T0181968

Lecture Notes in Computer Science 14187

Founding Editors

Gerhard Goos
Juris Hartmanis

Editorial Board Members

Elisa Bertino, *Purdue University, West Lafayette, IN, USA*
Wen Gao, *Peking University, Beijing, China*
Bernhard Steffen, *TU Dortmund University, Dortmund, Germany*
Moti Yung, *Columbia University, New York, NY, USA*

The series Lecture Notes in Computer Science (LNCS), including its subseries Lecture Notes in Artificial Intelligence (LNAI) and Lecture Notes in Bioinformatics (LNBI), has established itself as a medium for the publication of new developments in computer science and information technology research, teaching, and education.

LNCS enjoys close cooperation with the computer science R & D community, the series counts many renowned academics among its volume editors and paper authors, and collaborates with prestigious societies. Its mission is to serve this international community by providing an invaluable service, mainly focused on the publication of conference and workshop proceedings and postproceedings. LNCS commenced publication in 1973.

Gernot A. Fink · Rajiv Jain · Koichi Kise ·
Richard Zanibbi

Editors

Document Analysis
and Recognition –
ICDAR 2023

17th International Conference
San José, CA, USA, August 21–26, 2023
Proceedings, Part I

 Springer

Editors
Gernot A. Fink
TU Dortmund University
Dortmund, Germany

Rajiv Jain
Adobe
College Park, MN, USA

Koichi Kise
Osaka Metropolitan University
Osaka, Japan

Richard Zanibbi
Rochester Institute of Technology
Rochester, NY, USA

ISSN 0302-9743   ISSN 1611-3349 (electronic)
Lecture Notes in Computer Science
ISBN 978-3-031-41675-0   ISBN 978-3-031-41676-7 (eBook)
https://doi.org/10.1007/978-3-031-41676-7

This Springer imprint is published by the registered company Springer Nature Switzerland AG
The registered company address is: Gewerbestrasse 11, 6330 Cham, Switzerland

Foreword

We are delighted to welcome you to the proceedings of ICDAR 2023, the 17th IAPR International Conference on Document Analysis and Recognition, which was held in San Jose, in the heart of Silicon Valley in the United States. With the worst of the pandemic behind us, we hoped that ICDAR 2023 would be a fully in-person event. However, challenges such as difficulties in obtaining visas also necessitated the partial use of hybrid technologies for ICDAR 2023. The oral papers being presented remotely were synchronous to ensure that conference attendees interacted live with the presenters and the limited hybridization still resulted in an enjoyable conference with fruitful interactions.

ICDAR 2023 was the 17th edition of a longstanding conference series sponsored by the International Association of Pattern Recognition (IAPR). It is the premier international event for scientists and practitioners in document analysis and recognition. This field continues to play an important role in transitioning to digital documents. The IAPR-TC 10/11 technical committees endorse the conference. The very first ICDAR was held in St Malo, France in 1991, followed by Tsukuba, Japan (1993), Montreal, Canada (1995), Ulm, Germany (1997), Bangalore, India (1999), Seattle, USA (2001), Edinburgh, UK (2003), Seoul, South Korea (2005), Curitiba, Brazil (2007), Barcelona, Spain (2009), Beijing, China (2011), Washington, DC, USA (2013), Nancy, France (2015), Kyoto, Japan (2017), Sydney, Australia (2019) and Lausanne, Switzerland (2021).

Keeping with its tradition from past years, ICDAR 2023 featured a three-day main conference, including several competitions to challenge the field and a post-conference slate of workshops, tutorials, and a doctoral consortium. The conference was held at the San Jose Marriott on August 21–23, 2023, and the post-conference tracks at the Adobe World Headquarters in San Jose on August 24–26, 2023.

We thank our executive co-chairs, Venu Govindaraju and Tong Sun, for their support and valuable advice in organizing the conference. We are particularly grateful to Tong for her efforts in facilitating the organization of the post-conference in Adobe Headquarters and for Adobe's generous sponsorship.

The highlights of the conference include keynote talks by the recipient of the IAPR/ICDAR Outstanding Achievements Award, and distinguished speakers Marti Hearst, UC Berkeley School of Information; Vlad Morariu, Adobe Research; and Seiichi Uchida, Kyushu University, Japan.

A total of 316 papers were submitted to the main conference (plus 33 papers to the ICDAR-IJDAR journal track), with 53 papers accepted for oral presentation (plus 13 IJDAR track papers) and 101 for poster presentation. We would like to express our deepest gratitude to our Program Committee Chairs, featuring three distinguished researchers from academia, Gernot A. Fink, Koichi Kise, and Richard Zanibbi, and one from industry, Rajiv Jain, who did a phenomenal job in overseeing a comprehensive reviewing process and who worked tirelessly to put together a very thoughtful and interesting technical program for the main conference. We are also very grateful to the members of the Program Committee for their high-quality peer reviews. Thank you to our competition chairs, Kenny Davila, Chris Tensmeyer, and Dimosthenis Karatzas, for overseeing the competitions.

The post-conference featured 8 excellent workshops, four value-filled tutorials, and the doctoral consortium. We would like to thank Mickael Coustaty and Alicia Fornes, the workshop chairs, Elisa Barney-Smith and Laurence Likforman-Sulem, the tutorial chairs, and Jean-Christophe Burie and Andreas Fischer, the doctoral consortium chairs, for their efforts in putting together a wonderful post-conference program.

We would like to thank and acknowledge the hard work put in by our Publication Chairs, Anurag Bhardwaj and Utkarsh Porwal, who worked diligently to compile the camera-ready versions of all the papers and organize the conference proceedings with Springer. Many thanks are also due to our sponsorship, awards, industry, and publicity chairs for their support of the conference.

The organization of this conference was only possible with the tireless behind-the-scenes contributions of our webmaster and tech wizard, Edward Sobczak, and our secretariat, ably managed by Carol Doermann. We convey our heartfelt appreciation for their efforts.

Finally, we would like to thank for their support our many financial sponsors and the conference attendees and authors, for helping make this conference a success. We sincerely hope those who attended had an enjoyable conference, a wonderful stay in San Jose, and fruitful academic exchanges with colleagues.

August 2023

David Doermann
Srirangaraj (Ranga) Setlur

Preface

Welcome to the proceedings of the 17th International Conference on Document Analysis and Recognition (ICDAR) 2023. ICDAR is the premier international event for scientists and practitioners involved in document analysis and recognition.

This year, we received 316 conference paper submissions with authors from 42 different countries. In order to create a high-quality scientific program for the conference, we recruited 211 regular and 38 senior program committee (PC) members. Regular PC members provided a total of 913 reviews for the submitted papers (an average of 2.89 per paper). Senior PC members who oversaw the review phase for typically 8 submissions took care of consolidating reviews and suggested paper decisions in their meta-reviews. Based on the information provided in both the reviews and the prepared meta-reviews we PC Chairs then selected 154 submissions (48.7%) for inclusion into the scientific program of ICDAR 2023. From the accepted papers, 53 were selected for oral presentation, and 101 for poster presentation.

In addition to the papers submitted directly to ICDAR 2023, we continued the tradition of teaming up with the International Journal of Document Analysis and Recognition (IJDAR) and organized a special journal track. The journal track submissions underwent the same rigorous review process as regular IJDAR submissions. The ICDAR PC Chairs served as Guest Editors and oversaw the review process. From the 33 manuscripts submitted to the journal track, 13 were accepted and were published in a Special Issue of IJDAR entitled "Advanced Topics of Document Analysis and Recognition." In addition, all papers accepted in the journal track were included as oral presentations in the conference program.

A very prominent topic represented in both the submissions from the journal track as well as in the direct submissions to ICDAR 2023 was handwriting recognition. Therefore, we organized a Special Track on Frontiers in Handwriting Recognition. This also served to keep alive the tradition of the International Conference on Frontiers in Handwriting Recognition (ICFHR) that the TC-11 community decided to no longer organize as an independent conference during ICFHR 2022 held in Hyderabad, India. The handwriting track included oral sessions covering handwriting recognition for historical documents, synthesis of handwritten documents, as well as a subsection of one of the poster sessions. Additional presentation tracks at ICDAR 2023 featured Graphics Recognition, Natural Language Processing for Documents (D-NLP), Applications (including for medical, legal, and business documents), additional Document Analysis and Recognition topics (DAR), and a session highlighting featured competitions that were run for ICDAR 2023 (Competitions). Two poster presentation sessions were held at ICDAR 2023.

As ICDAR 2023 was held with in-person attendance, all papers were presented by their authors during the conference. Exceptions were only made for authors who could not attend the conference for unavoidable reasons. Such oral presentations were then provided by synchronous video presentations. Posters of authors that could not attend were presented by recorded teaser videos, in addition to the physical posters.

Three keynote talks were given by Marti Hearst (UC Berkeley), Vlad Morariu (Adobe Research), and Seichi Uchida (Kyushu University). We thank them for the valuable insights and inspiration that their talks provided for participants.

Finally, we would like to thank everyone who contributed to the preparation of the scientific program of ICDAR 2023, namely the authors of the scientific papers submitted to the journal track and directly to the conference, reviewers for journal-track papers, and both our regular and senior PC members. We also thank Ed Sobczak for helping with the conference web pages, and the ICDAR 2023 Publications Chairs Anurag Bharadwaj and Utkarsh Porwal, who oversaw the creation of this proceedings.

August 2023

Gernot A. Fink
Rajiv Jain
Koichi Kise
Richard Zanibbi

Organization

General Chairs

David Doermann   University at Buffalo, The State University of New York, USA

Srirangaraj Setlur   University at Buffalo, The State University of New York, USA

Executive Co-chairs

Venu Govindaraju   University at Buffalo, The State University of New York, USA

Tong Sun   Adobe Research, USA

PC Chairs

Gernot A. Fink   Technische Universität Dortmund, Germany (Europe)

Rajiv Jain   Adobe Research, USA (Industry)

Koichi Kise   Osaka Metropolitan University, Japan (Asia)

Richard Zanibbi   Rochester Institute of Technology, USA (Americas)

Workshop Chairs

Mickael Coustaty   La Rochelle University, France

Alicia Fornes   Universitat Autònoma de Barcelona, Spain

Tutorial Chairs

Elisa Barney-Smith   Luleå University of Technology, Sweden

Laurence Likforman-Sulem   Télécom ParisTech, France

Competitions Chairs

Kenny Davila   Universidad Tecnológica Centroamericana,
                                   UNITEC, Honduras
Dimosthenis Karatzas   Universitat Autònoma de Barcelona, Spain
Chris Tensmeyer   Adobe Research, USA

Doctoral Consortium Chairs

Andreas Fischer   University of Applied Sciences and Arts Western
                                   Switzerland
Veronica Romero   University of Valencia, Spain

Publications Chairs

Anurag Bharadwaj   Northeastern University, USA
Utkarsh Porwal   Walmart, USA

Posters/Demo Chair

Palaiahnakote Shivakumara   University of Malaya, Malaysia

Awards Chair

Santanu Chaudhury   IIT Jodhpur, India

Sponsorship Chairs

Wael Abd-Almageed   Information Sciences Institute USC, USA
Cheng-Lin Liu   Chinese Academy of Sciences, China
Masaki Nakagawa   Tokyo University of Agriculture and Technology,
                                   Japan

Industry Chairs

Andreas Dengel   DFKI, Germany
Véronique Eglin   Institut National des Sciences Appliquées (INSA)
                                   de Lyon, France
Nandakishore Kambhatla   Adobe Research, India

Publicity Chairs

Sukalpa Chanda   Østfold University College, Norway
Simone Marinai   University of Florence, Italy
Safwan Wshah   University of Vermont, USA

Technical Chair

Edward Sobczak   University at Buffalo, The State University of
                        New York, USA

Conference Secretariat

University at Buffalo, The State University of New York, USA

Program Committee

Senior Program Committee Members

Srirangaraj Setlur   Apostolos Antonacopoulos
Richard Zanibbi   Lianwen Jin
Koichi Kise   Nicholas Howe
Gernot Fink   Marc-Peter Schambach
David Doermann   Marcal Rossinyol
Rajiv Jain   Wataru Ohyama
Rolf Ingold   Nicole Vincent
Andreas Fischer   Faisal Shafait
Marcus Liwicki   Simone Marinai
Seiichi Uchida   Bertrand Couasnon
Daniel Lopresti   Masaki Nakagawa
Josep Llados   Anurag Bhardwaj
Elisa Barney Smith   Dimosthenis Karatzas
Umapada Pal   Masakazu Iwamura
Alicia Fornes   Tong Sun
Jean-Marc Ogier   Laurence Likforman-Sulem
C. V. Jawahar   Michael Blumenstein
Xiang Bai   Cheng-Lin Liu
Liangrui Peng   Luiz Oliveira
Jean-Christophe Burie   Robert Sabourin
Andreas Dengel   R. Manmatha
Robert Sablatnig   Angelo Marcelli
Basilis Gatos   Utkarsh Porwal

Program Committee Members

Harold Mouchere
Foteini Simistira Liwicki
Vernonique Eglin
Aurelie Lemaitre
Qiu-Feng Wang
Jorge Calvo-Zaragoza
Yuchen Zheng
Guangwei Zhang
Xu-Cheng Yin
Kengo Terasawa
Yasuhisa Fujii
Yu Zhou
Irina Rabaev
Anna Zhu
Soo-Hyung Kim
Liangcai Gao
Anders Hast
Minghui Liao
Guoqiang Zhong
Carlos Mello
Thierry Paquet
Mingkun Yang
Laurent Heutte
Antoine Doucet
Jean Hennebert
Cristina Carmona-Duarte
Fei Yin
Yue Lu
Maroua Mehri
Ryohei Tanaka
Adel M. M. Alimi
Heng Zhang
Gurpreet Lehal
Ergina Kavallieratou
Petra Gomez-Kramer
Anh Le Duc
Frederic Rayar
Muhammad Imran Malik
Vincent Christlein
Khurram Khurshid
Bart Lamiroy
Ernest Valveny
Antonio Parziale

Jean-Yves Ramel
Haikal El Abed
Alireza Alaei
Xiaoqing Lu
Sheng He
Abdel Belaid
Joan Puigcerver
Zhouhui Lian
Francesco Fontanella
Daniel Stoekl Ben Ezra
Byron Bezerra
Szilard Vajda
Irfan Ahmad
Imran Siddiqi
Nina S. T. Hirata
Momina Moetesum
Vassilis Katsouros
Fadoua Drira
Ekta Vats
Ruben Tolosana
Steven Simske
Christophe Rigaud
Claudio De Stefano
Henry A. Rowley
Pramod Kompalli
Siyang Qin
Alejandro Toselli
Slim Kanoun
Rafael Lins
Shinichiro Omachi
Kenny Davila
Qiang Huo
Da-Han Wang
Hung Tuan Nguyen
Ujjwal Bhattacharya
Jin Chen
Cuong Tuan Nguyen
Ruben Vera-Rodriguez
Yousri Kessentini
Salvatore Tabbone
Suresh Sundaram
Tonghua Su
Sukalpa Chanda

Mickael Coustaty
Donato Impedovo
Alceu Britto
Bidyut B. Chaudhuri
Swapan Kr. Parui
Eduardo Vellasques
Sounak Dey
Sheraz Ahmed
Julian Fierrez
Ioannis Pratikakis
Mehdi Hamdani
Florence Cloppet
Amina Serir
Mauricio Villegas
Joan Andreu Sanchez
Eric Anquetil
Majid Ziaratban
Baihua Xiao
Christopher Kermorvant
K. C. Santosh
Tomo Miyazaki
Florian Kleber
Carlos David Martinez Hinarejos
Muhammad Muzzamil Luqman
Badarinath T.
Christopher Tensmeyer
Musab Al-Ghadi
Ehtesham Hassan
Journet Nicholas
Romain Giot
Jonathan Fabrizio
Sriganesh Madhvanath
Volkmar Frinken
Akio Fujiyoshi
Srikar Appalaraju
Oriol Ramos-Terrades
Christian Viard-Gaudin
Chawki Djeddi
Nibal Nayef
Nam Ik Cho
Nicolas Sidere
Mohamed Cheriet
Mark Clement
Shivakumara Palaiahnakote
Shangxuan Tian

Ravi Kiran Sarvadevabhatla
Gaurav Harit
Iuliia Tkachenko
Christian Clausner
Vernonica Romero
Mathias Seuret
Vincent Poulain D'Andecy
Joseph Chazalon
Kaspar Riesen
Lambert Schomaker
Mounim El Yacoubi
Berrin Yanikoglu
Lluis Gomez
Brian Kenji Iwana
Ehsanollah Kabir
Najoua Essoukri Ben Amara
Volker Sorge
Clemens Neudecker
Praveen Krishnan
Abhisek Dey
Xiao Tu
Mohammad Tanvir Parvez
Sukhdeep Singh
Munish Kumar
Qi Zeng
Puneet Mathur
Clement Chatelain
Jihad El-Sana
Ayush Kumar Shah
Peter Staar
Stephen Rawls
David Etter
Ying Sheng
Jiuxiang Gu
Thomas Breuel
Antonio Jimeno
Karim Kalti
Enrique Vidal
Kazem Taghva
Evangelos Milios
Kaizhu Huang
Pierre Heroux
Guoxin Wang
Sandeep Tata
Youssouf Chherawala

Reeve Ingle
Aashi Jain
Carlos M. Travieso-Gonzales
Lesly Miculicich
Curtis Wigington
Andrea Gemelli
Martin Schall
Yanming Zhang
Dezhi Peng
Chongyu Liu
Huy Quang Ung
Marco Peer
Nam Tuan Ly
Jobin K. V.
Rina Buoy
Xiao-Hui Li
Maham Jahangir
Muhammad Naseer Bajwa

Oliver Tueselmann
Yang Xue
Kai Brandenbusch
Ajoy Mondal
Daichi Haraguchi
Junaid Younas
Ruddy Theodose
Rohit Saluja
Beat Wolf
Jean-Luc Bloechle
Anna Scius-Bertrand
Claudiu Musat
Linda Studer
Andrii Maksai
Oussama Zayene
Lars Voegtlin
Michael Jungo

Program Committee Subreviewers

Li Mingfeng
Houcemeddine Filali
Kai Hu
Yejing Xie
Tushar Karayil
Xu Chen
Benjamin Deguerre
Andrey Guzhov
Estanislau Lima
Hossein Naftchi
Giorgos Sfikas
Chandranath Adak
Yakn Li
Solenn Tual
Kai Labusch
Ahmed Cheikh Rouhou
Lingxiao Fei
Yunxue Shao
Yi Sun
Stephane Bres
Mohamed Mhiri
Zhengmi Tang
Fuxiang Yang
Saifullah Saifullah

Paolo Giglio
Wang Jiawei
Maksym Taranukhin
Menghan Wang
Nancy Girdhar
Xudong Xie
Ray Ding
Mélodie Boillet
Nabeel Khalid
Yan Shu
Moises Diaz
Biyi Fang
Adolfo Santoro
Glen Pouliquen
Ahmed Hamdi
Florian Kordon
Yan Zhang
Gerasimos Matidis
Khadiravana Belagavi
Xingbiao Zhao
Xiaotong Ji
Yan Zheng
M. Balakrishnan
Florian Kowarsch

Mohamed Ali Souibgui
Xuewen Wang
Djedjiga Belhadj
Omar Krichen
Agostino Accardo
Erika Griechisch
Vincenzo Gattulli
Thibault Lelore
Zacarias Curi
Xiaomeng Yang
Mariano Maisonnave
Xiaobo Jin
Corina Masanti
Panagiotis Kaddas
Karl Löwenmark
Jiahao Lv
Narayanan C. Krishnan
Simon Corbillé
Benjamin Fankhauser
Tiziana D'Alessandro
Francisco J. Castellanos
Souhail Bakkali
Caio Dias
Giuseppe De Gregorio
Hugo Romat
Alessandra Scotto di Freca
Christophe Gisler
Nicole Dalia Cilia
Aurélie Joseph
Gangyan Zeng
Elmokhtar Mohamed Moussa
Zhong Zhuoyao
Oluwatosin Adewumi
Sima Rezaei
Anuj Rai
Aristides Milios
Shreeganesh Ramanan
Wenbo Hu

Arthur Flor de Sousa Neto
Rayson Laroca
Sourour Ammar
Gianfranco Semeraro
Andre Hochuli
Saddok Kebairi
Shoma Iwai
Cleber Zanchettin
Ansgar Bernardi
Vivek Venugopal
Abderrhamne Rahiche
Wenwen Yu
Abhishek Baghel
Mathias Fuchs
Yael Iseli
Xiaowei Zhou
Yuan Panli
Minghui Xia
Zening Lin
Konstantinos Palaiologos
Loann Giovannangeli
Yuanyuan Ren
Shubhang Desai
Yann Soullard
Ling Fu
Juan Antonio Ramirez-Orta
Chixiang Ma
Truong Thanh-Nghia
Nathalie Girard
Kalyan Ram Ayyalasomayajula
Talles Viana
Francesco Castro
Anthony Gillioz
Huawen Shen
Sanket Biswas
Haisong Ding
Solène Tarride

Contents – Part I

Applications 1: Enterprise Documents (Medical, Legal, Finance)

Multi-stage Fine-Tuning Deep Learning Models Improves Automatic Assessment of the Rey-Osterrieth Complex Figure Test

Benjamin Schuster[1], Florian Kordon[1], Martin Mayr[1], Mathias Seuret[1], Stefanie Jost[2], Josef Kessler[2], and Vincent Christlein[1(✉)]

[1] Friedrich-Alexander-University, 91058 Erlangen, Germany
{benjamin.schuster,florian.kordon,martin.mayr,mathias.seuret,
vincent.christlein}@fau.de
[2] University Hospital Cologne, 50937 Cologne, Germany
{stefanie.jost,josef.kessler}@uk-koeln.de

Abstract. The Rey-Osterrieth Complex Figure Test (ROCFT) is a widely used neuropsychological tool for assessing the presence and severity of different diseases. It involves presenting a complex illustration to the patient who is asked to copy it, followed by recall from memory after 3 and 30 min. In clinical practice, a human rater evaluates each component of the reproduction, with the overall score indicating illness severity. However, this method is both time-consuming and error-prone. Efforts have been made to automate the process, but current algorithms require large-scale private datasets of up to 20,000 illustrations. With limited data, training a deep learning model is challenging. This study addresses this challenge by developing a fine-tuning strategy with multiple stages. We show that pre-training on a large-scale sketch dataset with initialized weights from ImageNet significantly reduces the mean absolute error (MAE) compared to just training with initialized weights from ImageNet, e.g., ReXNet-200 from 3.1 to 2.2 MAE. Additionally, techniques such as stochastic weight averaging (SWA) and ensembling of different architectures can further reduce the error to an MAE of 1.97.

Keywords: Rey-Osterrieth Complex Figure · Regression · Sketch pre-training · Ensembling

1 Introduction

The Rey-Osterrieth Complex Figure Test (ROCFT) was designed to examine the visuospatial ability and memory in patients who suffer from traumatic brain injury [1]. Additionally, the test is utilized to test for dementia and to evaluate children's cognitive development [1]. The test procedure starts with presenting the figure depicted in Fig. 1 to the patient, who is subsequently asked to copy it by drawing it, typically with a pen on paper. After 3 min, the patient is asked to reproduce the figure from memory. This procedure is repeated after 30 min.

© The Author(s), under exclusive license to Springer Nature Switzerland AG 2023
G. A. Fink et al. (Eds.): ICDAR 2023, LNCS 14187, pp. 3–19, 2023.
https://doi.org/10.1007/978-3-031-41676-7_1

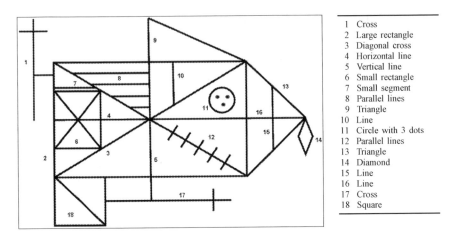

1	Cross
2	Large rectangle
3	Diagonal cross
4	Horizontal line
5	Vertical line
6	Small rectangle
7	Small segment
8	Parallel lines
9	Triangle
10	Line
11	Circle with 3 dots
12	Parallel lines
13	Triangle
14	Diamond
15	Line
16	Line
17	Cross
18	Square

Fig. 1. Rey-Osterrieth Complex Figure with annotated section numbers [23].

Those three steps are called *copy, immediate recall,* and *delayed recall.* While *copy* is always part of the procedure, sometimes only one of the steps *immediate recall* or *delayed recall* are carried out [1]. The figure can be subdivided into 18 separate sections, as annotated in Fig. 1. According to the Osterrieth scoring system, for each section, a score ranging from 0 to 2 is determined in the following way: If the unit is drawn and placed correctly, 2 points are assigned. In case it is placed poorly, this corresponds to only 1 point. In the case of a distorted section, which might be incomplete but still recognizable, 1 or 0.5 points are given depending on the placement quality. 0 points are assigned when the section is absent or unrecognizable. Once all sections are scored, the sum of them represents the score of the entire drawing and can range from 0 to 36.

The ROCFT has become one of the most widely applied neuropsychological tests for constructional and non-verbal memory [1] and can evaluate the patient's neuropsychological dysfunction [18]. There are several abilities necessary for good test performance, such as working memory, visuospatial abilities, planning [1], problem-solving, and visuomotor coordination [4]. Consequently, the test provides valuable data for evaluating the progress of patients through treatment and represents a tool for research into the organization of brain activity and its connection to behavior, brain disorders, and behavioral disabilities [2].

Manual scoring of the ROCFT represents a monotone and repetitive task. Moreover, the resulting score depends to some extent on the subjective judgments of the rater, thereby causing inter-rater variability. Hence, an automated scoring system could set a new standard to combat inter-rater variability. Given the popularity and wide acceptance of this test, automation holds great potential to reduce time and effort and ultimately save costs. In recent years, several machine learning methods have been developed to build automatic scoring systems for the ROCFT [5,10,13,19]. In most works that made use of deep learning, large training datasets with 2,000 to 20,000 images are used. However, such large datasets are typically not available for the average hospital. Moreover, to the best of our knowledge, there is no public dataset of ROCFT-drawings with annotated scores available.

In this work, we suggest the use of a multi-stage fine-tuning procedure for automatic ROCFT scoring given a small dataset of only about 400 ROCFT samples. In particular, we evaluate the effects of (1) sketch-based fine-tuning, i.e., pre-training on a large-scale sketch-dataset and then fine-tuning using the ROCFT dataset, (2) StochasticWeight Averaging (SWA) fine-tuning, and (3) ensembling.

The paper is organized as follows. Section 2 presents the related work in the field of recognizing ROCFT scores. We obtained a small ROCFT dataset, which we pre-processed, cf. Sect. 3. In Sect. 4, we present the deep learning architectures and the techniques for improving the automatic scoring. Our experiments and their evaluations are presented in Sect. 5. The paper is concluded in Sect. 6.

2   Related Work

An effort to automatically detect and score sections of the ROCFT with the use of traditional machine learning was made by several works [2,3,22].

Canham *et al.* [2,3] propose a method for localizing individual sections of the Rey-Osterrieth Complex Figure. They construct a relational graph from the vectorized binary image of the drawn figure and use it to identify the individual sections and to calculate geometric features like section orientation and size. The final score is calculated by combining the features using a weighted Yager intersection function. Li [22] presents a tablet-based testing procedure to support an automatic evaluation not only on the final image but already during the drawing process. For that purpose, strokes are first split up into line segments based on corner detection. Then, in a semi-automatic grading procedure, a clinician has to select a bounding box for each section of the figure. The program identifies the section based on the line segment information and calculates a grade. In contrast, Webb *et al.* [21] propose an tablet-based automatic scoring system for the conceptually similar OCS-Plus figure copy task. It identifies sections and calculates features by applying several pre-processing steps, e. g., circle identification, line segmentation/extraction, and star and cross identification.

The papers [5,10,13,19] approached the automatic assessment of the ROCFT with deep learning. Simfukwe *et al.* [19] propose a diagnosis system able to classify a drawing according to three clinical categories: normal, Mild Cognitive Impairment (MCI), or mild dementia. They built two datasets containing images from 2,232 patients, one for learning to distinguish normal from abnormal and one for learning to distinguish between all three categories. On the first dataset, an accuracy of 96% was achieved and on the second, 88%. In [5], the authors combined computer vision and deep learning to assign scores for each section in a drawing and used the results as features in a classification task for diagnosis of healthy, MCI, or dementia. The scoring system was changed to include four classes (omitted, distorted, misplaced, and correct) instead of a grade ranging from 0 to 2. Simple sections are assigned one of these classes using on computer vision techniques, while complex pattern sections are classified via deep learning. The average class accuracy is 70.7% and 67.5% for simple and complex pattern sections, respectively. Using the 18 features for diagnosis resulted in healthy vs. MCI (85%), healthy vs. dementia (91%), MCI vs. dementia (83%), and 3-class (73%).

Park *et al.* [15] evaluated a pre-trained DenseNet [10] for automatic scoring of the ROCFT using a single scalar output. From 6,680 subjects, three drawings each (*copy, immediate recall, delayed recall*) with corresponding scores are scanned, resulting in 20,040 images with a drawing. The authors evaluate model performance on this data using a cross-validation setup and report a Mean Absolute Error (MAE) of 0.95 and R^2 of 0.986. Similarly, Lander *et al.* [13] developed a Deep Learning model to automatically score the ROCFT using a dataset with 20,225 drawings. The authors empirically show that defining the evaluation task as a regression problem is preferable because it takes into account the distance between evaluations, a property not inherent to conventional classification. If the application scenario allows it, using a combination of both classification and regression tasks can further improve the scoring precision.

The last two presented papers are conceptually most related to our work and report convincing classification accuracy. However, both works had access to comparatively large datasets. In comparison, we explicitly tackle automatic scoring of ROCFT in scenarios where only little clinical data is available. For that purpose, we develop multiple stages of fine-tuning to improve a Deep Learning based scoring system able to compete with ones in large data regimes.

3   ROCFT Dataset

3.1   Data

The University Hospital Cologne provided data from 208 test sheets, with each sample consisting of three pages. Two pages contain a single drawing each, while the third page presents scores for both drawings in two columns. The scores include 18 rows for each section and an additional row for the overall sum. The individual scores are referred to as single scores and the total score, being the sum of all single scores for a drawing, is known as the total-score. An image of a sample is depicted in Fig. 2. The first page contains the *copy* drawing. The second page contains the *delayed recall* drawing. The tests carried out by the University Hospital Cologne do not include the immediate recall drawings.

For training a neural network to score automatically the ROCFT, we produced a dataset containing pairs of drawings and their corresponding scores. These scores are obtained by a semi-automatic method (image registration, number extraction, and recognition) where individual scores are detected and compared with the final sum. We only consider integer values, i. e., we truncate possible decimal places '.5'.

Fig. 2. Data sample consisting of 3 pages. Left: copy drawing of the patient, Middle: delayed recall drawing of the patient, Right: Scores for both copy and delayed recall.

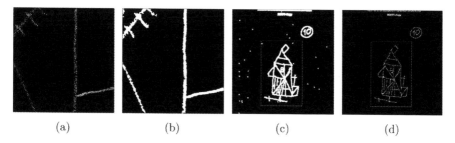

(a) (b) (c) (d)

Fig. 3. (a) Example of weak lines, (b) same lines after strengthening using mathematical morphological operators (dilation + closing), (c) bounding box around dilated drawing, (d) bounding box transferred to the original drawing.

3.2 Pre-processing

Because a drawing does not cover the whole page, its area is identified and cropped to extract it. This process can be subdivided into the following steps: (1) Binarization and inversion, (2) strengthening lines, (3) retrieving the drawing's bounding box, (4) and cropping and resizing the drawing. For binarization, we use local Gaussian thresholding. The provided test sheets contain drawings with well-defined lines, as well as drawings with weaker lines, which only include dots that are not connected (see Fig. 3a). The reason is that some figures were drawn with a pen, others with a pencil. Not well-defined lines in the drawing can disrupt the drawing identification and, therefore, possibly prevent a successful extraction. Hence, we strengthen lines by applying different mathematical morphological operators. In particular, we use *dilation* followed by *closing*, both with a kernel size of $(4, 4)$, cf. Fig. 3b. Next, the bounding box of the drawing is retrieved. Although the lines get strengthened, drawings are not always fully connected, i.e., cluttered into multiple pieces. Thus, simply taking the bounding box of the biggest component would not always yield satisfactory results because

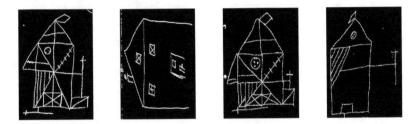

Fig. 4. Samples of the dataset with scores: 34, 0, 31, and 9.

parts of the drawings at the edge which are important for the score evaluation, might be missing. For instance, in Fig. 3d, the cross at the bottom of the drawing is not directly connected to the rest of the drawing. Therefore, heavy dilation (two times, with kernel size 12×12) is performed on a copy of the image, cf. Fig. 3c. Afterward, the drawing is usually fully connected such that the bounding box of the biggest component includes also parts of the drawings which are not directly connected to other parts of the drawing in the original image. To be independent from potentially varying image sizes, we use 2.5 % of the original image width for padding on all edges. The resulting bounding box gets cropped out of the original image (with only light dilatation) and is resized to the size of 354×500. The proportions are chosen experimentally by decreasing the sizes as much as possible without losing information that is relevant for scoring.

Subsequent to extracting both total scores and drawings, we combine corresponding pairs into a dataset which is further used for training and in the following referred to as *ROCFT dataset*. This dataset includes 416 samples, four of which are shown in Fig. 4. We only use total scores and not the vector of single scores, because given the small amount of data, we are concerned about the ability of the models to learn a proper mapping for the single scores.

4   Methodology

We suggest a multi-stage fine-tuning where we first fine-tune a model pre-trained on sketches and afterwards fine-tune once more using SWA. To further boost the performance, we combine different models in an ensemble.

4.1   Pre-training Using Sketch Dataset

Many models, which are already pre-trained for classification on ImageNet, are available. Yet, the ROCFT is a sketch. Sketches differ significantly in comparison to pictures of real-life objects, e. g., concerning color, precision, and shape. Therefore, we decided to pre-train the models on a dataset containing sketches with the aim of learning features that are specific to sketches. Another reason is to compensate for the small number of samples available in the ROCFT dataset. We chose the TU Berlin sketch dataset [6] for pre-training. It contains 20,000

(a) Snowman   (b) Giraffe   (c) Aeroplane

Fig. 5. Sketch Examples of the TU Berlin sketch dataset [6].

sketches made by humans evenly distributed over 250 object categories, i.e., there are 80 samples for each class, where every image is of size 1111×1111 pixels. Examples are shown in Fig. 5. A human classifies a given sample correctly in 73 % of the cases [6]. For pre-training, each image is inverted and downsampled to 500×500 pixels to match the ROCFT dataset. Several augmentations are performed: random horizontal flip, random rotation up to $10°$C, and randomly downsizing the image by up to 30 % to simulate different pen thicknesses. Every model gets initialized with downloaded weights that were acquired during pre-training on ImageNet. In initial experiments, we recognized that it helps to substantially speed up the model convergence. Since the problem on the TU Berlin sketch dataset is a classification problem, the last layer is configured to have 250 outputs. Predictions are derived from the *argmax* of these outputs. Cross-Entropy (CE)-loss is chosen as the loss function and accuracy as the metric. Afterward, the models are fine-tuned on the ROCFT dataset. To match the same dimensions, the ROCFT dataset images are zero-padded to the same image input size of the TU Berlin sketch dataset, i.e., 354×500 to 500×500.

Label predictions that are farther away from the correct label are more severe than those that are still wrong but closer. Therefore, we interpret the underlying task as a regression problem. We configured the last layer of each architecture to have one output, on which the logistic function is applied. To make a prediction, the output $o \in [0, 1]$ is mapped to the (integer) range 0 to 36.

4.2 Stochastic Weight Averaging

Stochastic Weight Averaging (SWA) computes the average of weights traversed by the optimizer [11]. It was demonstrated that SWA enhances generalization [11] and tends to find solutions in the center of wide flat loss regions making it less susceptible to the shift between train and test error surfaces [16]. We use SWA due to its potential of enhancing generalization at very low costs [11]. Especially important is that SWA is not used during the entire period of the training but only in the end, as an already trained model can be reloaded and trained for several more epochs with SWA in order to benefit from this technique [16].

4.3   Ensemble

Ensembling is implemented by computing a weighted average of the outputs (that range from 0 to 1) of the contributing members and then mapping the value to the desired range. The weights depend on the performance of the contributing models on the validation set. Given an ensemble with n members where each member achieves a validation MAE of $\{\mathrm{mae}_j | j \in \{1, \dots, n\}\}$, the weights w_j for each member are then calculated as follows:

$$
w_j = \left(\frac{1}{\mathrm{mae}_j \sum_{i=0}^{n} \frac{1}{\mathrm{mae}_i}} - \frac{1}{n} \right) + 1 \,. \tag{1}
$$

This formula ensures that $\sum_{i=0}^{n} w_i = n$. To make a prediction, the ensemble's output $o_{Ensemble} \in [0, 1]$ is calculated from the outputs of the contributing members o_j and their weights w_j:

$$
o_{Ensemble} = \frac{1}{n} \sum_{i=0}^{n} w_j \cdot o_i \,. \tag{2}
$$

This output is mapped to the (integer) range 0 to 36.

5   Evaluation

5.1   Metrics

Mean Absolute Error (MAE) is the main metric used for evaluation. Additionally, accuracy and Mean Squared Error (MSE) are employed, where accuracy refers to the average class accuracy. Moreover, we introduce *Max Error*, referring to the maximum absolute error made during the prediction of labels of an entire set.

5.2   Evaluation Protocol

Dataset split. We divided the ROCFT dataset into three disjoint sets for training, validation, and testing. Note that we ignore any patient information. Instead, we aimed to balance the labels. Therefore, the test set is assembled by randomly picking one to four samples from each label, with the constraint that the set has in total 74 samples, so on average, two per label. For the validation set, two samples are taken from each label. Since only one sample remains for labels 4 and 27, the validation set consists of 72 samples. All remaining samples are put into the training set, which subsequently has a size of 270.

Oversampling and Augmentation. The ROCFT dataset is quite imbalanced. The distributions for the training set is shown in Fig. 6. In order to counteract this imbalance, oversampling is performed.

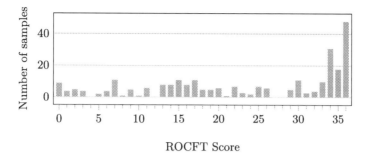

Fig. 6. ROCFT score distribution.

Fig. 7. Most left: Original, **others**: heavily augmented images of the original

Additionally, we apply augmentation for which we evaluate three different levels: *light, medium,* and *heavy*. The transformations applied for each level are given in the appendix (Table 4). Transformations are applied sequentially. While the augmentations *Opening, Closing, Erosion,* and *Dilation* are implemented by us, all other augmentations are used from the imgaug-library [12]. Results of *heavy* augmentations are shown in Fig. 7. The picture on the very left is the original image, while all others are augmented versions of the original.

Architectures. In this work, we tested several architectures: VGG [20], ReXNet [7], ResNet [9], TinyNet [8], and ConvNeXt [14]. In particular, we used the following models: *vgg16, rexnet_200, resnet50, tinynet_a,* and *convnext_small* of the timm-libary [17].

During training, MSE-loss serves as the loss function. Early stopping is used based on the MSE of the validation set. All intermediate results are obtained using the validation set while the test set is only used at the very end to evaluate the model based on the best method obtained on the validation set.

5.3 Results

Augmentation Effect Using No Pre-training. For each of the architectures, four training runs are executed, one for every augmentation level and one without any augmentations. The validation MSE achieved by these experiments is shown in Fig. 8.

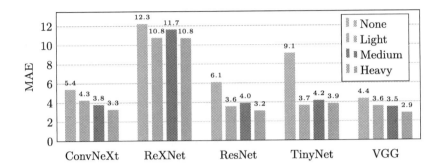

Fig. 8. Validation MMAE without pretraining using different augmentation levels

Table 1. Test accuracies on the TU Berlin sketch dataset.

Architecture	Accuracy↑ [%]
VGG16	69.5
TinyNet	74.8
ResNet50	76.0
ReXNet200	79.4
ConvNeXt	**80.5**

In all cases, the run without any augmentations performs worst. Using medium augmentations never performs the best. Heavy augmentations perform the best for ConvNeXt, ReXNet, ResNet and VGG, while light augmentations induced the best result for TinyNet. In subsequent fine-tuning experiments, we use the best augmentation level for each architecture.

TU Berlin Sketch Dataset Pre-training. Before continuing with the fine-tuning results, where we apply transfer learning from the TU Berlin sketch dataset to the ROCFT dataset, we give the pre-training results. The dataset of 20,000 images with 80 sketches for each of the 250 classes is divided into training, validation, and test sets by a stratified split with the ratio 70/15/15 %. Hence, all three sets are balanced, and while the validation and test set each contains 3,000 samples, the training set consists of 14,000 samples. To speed up convergence, the models use weights pre-trained on ImageNet. During network training, the accuracy gets tracked on the validation and training set. The model that performed best on the validation set is taken, and the accuracy on the test set is evaluated. The achieved test accuracies are presented in Table 1. The best-performing model is ConvNeXt with about 81 %, while VGG brings up the rear with 70 %. All models, except for VGG beat the human accuracy, which is at 73 % [6].

Training on this dataset is used as pre-training to learn sketch-related features. Outstanding performance is not of utmost importance, and thus satisfac-

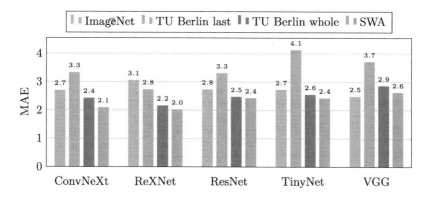

Fig. 9. Validation MAE after each fine-tuning step.

tory performance is achieved on all tested model architectures. It can be assumed that relevant features for hand-drawn sketches are learned by the models, and transfer learning to a similar domain can be applied successfully.

Effect of Multi-Stage Fine-Tuning. We start our multi-stage fine-tuning with the TU Berlin sketch dataset pre-trained weights. Then, we fine-tune with the ROCFT dataset in multiple steps. Therefore, we replace the last layer with a single output, which is initialized with random weights. Fine-tuning is performed in three subsequent training runs: (1) Last layer finetuning, (2) whole network finetuning, and (3) SWA finetuning.

First, the whole network is frozen except for the last layer and trained until convergence. In this way, the last layer learns how to use the pre-trained features to make predictions on the ROCFT dataset. The second step pursues the objective of slightly adjusting the pre-trained features and also the last layer. During the second step, the whole network is configured to be updated during training. The network is trained with early stopping and maximum epochs of 500. In the third step, we train for 30 epochs with SWA enabled as described in [16]. We compare these techniques with fine-tuning the five architectures just by using ImageNet pre-trained weights, i.e., omitting the TU Berlin sketch dataset and SWA fine-tuning steps. For each architecture, the best-performing augmentation level according to Fig. 8 is used. Figure 9 shows the resulting validation MAE for each step and architecture. We notice that just fine-tuning the head alone typically results in higher MAE. However, when fine-tuning then the whole network, all models (except VGG) improve upon the ImageNet baseline. The reason is almost certainly the similarity of sketches in the TU Berlin sketch dataset to sketches in the ROCFT dataset. Fine-tuning once more with SWA reduced the error further. Architecture-wise, ReXNet and ConvNext perform best while VGG performs worst.

Table 2. Overview of the best and average MAE for the different ensemble sizes. The performance of each combination can be found in Table 5.

	Best		Avg.	
	w. o. SWA↓	w. SWA↓	w. o. SWA↓	w. SWA↓
Single	2.18	2.04	2.51	2.34
Double	**2.14**	1.99	2.37	2.22
Triple	2.18	**1.97**	2.30	2.15
Quadrupel	2.18	2.06	**2.28**	**2.13**
All	2.19	2.11	—	—

Ensembling. In pursuit of further improvement, several ensembles with different architectures were formed. For each architecture combination (a detailed overview can be seen in Table 5), two ensembles are built and evaluated: (1) One where each architecture was fine-tuned with all three proposed steps, i. e., last layer fine-tuning, whole network fine-tuning and SWA fine-tuning and (2) one where only the first two fine-tuning steps were applied, i. e., without SWA. Most ensembles produced a MAE that was not better than the multi-step fine-tuned ReXNet model (2.042). Nevertheless, three ensembles exceeded this performance. The best performing ensemble was comprised of a ConvNeXt, ResNet, and ReXNet. Its MAE of 1.972 is 0.07 points better than ReXNet on its own. It can also be noted, that ensembles where the contributing members were not fine-tuned with SWA generally perform worse than those where the contributing members were also fine-tuned with SWA (with the exception of the combination of TinyNet + ReXNet). Additionally, Table 2 shows that on average a larger ensemble leads to improved results. However, when evaluating the best combination for each ensemble size, the best result without SWA was achieved by an ensemble of two models: ReXNet and ResNet. Whereas, with SWA the best result was accomplished by an ensemble of 3 models: ConvNeXt, ReXNet and ResNet.

Test Results. Since the previously mentioned ensemble achieves the smallest MAE on the validation set for our entire work, we additionally evaluate this model on our independent test set. Table 3 details all performance metrics on both datasets. Additionally, we give the error distribution on the test set in Fig. 10. It shows that the errors ranging from 0 to 3 are dominating. Over 70 % of predictions have an error smaller or equal to 3.

Table 3. All validation and test metrics of the ensemble (ConvNeXt, ResNet, ReXNet) with smallest MAE

	MAE↓	MSE↓	Max error↓	Accuracy↑
Validation	1.97	7.86	12	25.00
Test	3.10	17.04	13	14.19

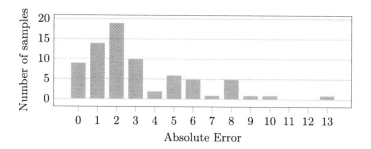

Fig. 10. Error distribution on the test set

6   Conclusion

In this work, we developed and validated an approach for the multi-stage fine-tuning of Deep Learning models for an automatic assessment of the Rey-Osterrieth Complex Figure Test. Specifically, we studied and combined (1) comprehensive data augmentation strategies, (2) pre-training on visually related sketch images, (3) SWA for improved model generalization to unseen image characteristics, and (4) ensembling with different model and augmentation combinations.

We could experimentally verify that using an application-driven fine-tuning strategy on visually similar but semantically contrasting sketch images can substantially improve the error rate. This improvement is not only shown in direct comparison with a fine-tuning strategy on ImageNet, but also holds for almost all network architectures studied. This means that additional data collection and curation can be avoided with relatively little effort.

Intriguingly, applying SWA shows a small but recognizable improvement. Since this technique is applied in a separate, very much shortened fine-tuning step, it imposes a negligible additional computational burden. In stark contrast, ensembling multiple network models represents a larger cut in the system's efficiency. Although we observe slight reductions in the error rate when using the best model combinations, we argue that this should be considered an optional component of the developed automatic ROCFT scoring system.

Overall, the low absolute errors show that an automatic assessment of the ROCFT is possible even if only a limited amount of data is available. The error difference of the test and validation set might indicate that there is an issue with overfitting, which needs to be investigated further. Future work should analyze if an augmentation technique that is more tailored to the ROCFT could further aid the scoring performance. For example, detecting one or more distinct sections in a sketch and then removing them and decreasing the score accordingly would substantially increase the number of possible structural compositions. Furthermore, given access to more data, integrating the prediction of section-wise single scores could prove to be a powerful auxiliary task that could not only increase the accuracy of the method but also provide additional information to the practitioner.

Acknowledgement. We thank the University Hospital Cologne for providing the data. The authors would also like to thank NVIDIA for their hardware donation.

A Augmentation Details

Table 4. Different augmentation levels (parameter ranges are uniformly distributed).

Augmentation	Probability	Parameters
Horizontal Flip	0.5	-
Vertical Flip	0.5	-
Rotation	1	$[-4, 4]$ (in degrees)
Closing	0.5	$[1, 3]$ (kernel size)
Jpeg Compression	0.4	$[40, 60]$ (compression level)

(a) Light augmentations

Augmentation	Probability	Parameters
Horizontal Flip AND Vertical Flip	0.2	-
Rotation	1	$[-6, 6]$ (in degrees)
Closing OR Opening	0.2	$[1, 4], [1, 2]$ (kernel size)
Max Pooling	0.2	$[1, 3]$ (kernel size)
Jpeg Compression	0.2	$[40, 70]$ (compression level)

(b) Medium augmentations

Augmentation	Probability	Parameters
Horizontal Flip AND Vertical Flip	0.2	-
Rotation	1	$[-10, 10]$ (in degrees)
Either 0 OR 1 of...		
Dilation	1	$[1, 5]$ (kernel size)
Erosion	1	$[1, 2]$ (kernel size)
Opening	1	$[1, 2]$ (kernel size)
Closing	1	$[1, 7]$ (kernel size)
Shrink and Pad to original size	0.4	$[0.6, 0.9]$ (shrinking factor)
Perspective Transform	0.3	$[0.01, 0.1]$ (scale)
Minimum of 1 and up to 3 of...		
Max Pooling	1	$[1, 3]$ (kernel size)
Jpeg Compression	0.7	$[60, 95]$ (compression level)
Elastic Transformation	0.7	$[1.0, 2.2]$ (alpha), 0.01 (sigma)

(c) Heavy augmentations

B Ensembling Combinations

Table 5. All different model combinations.

	ConvNeXt	ReXNet	ResNet	TinyNet	VGG	w.o. SWA↓	w. SWA↓
Single	✓					2.444	2.111
		✓				2.181	2.042
			✓			2.486	2.444
				✓		2.569	2.444
					✓	2.875	2.639
Double	✓	✓				2.208	1.986
	✓		✓			2.375	2.069
	✓			✓		2.361	2.264
	✓				✓	2.444	2.236
		✓	✓			**2.139**	2.083
		✓		✓		2.167	2.194
		✓			✓	2.361	2.167
			✓	✓		2.444	2.347
			✓		✓	2.556	2.472
				✓	✓	2.611	2.403
Triple	✓	✓	✓			2.194	**1.972**
	✓	✓		✓		2.194	2.028
	✓	✓			✓	2.375	2.181
	✓		✓	✓		2.292	2.028
	✓		✓		✓	2.389	2.181
	✓			✓	✓	2.403	2.250
		✓	✓	✓		2.181	2.111
		✓	✓		✓	2.319	2.222
		✓		✓	✓	2.319	2.292
			✓	✓	✓	2.389	2.264
Quadrupel	✓	✓	✓	✓		2.181	2.056
	✓	✓	✓		✓	2.264	2.083
	✓	✓		✓	✓	2.278	2.181
	✓		✓	✓	✓	2.389	2.153
		✓	✓	✓	✓	2.278	2.194
All	✓	✓	✓	✓	✓	2.194	2.111

References

1. Arango-Lasprilla, J.C., et al.: Rey-osterrieth complex figure - copy and immediate recall (3 minutes): normative data for Spanish-speaking pediatric populations. NeuroRehabilitation **41**(3), 593–603 (2017)
2. Canham, R.O., Smith, S.L., Tyrrell, A.M.: Automated scoring of a neuropsychological test: the Rey Osterrieth complex figure. In: Proceedings of the 26th Euromicro Conference. EUROMICRO 2000. Informatics: Inventing the Future, pp. 406–413. IEEE Comput. Soc (2000)
3. Canham, R.O., Smith, S.L., Tyrrell, A.M.: Location of structural sections from within a highly distorted complex line drawing. IEE Proc. Vis. Image Sig. Process. **152**(6), 741 (2005)
4. Conson, M., Siciliano, M., Baiano, C., Zappullo, I., Senese, V.P., Santangelo, G.: Normative data of the Rey-Osterrieth complex figure for Italian-speaking elementary school children. Neurol. Sci. **40**(10), 2045–2050 (2019)
5. Di Febbo, D., et al.: A decision support system for Rey-Osterrieth complex figure evaluation. Expert Syst. Appl. **213**, 119226 (2023)
6. Eitz, M., Hays, J., Alexa, M.: How do humans sketch objects? ACM Trans. Graph. (Proc. SIGGRAPH) **31**(4), 44:1–44:10 (2012)
7. Han, D., Yun, S., Heo, B., Yoo, Y.: Rethinking channel dimensions for efficient model design. In: Proceedings of the IEEE/CVF Conference on Computer Vision and Pattern Recognition (CVPR), pp. 732–741 (2021)
8. Han, K., Wang, Y., Zhang, Q., Zhang, W., XU, C., Zhang, T.: Model Rubik's cube: twisting resolution, depth and width for tinynets. In: Larochelle, H., Ranzato, M., Hadsell, R., Balcan, M., Lin, H. (eds.) Advances in Neural Information Processing Systems. vol. 33, pp. 19353–19364. Curran Associates, Inc. (2020)
9. He, K., Zhang, X., Ren, S., Sun, J.: Deep residual learning for image recognition. In: 2016 IEEE Conference on Computer Vision and Pattern Recognition (CVPR), pp. 770–778 (June 2016)
10. Huang, G., Liu, Z., van der Maaten, L., Weinberger, K.Q.: Densely connected convolutional networks. In: Staff, I. (ed.) 2017 IEEE Conference on Computer Vision and Pattern Recognition (CVPR), pp. 2261–2269. IEEE, Piscataway (July 2017)
11. Izmailov, P., Podoprikhin, D., Garipov, T., Vetrov, D., Wilson, A.G.: Averaging weights leads to wider optima and better generalization. In: Silva, R., Globerson, A., Globerson, A. (eds.) 34th Conference on Uncertainty in Artificial Intelligence 2018, UAI 2018, pp. 876–885. Association For Uncertainty in Artificial Intelligence (AUAI) (2018)
12. Jung, A.B., et al.: imgaug (2020)
13. Langer, N., et al.: The AI neuropsychologist: Automatic scoring of memory deficits with deep learning (2022). https://doi.org/10.1101/2022.06.15.496291
14. Liu, Z., Mao, H., Wu, C.Y., Feichtenhofer, C., Darrell, T., Xie, S.: A convnet for the 2020s. In: Proceedings of the IEEE/CVF Conference on Computer Vision and Pattern Recognition (CVPR), pp. 11976–11986 (2022)
15. Park, J.Y., Seo, E.H., Yoon, H.J., Won, S., Lee, K.H.: Automating Rey complex figure test scoring using a deep learning-based approach: A potential large-scale screening tool for congnitive decline (2022). https://www.researchsquare.com/article/rs-1973305/v1
16. Izmailov, P., Wilson, A.G.: Stochastic weight averaging in pytorch (29042019). https://pytorch.org/blog/stochastic-weight-averaging-in-pytorch/
17. Ross Wightman: Pytorch image models (2019)

18. Shin, M.S., Park, S.Y., Park, S.R., Seol, S.H., Kwon, J.S.: Clinical and empirical applications of the Rey-Osterrieth complex figure test. Nat. Protoc. **1**(2), 892–899 (2006)
19. Simfukwe, C., An, S.S., Youn, Y.C.: Comparison of RCF scoring system to clinical decision for the Rey complex figure using machine-learning algorithm. Dement. Neurocognitive Disord. **20**(4), 70–79 (2021)
20. Simonyan, K., Zisserman, A.: Very deep convolutional networks for large-scale image recognition. In: International Conference on Learning Representations (ICLR). San Diego (may 2015)
21. Webb, S.S., et al.: Validation of an automated scoring program for a digital complex figure copy task within healthy aging and stroke. Neuropsychology **35**(8), 847–862 (2021)
22. Li, Y.: Development of a Haptic-based Rey-Osterrieth Complex Figure Testing and Training System with Computer Scoring and Force-feedback Rehabilitation ... (2010). https://repository.lib.ncsu.edu/bitstream/handle/1840.16/6062/etd.pdf?sequence=1
23. Zhang, X., Lv, L., Min, G., Wang, Q., Zhao, Y., Li, Y.: Overview of the complex figure test and its clinical application in neuropsychiatric disorders, including copying and recall. Front. Neurol. **12**, 680474 (2021)

Structure Diagram Recognition
in Financial Announcements

Meixuan Qiao[1], Jun Wang[2(✉)], Junfu Xiang[2], Qiyu Hou[2], and Ruixuan Li[1]

[1] Huazhong University of Science and Technology, Wuhan, China
{qiaomeixuan,rxli}@hust.edu.cn
[2] iWudao Tech, Wuhan, China
{jwang,xiangjf,houqy}@iwudao.tech

Abstract. Accurately extracting structured data from structure diagrams in financial announcements is of great practical importance for
building financial knowledge graphs and further improving the efficiency
of various financial applications. First, we proposed a new method for
recognizing structure diagrams in financial announcements, which can
better detect and extract different types of connecting lines, including
straight lines, curves, and polylines of different orientations and angles.
Second, we developed a semi-automated, two-stage method to efficiently
generate the industry's first benchmark of structure diagrams from Chinese financial announcements, where a large number of diagrams were
synthesized and annotated using an automated tool to train a preliminary recognition model with fairly good performance, and then a high-
quality benchmark can be obtained by automatically annotating the real-
world structure diagrams using the preliminary model and then making
few manual corrections. Finally, we experimentally verified the significant
performance advantage of our structure diagram recognition method over
previous methods.

Keywords: Structure Diagram Recognition · Document AI ·
Financial Announcements

1 Introduction

As typical rich-format business documents, financial announcements contain
not only textual content, but also tables and graphics of various types and
formats, which also contain a lot of valuable financial information and data.
Document AI, which aims to automatically read, understand, and analyze rich-
format business documents, has recently become an important area of research
at the intersection of computer vision and natural language processing [3]. For
example, layout analysis [3], which detects and identifies basic units in documents (such as headings, paragraphs, tables, and graphics), and table structure recognition [3], which extracts the semantic structure of tables, have been
widely used in the extraction of structured data from various rich-format documents, including financial announcements, and a number of benchmark datasets

G. A. Fink et al. (Eds.): ICDAR 2023, LNCS 14187, pp. 20–44, 2023.
https://doi.org/10.1007/978-3-031-41676-7_2

have been established [11,32–34]. Nevertheless, research on the recognition and understanding of graphics in rich-format documents remains in the early stages. Although some studies have focused on recognizing flowcharts [7,18–20], statistical charts [8,9,14,23,30] and geometry problem [2,13,21], to the best of our knowledge, there is no existing work explicitly dedicated to recognizing and understanding various types of structure diagrams in financial announcements.

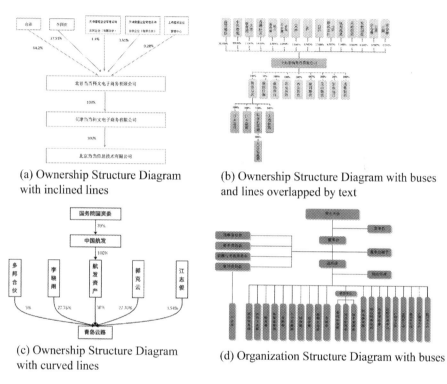

(a) Ownership Structure Diagram with inclined lines

(b) Ownership Structure Diagram with buses and lines overlapped by text

(c) Ownership Structure Diagram with curved lines

(d) Organization Structure Diagram with buses

Fig. 1. Typical examples of Structure Diagrams in Chinese Financial Announcements

Figure 1 (a), (b), and (c) are typical ownership structure diagrams extracted from Chinese financial announcements, where the nodes represent institutional or individual entities, and the connecting lines represent the ownership relationship and proportion between a pair of entities. Although the ownership structure information can also be obtained from a business registration database, generally there is a timeliness gap, and it often takes some time for the latest ownership changes disclosed in financial announcements to be updated and reflected in the business registration database. Figure 1 (d) shows a typical organization structure diagram in a Chinese financial announcement, where the nodes represent the departments or occupations in an institution, and the connecting lines between the nodes represent the superior-subordinate relationships. Recognizing organization structure diagrams can help understand the decision-making structure of an institution and analyze its business planning for future development. Currently, many practitioners in the financial industry still rely on manual data extraction from structure diagrams in Chinese financial announcements, which is time-consuming and error-prone. Therefore, it is of great practical significance to improve the efficiency of various related financial applications by automatically extracting structured data from structure diagrams in financial announcements and constructing corresponding financial knowledge graphs in a timely and accurate manner.

Although the recent deep learning-based object detection methods [7, 18–20, 24] have improved diagram recognition compared to traditional image processing-based methods [15, 17, 25], they mainly focus on flowchart recognition and are not very effective at structure diagram recognition, especially not good at detecting various connecting lines in structure diagrams.

First, to address the above problem, we proposed a new method called SDR by extending the Oriented R-CNN [26] with key point detection [6], which is particularly good at detecting various connecting lines with different orientations and angles, including straight lines, curves, and polylines.

Second, to overcome the lack of training data and the high cost of annotation, a two-stage method was developed to efficiently build a benchmark with high-quality annotations. (1) An automated tool has been built to efficiently synthesize and annotate a large number of structure diagrams of different styles and formats, which can be used to train a preliminary structure diagram recognition model. (2) The preliminary model can automatically annotate real-world structure diagrams with very reasonable quality, so that only a very limited number of manual corrections are required.

Third, to evaluate the effectiveness of our methods in real scenarios, we used the above two-stage method to build the industry's first benchmark containing 2216 real ownership structure diagrams and 1750 real organization structure diagrams extracted from Chinese financial announcements, and experimentally verified the significant performance advantage of our SDR method over previous methods.

2   Related Works

The work on flowchart recognition is the closest to the structure diagram recognition studied in this paper. Early work mainly used traditional image processing methods based on Connected Components Analysis and various heuristic rules [15, 17, 25], which usually had poor performance in detecting dashed and discontinuous lines and were easily disturbed by noise on poor quality scanned images.

Recently, some deep learning-based methods have been applied to flowchart recognition, and have made some progress compared to traditional image processing methods [7, 18–20, 24]. Julca-Aguilar et al. [7] first used Faster R-CNN [16] to detect symbols and connecting lines in handwritten flowcharts.

Subsequently, Schäfer et al. proposed a series of models [18–20], such as Arrow R-CNN [19], to extend the Faster R-CNN based on the characteristics of handwritten flowcharts. The Arrow R-CNN, like the Faster R-CNN, is constrained to generating horizontal rectangular bounding boxes. Furthermore, Arrow R-CNN was initially developed to detect simple and direct connecting lines between nodes, and it treats the entire path between each pair of linked nodes as a single connecting line object. These two factors make it difficult to correctly detect certain connecting lines in the structure diagrams. As shown in Fig. 3 (a), the horizontal bounding boxes of the two middle inclined lines are almost completely covered by the horizontal bounding boxes of the two outer inclined lines, resulting in one of the middle inclined lines not being correctly detected. When nodes on different layers are connected via buses, as shown in Fig. 5 (a), Arrow R-CNN can only annotate the entire multi-segment polyline connecting each pair of nodes as a single object, and causes the bounding boxes of the middle polylines to completely overlap with the bounding boxes of the outer polylines. And we can see that most of the connecting polylines containing the vertical short line segments in Fig. 5(a) are not detected. Sun et al. [24] created a new dataset containing more than 1000 machine-generated flowcharts, and proposed an end-to-end multi-task model called FR-DETR by merging DETR [1] for symbol detection and LETR [28] for line segment detection. FR-DETR assumed that the connection path between each pair of connected symbols in flowcharts was all composed of straight line segments, and defined only straight line segments as the object of connecting line detection. FR-DETR detects straight line segments by directly regressing the coordinates of the start and end points for each line segment, and thus can handle inclined line segments naturally. Although FR-DETR may be able to fit certain curves with straight lines, it cannot correctly detect connecting lines with high curvature. As shown in Fig. 4 (a), all four curves are not detected by FR-DETR. Also, in some ownership structure diagrams, connecting lines between nodes are overlapped with corresponding text, and this often causes FR-DETR to fail to detect these connecting lines with text. As shown in Fig. 6 (a), almost all of the short vertical lines overlaid with numbers are not detected. FR-DETR's Transformer-based model also results in a much slower recognition speed. Our SDR method can not only better detect various connecting lines in structure diagrams, but also further parse the corresponding semantic structure and extract the structured data based on the obtained connecting relationships.

3   System Framework

3.1   Structure Diagram Detection

An ownership or organization structure diagram usually appears in a specific section of a financial announcement, so the text in the announcement, including the section titles, can be analyzed to determine the page range of the section where the structure diagram is located, and then the layout of the candidate page can be further analyzed to locate the bounding boxes of the required structure

diagram. We used VSR [31] to implement layout analysis because it adds textual
semantic features to better distinguish different types of diagrams with similar
visual appearance (e.g., ownership and organizational structure diagrams) compared to models that use only visual features, such as Mask R-CNN [6], and
has a much lower training cost compared to the pre-trained language models,
such as LayoutLM [10,27,29]. The cyan area in Fig. 2 (a) and the pink area in
Fig. 2 (b) are two examples of the ownership structure diagram and organization
structure diagram, respectively, detected using layout analysis.

 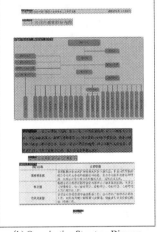

(a) Ownership Structure Diagram   (b) Organization Structure Diagram

Fig. 2. Examples of structure diagram detection using layout analysis.

3.2   Detection of Connecting Lines and Nodes

This paper proposed a new method called SDR for structure diagram recognition, which can better handle various connection structures between nodes in
particular. Figure 7 is an illustration of the network structure of our SDR model.
First, our SDR model extended the Oriented R-CNN model [26] to support the
detection of oriented objects, and this allows the output bounding boxes to be
rotated by a certain angle to achieve better detection of inclined or curved lines,
as shown in Fig. 3 (b) and Fig. 4 (b). In the first stage of SDR, we extend the
regular RPN in Faster R-CNN to produce oriented proposals, each correspond-
ing to a regression output $(x, y, w, h, \Delta\alpha, \Delta\beta)$ [26]. (x, y) is the center coordinate
of the predicted proposal, and w and h are the width and height of the external
rectangle box of the predicted oriented proposal. $\Delta\alpha$ and $\Delta\beta$ are the offsets relative to the midpoints of the top and right sides of the external rectangle. Each
oriented proposal generated by the oriented RPN is usually a parallelogram,
which can be transformed by a simple operation into an oriented rectangular

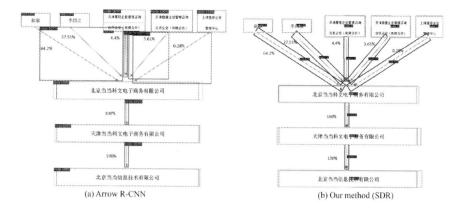

(a) Arrow R-CNN   (b) Our method (SDR)

Fig. 3. Recognition of Ownership Diagram with inclined lines

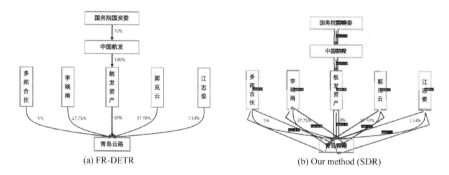

(a) FR-DETR   (b) Our method (SDR)

Fig. 4. Recognition of Ownership Diagram with curved lines

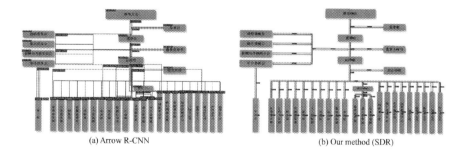

(a) Arrow R-CNN   (b) Our method (SDR)

Fig. 5. Recognition of Ownership Diagram with bus structure

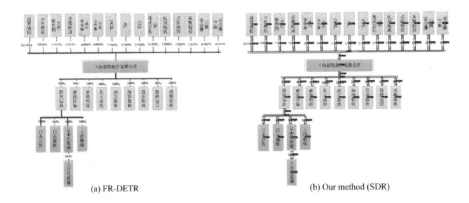

(a) FR-DETR   (b) Our method (SDR)

Fig. 6. Recognition of Ownership Diagram with lines overlapped by text

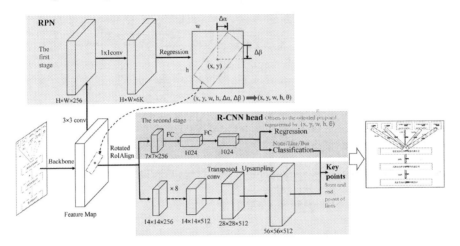

Fig. 7. Structure Diagram Recognition (SDR) Model

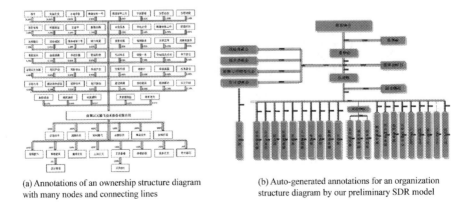

(a) Annotations of an ownership structure diagram   (b) Auto-generated annotations for an organization
with many nodes and connecting lines   structure diagram by our preliminary SDR model

Fig. 8. Some annotation examples for structure diagrams

proposal represented by (x, y, w, h, θ). The oriented rectangular proposal is projected onto the feature map for extraction of a fixed size feature vector using rotated RoIAlign [26]. In the second stage of SDR, each feature vector is fed into a series of fully-connected layers with two branches: one for classifying each proposal, and another one for regressing the offsets to the oriented rectangle corresponding to each proposal. In addition, unlike the regression method used in Arrow R-CNN, we integrate the keypoint detection method based on image segmentation used in Mask R-CNN [6] to detect the start and end points of the connecting lines with arrowheads, which facilitates the aggregation of the lines in post-processing.

To better handle the common bus structures in structure diagrams and the corresponding multi-segment polylines between the nodes connected through buses, as shown in Fig. 1 (b) and (d), SDR has defined the buses as a special type of detection object in addition to the regular connecting lines as a type of detection object. In the post-processing, the regular connecting lines and buses can be aggregated into the multi-segment polylines, which establish the corresponding connection relationship between the nodes. Figure 5 (b) showed that the SDR can accurately detect all the bus (colored in red) and the attached regular connecting lines (colored in blue). They can be easily aggregated into corresponding multi-segment polylines in post-processing, and the problem of Arrow R-CNN shown in Fig. 5 (a) can be avoided. Compared with the regression method of FR-DETR, our SDR is also very reliable at detecting line segments overlapped by text, as shown in Fig. 6 (b), and all the line segments that were not detected in Fig. 6 (a) were correctly detected by SDR. The shapes of the nodes are relatively limited compared to the connecting lines, so the accuracy of node detection is high for all the above methods, and the bounding boxes of the nodes are shown in green in all the figures.

3.3   Post-processing

The text blocks within a structural diagram are initially detected by an OCR system that utilizes DBNet [12] and CRNN [22] and is implemented in PaddleOCR[1]. This OCR system was trained on a dataset of Chinese financial announcements. After detection, the text blocks are merged into previously identified nodes or connected lines based on their respective coordinates. For example, if the bounding box of the current text block is located inside the bounding box of a node, the text recognized by OCR is used as the corresponding entity name of that node. Another example is that in an ownership structure diagram if the nearest object to the current text block is a line and the text is a number, that text is extracted as the ownership percentage corresponding to the line. For organization diagrams, text blocks are located only inside nodes and are recognized as names of the departments or occupations.

For each line other than the bus type, its start and end points must be connected to nodes or other lines, so that each line can always find the two

---

[1] https://github.com/PaddlePaddle/PaddleOCR/.

bounding boxes closest to its start and end points, respectively. For the two bounding boxes that are closest to the start and end of the current line, if both are the node type, the ownership relationship between these two nodes is created based on the ownership percentage corresponding to the current line if they are in an ownership structure diagram, or the subordination relationship between these two nodes is determined based on the current line if they are in an organization structure diagram. If one of the two bounding boxes closest to the start and end of the current line corresponds to a bus or a line, it is necessary to extend the current line by merging it with the current line until both bounding boxes connected to the current line are of node type. Not all lines have arrows. For lines with no clear direction, the default direction is top to bottom or left to right.

3.4   Structured Data Extraction from Diagrams

Although the ultimate goal of conducting diagram recognition is to construct financial knowledge graphs, this paper focuses mainly extracting structured data from diagrams, leaving aside tasks such as entity alignment and disambiguation for the time being. Each node pair detected and extracted from ownership structure diagrams, along with the corresponding ownership relationship, is output as a relation tuple of **(Owner, Percentage, Owned)**. Similarly, each node pair detected and extracted from organization structure diagrams, and the corresponding hierarchical relationship, is output as a relation tuple of **(Supervisor, Subordinate)**.

4   Semi-automated Two-Stage Method for Structure Diagram Annotation

Some structure diagrams often have a large number of nodes and dense connecting lines, as shown in Fig. 8 (a), and annotating all these relatively small object areas also requires more delicate operations, which can be time-consuming and labor-intensive if we rely solely on manual work. To address this problem, we developed a semi-automated, two-stage method to generate high-quality annotations for real-world structure diagrams.

In the first stage, we created an automated tool exploiting the structural properties of structure diagrams to generate structure diagrams for different scenarios and corresponding annotations for training structure diagram recognition models. In our previous work on extracting information from the textual content of financial announcements, we have built and accumulated a knowledge base of relevant entities of individuals, departments and institutions. A random integer n can be generated as the number of nodes from a given range, and then n entities can be randomly selected as nodes from the knowledge base according to the type of diagram being synthesized. The number of levels m is randomly chosen from a certain range according to the constraint on the number of nodes n, and then the nodes between two adjacent levels are connected in different patterns of one-to-one, one-to-many and many-to-one with certain probabilities, and a small number of shortcut connections between nodes that are not located in adjacent levels can be further randomly generated. Each of these connections corresponds to a specific relationship in a specific type of structure diagram, reflecting the topology in the structure diagram to be synthesized. A topology generated above is then imported into the visualization tool Graphviz [5] to automatically draw a corresponding structure diagram according to the settings. In the settings, we can choose the shape and color of the nodes and lines, as well as the orientation and angle of the lines, the font and the position of the text attached to the lines, and whether to use a bus structure to show one-to-many or many-to-one connections. Flexible settings allow the automated tool to synthesize a wide variety of structure diagrams covering a wide range of real-world scenarios (See Appendix 7.1 for more examples of synthesized structure diagrams). Graphviz is able to export the synthesized structure diagrams into SVG format, so we can easily obtain the coordinates of each object in the diagrams for corresponding annotations, and finally convert them into DOTA format [4] as training data. Then a preliminary model can be trained based on the automatically synthesized training data.

In the second stage, the preliminary model can be used to automatically annotate the real-world structure diagrams extracted from financial announcements. Our experiments in Sect. 5.2 show that the preliminary model usually has a reasonably good performance. A typical example in Fig. 8 (b) shows the results of the automatic annotation of a diagram using the preliminary model, where only a very small number of short vertical line segments are not correctly detected, and generally, we only need to make a few corrections to the automatic annotation results (See Appendix 7.2 for more examples of structure diagrams automatically annotated by the preliminary model). The manual correction tool first converts the auto-annotated DOTA data into the COCO format, then imports it into the COCO Annotator for correction, and finally converts it back to the DOTA format.

5   Experiments

5.1   Evaluation Metrics

Average Precision (AP) and mean Average Precision (mAP) are the most common metrics used to evaluate object detection [16]. Arrow key points do not participate in the mAP calculation because they are predicted by a separate head that is different from nodes, buses, and lines. We also use $Precision/Recall/F1$ as metrics for object detection, which are commonly used in flowchart recognition [19,24]. The IoU threshold for the above metrics is 50%.

However, in the task of structure diagram recognition, object detection is only one part of the process, and the ultimate goal is to obtain the topology of the structure diagram and extract the tuples of structured data for import into the knowledge base. Therefore, a method cannot only be measured in terms of object detection, but must also be evaluated on the structured data extracted through aggregation and post-processing. For an ownership relationship tuple **(Owner, Percentage, Owned)** or a subordinate relationship tuple **(Supervisor, Subordinate)**, a tuple is correct only if all elements contained in that tuple are correctly extracted. By checking the extracted tuples in each structure diagram, the *Precision/Recall/F*1 of the extracted tuples can be counted as metrics for evaluating structured data extraction.

5.2   Datasets

Synthesized Dataset. Based on the observation of the real data distribution, we automatically synthesized and annotated 8050 ownership structure diagrams and 4450 organization structure diagrams using the automated tool introduced in Sect. 4. Based on this synthesized dataset, we can train a preliminary SDR model for structure diagram recognition.

Real-World Benchmark Dataset. Following the method presented in Sect. 3.1, we first used layout analysis to automatically detect and extract some structure diagrams from publicly disclosed Chinese financial announcements such as prospectuses, and then invited several financial professionals to manually review and finally select 2216 ownership structure diagrams and 1750 organization diagrams. The principle is to cover as wide a range of different layout structures and styles as possible, including nodes of different shapes, colors, and styles, as well as connecting lines of different patterns, directions, and angles, in an attempt to maintain the diversity and complexity of the structure diagrams in a real-world scenario. Then, using the two-stage method presented in Sect. 4, we applied the preliminary SDR model trained on the synthesized dataset to automatically annotate all diagrams in the above real-world dataset, and then manually corrected the automatic annotations to create the industry's first structure diagram benchmark.

Comparing the results of automatic annotation with the results after manual corrections, the preliminary SDR model showed a fairly good performance, as shown in Table 1, and only a small number of corrections are needed. So actual experiments verify that the two-stage method can significantly improve efficiency and reduce costs.

As introduced in Sect. 3.2, Arrow RCNN, FR-DETR, and our SDR have different definitions of connecting lines, so the corresponding annotations are different as well. However, we can easily convert the annotation data used in our SDR to the annotation data used in Arrow R-CNN and FR-DETR automatically. A preliminary Arrow R-CNN model and a preliminary FR-DETR model were also trained on the synthesized dataset, respectively, and used to automatically annotate all the above real dataset. Table 1 shows that the preliminary Arrow R-CNN model is poor. However, the preliminary FR-DETR model is also reasonably good.

Table 1. Evaluation of the different preliminary models on the entire benchmark dataset

Datasets		Ownership				Organization			
Category	Metrics(%)	Arrow R-CNN	FR-DETR	Our SDR (R50)	Our SDR (Swin-S)	Arrow R-CNN	FR-DETR	Our SDR (R50)	Our SDR (Swin-S)
Node	Precision	**97.5**	96.1	92.7	92.5	97.2	95.6	98.9	**99.4**
	Recall	98.6	98.9	**99.3**	99.2	98.1	99.7	**99.9**	99.8
	F1	**98.0**	97.5	95.9	95.7	97.6	97.6	99.4	**99.6**
	AP	96.8	95.3	**98.8**	98.7	96.2	98.7	98.9	**99.9**
Line	Precision	49.0	71.9	79.0	**83.8**	52.7	82.0	85.7	**88.6**
	Recall	53.8	76.5	83.0	**88.8**	54.8	**86.7**	81.8	82.2
	F1	51.3	74.1	81.0	**86.2**	53.7	84.3	83.7	**85.3**
	AP	47.4	67.0	77.9	**85.3**	48.3	81.3	79.3	**81.0**
Bus	Precision	N/A	N/A	66.9	**75.5**	N/A	N/A	54.6	**70.2**
	Recall	N/A	N/A	80.6	**89.4**	N/A	N/A	94.4	**98.2**
	F1	N/A	N/A	73.1	**81.9**	N/A	N/A	69.2	**81.9**
	AP	N/A	N/A	72.9	**84.5**	N/A	N/A	91.4	**97.0**
mAP		72.1	81.2	83.2	**89.5**	72.3	90.0	89.9	**92.3**
Arrow keypoints	Precision	49.5	N/A	86.8	**91.6**	41.3	N/A	74.9	**75.3**
	Recall	52.4	N/A	95.5	**96.8**	44.8	N/A	**96.1**	95.6
	F1	50.9	N/A	90.9	**94.1**	43.0	N/A	**84.2**	84.2
	AP	46.9	N/A	83.0	**88.7**	39.6	N/A	71.6	**72.2**

5.3 Implementation of Baselines and Our SDR

As discussed in the previous sections, Arrow R-CNN and FR-DETR are the methods closest to our work, so they are selected as the baselines. Arrow R-CNN mainly extended Faster R-CNN for better handwritten flowchart recognition. It is not yet open source, so we modify the Faster R-CNN model obtained from Detectron2[2] to reproduce Arrow R-CNN according to the description in the original paper [19]. FR-DETR simply merged DETR and LETR into a multi-task model, and the experiments [24] showed that the multi-task FR-DETR performed slightly worse than the single-task of DETR in symbol detection and slightly worse than the single-task LETR in line segment detection, respectively. So even though FR-DETR is not open source, we can use separate DETR to get FR-DETR's upper bounds on node detection performance, and separate LETR to get FR-DETR's upper bounds on line segment detection performance. Our DETR codes are from the official implementation in Detectron2, and our LETR codes are from the official implementation of the paper [28]. Our SDR model mainly extended Oriented R-CNN implementation in MMRotate [35]. We tried two different backbones based on ResNet50 and Swin-Transformer-Small, referred to in all the tables as **R50** and **Swin-S**, respectively. See Appendix 7.3 for more implementation details.

5.4 Evaluation on Real-World Benchmark

After shuffling the real-world benchmark, we selected 1772 diagrams as the training set and 444 diagrams as the test set for the ownership structure diagrams, and 1400 diagrams as the training set and 350 diagrams as the test set for the organization structure diagrams.

---

[2] https://github.com/facebookresearch/detectron2.

Table 2. Evaluation on the recognition of ownership structure diagrams (Pre: Preliminary models, FT: Fine-tuned models)

Category	Metrics(%)	Arrow R-CNN		FR-DETR		Our SDR(R50)		Our SDR(Swin-S)	
		Pre	FT	Pre	FT	Pre	FT	Pre	FT
Node	Precision	97.6	98.7	96.9	**99.5**	93.3	99.0	92.9	99.1
	Recall	98.8	99.0	98.8	**99.8**	99.7	**99.8**	99.5	**99.8**
	F1	98.2	98.8	97.8	**99.6**	96.4	99.4	96.1	99.4
	AP	96.7	98.9	96.3	98.8	98.8	**98.9**	98.7	**98.9**
Line	Precision	50.9	67.6	73.2	88.5	79.4	92.8	84.5	**93.7**
	Recall	53.6	70.7	76.7	90.3	83.8	98.3	90.5	**98.8**
	F1	52.2	69.1	74.9	89.4	81.5	95.5	87.4	**96.2**
	AP	48.1	65.5	70.9	87.8	78.3	97.3	86.5	**98.2**
Bus	Precision	N/A	N/A	N/A	N/A	66.7	79.8	74.3	**85.3**
	Recall	N/A	N/A	N/A	N/A	81.6	97.7	90.1	**98.4**
	F1	N/A	N/A	N/A	N/A	73.4	87.8	81.4	**91.4**
	AP	N/A	N/A	N/A	N/A	73.2	96.5	85.0	**97.6**
mAP		72.4	82.2	83.6	93.3	83.4	97.6	90.1	**98.2**
Arrow keypoints	Precision	50.4	54.8	N/A	N/A	86.7	96.9	91.3	**97.0**
	Recall	49.1	57.6	N/A	N/A	96.0	**98.7**	97.0	98.5
	F1	47.2	56.2	N/A	N/A	91.1	**97.8**	94.1	97.7
	AP	44.0	52.7	N/A	N/A	83.7	**95.1**	88.9	95.0

Table 3. Evaluation on the recognition of organization structure diagrams (Pre: Preliminary models, FT: Fine-tuned models)

Category	Metrics(%)	Arrow R-CNN		FR-DETR		Our SDR(R50)		Our SDR(Swin-S)	
		Pre	FT	Pre	FT	Pre	FT	Pre	FT
Node	Precision	96.7	98.9	98.0	99.6	99.4	**99.8**	99.6	99.7
	Recall	98.9	99.3	98.8	**99.9**	99.9	99.9	99.8	**99.9**
	F1	97.8	99.7	98.4	99.7	99.6	**99.8**	99.7	**99.8**
	AP	96.5	98.7	97.9	99.0	98.8	**99.0**	98.9	**99.0**
Line	Precision	51.5	77.4	89.0	95.5	87.3	95.9	88.0	**98.4**
	Recall	53.2	73.5	86.4	**96.8**	79.2	95.4	79.9	95.3
	F1	52.3	75.4	87.7	96.1	83.1	95.6	83.8	**96.8**
	AP	49.0	68.3	81.6	94.0	77.4	**95.0**	78.0	**95.0**
Bus	Precision	N/A	N/A	N/A	N/A	53.8	90.0	69.9	**95.3**
	Recall	N/A	N/A	N/A	N/A	93.9	98.3	97.9	**98.5**
	F1	N/A	N/A	N/A	N/A	68.4	94.0	81.6	**96.9**
	AP	N/A	N/A	N/A	N/A	91.1	**98.0**	96.4	97.9
mAP		72.8	83.5	89.8	96.5	89.1	**97.3**	91.1	**97.3**
Arrow keypoints	Precision	42.0	51.0	N/A	N/A	70.3	**97.7**	72.5	96.3
	Recall	44.3	54.2	N/A	N/A	95.9	**99.6**	94.3	**99.6**
	F1	43.1	52.6	N/A	N/A	81.1	**98.6**	82.0	97.9
	AP	40.2	48.1	N/A	N/A	67.7	**96.7**	68.6	95.7

Table 4. Evaluation of extracting structured data from the ownership structure diagrams (Pre: Preliminary models, FT: Fine-tuned models)

Metrics(%)	Arrow R-CNN		FR-DETR		Our SDR(R50)		Our SDR(Swin-S)	
	Pre	FT	Pre	FT	Pre	FT	Pre	FT
Precision	45.6	61.2	78.7	85.6	80.2	**91.5**	85.2	91.4
Recall	47.9	64.8	72.2	82.8	75.6	87.0	81.3	**87.6**
F1	46.7	62.9	75.3	84.2	77.8	89.2	83.2	**89.5**

Table 5. Evaluation of extracting structured data from the organization structure diagrams (Pre: Preliminary models, FT: Fine-tuned models)

Metrics(%)	Arrow R-CNN		FR-DETR		Our SDR(R50)		Our SDR(Swin-S)	
	Pre	FT	Pre	FT	Pre	FT	Pre	FT
Precision	46.5	66.1	81.4	90.9	81.8	**91.9**	82.3	91.8
Recall	50.6	69.0	78.5	86.6	76.4	87.4	77.9	**87.5**
F1	48.5	67.5	79.9	88.7	79.0	**89.6**	80.0	**89.6**

Table 6. Parameter number and inference time of different models.

Network	params	seconds per image
Arrow R-CNN	59.0M	0.18
FR-DETR	59.5M	1.22
Our SDR(R50)	59.1M	**0.16**
Our SDR(Swin-S)	83.8M	0.21

For the input size of diagram images, the longer side of the image is scaled to 1024 and the shorter side is subsequently scaled according to the aspect ratio of the original image before being input to the model. For all models in the experiments, we followed the above configuration.

Table 2 and Table 3 showed the evaluations performed on the test set for the ownership structure diagrams and the test set for the organization structure diagrams, respectively. As discussed in Sect. 3.2, the shapes of the nodes are relatively limited compared to the connecting lines, so the performance of node detection is good for all the above methods, and we focus on the detection of connecting lines in this section. Although the synthesized data tried to capture the real data distribution as much as possible, there are always some complex and special cases in the real data that differ from the regular scenarios, which inevitably lead to some differences between the two. Therefore, fine-tuning on real training set can allow the model to better adapt to the real data distribution and thus further improve the performance of the model, and the experimental results in Table 2 and Table 3 showed the effectiveness of fine-tuning.

Schäfer et al. [19] also observed that when some connecting lines are very close together in handwritten flowcharts, their horizontal bounding boxes overlap each other to a large extent, so they tried to increase the IoU threshold of NMS from 0.5 used in Faster R-CNN to 0.8 to improve the detection of lines with overlapping bounding boxes in Arrow R-CNN. After increasing the threshold, the detection performance of the connection lines did improve significantly, with F1 increasing from 51.8% to 69.1% on the ownership test set and from 57.7% to 73.5% on the organization test set, but Arrow R-CNN is still difficult to correctly detect many common connecting lines whose horizontal rectangular bounding boxes completely overlap each other, as shown in Fig. 3 (a) and Fig. 5 (a). Therefore, the performance of the Arrow R-CNN in the detection of connecting lines is still not so good.

FR-DETR does not do a good job of detecting curves, but curves occur in a relatively small percentage of the current dataset of structure diagrams, so this has less impact on its overall performance. FR-DETR only detects each individual straight line segment as an object, instead of detecting each complex ployline containing multiple line segments as an object like Arrow R-CNN. Therefore, its task is simpler than Arrow R-CNN, and the corresponding performance is much better. As shown in Fig. 6 (a), there are often text overlays on connecting lines in ownership structure diagrams, and LETR in FR-DETR is not very robust and can cause many connecting lines to be missed. Therefore, as shown in Table 2, FR-DETR's performance is not as good as SDRs. However, for organization structure diagrams, the connecting lines are generally not covered by text, so FR-DETR's detection of connecting lines by FR-DETR is not affected, and its performance is very close to that of SDRs, as shown in Table 3.

Consistent with Table 1, Table 2 and Table 3 show that the SDR (Swin-S) has a significant advantage over the SDR (R50) on the preliminary models before fine tuning on real data. This may be due to the fact that Swin-Transformer has better feature extraction and representation capabilities, and thus better generalization capabilities. The advantage of the SDR (Swin-S) over the SDR (R50) is relatively small after fine-tuning on real data.

We also examined the SDR failure cases, which are typically due to very short lines connected to the bus, complex structures containing intersecting lines, and interference from dashed line boxes. See Appendix 7.4 for some examples.

To facilitate the extraction of topology and structured data, our SDR defines buses as a special type of object that allows for easier and more reliable aggregation in post-processing. The relationship tuples obtained from structured data extraction were evaluated on the test sets of the ownership structure diagrams and the organization structure diagrams, respectively, according to the method introduced in Sect. 3.4. Table 4 and Table 5 show that SDR is also better than FR-DETR, and Arrow R-CNN is still the worst, and the fine-tuning also has significant effects.

Table 6 compares the number of parameters and inference speed of the above models. SDR (R50) has the fastest speed and its number of parameters is comparable to Arrow R-CNN, but its recognition performance is much better than Arrow R-CNN. The number of parameters of FR-DETR is comparable to that of SDR (R50), and FR-DETR's overall recognition performance is also not far from that of SDR (R50), but SDR (R50)'s inference speed is more than 7 times FR-DETR's. FR-DETR's speed bottleneck is mainly the LETR part. SDR (Swin-S) has the largest number of parameters and the best recognition performance, and its inference speed is 30% slower than SDR(R50), and but still more than 4 times faster than FR-DETR.

All the above experimental results show that the SDR method proposed in this paper has significantly improved compared to the previous methods, especially the detection of various connecting lines, which in turn leads to an improvement in the structured data extraction corresponding to the connection relationships.

6   Conclusion

In this paper, we proposed a new method for structure diagram recognition that can better detect various complex connecting lines. We also developed a two-stage method to efficiently generate high-quality annotations for real-world diagrams, and constructed the industry's first structure diagram benchmark from real financial announcements. Empirical experiments validated the significant performance advantage of our proposed methods over existing methods. In the future, we plan to extend the methods in this paper and apply it to the recognition of more types of diagrams, such as process diagrams.

7   Appendix

7.1   Examples of Synthesized Structure Diagrams

Figure 9 and Fig. 10 show some examples of synthesized ownership structure diagrams and organization structure diagrams, respectively.

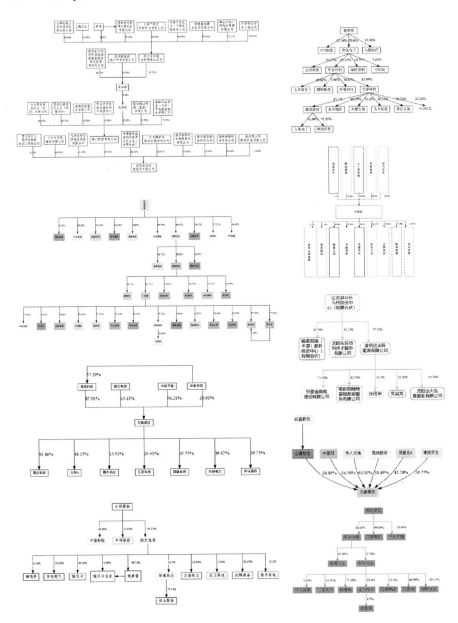

Fig. 9. Some examples of Synthesized Ownership Structure Diagrams

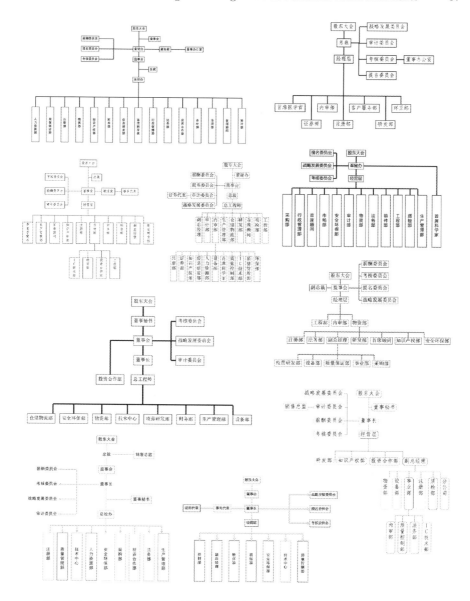

Fig. 10. Some examples of Synthesized Organization Structure Diagrams

7.2 Examples of Structure Diagrams Automatically Annotated by the Preliminary SDR Models

Figure 11 and Fig. 12 show some examples of ownership structure diagrams and organization structure diagrams, respectively, automatically annotated by the preliminary SDR models.

7.3   Implementation details for Arrow R-CNN, FR-DETR and SDR

Arrow R-CNN. The backbone is ResNet50 with FPN, and is initialized with the ImageNet-based pre-training model[3]. The model was trained using Momentum SGD as the optimizer, with a batch size of 2, a maximum number of iterations of 38000, an initial learning rate of 0.0025, and dividing by 10 at the 32000th and 36000th iteration. Horizontal and vertical flips and rotations were used for data augmentation during training.

FR-DETR. For DETR, the backbone is ResNet50 with pretrained weights[4]. The model was trained with a batch size of 2 and a total of 300 epochs, and the learning rate was divided by 10 at the 200th epochs. The model was optimized using AdamW with the learning rate of 1e−04, and the weight decay of 1e−04. Only horizontal flips were used for data augmentation during training.

For LETR, The backbone is ResNet50[5]. The model was trained with a batch size of 1, and 25 epochs for focal-loss fine-tuning on the pre-training model. The model was optimized using AdamW with the learning rate of 1e−05, and the weight decay of 1e−4. Horizontal and vertical flips were used for data augmentation during training.

SDR. The codes are modified and extended from MMRotate[6]. The model was trained with a batch size of 1 and a total of 12 epochs, and the learning rate was divided by 10 at the 8th and 11th epochs. Horizontal and vertical flips and rotations were used for data augmentation.

One option for the backbone is ResNet50, and the network was optimized using the SGD algorithm with a momentum of 0.9, a weight decay of 0.0001, and an initial learning rate set to 1.25e−03.

Another backbone option is Swin-Transformer-small, and the network was optimized using AdamW with the learning rate of 2.5e−05, and the weight decay of 0.05.

Our SDR provided seven customized anchor box aspect ratios (0.02, 0.1, 0.5, 1.0, 2.0, 4.0, 10.0) to accommodate different types of objects with different sizes and shapes in structure diagrams.

---

[3] https://github.com/facebookresearch/detectron2.
[4] https://github.com/facebookresearch/detr.
[5] https://github.com/mlpc-ucsd/LETR.
[6] https://github.com/open-mmlab/mmrotate.

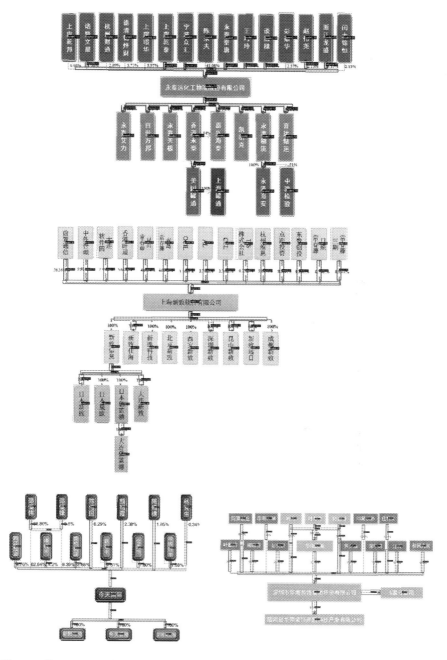

Fig. 11. Some examples of Ownership Structure Diagrams Automatically Annotated by the Preliminary SDR Model

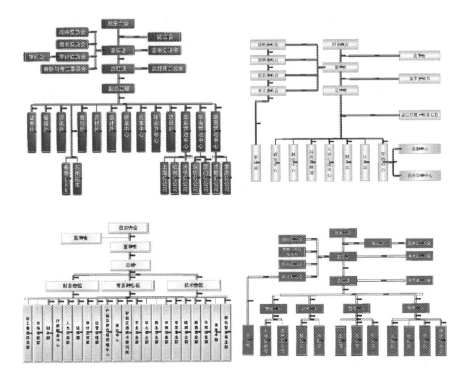

Fig. 12. Some examples of Organization Structure Diagrams Automatically Annotated by the Preliminary SDR Model

7.4   Examples of SDR Failure Cases

Figure 13 shows some examples of SDR failure cases of detecting connecting lines.

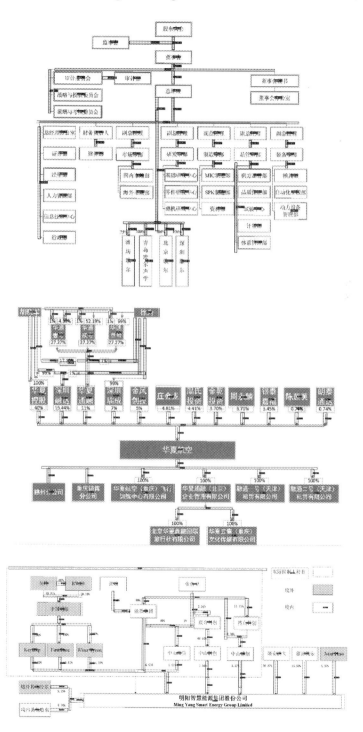

Fig. 13. Some examples of SDR failure cases of detecting connecting lines

References

1. Carion, N., Massa, F., Synnaeve, G., Usunier, N., Kirillov, A., Zagoruyko, S.: End-to-end object detection with transformers. In: Vedaldi, A., Bischof, H., Brox, T., Frahm, J.-M. (eds.) ECCV 2020. LNCS, vol. 12346, pp. 213–229. Springer, Cham (2020). https://doi.org/10.1007/978-3-030-58452-8_13
2. Chen, J., et al.: GeoQA: a geometric question answering benchmark towards multimodal numerical reasoning (2021)
3. Cui, L., Xu, Y., Lv, T., Wei, F.: Document AI: benchmarks, models and applications. CoRR abs/2111.08609 (2021). https://arxiv.org/abs/2111.08609
4. Ding, J., et al.: Object detection in aerial images: a large-scale benchmark and challenges. IEEE Trans. Pattern Anal. Mach. Intell. 1–1 (2021). https://doi.org/10.1109/TPAMI.2021.3117983
5. Ellson, J., Gansner, E.R., Koutsofios, E., North, S.C., Woodhull, G.: Graphviz and dynagraph - static and dynamic graph drawing tools. In: Jünger, M., Mutzel, P. (eds.) Graph Drawing Software, pp. 127–148. Springer, Heidelberg (2003). https://doi.org/10.1007/978-3-642-18638-7_6
6. He, K., Gkioxari, G., Dollár, P., Girshick, R.: Mask R-CNN. In: 2017 IEEE International Conference on Computer Vision (ICCV), pp. 2980–2988 (2017). https://doi.org/10.1109/ICCV.2017.322
7. Julca-Aguilar, F., Hirata, N.: Symbol detection in online handwritten graphics using faster R-CNN. In: 13th IAPR International Workshop on Document Analysis Systems (DAS) (2018)
8. Kantharaj, S., et al.: Chart-to-text: a large-scale benchmark for chart summarization. In: Proceedings of the 60th Annual Meeting of the Association for Computational Linguistics (Volume 1: Long Papers), Dublin, Ireland, May 2022, pp. 4005–4023. Association for Computational Linguistics (2022). https://aclanthology.org/2022.acl-long.277
9. Kato, H., Nakazawa, M., Yang, H.K., Chen, M., Stenger, B.: Parsing line chart images using linear programming. In: 2022 IEEE/CVF Winter Conference on Applications of Computer Vision (WACV), pp. 2553–2562 (2022). https://doi.org/10.1109/WACV51458.2022.00261
10. Li, C., et al.: StructuralLM: structural pre-training for form understanding. In: Proceedings of the 59th Annual Meeting of the Association for Computational Linguistics and the 11th International Joint Conference on Natural Language Processing (Volume 1: Long Papers), August 2021, pp. 6309–6318. Association for Computational Linguistics (2021). https://doi.org/10.18653/v1/2021.acl-long.493, https://aclanthology.org/2021.acl-long.493
11. Li, M., et al.: DocBank: a benchmark dataset for document layout analysis. In: Proceedings of the 28th International Conference on Computational Linguistics, Barcelona, Spain, December 2020, pp. 949–960. International Committee on Computational Linguistics (2020). https://doi.org/10.18653/v1/2020.coling-main.82, https://aclanthology.org/2020.coling-main.82
12. Liao, M., Zou, Z., Wan, Z., Yao, C., Bai, X.: Real-time scene text detection with differentiable binarization and adaptive scale fusion. IEEE Trans. Pattern Anal. Mach. Intell. **45**, 919–931 (2022)
13. Lu, P., et al.: Inter-GPS: interpretable geometry problem solving with formal language and symbolic reasoning. In: Annual Meeting of the Association for Computational Linguistics (2021)

14. Ma, W., et al.: Towards an efficient framework for data extraction from chart images. In: Lladós, J., Lopresti, D., Uchida, S. (eds.) ICDAR 2021. LNCS, vol. 12821, pp. 583–597. Springer, Cham (2021). https://doi.org/10.1007/978-3-030-86549-8_37

15. Mörzinger, R., Schuster, R., Horti, A., Thallinger, G.: Visual structure analysis of flow charts in patent images. In: Forner, P., Karlgren, J., Womser-Hacker, C. (eds.) CLEF 2012 Evaluation Labs and Workshop, Online Working Notes, Rome, Italy, September 17–20, 2012. CEUR Workshop Proceedings, vol. 1178. CEUR-WS.org (2012). https://ceur-ws.org/Vol-1178/CLEF2012wn-CLEFIP-MorzingerEt2012.pdf

16. Ren, S., He, K., Girshick, R., Sun, J.: Faster R-CNN: towards real-time object detection with region proposal networks. In: Cortes, C., Lawrence, N., Lee, D., Sugiyama, M., Garnett, R. (eds.) Advances in Neural Information Processing Systems, vol. 28. Curran Associates, Inc. (2015). https://proceedings.neurips.cc/paper/2015/file/14bfa6bb14875e45bba028a21ed38046-Paper.pdf

17. Rusiñol, M., de las Heras, L.P., Terrades, O.R.: Flowchart recognition for nontextual information retrieval in patent search. Inf. Retrieval **17**(5), 545–562 (2014). https://doi.org/10.1007/s10791-013-9234-3

18. Schäfer, B., van der Aa, H., Leopold, H., Stuckenschmidt, H.: Sketch2BPMN: automatic recognition of hand-drawn BPMN models. In: La Rosa, M., Sadiq, S., Teniente, E. (eds.) CAiSE 2021. LNCS, vol. 12751, pp. 344–360. Springer, Cham (2021). https://doi.org/10.1007/978-3-030-79382-1_21

19. Schäfer, B., Keuper, M., Stuckenschmidt, H.: Arrow R-CNN for handwritten diagram recognition. Int. J. Doc. Anal. Recogn. (IJDAR) **24**(1), 3–17 (2021). https://doi.org/10.1007/s10032-020-00361-1

20. Schäfer, B., Stuckenschmidt, H.: DiagramNet: hand-drawn diagram recognition using visual arrow-relation detection. In: Lladós, J., Lopresti, D., Uchida, S. (eds.) ICDAR 2021. LNCS, vol. 12821, pp. 614–630. Springer, Cham (2021). https://doi.org/10.1007/978-3-030-86549-8_39

21. Seo, M.J., Hajishirzi, H., Farhadi, A., Etzioni, O.: Diagram understanding in geometry questions. In: Proceedings of the AAAI Conference on Artificial Intelligence, vol. 28 (2014)

22. Shi, B., Bai, X., Yao, C.: An end-to-end trainable neural network for image-based sequence recognition and its application to scene text recognition. IEEE Trans. Pattern Anal. Mach. Intell. **39**(11), 2298–2304 (2017). https://doi.org/10.1109/TPAMI.2016.2646371

23. Sohn, C., Choi, H., Kim, K., Park, J., Noh, J.: Line chart understanding with convolutional neural network. Electronics **10**(6) (2021). https://doi.org/10.3390/electronics10060749, https://www.mdpi.com/2079-9292/10/6/749

24. Sun, L., Du, H., Hou, T.: FR-DETR: end-to-end flowchart recognition with precision and robustness. IEEE Access **10**, 64292–64301 (2022). https://doi.org/10.1109/ACCESS.2022.3183068

25. Thean, A., Deltorn, J., Lopez, P., Romary, L.: Textual summarisation of flowcharts in patent drawings for CLEF-IP 2012. In: Forner, P., Karlgren, J., Womser-Hacker, C. (eds.) CLEF 2012 Evaluation Labs and Workshop, Online Working Notes, Rome, Italy, September 17–20, 2012. CEUR Workshop Proceedings, vol. 1178. CEUR-WS.org (2012). https://ceur-ws.org/Vol-1178/CLEF2012wn-CLEFIP-TheanEt2012.pdf

26. Xie, X., Cheng, G., Wang, J., Yao, X., Han, J.: Oriented R-CNN for object detection. In: Proceedings of the IEEE/CVF International Conference on Computer Vision (ICCV), pp. 3520–3529 (October 2021)

27. Xu, Y., et al.: LayoutLMv2: multi-modal pre-training for visually-rich document understanding. In: Proceedings of the 59th Annual Meeting of the Association for Computational Linguistics and the 11th International Joint Conference on Natural Language Processing (Volume 1: Long Papers), August 2021, pp. 2579–2591. Association for Computational Linguistics (2021). https://doi.org/10.18653/v1/2021.acl-long.201, https://aclanthology.org/2021.acl-long.201

28. Xu, Y., Xu, W., Cheung, D., Tu, Z.: Line segment detection using transformers without edges. In: 2021 IEEE/CVF Conference on Computer Vision and Pattern Recognition (CVPR), pp. 4255–4264 (2021). https://doi.org/10.1109/CVPR46437.2021.00424

29. Xu, Y., et al.: Layoutxlm: multimodal pre-training for multilingual visually-rich document understanding. CoRR abs/2104.08836 (2021). https://arxiv.org/abs/2104.08836

30. Zhang, H., Ma, W., Jin, L., Huang, Y., Ding, K., Wu, Y.: DeMatch: towards understanding the panel of chart documents. In: Lladós, J., Lopresti, D., Uchida, S. (eds.) ICDAR 2021. LNCS, vol. 12823, pp. 692–707. Springer, Cham (2021). https://doi.org/10.1007/978-3-030-86334-0_45

31. Zhang, P., et al.: VSR: a unified framework for document layout analysis combining vision, semantics and relations. In: Lladós, J., Lopresti, D., Uchida, S. (eds.) ICDAR 2021. LNCS, vol. 12821, pp. 115–130. Springer, Cham (2021). https://doi.org/10.1007/978-3-030-86549-8_8

32. Zheng, X., Burdick, D., Popa, L., Zhong, P., Wang, N.X.R.: Global table extractor (GTE): a framework for joint table identification and cell structure recognition using visual context. In: Winter Conference for Applications in Computer Vision (WACV) (2021)

33. Zhong, X., ShafieiBavani, E., Yepes, A.J.: Image-based table recognition: data, model, and evaluation. arXiv preprint arXiv:1911.10683 (2019)

34. Zhong, X., Tang, J., Yepes, A.J.: PubLayNet: largest dataset ever for document layout analysis. In: 2019 International Conference on Document Analysis and Recognition (ICDAR), September 2019, pp. 1015–1022. IEEE (2019). https://doi.org/10.1109/ICDAR.2019.00166

35. Zhou, Y., et al.: MMRotate: a rotated object detection benchmark using pytorch. In: Proceedings of the 30th ACM International Conference on Multimedia (2022)

TransDocAnalyser: A Framework for Semi-structured Offline Handwritten Documents Analysis with an Application to Legal Domain

Sagar Chakraborty[1,2]([✉]), Gaurav Harit[2], and Saptarshi Ghosh[3]

[1] Wipro Limited, Salt Lake, Kolkata, India
[2] Department of Computer Science and Engineering, Indian Institute of Technology, Jodhpur, Rajasthan, India
{chakraborty.4,gharit}@iitj.ac.in
[3] Department of Computer Science and Engineering, Indian Institute of Technology, Kharagpur, West Bengal, India
saptarshi@cse.iitkgp.ac.in

Abstract. State-of-the-art offline Optical Character Recognition (OCR) frameworks perform poorly on semi-structured handwritten domain-specific documents due to their inability to localize and label form fields with domain-specific semantics. Existing techniques for semi-structured document analysis have primarily used datasets comprising invoices, purchase orders, receipts, and identity-card documents for benchmarking. In this work, we build the first semi-structured document analysis dataset in the legal domain by collecting a large number of First Information Report (FIR) documents from several police stations in India. This dataset, which we call the FIR dataset, is more challenging than most existing document analysis datasets, since it combines a wide variety of handwritten text with printed text. We also propose an end-to-end framework for offline processing of handwritten semi-structured documents, and benchmark it on our novel FIR dataset. Our framework used Encoder-Decoder architecture for localizing and labelling the form fields and for recognizing the handwritten content. The encoder consists of Faster-RCNN and Vision Transformers. Further the Transformer-based decoder architecture is trained with a domain-specific tokenizer. We also propose a post-correction method to handle recognition errors pertaining to the domain-specific terms. Our proposed framework achieves state-of-the-art results on the FIR dataset outperforming several existing models.

Keywords: Semi-structured document · Offline handwriting recognition · Legal document analysis · Vision Transformer · FIR dataset

1 Introduction

Semi-Structured documents are widely used in many different industries. Recent advancement in digitization has increased the demand for analysis of scanned or

G. A. Fink et al. (Eds.): ICDAR 2023, LNCS 14187, pp. 45–62, 2023.
https://doi.org/10.1007/978-3-031-41676-7_3

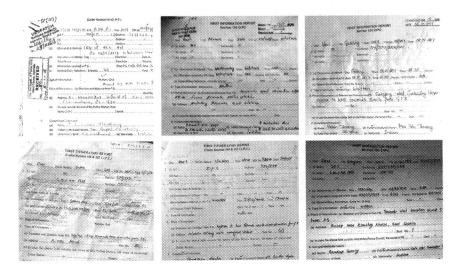

Fig. 1. Examples of First Information Report (FIR) documents from different police stations in India. The FIR dataset developed in this paper consists of a wide variety of such semi-structured FIR documents containing both printed and handwritten text.

mobile-captured semi-structured documents. Many recent works have used different deep learning techniques to solve some of the critical problems in processing and layout analysis of semi-structured documents [16,23,34]. Semi-structured documents consist of printed, handwritten, or hybrid (both printed and handwritten) text forms. In particular, hybrid documents (see Fig. 1) are more complex to analyze since they require segregation of printed and handwritten text and subsequent recognition. With recent advancements, the OCR accuracy has improved for printed text; however, recognition of handwritten characters is still a challenge due to variations in writing style and layout.

Earlier works have focused on techniques for layout analysis, named-entity recognition, offline handwriting recognition, etc., but sufficient work has *not* been done on developing an end-to-end framework for processing semi-structured documents. A general end-to-end framework can be easily fine-tuned for domain-specific requirements. In this paper we present the first framework for semi-structured document analysis applied to legal documents.

There have been many works on legal documents, such as on case document summarization [6], relevant statute identification from legal facts [31], pretraining language models on legal text [32] and so on. But almost all prior research in the legal domain has focused on textual data, and *not* on document images. In particular, the challenges involved in document processing and layout analysis of legal documents is unattended, even though these tasks have become important due to the increasing availability of scanned/photographed legal documents.

In this work, we build the first dataset for semi-structured document analysis in the legal domain. To this end, we focus on **First Information Report** (FIR) documents from India. An FIR is usually prepared by police stations in some

South Asian countries when they first get a complaint by the victim of a crime (or someone on behalf of the victim).[1] An FIR usually contains a lot of details such as the date, time, place, and details of the incident, the names of the person(s) involved, a list of the statutes (written laws, e.g., those set by the Constitution of a country) that might have been violated by the incident, and so on. The FIRs are usually written on a printed form, where the fields are filled in by hand by police officials (see examples in Fig. 1). It is estimated that more than 6 million FIRs are filed every year across thousands of police stations in various states in India. Such high volumes lead to inconsistent practices in-terms of handwriting, layout structure, scanning procedure, scan quality, etc., and introduce huge noise in the digital copies of these documents.

Our target fields of interest while processing FIR documents are the hand-written entries (e.g., name of the complainant, the statutes violated) which are challenging to identify due to the wide variation in handwriting. To form the dataset, which we call the **FIR dataset**, we created the meta-data for the target fields by collecting the actual text values from the police databases, and also annotated the documents with layout positions of the target fields. The FIR dataset is made publicly available at https://github.com/LegalDocumentProcessing/FIR_Dataset_ICDAR2023.

The FIR dataset is particularly challenging since its documents are of mixed type, with both printed and handwritten text. Traditional OCR identifies blocks of text strings in documents and recognizes the text from images by parsing from left to right [19]. NLP techniques like named-entity recognition (NER), which uses raw text to find the target fields, cannot be applied easily, since traditional OCRs do not work well in recognition of mixed documents with handwritten and printed characters occurring together. Another drawback of traditional OCRs in this context is their inability to recognise domain-specific words due to their general language-based vocabulary. In this work, we propose a novel framework for analysing such domain-specific semi-structured documents. The contributions of the proposed framework as follows:

1. We use a FastRCNN + Vision Transformer-based encoder trained for target field localization and classification. We also deploy a BERT-based text decoder that is fine-tuned to incorporate legal domain-specific vocabulary.
2. We use a domain-specific pretrained language model [32] to improve the recognition of domain-specific text (legal statutes, Indian names, etc.). This idea of using a domain-specific language model along with OCR is novel, and has a wider applicability over other domains (e.g., finance, healthcare, etc.) where this technique can be used to achieve improved recognition from domain-specific documents.
3. We improve the character error rate (CER) by reducing the ambiguities in OCR through a novel domain-specific post-correction step. Using domain knowledge, we created a database for each target field (such as Indian names, Indian statutes, etc.) to replace the ambiguous words from OCR having low

---

[1] https://en.wikipedia.org/wiki/First_information_report.

confidence using a combination of TF-IDF vectorizer and K-Nearest Neighbour classifier. This novel post-correction method to handle recognition errors pertaining to proper nouns, enables our proposed framework to outperform state-of-the-art OCR models by large margins.

To summarize, in this work we build the first legal domain-specific dataset for semi-structured document analysis. We also develop a framework to localise the handwritten target fields, and fine-tune a transformer-based OCR (TrOCR) to extract handwritten text. We further develop post-correction techniques to improve the character error rate. To our knowledge, the combination of Faster-RCNN and TrOCR with other components, such as Vision Transformer and legal domain-specific tokenizers, to create an end-to-end framework for processing offline handwritten semi-structured documents is novel, and can be useful for analysis of similar documents in other domains as well.

2   Related Work

We briefly survey four types of prior works related to our work – (i) related datasets, (ii) works addressing target field localization and classification, (iii) handwritten character recognition, and (iv) works on post-OCR correction methods.

Related Datasets: There exist several popular datasets for semi-structured document analysis. FUNSD [22] is a very popular dataset for information extraction and layout analysis. FUNSD dataset is a subset of RVL-CDIP dataset [17], and contains 199 annotated financial forms. The SROIE dataset [21] contains 1,000 annotated receipts having 4 different entities, and is used for receipt recognition and information extraction tasks. The CloudSCan Invoice dataset [29] is a custom dataset for invoice information extraction. The dataset contained 8 entities in printed text.

Note that no such dataset exists in the legal domain, and our FIR dataset is the first of its kind. Also, the existing datasets contain only printed text, while the dataset we build contains a mixture of printed and hand-written text (see Table 2 for a detailed comparison of the various datasets).

Localization and Labelling of Field Components: Rule-based information extraction methods (such as the method developed by Kempf et al. [10] and many other methods) could be useful when documents are of high quality and do not contain handwritten characters. But when document layouts involve huge variations, noise and handwritten characters, keyword-based approaches fail to provide good results. Template-based approaches also fail due to scanning errors and layout variability [1,2,36].

Srivastava et al. [12] developed a graph-based deep network for predicting the associations between field labels and field values in handwritten form images. They considered forms in which the field label comprises printed text and field value can be handwritten text; this is similar to what we have in the FIR dataset developed in this work. To perform association between the target field labels and values, they formed a graphical representation of the textual scripts using their associated layout position.

In this work, we tried to remove the dependency on OCR of previous works [12] by using layout information of images to learn the positions of target fields and extract the image patches using state-of-the-art object detection models such as [33,35,37].

Zhu et al. [37] proposed attention modules that only attend to a small set of key sampling points around a reference, which can achieve better performance than baseline model [8] with $10\times$ less training epochs. Tan et al. [35] used weighted bi-directional feature pyramid network (BiFPN), which allows easy and fast multi-scale feature fusion. Ren et al [33] proposed an improved version of their earlier work [14] provides comparative performances with [35,37] with lower latency and computational resources on FIR dataset. Hence, we use Faster RCNN model in this framework for localization and classification of the field component.

Handwritten Character Recognition: Offline handwriting recognition has been a long standing research interest. The works [3–5] presented novel features based on structural features of the strokes and their spatial relations with a character, as visible from different viewing directions on a 2D plane. Diesendruck et al. [11] used Word Spotting to directly recognise handwritten text from images. The conventional text recognition task is usually framed as an encoder-decoder problem where the traditional methods [19] leveraged CNN-based [24] encoder for image understanding and LSTM-based [20] decoder for text recognition.

Chowdhury et al. [9] combined a deep convolutional network with a recurrent Encoder-Decoder network to map an image to a sequence of characters corresponding to the text present in the image. Michael, Johannes et al. [28] proposed a sequence-to-sequence model combining a convolutional neural network (as a generic feature extractor) with a recurrent neural network to encode both the visual information, as well as the temporal context between characters in the input image. Further, Li et al. [25] used for the first time an end-to-end Transformer-based encoder-decoder OCR model for handwritten text recognition and achieved SOTA results. The model [25] is convolution-free unlike previous methods, and does not rely on any complex pre/post-processing steps. The present work leverages this work and extends its application in legal domain.

Post-OCR Correction: Rectification of errors in the recognised text from the OCR would require extensive training which is computation heavy. Further, post-OCR error correction requires a large amount of annotated data which may not always be available. After the introduction of the Attention mechanism and BERT model, many works have been done to improve the results of the

OCR using language model based post-correction techniques. However, Neural Machine Translation based approaches as used by Duong et al. [13] are not useful in the case of form text due to the lack of adequate context and neighbouring words. We extend the idea used in the work of Trstenjak et al. [7] where they used edit distance and cosine similarity to find the matching words. In this paper we used K-nearest neighbour with edit distance to find best matches for the words predicted with low confidence score by the OCR.

3   The FIR Dataset

First Information Report (FIR) documents contain details about incidents of cognisable offence, that are written at police stations based on a complaint. FIRs are usually filed by a police official filling up a printed form; hence the documents contain both printed and handwritten text. In this work, we focus on FIR documents written at police stations in India. Though the FIR forms used across different Indian states mostly have a common set of fields, there are some differences in their layout (see examples in Fig. 1). To diversify the dataset, we included FIR documents from the databases of various police stations across several Indian states – West Bengal[2], Rajasthan[3], Sikkim[4], Tripura[5] and Nagaland[6].

As stated earlier, an FIR contains many fields including the name of the complainant, names of suspected/alleged persons, statutes that may have been violated, date and location of the incident, and so on. In this work, we selected *four target fields* from FIR documents for the data annotation and recognition task – (1) *Year* (the year in which the complaint is being recorded), (2) *Complainant's name* (name of the person who lodged the complaint), (3) *Police Station* (name of the police station that is responsible for investigating the particular incident), and (4) *Statutes* (Indian laws that have potentially been violated in the reported incident; these laws give a good indication of the type of the crime). We selected these four target fields because we were able to collect the gold standard for these four fields from some of the police databases. Also, digitizing these four fields would enable various societal analysis, such as analysis of the nature of crimes in different police stations, temporal variations in crimes, and so on.

Annotations: We manually analysed more than 1,300 FIR documents belonging to different states, regions, police stations, etc. We found that FIR documents from the same region / police station tend to have the similar layout and form structure. Hence we selected a subset of 375 FIR documents with reasonably varying layouts / form structure, so that this subset covers most of the different variations. These 375 documents were manually annotated. Annotations were

---

[2] http://bidhannagarcitypolice.gov.in/fir_record.php.
[3] https://home.rajasthan.gov.in/content/homeportal/en.html.
[4] https://police.sikkim.gov.in/visitor/fir.
[5] https://tripurapolice.gov.in/west/fir-copies.
[6] https://police.nagaland.gov.in/fir-2/.

Fig. 2. Sample of various entities present in First Information Reports with different writing styles, distortions and scales.

Table 1. FIR Dataset statistics

Split	Images	Layout	Words	Labels
Training	300	61	1,830	1,230
Testing	75	18	457	307

done on these documents using LabelMe annotation tool[7] to mark the bounding boxes of the target fields.

Figure 2 shows some samples of various entities present in our dataset, and Fig. 3 shows examples of ground truth annotations for two of the entities in Fig. 2. In the ground truth, each bounding box has four co-ordinates (X_left, X_width, Y_right, Y_height) which describe the position of the rectangle containing the field value for each target field.

Train-test Split: During the annotation of our dataset, we identified 79 different types of large scale variations, layout distortions/deformations, which we split into training and testing sets. We divided our dataset (of 375 document images) such that 300 images are included in the training set and the other 75 images are used as the test set. During training, we used 30% of training dataset as a validation set. Table 1 shows the bifurcation statistics for training and test sets.

Preprocessing the Images: For Faster-RCNN we resized the document images to a size of 1180 × 740, and used the bounding boxes and label names to train the model to predict and classify the bounding boxes. We convert the dataset into IAM Dataset format [27] to fine-tune the transformer OCR.

---

[7] https://github.com/wkentaro/labelme

```
{
    "First Information Report": [
        {
            "id": 0,
            "image_id": 127,
            "text": "Randeep Gurung",
            "box": [
                [
                    293.56, 64.76, 057.43, 108.84
                ]
            ],
            "label": "complaint_name"
        },
        {
            "id": 1,
            "image_id": 127,
            "text": "122/269/270",
            "box": [
                [
                    469.47, 104.65, 628.63, 130.93
                ]
            ],
            "label": "section"
        }
    ]
}
```

Fig. 3. Examples of ground truth annotations for two of the entities shown in Fig. 2

Table 2. Comparison of the FIR dataset with other similar datasets

Dataset	Category	#Images	Text Type		#Entites
			Printed	**Handwritten**	
FUNSD [22]	Form	199	✓	x	4
SROIE [21]	Receipt	1000	✓	x	4
Cloud Invoice [29]	Invoice	326571	✓	x	8
FIR (**Ours**)	Form	375	✓	✓	4

Novelty of the FIR Dataset: We compare our FIR dataset[8] with other datasets for semi-structure document analysis in Table 2. The FIR dataset contains both printed and handwritten information which makes it unique and complex compared to several other datasets. Additionally, the FIR dataset is the first dataset for semi-structured document analysis in the legal domain.

4   The TransDocAnalyser Framework

We now present TransDocAnalyser, a framework for offline processing of handwritten semi-structured documents, by adopting Faster-RCNN and Transformer-based encoder-decoder architecture, with post-correction to improve performance.

4.1   The Faster-RCNN Architecture

Faster-RCNN [33] is a popular object detection algorithm that has been adopted in many real-world applications. It builds upon the earlier R-CNN [15] and

---

[8] https://github.com/LegalDocumentProcessing/FIR_Dataset_ICDAR2023.

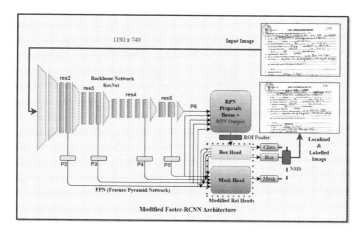

Fig. 4. Modified Faster-RCNN based architecture for target field localization and labelling

Fast R-CNN [33] architectures. We pass the input images through the Faster-RCNN network to get the domain-specific field associations and extract the image patches from the documents.

Our modified Faster-RCNN architecture consists of three main components (as schematically shown in Fig. 4)– (1) Backbone Network , (2) Region Proposal Network (RPN), and (3) ROI Heads as detailed below.

(1) Backbone Network: ResNet-based backbone network is used to extract multi-scaled feature maps from the input – that are named as P2, P3, P4 , P8 and so on – which are scaled as 1/4th, 1/8th, 1/16th and so on. This backbone network is FPN-based (Feature Pyramid network) [26] which is multi-scale object detector invariant to the object size.

(2) Region Proposal Network (RPN): Detects ROI (regions of interest) along with a confidence score, from the multi-scale feature maps generated by the backbone network. A fixed-size kernel is used for region pooling. The regions detected by the RPN are called *proposal boxes.*

(3) ROI Heads: The input to the box head comprises (i) the feature maps generated by a Fully Connected Network (FCN), (ii) the *proposed boxes* which come from the RPN. These are 1,000 boxes with their predicted labels. Box head uses the bounding boxes proposed by the RPN to crop and prepare the feature maps. (iii) ground truth bounding boxes from the annotated training datasets. The ROI pooling uses the proposed boxes detected by RPN, crops the rectangular areas of the feature maps, and feeds them into the head networks. Using Box head and mask head together in Faster-RCNN network, inspired by He et al. [18] improves the overall performance.

During training, the box head makes use of the ground truth boxes to accelerate the training. The mask head provides the final predicted bounding boxes and confidence scores during the training. At the time of inference the head

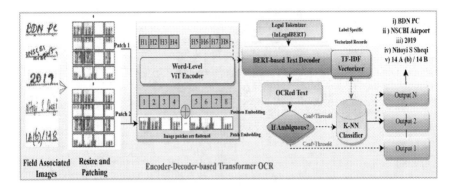

Fig. 5. TrOCR architecture with custom enhancements. The Text decoder uses a domain-specific InLegalBert [32] based tokenizer. OCR predictions go for post-correction if the confidence score is less than the threshold. We convert the OCR prediction into a TF-IDF vector and search in the domain-specific field database to find the Nearest Match.

network uses non-maximum suppression (NMS) algorithm to remove the overlapping boxes and selects the top-k results as the predicted output based on thresholds on their confidence score and intersection over union (IOU).

4.2   The TrOCR Architecture

Once the localized images are generated for a target field (e.g., complainant name) by Faster-RCNN, the image patches are then flattened and sent to the Vision Transformer (ViT) based encoder model. We use TrOCR [25] as the backbone model for our finetuning (see Fig. 5). TrOCR [25] is a Transformer-based OCR model which consists of a pretrained vision Transformer encoder and a pretrained text decoder. The ViT encoder is trained on the IAM handwritten dataset, which we fine-tune on our FIR dataset. We use the output patches from the Faster-RCNN network as input to the ViT encoder, and fine-tune it to generate features. As we are providing the raw image patches received from Faster-RCNN into the ViT encoder, we did not apply any pre-processing or layout enhancement technique to improve the quality of the localised images. On the contrary, we put the noisy localised images cropped from the *form fields* directly, which learns to suppress noise features by training.

We also replace the default text decoder (RoBERTa) with the Indian legal-domain specific BERT based text decoder InLegalBERT [32] as shown in Fig. 5. InLegalBert [32] is pre-trained with a huge corpus of about 5.4 million Indian Legal documents, including court judgements of the Indian Supreme Court and other higher courts of India, and various Central Government Acts.

To recognize characters in the cropped image patches, the images are first resized into square boxes of size 384 × 384 pixels and then flattened into a sequence of patches, which are then encoded by ViT into high-level representations and decoded by InLegalBERT into corresponding characters step-by-step.

Table 3. Excerpts from field-specific databases used to prepare TF-IDF vectorized records for KNN search. All databases contain India-specific entries.

Names	Surnames	Police Stations	Statutes / Acts
Anamul	Haque	Baguiati	IPC (Indian Penal Code)
Shyam	Das	Airport	D.M. Act (Disaster Management Act)
Barnali	Pramanik	Newtown	D.C. Act (Drug and Cosmetics Act)
Rasida	Begam	Saltlake	NDPS Act

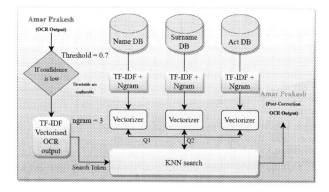

Fig. 6. Term Frequency and Inverse Document frequency (TF-IDF) Vectorizer based K-Nearest Neighbour model for post-correction on OCR output

We evaluate and penalise the model based on the Character Error Rate (CER). CER calculation is based on the concept of Levenshtein distance, where we count the minimum number of character-level operations required to transform the ground truth text into the predicted OCR output. CER is computed as $CER = (S + D + I)/N$ where S is the number of substitutions, D is the number of deletions, I is the number of Insertions, and N is the number of characters in the reference text.

4.3 KNN-Based OCR Correction

For each predicted word from OCR, if the confidence score is less than a threshold 0.7, we consider the OCR output to be ambiguous for that particular word. In such cases, the predicted word goes through a post-correction step which we describe now (see Fig. 6).

For each target field, we create a database of relevant values and terms (which could be written in the field) from various sources available on the Web. Table 3 shows a very small subset of some of the field-specific databases such as Indian names, Indian surnames, Indian statutes (Acts and Sections), etc. We converted each database into a set of TF-IDF vectors (see Fig. 6). Here TF-IDF stands for Term Frequency times Inverse Document Frequency. The TF-IDF scores

Table 4. Faster-RCNN model training parameters

Base Model	Base Weights	Learning Rate	Epoch #	# of Class	IMS/batch	Image Size
ResNet 50	Mask RCNN	0.00025	2500	4	4	1180 × 740

Table 5. Transformer OCR (TrOCR) parameters used for model fine-tuning

Feature Extractor	Tokenizer	Max Len	N-gram	Penalty	# of Beam	Optimizer
google-vit-patch16-384	InLegalBERT	32	3	2.0	4	AdamW

are computed using n-grams of groups of letters. In our work we used $n = 3$ (trigrams) for generating the TF-IDF vectors for OCR predicted words as well as for the entities in the databases.

For a given OCR output, based on the associated field name which is already available from the field classification by Faster-RCNN, we used the K-Nearest Neighbour (KNN) classifier to select the appropriate vectorized database. KNN returns best matches with a confidence score based on the distance between the search vector (OCR output) and the vectors in the chosen database. If the confidence score returned by KNN is greater than 0.9, then the OCR predicted word gets replaced with the word predicted by the K-Nearest Neighbour search.

5 Experimental Settings

We ran all experiments on a Tesla T4 GPU with CUDA version 11.2. We used CUDA enabled Torch framework 1.8.0.

In the first stage of the TransDocAnalyser framework, we trained the Faster RCNN from scratch using the annotated dataset (the training set). Table 4 shows the settings used for training the Faster-RCNN model. Prior to the training, input images are resized in 1180 × 740. For memory optimization, we run the model in two steps, first for 1500 iteration and then for 1000 iteration on the stored model. We tried batch sizes (BS) of 16, 32 and 64, and finalized BS as 64 because of the improvement in performance and training time. We used the trained model Faster-RCNN model to detect and crop out the bounding boxes of each label from the original document (as shown in Fig. 2) and created our dataset to fine-tune the ViT encoder.

We also created a metadata file mapping each cropped image (as shown in Fig. 2) with its corresponding text as described in [27] to fine-tune the decoder.

Table 5 shows the parameter settings used for fine-tuning the TrOCR model. Image patches are resized to 384 × 384 dimension to fine-tune ViT encoder. In the TrOCR model configuration, we replaced the tokenizer and decoder settings based on InLegalBert. We tried with batch size (BS) of 2, 4, 8, 16, 32, 64, and BS = 8 provided the best result on the validation set. We fine-tuned the Encoder and Decoder of the OCR for 40 epochs and obtained the final results.

Table 6. Performance of field labelling on the FIR dataset (validation set and test set). Re: Recall, Pr: Precision, F1: F1-score, mAP: mean average precision.

Results on dataset	Target field	Faster R-CNN			
		Re ↑	Pr ↑	F1 ↑	mAP ↑
Validation	Year	0.98	0.96	0.97	0.97
	Statute	0.85	0.82	0.83	0.84
	Police Station	0.96	0.90	0.93	0.93
	Complainant Name	0.84	0.76	0.80	0.77
Test	Year	0.97	0.96	0.97	0.96
	Statute	0.84	0.87	0.86	0.80
	Police Station	0.93	0.88	0.91	0.91
	Complainant Name	0.80	0.81	0.81	0.74

Fig. 7. Examples of localization and labelling of target fields by Faster-RCNN. The predicted bounding boxes are highlighted in green on the images. The associated class labels are highlighted in red. (Color figure online)

The KNN-based OCR correction module used n-grams with $n = 1, 2, 3, 4$ to generate the TF-IDF vectors of the field-specific databases. Using $n = 3$ (trigrams) and KNN with $K = 1$ provided the best results.

6   Results

In this section, we present the results of the proposed framework TransDocAnalyser in three stages – (i) The performance of Faster-RCNN on localization and labelling of the target fields (Table 6); (ii) Sample of OCR results with Confidence Scores (Table 7); and (iii) Comparison of the performance of the proposed framework with existing OCR methods (Table 8).

Table 6 shows the results of field label detection using Faster-RCNN on both test and validation sets of the FIR dataset. The performance is reported in terms of Recall (Re), Precision (Pr), F1 (harmonic mean of Recall and Precision) and mean Average Precision (mAP). For the localization and labelling, a prediction

Table 7. Finetuned (TrOCR) predictions on the generated image patches shown below

Image Patches	OCR Results	Confidence Score
2019	2019	0.89
Lian Min Thang	Lian Min Thang	0.77
NSCBI Airport	Nscbi Airport	0.79
Amar Prakash	Amar Prakesh	0.63
379	379	0.96

is considered correct if both the IOU (with the ground truth) and the confidence threshold are higher than 0.5. The results show that our model is performing well, with the best and worst results for the fields 'Year' (F1 = 0.97) and 'Name' (F1 = 0.8) respectively. This variation in the results is intuitive, since names have a lot more variation than the year.

Figure 7 shows examples of outputs of Faster-RCNN on some documents from the test set of the FIR dataset. The predicted bounding boxes are highlighted in green rectangles, and the predicted class names are marked in red on top of each bounding box.

The output of Faster-RCNN provides bounding boxes and field names for each image, using which image patches are generated and sent to the Encoder-Decoder architecture. Table 7 shows some examples of image patches and the finetuned TrOCR predictions for those image patches. It is seen that the name "Amar Prakash" is predicted as 'Amar Prakesh" with confidence score below a threshold of 0.7 (which was decided empirically). As the prediction confidence is below the threshold, this output goes to the post-correction method proposed in this work.

Table 8 compares the final performance of our proposed framework TransDoc-Analyser, and compares our model with Google-Tesseract and Microsoft-TrOCR for handwritten recognition on proposed FIR dataset.[9] The performances are reported in terms of Character Error Rate (CER), Word Error Rate (WER), and BLEU scores [30]. Lower values of CER and WER indicate better performance, while higher BLEU scores are better.

We achieve state-of-the-art results using the proposed TransDocAnalyser framework which outperforms the other models with quite a good margin (see Table 8). While the TrOCR + InLegalBert model also performed well, our proposed framework TransDocAnalyser (consisting of vision transformer-based encoder, InLegalBert tokenizer and KNN-based post-correction) achieved the best results across all the four target fields of the FIR dataset.

---

[9] We initially compared Tesseract with TrOCR-Base, and found TrOCR to perform much better. Hence subsequent experiments were done with TrOCR only.

Table 8. Benchmarking state-of-the-art TrOCR and our proposed framework Trans-DocAnalyser on the FIR dataset (best values in boldface)

OCR models	Target Field	Evaluation Metrics		
		CER ↓	WER ↓	BLEU ↑
Tesseract-OCR	Year	0.78	0.75	0.14
	Statute	0.89	0.83	0.12
	Police Station	0.91	0.89	0.10
	Complainant Name	0.96	0.87	0.9
TrOCR-Base	Year	0.38	0.32	0.72
	Statute	0.42	0.38	0.68
	Police Station	0.50	0.44	0.62
	Complainant Name	0.62	0.56	0.56
TrOCR-Large	Year	0.33	0.32	0.75
	Statute	0.34	0.33	0.73
	Police Station	0.36	0.38	0.65
	Complainant Name	0.51	0.50	0.57
TrOCR-**InLegalBert**	Year	0.17	0.17	0.84
	Statute	0.19	0.21	0.92
	Police Station	0.31	0.26	0.78
	Complainant Name	0.45	0.39	0.72
TransDocAnalyser (proposed)	Year	**0.09**	**0.02**	**0.96**
	Statute	**0.11**	**0.10**	**0.93**
	Police Station	**0.18**	**0.20**	**0.83**
	Complainant Name	**0.24**	**0.21**	**0.78**

7   Conclusion

In this work, we (i) developed the first dataset for semi-structured handwritten document analysis in the legal domain, and (ii) proposed a novel framework for offline analysis of semi-structured handwritten documents in a particular domain. Our proposed TransDocAnalyser framework including Faster-RCNN, TrOCR, a domain-specific language model/tokenizer, and KNN-based post-correction outperformed existing OCRs.

We hope that the FIR dataset developed in this work will enable further research on legal document analysis which is gaining importance world-wide and specially in developing countries. We also believe that the TransDocAnalyser framework can be easily extended to semi-structured handwritten document analysis in other domains as well, with a little fine-tuning.

Acknowledgement. This work is partially supported by research grants from Wipro Limited (www.wipro.com) and IIT Jodhpur (www.iitj.ac.in).

References

1. Amano, A., Asada, N.: Complex table form analysis using graph grammar. In: Lopresti, D., Hu, J., Kashi, R. (eds.) DAS 2002. LNCS, vol. 2423, pp. 283–286. Springer, Heidelberg (2002). https://doi.org/10.1007/3-540-45869-7_32
2. Amano, A., Asada, N., Mukunoki, M., Aoyama, M.: Table form document analysis based on the document structure grammar. Int. J. Doc. Anal. Recogn. (IJDAR) **8**, 210–213 (2006). https://doi.org/10.1007/s10032-005-0008-3
3. Bag, S., Harit, G.: A medial axis based thinning strategy and structural feature extraction of character images. In: Proceedings of IEEE International Conference on Image Processing, pp. 2173–2176 (2010)
4. Bag, S., Harit, G.: An improved contour-based thinning method for character images. Pattern Recogn. Lett. **32**(14), 1836–1842 (2011)
5. Bag, S., Harit, G.: Topographic feature extraction for Bengali and Hindi character images. Sig. Image Process. Int. J. **2**, 2215 (2011)
6. Bhattacharya, P., Hiware, K., Rajgaria, S., Pochhi, N., Ghosh, K., Ghosh, S.: A comparative study of summarization algorithms applied to legal case judgments. In: Proceedings of European Conference on Information Retrieval (ECIR), pp. 413–428 (2019)
7. Bruno, T., Sasa, M., Donko, D.: KNN with TF-IDF based framework for text categorization. Procedia Eng. **69**, 1356–1364 (2014)
8. Carion, N., Massa, F., Synnaeve, G., Usunier, N., Kirillov, A., Zagoruyko, S.: End-to-end object detection with transformers. In: Vedaldi, A., Bischof, H., Brox, T., Frahm, J.-M. (eds.) ECCV 2020. LNCS, vol. 12346, pp. 213–229. Springer, Cham (2020). https://doi.org/10.1007/978-3-030-58452-8_13
9. Chowdhury, A., Vig, L.: An efficient end-to-end neural model for handwritten text recognition. In: British Machine Vision Conference (2018)
10. Constum, T., et al.: Recognition and information extraction in historical handwritten tables: Toward understanding early 20th century Paris census. In: Proceedings of IAPR Workshop on Document Analysis Systems (DAS), pp. 143–157 (2022)
11. Diesendruck, L., Marini, L., Kooper, R., Kejriwal, M., McHenry, K.: A framework to access handwritten information within large digitized paper collections. In: Proceedings of IEEE International Conference on E-Science, pp. 1–10 (10 2012)
12. Divya, S., Gaurav, H.: Associating field components in heterogeneous handwritten form images using graph autoencoder. In: 2019 International Conference on Document Analysis and Recognition Workshops (ICDARW), vol. 5, pp. 41–46 (2019)
13. Duong, Q., Hämäläinen, M., Hengchen, S.: An unsupervised method for OCR post-correction and spelling normalisation for Finnish. In: Proceedings of the 23rd Nordic Conference on Computational Linguistics (NoDaLiDa), pp. 240–248 (2021)
14. Girshick, R.: Fast R-CNN. In: Proceedings of IEEE International Conference on Computer Vision (ICCV), pp. 1440–1448 (2015)
15. Girshick, R., Donahue, J., Darrell, T., Malik, J.: Rich feature hierarchies for accurate object detection and semantic segmentation. In: Proceedings of IEEE Conference on Computer Vision and Pattern Recognition, pp. 580–587 (2014)
16. Ha, H.T., Medved', M., Nevěřilová, Z., Horák, A.: Recognition of OCR invoice metadata block types. In: Proceedings of Text, Speech, and Dialogue, pp. 304–312 (2018)
17. Harley, A.W., Ufkes, A., Derpanis, K.G.: Evaluation of deep convolutional nets for document image classification and retrieval. In: 2015 13th International Conference on Document Analysis and Recognition (ICDAR), pp. 991–995 (2015)

18. He, K., Gkioxari, G., Dollár, P., Girshick, R.: Mask R-CNN. In: Proceedings of IEEE International Conference on Computer Vision (ICCV), pp. 2980–2988 (2017)
19. Hegghammer, T.: OCR with tesseract, Amazon textract, and google document AI: a benchmarking experiment. J. Comput. Soc. Sci. **5**(1), 861–882 (2022)
20. Hochreiter, S., Schmidhuber, J.: Long short-term memory. Neural Comput. **9**(8), 1735–1780 (1997)
21. Huang, Z., et al.: Competition on scanned receipt OCR and information extraction. In: Proceedings of International Conference on Document Analysis and Recognition (ICDAR), pp. 1516–1520 (2019)
22. Jaume, G., Kemal Ekenel, H., Thiran, J.P.: FUNSD: a dataset for form understanding in noisy scanned documents. In: 2019 International Conference on Document Analysis and Recognition Workshops (ICDARW), vol. 2, pp. 1–6 (2019)
23. Kim, G., et al.: OCR-free document understanding transformer. In: Avidan, S., Brostow, G., Cissé, M., Farinella, G.M., Hassner, T. (eds.) Computer Vision - ECCV 2022. ECCV 2022. LNCS, vol. 13688, pp. 498–517. Springer, Cham (2022). https://doi.org/10.1007/978-3-031-19815-1_29
24. Krizhevsky, A., Sutskever, I., Hinton, G.E.: ImageNet classification with deep convolutional neural networks. In: Pereira, F., Burges, C., Bottou, L., Weinberger, K. (eds.) Advances in Neural Information Processing Systems, vol. 25. Curran Associates, Inc. (2012)
25. Li, M., et al.: TrOCR: transformer-based optical character recognition with pre-trained models. In: Proceedings of AAAI (2023)
26. Lin, T.Y., Dollár, P., Girshick, R., He, K., Hariharan, B., Belongie, S.: Feature pyramid networks for object detection. In: Proceedings of IEEE Conference on Computer Vision and Pattern Recognition (CVPR), pp. 936–944 (2017)
27. Marti, U.V., Bunke, H.: The IAM-database: an English sentence database for offline handwriting recognition. Int. J. Doc. Anal. Recogn. **5**, 39–46 (2002)
28. Michael, J., Labahn, R., Grüning, T., Zöllner, J.: Evaluating sequence-to-sequence models for handwritten text recognition. In: 2019 International Conference on Document Analysis and Recognition (ICDAR), pp. 1286–1293 (2019)
29. Palm, R.B., Winther, O., Laws, F.: CloudScan - a configuration-free invoice analysis system using recurrent neural networks. In: Proceedings of IAPR International Conference on Document Analysis and Recognition (ICDAR), pp. 406–413 (2017)
30. Papineni, K., Roukos, S., Ward, T., Zhu, W.J.: BLEU: a method for automatic evaluation of machine translation. In: Proceedings of the 40th Annual Meeting on Association for Computational Linguistics, pp. 311–318 (2002)
31. Paul, S., Goyal, P., Ghosh, S.: LeSICiN: a heterogeneous graph-based approach for automatic legal statute identification from Indian legal documents. In: Proceedings of the AAAI Conference on Artificial Intelligence, vol. 36 (2022)
32. Paul, S., Mandal, A., Goyal, P., Ghosh, S.: Pre-trained language models for the legal domain: a case study on Indian law. In: Proceedings of the International Conference on Artificial Intelligence and Law (ICAIL) (2023)
33. Ren, S., He, K., Girshick, R., Sun, J.: Faster R-CNN: towards real-time object detection with region proposal networks. In: Proceedings of the International Conference on Neural Information Processing Systems - vol. 1, pp. 91–99. MIT Press (2015)
34. Subramani, N., Matton, A., Greaves, M., Lam, A.: A survey of deep learning approaches for OCR and document understanding. CoRR abs/2011.13534 (2020)
35. Tan, M., Pang, R., Le, Q.V.: EfficientDet: scalable and efficient object detection. In: Proceedings of IEEE/CVF Conference on Computer Vision and Pattern Recognition (CVPR), pp. 10778–10787 (2020)

36. Watanabe, T., Luo, Q., Sugie, N.: Layout recognition of multi-kinds of table-form documents. IEEE Trans. Pattern Anal. Mach. Intell. **17**(4), 432–445 (1995)
37. Zhu, X., Su, W., Lu, L., Li, B., Wang, X., Dai, J.: Deformable DETR: deformable transformers for end-to-end object detection. In: Proceedings of International Conference on Learning Representations (ICLR) (2021)

Context Aware Document Binarization and Its Application to Information Extraction from Structured Documents

Ján Koloda$^{(\boxtimes)}$ ⓘ and Jue Wang

Gini GmbH, Department Computer Vision & Information Extraction, Munich, Germany
{jan,jue}@gini.net

Abstract. Document binarization plays a key role in information extraction pipelines from document images. In this paper, we propose a robust binarization algorithm that aims at obtaining highly accurate binary maps with reduced computational burden. The proposed technique exploits the effectiveness of the classic Sauvola thresholding set within a DNN environment. Our model learns to combine multi-scale Sauvola thresholds using a featurewise attention module that exploits the visual context of each pixel. The resulting binarization map is further enhanced by a spatial error concealment procedure to recover missing or severely degraded visual information. Moreover, we propose to employ an automatic color removal module that is responsible for suppressing any binarization irrelevant information from the image. This is especially important for structured documents, such as payment forms, where colored structures are used for better user experience and readability. The resulting model is compact, explainable and end-to-end trainable. The proposed technique outperforms the state-of-the-art algorithms in terms of binarization accuracy and successfully extracted information rates.

Keywords: Binarization · Information extraction · Structured documents

1 Introduction

Document binarization refers to the process of assigning a binary value to every pixel. The objective is to distinguish the foreground pixels from the background. In other words, the binarization aims at identifying pixels of interest, typically pixels comprising relevant characters and symbols, and suppressing the rest. The goal is to enhance the overall document readability. Note that in many practical applications the purpose of binarization is not just to obtain the binary mask but rather to aid the subsequent processing, e.g., text detection, optical character recognition (OCR) or named entity recognition (NER). This aspect, however, is generally not evaluated in the literature and only the accuracies of the obtained binary masks are directly measured [1, 14, 16, 29]. Therefore, in this paper, we will not only evaluate the binarization performance in terms of correctly classified

G. A. Fink et al. (Eds.): ICDAR 2023, LNCS 14187, pp. 63–78, 2023.
https://doi.org/10.1007/978-3-031-41676-7_4

pixels but we will also evaluate the influence of the binarization quality on the information extraction results, concretely on NER tasks. An example of a typical information extraction pipeline is shown in Fig. 1. Throughout the paper, we will use the Gini document processing pipeline [34] as reference for information extraction. The binarization task is involved in processing of millions of payment forms and other financial documents that are run through the pipeline on weekly basis [35].

Additionally, the widely used binarization datasets typically refer to manuscript-like documents, either machine typed [2, 4, 6, 9] or handwritten [5, 7, 8, 10]. Such datasets, however, are not suitable for evaluation of modern structured documents, such as payment forms or invoices. The current growth of various document based information extraction services [34] requires the use of more complex document layouts and visual characteristics. In fact, modern structured documents contain visual features to better guide the user and improve the user experience. These complex visual contexts include colored regions that tend to confuse the binarization process. As a consequence, different background structures, logos and other irrelevant objects are highlighted during the binarization. This binarization noise, in turn, will confuse the subsequent processing and it can mask relevant information. Figure 3 shows an example of a typical German payment form. It is clear that pure gray scale processing has a tendency to mask important characters or produce excessive binarization masks that will have negative effect on subsequent processing (e.g. information extraction). In addition, such masks would also add to the overall processing time of a document.

In this paper, we introduce a robust and light-weight binarization convolutional neural network (CNN). The objective is to achieve a high binarization performance on manuscript-like documents as well as modern structured financial documents with non-negligible color content. The proposed approach exploits the advantages of the classic Sauvola thresholding algorithm [17] and the recent SauvolaNet architecture [1]. Unlike many binarization approaches [1, 14, 29], we will not only rely on gray scale inputs but also leverage the color information. We will show the importance of document color processing and its effect on the binarization results. The resulting architecture includes a restoration subnet to reconstruct objects and characters affected by noise or missing ink. Furthermore, the training process and the information propagation is aided by means of a featurewise attention module that leverages the visual context of every pixel. Finally, the proposed architecture is end-to-end trainable and does not require any pre-processing or post-processing steps.

The paper is structured as follows. In Sect. 2 the relevant document binarization techniques are reviewed. In Sect. 3, the proposed context aware document binarization method is presented. Experimental results are discussed in Sect. 4. The last section is devoted to conclusions.

2   Related Work

Document binarization converts the input image into a binary map. The objective is to detect the characters, or any other relevant objects, and suppress the

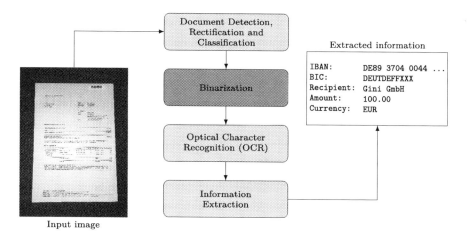

Input image

Fig. 1. Document information extraction pipeline.

irrelevant background. Let I be a document image with size $M \times N$ pixels. The resulting binarized image $\hat{B}$ is then obtained by means of the binarization function β as

$$\hat{B} = \beta\left(I\right). \tag{1}$$

Typically, the binarization function β is not constructed directly but via pixel thresholding. The thresholding function f binarizes the input image using the threshold T as

$$\hat{B} = \beta\left(I\right) = f\left(I, T\right) \tag{2}$$

with the thresholding function being computed as

$$\hat{B}(i,j) = f\left(I(i,j), T(i,j)\right) = \begin{cases} +1, & \text{if } I(i,j) \geq T(i,j) \\ -1, & \text{otherwise} \end{cases}, \tag{3}$$

where $i \in [0, M-1]$ and $j \in [0, N-1]$ denote pixel spatial indices.

The main binarization goal is to estimate a suitable threshold T, local or global, that is used to cluster the pixels into foreground and background. In fact, any binarization algorithm can be considered as a thresholding approach. The main differences lie in the estimation of T. Let τ denote a function that estimates T. Typically, this estimation is done by optimizing a model consisting of a set of parameters Θ that actuates on the input image I, i.e.,

$$\hat{B} = f\left(I, T\right) = f\left(I, \tau\left(I|\Theta\right)\right). \tag{4}$$

The most straightforward approaches employ simple global thresholding, where all pixels below a certain scalar threshold are highlighted as foreground. In controlled environments, the threshold can be set manually, however, adaptive global thresholding yields, in general, more stable results. For instance, the

Otsu algorithm [22] uses a simple optimization model that computes a global threshold that minimizes the intra-class variance for foreground and background clusters. Nevertheless, given the complexity and variance of documents and the differences in imaging conditions, global thresholding approaches usually lead to poor performance and more thresholding flexibility is necessary.

Therefore, methods that adaptively estimate local thresholds are introduced. The most popular ones are Niblack [15] and Sauvola [17]. The Sauvola method computes the pixelwise thresholds, $\boldsymbol{T}_S$, by inspecting local statistics around each pixel, i.e.,

$$\boldsymbol{T}_S(i,j|\boldsymbol{\Theta}) = \boldsymbol{\mu}(i,j)\left(1 + k\left(\frac{\boldsymbol{\sigma}(i,j)}{r} - 1\right)\right) \tag{5}$$

where $\boldsymbol{\Theta} = \{w, k, r\}$. The parameters r and k control the influence of local luminance and contrast, respectively. The first and second order local statistics, the mean $\boldsymbol{\mu}$ and the standard deviation $\boldsymbol{\sigma}$, are computed in the $w \times w$ vicinity around every image pixel. This method, although computationally efficient, is based on parameter heuristics and tends to fail if the heuristics do not hold. Therefore, further improvements on adaptively selecting the hyperparameters $\boldsymbol{\Theta}$ have been introduced [18,27,30]. The binarization robustness can be further enhanced by employing input prefiltering [23] and output postprocessing [24].

The state-of-the-art binarization approaches rely on deep neural networks (DNN) to estimate the threshold matrix $\boldsymbol{T}$. Many of the DNN-based approaches consider the binarization problem as a semantic segmentation task [23,25,26]. This allows to threshold the document image based on the higher level features extracted from the visual information. In [14], a selectional convolutional autoencoder is used to compute local activations and binarize them by means of global thresholding. On the other hand, the approach in [16] focuses on iteratively enhancing the input image, by means of a DNN sequence, and binarize the result using the traditional Otsu algorithm. The SauvolaNet approach [1] implements a multi-scale Suauvola thresholding within a DNN framework. It further introduces a pixelwise attention module that helps to remove the hard dependency on hyperparameters that the original Sauvola algorithm suffers from.

Finally, generative adversarial networks (GAN) allow to formulate the binarization problem as image-to-image generation task. The methods in [19,20] jointly process global and local pixel information to improve the model robustness against various visual degradations. The performance can be further enhanced by incorporating text recognition into the generative pipeline [21].

3   Proposed Algorithm

Similar to SauvolaNet [1], our proposal exploits the modeling abilities of the adaptive Sauvola thresholding in a multi-scale setting. In [1], a set of Sauvola threshold matrices is computed using different window sizes w (see Eq. (5)). These multi-scale thresholds are then combined via a DNN-based pixelwise attention mechanism which yields the final threshold value for each input pixel. The main goal of [1] is to learn and adapt the hyperparameters $\{w, k, r\}$ (see Eq. (5)) to every image by learning the relevant image features. This leads to a very compact and explainable model architecture. This architecture is well adapted to historical and rather monochrome document images. However, as will be shown in Sects. 3 and 4, it exhibits significant drawbacks when applied to more complex modern structured documents. Therefore, and in order to overcome those drawbacks, our proposal introduces three novel components, namely adaptive color removal, featurewise attention window and spatial error concealment. In addition, we leverage the SauvolaNet multi-scale pipeline which makes the proposed method compact, explainable and end-to-end trainable. The proposed context aware document binarization (CADB) algorithm is depicted in Fig. 2. The aforementioned three components as well as the overall proposed architecture are presented in detail in the following subsections.

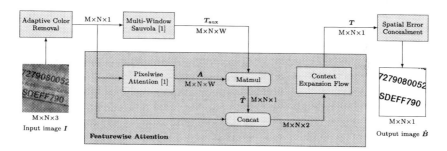

Fig. 2. Overview of the proposed CADB architecture.

3.1   Adaptive Color Removal

Many binarization benchmarking datasets comprise images of documents typed with a certain ink on a relatively homogeneous paper [2–10]. The visual properties are further affected by various imaging conditions and physical degradations of the documents. Therefore, many binarization approaches convert the input image to grayscale before starting the binarization pipeline [1,14,22]. On the other hand, the methods that do consider the color information assume that the color is an overall property of the whole document [37]. However, in many modern documents, color is used for layout highlighting, to improve the user experience or for aesthetic reasons (see Fig. 3). This implies that the color influence is oftentimes local and cannot be expanded to the document as a whole.

This color dependency can be controlled by employing scanners with a specific light composition. Such a controlled imaging process allows for applying blind color removal algorithms [36]. In fact, blind color removal is an important step in the automatic processing of financial documents such as payment forms [36]. However, such an imaging pipeline requires controlled environment and specific hardware. Hence, it is unviable for large scale deployment.

Our objective is to keep the color removal process hardware independent and robust against different lighting conditions. This will allow the use of relatively cheap smartphone camera sensors instead of dedicated scanners. Therefore, we propose to employ an adaptive color removal (ACR) module as a first step before the actual luminance inspection is launched (see Fig. 2). In order to keep the color removal process simple and efficient, the ACR is proposed to employ a stack of shallow convolutional layers to process the input image. The architecture of ACR is summarized in Fig. 4. The output of ACR is a monochromatic image where the irrelevant (even though colorful) regions are suppressed in order to aid the subsequent threshold estimation. The ACR is trained jointly with the rest of the network and does not require any special considerations.

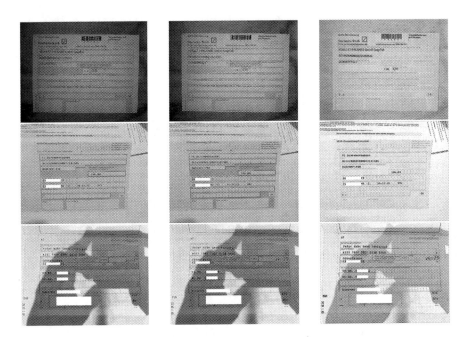

Fig. 3. Examples of adaptive color removal. Original images (left), the corresponding gray scale images (middle) and the ACR results (right). Sensitive personal information is covered by white boxes. (Color figure online)

Figure 3 shows examples of commonly used German and Austrian payment forms using different color palettes. These documents, in addition, are captured under various lighting conditions. The comparison between the classic grayscale image and the ACR result is also shown in Fig. 3. It is observed that the ACR module helps to suppress the irrelevant pixels much more efficiently than a simple grayscale conversion. The ACR does not only suppress the colored regions related to the document visual layout (otherwise irrelevant for the binarization purposes) but also reduces the effect of strong shadows. As it will be shown in Sect. 4, the

ACR module plays a key role in processing and extracting information from documents with significant color content. Also note that the ACR does not require any specific training pipeline nor data annotation. It is trained implicitly as part of the CADB end-to-end training (see Sect. 4).

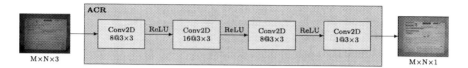

Fig. 4. Architecture of the automatic color removal (ACR) module. Each Conv2D layer is denoted as `filters@ksize`×`ksize`. The last Conv2D layer uses the sigmoid activation. The structure of context expansion flow (CEF) module, described in Sect. 3.2, is identical.

3.2   Featurewise Attention Window

As shown in Eq. (5), the Sauvola threshold is computed using the local mean and standard deviation of intensity values around each pixel, respectively. The problem of such an approach is that the local statistics may also capture irrelevant structures that can heavily influence the computed threshold. The SauvolaNet tries to overcome this issue by computing a set of auxiliary thresholds T_{aux} using the multi-window Sauvola module (MWS) [1]. These thresholds correspond to different processing scales and are computed by varying the local window size w (see Eq. (5)). The resulting SauvolaNet threshold matrix $\hat{T}$ is finally computed by linear combination of the auxiliary thresholds. This combination is controlled by the pixelwise attention (PWA) mechanism that effectively selects the optimal scale for each pixel [1]. Let A denote the attention tensor estimated by PWA and let W be the number of different window sizes w used in MWS. The threshold matrix $\hat{T}$ is then computed as

$$\hat{T}(i,j) = \sum_{k=0}^{W-1} A(i,j,k) T_{\text{aux}}(i,j,k). \tag{6}$$

The concrete signal flow is also depicted in Fig. 2. The estimation of $\hat{T}$ alleviates the rigidity of the original Sauvola thresholding but the auxiliary thresholds are still ignorant of the local structure of the document.

In fact, the auxiliary threshold estimation for each scale, performed by MWS, is purely pixelwise and the threshold for a given pixel does not influence, nor is influenced by, the estimated thresholds of its neighboring pixels. This lack of context information hinders an efficient training since the PWA not only has to learn the adequate pixelwise attention for auxiliary threshold weighting but also to compensate for any irregularities in the input image. This approach also ignores the a priori information that natural images, including documents, are mainly low-pass signals where the spatial correlation among pixels is high [32].

We propose to overcome this issue by encapsulating the PWA into a generic featurewise attention window (FWA) module. The FWA preserves the pixel-wise attention and allows for information flow among neighboring pixels and the corresponding auxiliary thresholds. This is achieved by refining the computed threshold matrix $\hat{T}$ by a sequence of shallow convolutional layers, similar to ACR. We denote this module as context expansion flow (CEF) and its structure is depicted in Fig. 4. The FWA additionally includes a residual connection that bypasses the PWA. The objective of this bypass is twofold. First, it helps to avoid vanishing gradients and, in fact, the experiments have shown a faster learning curve. Second, it avoids losing higher order statistics through MWA, since it establishes an additional direct connection between ACR and the pixel-wise attentions and auxiliary thresholds. The structure of the FWA module is illustrated in Fig. 2.

At this point, the CADB architecture can already be trained using a modified hinge loss [1], which is computed as

$$\mathcal{L}(i,j) = \max\left(1 - \alpha\left(\boldsymbol{I}(i,j) - \boldsymbol{T}(i,j)\right)\boldsymbol{B}(i,j), 0\right) \tag{7}$$

where the parameter α controls the margin of the decision boundary. This implies that only pixels close to the boundary will be used in gradient back-propagation. As in [1], α is set to 16.

3.3 Spatial Error Concealment

Document images are also affected by physical degradations of the documents themselves. Oftentimes, documents contain printing errors, noise or missing ink. Therefore, incomplete or otherwise degraded characters are not fully detected by the binarizer. This, on one hand, may be considered as a desirable behavior since, in fact, there is physically no information to be binarized. However, such incomplete results will negatively affect the subsequent higher level processing. In order to overcome this issue, we propose to include a restoration subnet to recover the errors. Spatial error concealment (SEC) techniques are especially designed to leverage the spatial information of the neighboring pixels and reconstruct the missing information [31]. The concealment strategy can be moved from the pixel domain to feature domain where the calculated thresholds are adjusted based on the higher level information that is not considered during the estimation of the auxiliary thresholds.

We propose to perform the SEC by means of a downsampled U-Net architecture [28] where the number of filters of each layer is reduced by 16 at each stage. This light-weight restoration subnet is refining the thresholds estimated by FWA and delivers the final binarization results, recovering the missing information (if any). To train the restoration subnet, the eroded binary cross-entropy loss is proposed. Since the class distribution in document binarization tends to be highly imbalanced, we compensate for this issue by eroding the over-represented background pixels. This is achieved by dilating the binarization mask with a 7×7 kernel and all pixels not included in the dilation are excluded from the loss computation. The proposed eroded binary cross-entropy (ECE) loss is then computed as

$$ECE = \frac{1}{MN} \sum_{i=0}^{M-1} \sum_{j=0}^{N-1} \left(\boldsymbol{y}_{ij} \log\left(\boldsymbol{p}_{ij}\right) + (1 - \boldsymbol{y}_{ij}) \log\left(1 - \boldsymbol{p}_{ij}\right) \right) \boldsymbol{e}_{ij} \qquad (8)$$

where $\boldsymbol{y}$ corresponds to the binarization ground truth and $\boldsymbol{p}$ is the corresponding SEC prediction. The background erosion mask $\boldsymbol{e}$ is computed as the foreground dilation with a 7×7 morphological kernel $\boldsymbol{k}$, i.e.,

$$\boldsymbol{e}_{ij} = \boldsymbol{y}_{ij} \oplus \boldsymbol{k} \qquad (9)$$

where $\oplus$ denotes the dilation operation.

The ECE, however, completely disregards larger background areas which can have significantly negative effect on the final binarization results. Therefore, during the training, we will alternate between the ECE loss and the conventional cross-entropy with a certain probability distribution. In our experiments, the ECE is employed with the probability of 0.9. Note that the mask erosion can be considered as a special case of data augmentation and can be efficiently implemented by introducing an additional *don't care* class, similar to [33]. Simulations have revealed that the proposed loss yields higher binarization accuracies and better extraction rates than a simple class weighting used to compensate for the class imbalance between foreground and background pixels.

Finally, the complete CADB network architecture is shown in Fig. 2. Note that, unlike some others segmentation-based approaches [14], the proposed restoration subnet is fully convolutional and, therefore, does not require to split the input image into regions with predefined shapes and reconstruct the output afterwards.

4   Experimental Results

In order to thoroughly evaluate the performance of the proposed algorithm, we have conducted series of tests on complex modern structured documents. In addition, in order to demonstrate the generalization abilities and robustness of the proposed technique, we first conduct evaluations on standard handwritten and machine typed benchmarking datasets. Detailed dataset descriptions and the obtained results are presented in Sects. 4.1 and 4.2, respectively.

The proposed technique is compared to other state-of-the-art binarization algorithms, namely SauvolaNet [1], ISauvola [18], Wan [29], Sauvola thresholding [17] and Otsu binarization [22]. The proposed algorithm will be evaluated in three different configurations. The goal is to assess the individual effect of the three proposed components to the overall binarization performance. First, the full CADB architecture, as described in Fig. 2, is evaluated. Second, we switch off the SEC module to better assess the importance of spatial information flow in the feature domain. This configuration will be referred to as $\text{CADB}_{\text{ACR/FWA}}$. Finally, we switch both SEC and FWA off. This configuration will be denoted as CADB_{ACR} and will allow to evaluate the influence of adaptive color removal on the binarization results.

We implement the CADB pipeline in the PyTorch framework. Throughout the experiments, image patches of size 256×256 pixels are used for training. The training batch size is set to 32 and the Adam optimizer with the initial learning rate of 10^{-4} is employed.

4.1   Standard Benchmarking

For these experiments, and in order to achieve a fair comparison, the same setup as described in [1] is employed. There are 13 document binarization datasets, namely (H-)DIBCO 2009 [2], 2010 [3], 2011 [4], 2012 [5], 2013 [6], 2014 [7], 2016 [8], 2017 [9], 2018 [10]; PHIDB [11], Bickely-diary dataset [12], Synchromedia Multispectral dataset [13], and Monk Cuper Set [16]. The data splits are the same as in [1]. All approaches in this section have been studied using the same evaluation protocol.

Table 1. Average binarization metrics obtained by the tested algorithms. The best performances are in bold face.

	PSNR [dB]	ACC [%]	FM [%]	DRD
Otsu	17.13	96.93	85.99	7.11
Sauvola	16.32	96.82	83.30	10.17
Wan	13.73	93.07	75.09	29.01
ISauvola	17.32	97.40	86.23	6.55
SauvolaNet	18.23	98.03	88.50	5.34
CADB_{ACR}	**19.58**	98.43	90.98	4.53
$\text{CADB}_{\text{ACR/FWA}}$	19.17	98.38	89.96	4.38
CADB	19.53	**98.53**	**91.35**	**3.61**

For evaluation, the standard binarization metrics [1,4–6] are adopted, i.e., peak signal-to-noise ratio (PSNR), accuracy (ACC), F-measure (FM) and distance reciprocal distortion metric (DRD). The binarization results are shown in Table 1. It follows that the proposed algorithm considerably outperforms other state-of-the-art techniques. In fact, the different CADB configurations achieve PSNR gains of more than 1 dB with respect to the SauvolaNet. Subjective comparisons are provided in Fig. 5. It is observed that, unlike the tested state-of-the-art techniques, our proposal can successfully binarize strongly faded characters.

4.2   Experiments with Structured Documents

In this section, experiments with modern structured documents are conducted. In particular, images of filled in German and Austrian payment forms are used.

Original image	Otsu	SauvolaNet	CADB$_{ACR/FWA}$

Ground truth	Sauvola	CADB$_{ACR}$	CADB

Fig. 5. Example of binarization performance on a handwritten document [8].

Examples of such documents can be found in Fig. 3. During the training, various payment form templates are filled with random data and the corresponding binarization mask are generated. The results are further augmented to achieve photorealistic visual properties. This data generation allows for building large amounts of training data without the need of laborious annotations. Examples of automatically generated training data are shown in Fig. 6. These data have been used to train all the methods evaluated in this section.

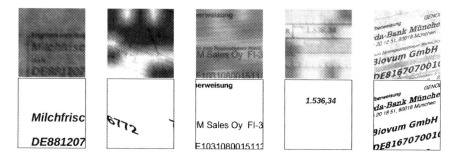

Fig. 6. Examples of payment form cuts (top) and the corresponding binarization masks (bottom).

The performance is first assessed using the metrics from the previous section. The evaluation dataset consists of 1000 document cuts with size 256×256 pixels and their corresponding binarization masks. The evaluation set is available in [38]. The evaluation results are shown in Table 2. It is observed that our proposal considerably outperforms the evaluated methods in all considered metrics. The PSNR improvements with respect to SauvolaNet are more than 9 dB and the DRD metric is orders of magnitude lower. It is also shown that, for structured documents, a significant part of the improvement is due to ACR. The full scale CADB then improves the results even further.

In addition, the binarization performance is also evaluated on real production data from the Gini information extraction pipeline [34]. For such a purpose, 1000 payment form documents, taken by non-expert users, are employed. In this case, we will evaluate the performance of the different binarization approaches

Table 2. Average binarization metrics obtained by the tested algorithms. The best performances are in bold face.

	PSNR [dB]	ACC [%]	FM [%]	DRD
Otsu	4.96	62.26	16.27	978.44
Sauvola	12.22	89.86	27.38	155.01
Wan	8.28	72.27	21.76	448.52
ISauvola	12.49	90.44	28.34	147.42
SauvolaNet	14.61	95.96	45.99	112.18
CADB$_{ACR}$	22.93	99.03	79.34	5.76
CADB$_{ACR/FWA}$	23.63	99.20	83.70	**5.72**
CADB	**23.70**	**99.27**	**84.54**	6.88

indirectly by measuring the information extraction rates from the documents. Classification accuracies and reconstruction error measures, such as PSNR, do not always reflect the perceptual quality of the final results [31]. In the case of processing of structured financial document, our ultimate goal is the accurate information extraction from the images. Therefore, we run the evaluation set through the whole information extraction pipeline [34] and evaluate the extraction rates for the different binarization approaches. The rest of the pipeline is kept intact and is not adapted to any binarization in particular. The best performing binarization approaches from Sect. 4.1 are considered here. The extraction rates are measured on 4 named entities, namely the international bank account number (IBAN), bank identifier code (BIC), the amount to be paid and the recipient name. The extraction rates are computed as the ratio of successfully delivered labels with respect to the total amount of documents. The results are shown in Table 3. It follows that the CADB variants outperform the SauvolaNet by up to 15 percentual points. In addition, the SEC restoration adds further enhancements of up to 4 percentual points and, on average, improves the extraction rates for most of the fields.

Table 3. Extraction rates (in %) for the fields of IBAN, BIC, amount and recipient. The best performances are in bold face.

	IBAN	BIC	Amount	Recipient
SauvolaNet	79.00	86.10	57.70	92.30
CADB$_{ACR}$	88.80	92.90	67.40	93.10
CADB$_{ACR/FWA}$	88.59	93.69	**73.27**	94.89
CADB	**92.40**	**95.60**	72.90	**95.70**

Subjective comparisons are shown in Fig. 7. It is observed that the proposed method renders robust results and does not suffer from binarization artifacts that make the resulting binarized text partially, or even fully, illegible. This is also corroborated by the examples in Fig. 8. It is shown that the proposed CADB method is robust against adversarial imaging conditions and is able to handle different color pallettes and camera devices.

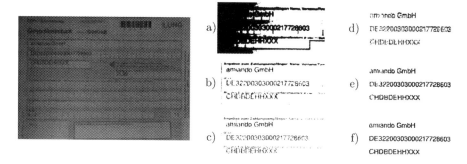

Fig. 7. Example of binarization performance on a payment form. The results for (a) Otsu, (b) Sauvola, (c) SauvolaNet, (d) CADB$_{ACR}$, (e) CADB$_{ACR/FWA}$, and (f) CADB are shown.

Additionally, due to the shallow convolutions in ACR and FWA, the computational overhead with respect to SauvolaNet is relatively small. In fact, ACR and FWA (including the residual bypass), increment both the network size and the inference time by around 7%. The inference time is measured on Nvidia RTX 2070 with 6GB of RAM. The information extraction gains of SEC come at a price of increased model size. In fact, the full CADB architecture, including SEC, comprises 1.34 millions parameters. The processing time is increased by 54.5% and is about 20 ms for a high-definition document image. Nevertheless, CADB in full configuration is still more compact than other state-of-the-art architectures [9,20] and order of magnitude smaller than [14,16]. Depending on the use case, the high modularity of CADB offers the user the possibility to select the best fitting configuration.

Fig. 8. Comparison of binarization results. Original images (left), SauvolaNet (middle) and the proposed CADB (right).

5 Conclusions

In this paper, we have proposed a new binarization model based on Sauvola thresholding principles. We have leveraged the SauvolaNet approach, a DNN-based Sauvola approximation, and discussed the drawbacks when applied to modern structured documents. In order to overcome those drawbacks, we have proposed three compact and explainable modules. The objective is not only to improve the accuracy of the computed binary maps but also to enhance the information extraction quality from the documents. The proposed CADB architecture is an end-to-end trainable architecture that is able to remove from the document visually prominent but otherwise irrelevant structures. Furthermore, we have designed a featurewise attention mechanism that allows the information flow both in the pixel and in the feature domain. Finally, we have taken advantage of the spatial error concealment methods to restore the missing or damaged information. Simulations reveal that our proposal significantly outperforms the state-of-the-art techniques both in terms of binarization accuracy and information extraction rates from the binarized documents.

References

1. Li, D., Wu, Y., Zhou, Y: SauvolaNet: learning adaptive sauvola network for degraded document binarization. In: Proceedings of ICDAR, Lausanne (2021)
2. Gatos, B., Ntirogiannis, K., Pratikakis, I.: ICDAR 2009 document image binarization contest (DIBCO 2009). In: Proceedings of ICDAR, Barcelona (2009)
3. Pratikakis, I., Gatos, B., Ntirogiannis, K.: H-DIBCO 2010 - handwritten document image binarization competition. In: Proceedings of ICFHR, Kolkata (2010)
4. Pratikakis, I., Gatos, B., Ntirogiannis, K.: ICDAR 2011 document image binarization contest (DIBCO 2011). In: Proceedings of ICDAR, Beijing (2011)
5. Pratikakis, I., Gatos, B., Ntirogiannis, K.: ICFHR 2012 competition on handwritten document image binarization (H-DIBCO 2012). In: Proceedings of ICFHR, Bari (2012)
6. Pratikakis, I., Gatos, B., Ntirogiannis, K.: ICDAR 2013 document image binarization contest (DIBCO 2013). In: Proceedings of ICDAR, Washington (2013)
7. Ntirogiannis, K., Gatos, B., Pratikakis, I.: ICFHR2014 competition on handwritten document image binarization (H-DIBCO 2014). In: Proceedings of ICFHR, Hersonissos (2014)
8. Pratikakis, I., Zagoris, K., Barlas, G., Gatos, B.: ICFHR2016 handwritten document image binarization contest (H-DIBCO 2016). In: Proceedings of ICFHR, Shenzhen (2016)
9. Pratikakis, I., Zagoris, K., Barlas, G., Gatos, B.: ICDAR2017 competition on document image binarization (DIBCO 2017). In: Proceedings of ICDAR, Kyoto (2017)
10. Pratikakis, I., Zagoris, K., Kaddas, P., Gatos, B.: ICFHR2018 competition on handwritten document image binarization (H-DIBCO 2018). In: Proceedings of ICFHR, Niagara Falls (2018)
11. Nafchi, H.Z., Ayatollahi, S.M., Moghaddam, R.F., Cheriet, M.: An efficient ground truthing tool for binarization of historical manuscripts. In: Proceedings of ICDAR (2013)

12. Deng, F., Wu, Z., Lu, Z., Brown, M.S.: Binarizationshop: a user-assisted software suite for converting old documents to black-and-white. In: Proceedings of Annual Joint Conf. on Digital Libraries (2010)
13. Hedjam, R., Nafchi, H.Z., Moghaddam, R.F., Kalacska, M., Cheriet, M.: ICDAR 2015 contest on multispectral text extraction (MS-Tex 2015). In: Proceedings of ICDAR (2015)
14. Calvo-Zaragoza, J., Gallego, A.J.: A selectional auto-encoder approach for document image binarization. Pattern Recogn. **86**, 34–47 (2019)
15. Niblack, W.: An Introduction to Digital Image Processing, Strandberg Publishing Company (1985)
16. He, S., Schomaker, L.: Deepotsu: document enhancement and binarization using iterative deep learning. Pattern Recogn. **91**, 379–390 (2019)
17. Sauvola, J., Pietikäinen, M.: Adaptive document image binarization. Pattern Recogn. **33**, 225–236 (2000)
18. Hadjadj, Z., Meziane, A., Cherfa, Y., Cheriet, M., Setitra, I.: Isauvola: improved sauvola's algorithm for document image binarization. In: Proceedings of ICIAR, Póvoa de Varzim (2016)
19. De, R., Chakraborty, A., Sarkar, R.: Document image binarization using dual discriminator generative adversarial networks. IEEE Signal Process. Lett. **27**, 1090–1094 (2020)
20. Zhao, J., Shi, C., Jia, F., Wang, Y., Xiao, B.: Document image binarization with cascaded generators of conditional generative adversarial networks. Pattern Recogn. **96**, 106968 (2019)
21. Jemni, S.K., Souibgui, M.A., Kessentini, Y., Fornés, A.: Enhance to read better: a multi-task adversarial network for handwritten document image enhancement. Pattern Recogn. 123, 108370 (2021)
22. Otsu, N.: A threshold selection method from gray-level histograms. Automatica (1975)
23. Vo, G.D., Park, C.: Robust regression for image binarization under heavy noise and nonuniform background. Pattern Recogn. **81**, 224–239 (2018)
24. Moghaddam, R.F., Cheriet, M.: A multi-scale framework for adaptive binarization of degraded document images. Pattern Recogn. **43**, 2186–2198 (2010)
25. Peng, X., Wang, C., Cao, H.: Document binarization via multi-resolutional attention model with DRD loss, In: Proceedings of ICDAR, Sydney (2020)
26. Tensmeyer, C., Martínez, T.: Document image binarization with fully convolutional neural networks. In: Proceedings of ICDAR, Kyoto (2017)
27. Lazzara, G., Géraud, T.: Efficient multiscale Sauvola's binarization. Int. J. Document Anal. Recogn. **17**, 105–123 (2014)
28. Ronneberger, O., Fischer, P., Brox, T.: U-Net: convolutional networks for biomedical image segmentation. Medical Image Computing and Computer-Assisted Intervention (2015)
29. Wan, A.M., Mohamed, M.M.A.K.: Binarization of document image using optimum threshold modification. J. Phys. **1019**, 012022 (2018)
30. Kaur, A., Rani, U., Gurpreet, S.J.: Modified Sauvola Binarization for Degraded Document Images. Engineering Applications of Artificial Intelligence (2020)
31. Koloda, J., Peinado, A.M., Sánchez, V.: Kernel-based MMSE multimedia signal reconstruction and its application to spatial error concealment. IEEE Trans. Multimed. (2014)
32. Koloda, J., Seiler, J., Peinado, A.M., Kaup, A.: Scalable kernel-based minimum mean square error estimator for accelerated image error concealment. IEEE Trans. Broadcasting **63**, 59–70 (2017)

33. Geiger, A., Lenz, P., Urtasun, R.: Are We Ready for Autonomous Driving? The KITTI Vision Benchmark Suite. In: Proceedings of CVPR, Providence (2012)
34. Gini GmbH. https://gini.net/en/products/extract/gini-smart. Accessed 15 Jan 2023
35. Gini Photo Payment. https://gini.net/en/gini-now-processes-over-7-million-photo-payments-per-month. Accessed 15 Jan 2023
36. Document Data Capture. https://www.bitkom.org/sites/default/files/file/import/130302-Document-Data-Capture.pdf. Accessed 15 Jan 2023
37. Lin, Y.-S., Ju, R.-Y., Chen, C.-C., Lin, T.-Y., Chiang, J.-S.: Three-Stage Binarization of Color Document Images Based on Discrete Wavelet Transform and Generative Adversarial Networks, arXiv preprint (2022)
38. CADB Testset. https://github.com/gini/vision-cadb-testset.git. Accessed 15 Jan 2023

Frontiers in Handwriting Recognition 1 (Online)

SET, SORT! A Novel Sub-stroke Level Transformers for Offline Handwriting to Online Conversion

Elmokhtar Mohamed Moussa[1,2]([✉]) [ID], Thibault Lelore[1] [ID], and Harold Mouchère[2] [ID]

[1] MyScript SAS, Nantes, France
{elmokhtar.mohamed.moussa,thibault.lelore}@myscript.com
[2] Nantes Université, École Centrale Nantes, CNRS, LS2N, UMR 6004, 44000 Nantes, France
{elmokhtar.mohamedmoussa,harold.mouchere}@univ-nantes.fr

Abstract. We present a novel sub-stroke level transformer approach to convert offline images of handwriting to online. We start by extracting sub-strokes from the offline images by inferring a skeleton with a CNN and applying a basic cutting algorithm. We introduce sub-stroke embeddings by encoding the sub-stroke point sequence with a **Su**b-stroke **E**ncoding **T**ransformer (SET). The embeddings are then fed to the **Su**b-strokes **OR**dering **T**ransformer (SORT) which predicts the discrete sub-strokes ordering and the pen state. By constraining the Transformer input and output to the inferred sub-strokes, the recovered online is highly precise. We evaluate our method on Latin words from the IRONOFF dataset and on maths expressions from CROHME dataset. We measure the performance with two criteria: fidelity with Dynamic Time Warping (DTW) and semantic coherence using recognition rate. Our method outperforms the state-of-the-art in both datasets, achieving a word recognition rate of 81.06% and a 2.41 DTW on IRONOFF and an expression recognition rate of 62.00% and a DTW of 13.93 on CROHME 2019. This work constitutes an important milestone toward full offline document conversion to online.

Keywords: offline handwriting · transformer · online recovery

1 Introduction

In today's highly virtual and automated world, note-taking is still a manual procedure. It is a ground to express our volatile thoughts and ideas, allowing their organization and the emergence of our creativity afterward. While pen and paper still offer unmatched comfort and efficient input methods for handwritten notes, it disables their exploitation to their full potential. They are usually digitized as offline documents by capturing images with a scanner or camera. This is an inconvenient step for most users and it also adds noise that affects offline processing systems. Online documents - the offline counterpart - are recorded on touch-sensitive surface devices with an e-pen, enabling a more powerful machine-automated organization and edition of handwritten documents, with intuitive pen gestures. Many commercial software specializing in note-taking exists, proposing a plethora of functionalities such as recognition, note indexing, collaborative note-taking, *etc.* The online domain is ever-evolving as the offline is already far behind. By developing an offline-to-online conversion system we allow the users to take to their advantage the best of the two modalities: ergonomic note-taking with a pen and paper and powerful editing and processing of the digital ink. Recently, attractive hybrid devices are surfacing. Their hardware closely mimics a pencil offering a more ergonomic input method while still proposing online processing tools. However, in the quest for paper-like hardware, the devices are still today limited in computational resources compared to other touch devices. Research efforts [17] have been conducted in the document analysis domain to automatically recover online from offline documents by retrieving the pen trajectory. Thus allowing for the direct exploitation of paper and pen notes in the existing online processing systems.

However, as datasets coupling online with offline are scarce [18,20], the applications of data-driven approaches remain limited. To overcome this issue, rasterization or online data to offline conversion is commonly used for training multi-modal systems. Converting online signals to realistic raster images often involves adding noise and simulating pen tip width and movement speed [7]. Other advanced applications use generative adversarial networks [11] to generate artificial papyrus and other historic documents. Multi-modal systems utilize both online and offline, combining temporality with spatial clues for better performances. For instance, handwriting recognition is typically classified into two types: offline and online systems. Multi-modal Handwriting recognition systems [21,22] are shown to outperform their mono-modal counterparts. In this paper, we focus on the reverse problem, which is offline to online conversion. Vectorization similarly tries to model a line drawing image as a set of geometric primitives (polygons, parametric curves, *etc.*) corresponding to elements found in Scalable Vector Graphic (SVG) format. It is mainly applied to technical drawing [6] and 2D animation sketching [7]. In this particular application, retrieving temporal ordering between the extracted vector elements is less relevant.

For handwriting applications, we are more involved in the recovery of pen trajectory from images. The availability of temporal information in online systems often makes them better performing than their offline analog [16]. In 2019, the Competition on Recognition of Online Handwritten Mathematical Expressions (CROHME) [12] included for the first time an offline recognition task. It has since sparked a great interest in offline to online conversion [4] in this specific domain. Classical approaches are rule-based systems. They usually operate on a topology to detect regions where the drawing direction is ambiguous (*e.g.* junctions) and employ a set of handcrafted heuristics to simplify and resolve them. However, they are very hard to maintain and do not generalize to different languages or content. Recently, many data-driven approaches have been proposed in the literature to recover online from offline. However, CRNNs models [2,3] rely on fixed-size feature maps of the whole offline image, regardless of the ink density, to predict all the underlying intricacies in the temporality of the different strokes resulting in the partial or full omission of strokes. In this paper, we propose the following contributions:

- We propose novel sub-stroke level Transformers (SET and SORT) to recover the online from offline (see Fig. 1) instead of CRNNs architectures [2,3].
- We move from the image to sequence framework to operate on the sub-stroke level to perform a local and global analysis of the different junctions as is adopted in classical approaches [4].
- Our SET and SORT approach outperforms prior online recovery work on the handwritten text of the IRONOFF dataset. We also extend our work to more complex maths equations of the CROHME dataset.

(a) Input offline image.

(b) Extracted sub-strokes from the CNN inferred skeleton. A sub-stroke can be drawn in both directions.

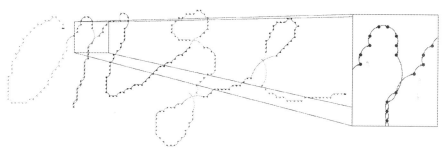

(c) Network online prediction. *start* and *end* nodes are annotated. The zoom box shows the predicted sub-stroke direction and ordering as illustrated by arrows. The edge color is the destination node. For double-traced sub-strokes, edges are read in the clockwise orientation as in [9].

Fig. 1. Given an input offline image (a), sub-strokes incident on the same junctions are extracted (b). Our Network predicts the sub-stroke's order and directions (c). The trivial longest path obtained by following the outgoing edges from *start* node to *end* node is the predicted online signal.

2   Related Works

Line Drawing Vectorization is a crucial step in the creation of 2D animations and sketches. It involves converting drawing images into vector graphics.

Artists often begin by sketching their work on paper and then manually vectorizing it digitally for finalization. However, vectorizing rough and complex real-world sketches can be challenging, as multiple overlapping lines need to be combined into a single line and extraneous lines and background noise need to be removed. [19] proposed a fully convolutional simplification network with a discriminator network to clean high-resolution sketches. [7] developed a two-step system employing two neural networks to vectorize clean line drawings. The first multi-task CNN predicts skeleton and junction images, and the second CNN resolves the segment connectivity around the junctions. They demonstrate state-of-the-art results on the public *Quick, Draw!* [8] dataset. However, their method is restricted to relatively small junctions of degrees 3 to 6 that fit in a 32×32 window.

Pen Trajectory Recovery. Throughout the years, numerous methods have been proposed by researchers to tackle the task of pen trajectory recovery from offline images. The steps involved in these methods typically include extraction of topology and detection of local ambiguous regions such as junctions and double-traced strokes. These ambiguities are then resolved using hand-designed rules. The existing methods can be broadly categorized into three types: recognition-based, topology-based, and tracking-based. Recognition-based [5] methods, which were first introduced for drawings composed of regular shapes such as diagrams and engineering drawings, detect these shapes by fitting geometric primitives. This approach is not ideal for handwritten text due to limitations in the possible graphical representation. Topology-based methods [10] construct a representation using topological information from the image (skeleton, contour, *etc.*) and view pen trajectory recovery as a global or local optimization problem. [17] developed a weighted graph approach that finds the best matching paths for pen trajectory recovery and demonstrated good performance on English characters. The tracking-based approach estimates the pen's relative direction iteratively. [24] proposed an image-to-sequence framework to generate pen trajectories using a CNN and fully connected layers without any RNN. This approach showed good results on Chinese and English handwriting datasets but the model's complexity is directly proportional to the image resolution. Moreover, their method requires a skeleton as input and inferred skeletons can be noisy and very different from the perfect skeletons their network was trained on, leading to unexpected failures at test time. [14] investigated the generalization of the previous approach to arbitrary-size images of math equations. They suggest using a fully convolutional neural network trained on noisy offline images. The network learns to predict both a skeleton and the next pen positions. However, the lack of temporal modeling causes over-segmentation of long strokes. Other lines of research followed a sequence modeling approach with CRNNs. In [3] a CNN-BLSTM network was proposed. They obtain good results on Tamil, Telugu, and Devanagari characters. However, this approach is limited to single isolated characters and requires separate models for each script. [2] extended the same CRNNs network to the text line scale [13] by applying a variety of data-augmentation techniques and an adaptive ground-truth loss to counter pathological strokes impact on the model learning. Their system is shown to recover a great portion of the online signal but still tends to omit some small strokes or even to over-simplify complex long strokes. Moreover, this approach is not well suited for 2D content such as math equations. In fact, resizing larger images of equations to a small fixed height (61 pixels) can lead to illegible content.

Stroke Embeddings and Transformers. Most of the proposed aforementioned approaches rely on the sequence-based networks to recognize drawing [8] or handwritten text [3]. [1] propose stroke-level Transformers to embed strokes into fixed lengths representations that are used to generate auto-completion of diagrams drawings. They show that Transformers outperform the sequential RNN approach [8]. However, they conclude that cursive handwriting strokes are challenging and longer strokes can't be correctly encoded in a fixed-size embedding. In this work, we model sub-stroke as embedding. Sub-strokes are much simpler shapes (straight lines, short open curves, *etc.*) that are far easier to model.

3   SET, SORT: Sub-stroke Level Transformers

After an overview of the proposed system, we present the sub-stroke extraction algorithm, the **S**ub-stroke **E**ncoding **T**ransformer (SET) and the **S**ub-strokes **OR**dering **T**ransformer (SORT).

3.1   Overview

We propose a novel sub-stroke ordering Transformer model to reconstruct the online signal from offline images. We start by using the FCNN from [14] to extract a skeleton (1-pixel thick outline) from the input offline image. A sub-stroke cutting algorithm based on junction detection is then applied to the extracted skeleton. We use a Transformer auto-encoder to learn sub-stroke embedding [1]. Finally, an auto-regressive Transformer decoder is used to predict the sub-strokes ordering using their embeddings. Figure 2 shows an overview of our pipeline. More formally, given a set of sub-strokes $V = \{ss_1, ss_2, \ldots, ss_N\}$, with a sub-stroke defined as a sequence of coordinates $ss_i = (x_k, y_k)_{k=1}^m$. Each sub-stroke from the skeleton appears twice, in both directions. The goal is to predict the sequence indicating the writing order of the different sub-strokes $S = (o_1, o_i, \ldots, o_M)$ $o_i \in \{1, \ldots, N\}$ and how they should be merged to form strokes. This is achieved by predicting a pen-up to indicate the end of the stroke. We note that V and S can be of unequal lengths, for instance, sub-strokes can be ignored (as noise), used several times (redrawing), and most of the time sub-strokes are drawn in one direction only, therefore the opposite sub-stroke is omitted.

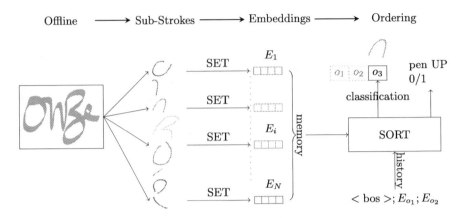

Fig. 2. Overview of our approach for the generation of the first three sub-strokes. After sub-stroke extraction, the SET encoder provides embedding for each sub-stroke. The SORT network uses the memory and the history to predict the next sub-stroke and the pen state. The sequence begins with a *bos* token and will end after M tokens with the eos token. Here we show the first three timesteps.

3.2 Sub-strokes Extraction

After extracting a skeleton using an already trained FCNN [14], we apply a thinning algorithm [23] to remove the few small remaining ambiguities in the skeleton, obtaining I_{thin}. We cut the skeleton into sub-strokes by removing the different junctions pixels and computing the resulting connected components. A junction pixel is defined as a skeleton pixel with 3 or more 8-connected skeleton pixels. Each connected component will have two extremities, the skeleton pixels with exactly one 8-connected skeleton neighbor (see Fig. 3). We compute the path from one extremity to another to define a sub-stroke. The opposite traversal path is also included as a distinct sub-stroke. Using this simple heuristic-free algorithm allows us to generalize to any handwritten content. Our sub-stroke cutting algorithm results in a normalization of stroke drawing. Partial inconspicuous stroke retracing is removed.

3.3 SET: Sub-stroke Embedding Transformer

We adapt the stroke embedding from [1] to the lower level of a sub-stroke. In fact, learning meaningful fixed-size vector representation for a stroke of arbitrary size and complexity can prove to be especially difficult for cursive handwriting. Sub-strokes are usually much simpler geometric primitives that are far easier to model. We define the sub-stroke auto-encoder as a Transformer followed by a sub-stroke reconstruction MLP, as shown in Fig. 4. Before being embedded by the encoder, the input sub-stroke points are shifted to start at the origin and

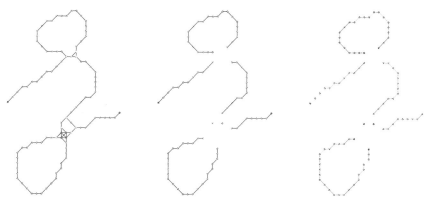

(a) Skeleton pixels 8-connectivity, in yellow and blue junction pixels.

(b) Junctions pixels are removed, splitting the skeleton into segments.

(c) Connected components are labeled to obtain the sub-strokes. Extremities are marked by a cross.

Fig. 3. Illustration of the sub-strokes cutting algorithm.

normalized w.r.t. to the offline image dimensions. This ensures that embedding only captures important local geometric features.

Encoder. Given an input sub-stroke defined as a sequence of points $ss_i = ((x_1, y_1), \ldots, (x_m, y_m))$, the points are first linearly projected to vectors of size 64 and summed with a sinusoidal positional encoding of each timestep. The input embedding is then fed through a Transformer with a stack of 6 layers, and 4 attention heads, with a model dimension of 64 and a feed-forward size of 256. The decoder output vector for the last timestep n of ss_i is projected linearly to a vector of size 8 corresponding to the sub-stroke embedding E_i.

Decoder. The sub-stroke reconstruction $F(E_i, t) \in \mathbb{R}^2, t \in [0, 1]$ is a parametric approximation of the sub-stroke curve using a two-layer MLP. It estimates the coordinates of the sub-stroke curve at every timestamp t. It's composed of a hidden layer of size 512 followed by $ReLU$ and an output layer of size 2 corresponding to the coordinates (x_t, y_t) of a point. The auto-encoder stroke embedding network objective is to reconstruct accurately the input sub-stroke.

3.4 SORT: Sub-stroke Ordering Transformer

We present a novel sub-stroke ordering auto-regressive transformer based on the sub-stroke embedding. Each sub-stroke embedding is concatenated with the positional embedding of its starting point $[f(ss_i[0]); E_i]$ to add global information of the sub-stroke spatial arrangement in the offline image. We use a stack of

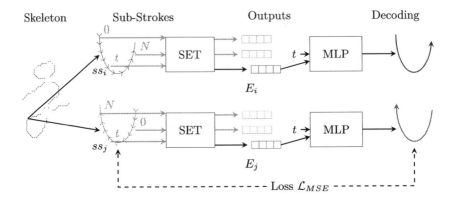

Fig. 4. Sub-stroke encoding transformer.

n_{layers} encoder-decoder Transformer with a model size of d_m and n_{heads} attention heads. The last transformer decoder layer employs a single attention head to compute the cross-attention between the encoder's output keys K and values V and the decoder self-attention output queries Q. This layer's output attention scores $\hat{A}_i$ over the sub-strokes set are used as predictions for the next sub-stroke ss_{i+1} probability distribution. As we can see in Eq. 1, the SORT outputs two decisions. On one hand, the attention scores over the sub-strokes set are used as predictions for the next sub-stroke ss_{i+1} probability distribution $\hat{A}_i$. On the other hand, the values O_i are used to predict the pen up state $\hat{P}_i$ with a small classification MLP.

$$\begin{aligned}
Q_i, K_i, V_i &= QW_i^Q, KW_i^K, VW_i^V \\
\hat{A}_i &= \text{softmax}\left(\frac{Q_i K_i^\top}{\sqrt{d_k}}\right) \in \mathbb{R}^{T \times L} \\
O_i &= \hat{A}_i V_i \\
\hat{P}_i &= \text{MLP}\left(O_i\right) \\
\text{with } & W_i^Q \in \mathbb{R}^{d_m \times d_k}, W_i^K \in \mathbb{R}^{d_m \times d_k}, W_i^V \in \mathbb{R}^{d_m \times d_v}
\end{aligned} \tag{1}$$

To alleviate the lack of coverage we employ the Attention Refinement Module (ARM) [25].

3.5   Training

The SET network is trained separately from the SORT. We sample five points at random $t \in [0, 1]$ from the sub-stroke latent representation E_i by using $F(E_i, t)$. The network is trained with an *MSE* loss between the reconstructed points and the ground-truth sub-stroke points as in Eq. 2.

$$\mathcal{L}_{MSE} = \frac{1}{5} \sum_{n=1}^{5} (F(E_{ss_i}, t_n) - ss_{i_{t_n}})^2 \tag{2}$$

We use teacher forcing to train the SORT network, it predicts the $i + 1$ sub-stroke probability distribution with $\hat{A}_i$ and the associated pen state $\hat{P}_i$ given the i sub-stroke. The network is trained with a multi-task loss $\mathcal{L}$ combining a cross-entropy classification loss for sub-stroke ordering $\mathcal{L}_O$ and a binary cross entropy loss for pen state classification $\mathcal{L}_P$.

$$\mathcal{L} = \lambda_1 \mathcal{L}_O + \lambda_2 \mathcal{L}_P,$$

$$\mathcal{L}_O(\hat{A}, A) = \frac{1}{|V|} \sum_{y=0}^{|V|} -A_y \log\left(\hat{A}_y\right), \tag{3}$$

$$\mathcal{L}_P(\hat{P}, P) = \frac{1}{2} \sum_{y \in \{ \text{down,up} \}} -\left(P_y \log\left(\hat{P}_y\right) + (1 - P_y) \log\left(1 - \hat{P}_y\right)\right),$$

where $\lambda_1, \lambda_2 \in \mathbb{R}$ and P, A are respectively the ground-truth pen-state and sub-stroke successor. Sub-strokes are extracted from an accurate but still not perfect inferred skeleton. They are ordered using the ground-truth online signal to obtain A used to train our network. This ordering is defined as the oracle machine's solution to the sub-strokes ordering problem. The oracle ordering is obtained by using the original online to map each extracted sub-stroke from the offline image independently to a sub-section of the online. They are then ordered using their time of apparition in the online signal. We note that this oracle answer is a satisfying approximation of the original online. However, it can still introduce a small disparity from the original online, particularly in cases of invisible pen-ups or erroneous skeletons.

3.6 Inference

At inference time, we follow the same pipeline to extract sub-strokes from an offline image as explained in Sect. 3.2. Sub-stroke embeddings are then produced using the SET network. The SORT network then iteratively predicts the next sub-stroke and corresponding pen state. We select the sub-stroke with the highest predicted probability as the next one which will be fed as input for the next timestep. Inference ends when a special *eos* token is predicted as the next sub-stroke. The result is a sequence of sub-strokes that we linearly interpolate to fill in the void left between two consecutive sub-stroke extremities (see Fig. 3), only when the pen state is "down" (i.e. $\hat{P}_i < 0.5$).

4 Evaluation

The goal of our method is to reconstruct accurately the pen trajectory reflected by a user's offline drawing. To quantitatively evaluate the quality of the online reconstructions, we employ two evaluation metrics DTW and handwriting recognition rate. While the DTW strictly measures geometric reconstruction fidelity, the recognition rate is a more lenient metric that measures semantic coherence.

4.1   DTW Point-Wise and DTW Point-to-Segment-Wise

We compute a DTW distance between the inferred online signal and the ground truth signal to measure the accuracy of the network prediction. We also employ a modified DTW with a point-to-segment distance DTW_{seg} by [15] which is less sensitive to the sampling rate. We also evaluate the stroke extraction by using a DTW on the stroke level. Similar to the offline Stroke IoU proposed by [7], we use an online stroke DTW defined as :

$$\text{SDTW} = \frac{1}{n} \sum_{i=1,\dots,n} \min_{j=1,\dots,m} \text{DTW}(S_i, \hat{S}_j), \tag{4}$$

where S_i are ground-truth strokes and $\hat{S}_j$ are predicted strokes. This metric is useful to detect under/over-segmentation issues of strokes which are otherwise not taken into account by DTW.

4.2   Handwriting Recognition Rate

The natural variability in writing styles makes it so that different reconstructions are plausible. DTW-based metrics continuity constraint strictly matches two online signals, which leads to high-cost alignment in some cases such as delayed strokes, interchangeable strokes and reversed strokes. An online automatic handwriting recognition system can be used to recognize the retrieved online signal. The recognition results can be compared with a ground-truth text label, computing a word and character recognition rate (WRR and CRR) for handwritten text and expression recognition rate for handwritten math. This results in higher-level evaluation which is far less sensitive to writing styles. However, we note that powerful state-of-art recognition systems can correctly predict the text even if some symbols are approximated roughly. In our case, this is problematic since the predicted signal is no longer loyal to the user's handwriting. For this reason, it is important to supplement the recognition rate with the DTW to also account for visual accuracy. We use the MyScript interactive ink recognition engine version 2.0[1] to evaluate the recognition.

5   Experiments

In this section, we present the training protocol and the evaluation results of our approach using online metrics.

---

[1] MyScript iink SDK is available at https://developer.myscript.com/docs/interactive-ink/2.0/overview/about/.

5.1   Datasets and Training

Our networks are trained and evaluated on IRONOFF [20] and CROHME [12] datasets. We follow the same procedure as [14] to render synthetic offline images from their online counterpart. Our rendered offline images are noisier than the constant stroke width rendering proposed in [12], as we want to better mimic the end goal real word noisier offline images (cf. Fig. 5). The training set of IRONOFF and CROHME contains respectively $48K$ and $10K$ samples, roughly equating to a total of $100K$ strokes each. We supplement CROHME with $15K$ equations from our private proprietary dataset.

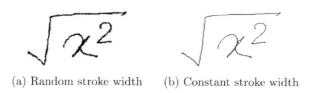

(a) Random stroke width   (b) Constant stroke width

Fig. 5. Comparison between variable and constant thickness stroking.

We first train the SET network on IRONOFF and freeze it during the training of the SORT network. The SORT network is trained on IRONOFF and fine-tuned on CROHME and our private datasets. We use the Adam optimizer with a learning rate of 0.001 and a batch size of 10. The training is performed on a single NVIDIA GeForce RTX 2080 Ti GPU with 24GB of memory and takes 20 h to be completed.

5.2   Results

We evaluate and benchmark our method against state-of-art offline to online conversion systems. Table 1 shows the results on the test set of *IRONOFF* containing $17K$ test samples. Row (d) shows that the oracle approximation is very close to the original online (e). The small difference reflects the previously mentioned errors and simplifications. Our method (c) outperforms other state-of-the-art approaches (a) and (b) while using a relatively lighter model compared to the other data-driven approach of [2]. However, as shown by (d) there is still a margin for progression.

We also evaluate our method on the CROHME 2014 and 2019 test sets. The evaluation results of [4] and our method are presented in Table 2 and 3.

Our approach achieves a better stroke extraction resulting in higher expression recognition rates. As reported by rows (d) of Tables 2 and 3 respectively, better online level DTW is not always synonymous with more precise stroke segmentation and more accurate recognition. In fact, [4] obtains a slightly better DTW of 14.29 as reported by Table 3 however a fairly lesser stroke DTW 7.19 compared to ours of 3.85. The same applies to ExpRate as well, 57.01% compared to 62.00% of our approach.

Table 1. Results on *IRONOFF* test set.

Method	Parameters	DTW↓	DTW$_{seg}$ ↓	CRR↑	WRR↑
(a) CNN-BiLSTM [2]	7M	7.09	7.45	59.22	41.43
(b) Chungkwong et al. [4]	–	5.75	5.06	73.45	60.00
(c) Sub-stroke Transformer (Ours)	2M	**3.25**	**2.72**	**90.85**	**81.06**
(d) Oracle	–	0.33	0.32	92.56	83.45
(e) online GT	–			93.03	83.81

Table 2. Results on CROHME 2014 test set.

Method	DTW↓	DTW$_{seg}$ ↓	SDTW↓	ExpRate↑
(a) Chungkwong et al	16.30	16.13	6.54	52.43
(b) Sub-stroke Transformer (Ours)	24.54	24.37	12.29	29.37
(c) fine-tune (b) on CROHME	**13.75**	**13.59**	4.43	53.75
(d) fine-tune (c) on private datasets	13.93	13.80	**2.93**	**59.31**
(f) Oracle	0.24	0.22	0.50	66.63
(g) GT online	—	—	—	69.77

Table 3. Results on CROHME 2019 test set. Row (d) reports the results of the fine-tuned model on equations from CROHME and our private dataset.

Method	DTW↓	DTW$_{seg}$ ↓	SDTW↓	ExpRate↑
(a) Chungkwong et al	**14.29**	**14.14**	7.19	57.01
(d) fine-tune on private dataset	14.98	14.80	**3.85**	**62.00**
(f) Oracle	0.26	0.24	0.69	70.19
(g) GT online	—	—	—	73.13

Figure 6 shows a visual comparison between our approaches and other state-of-the-art methods on IRONOFF. Our approach Fig. 6c is observed to cover very closely most of the offline image compared to Fig. 6a and 6b. In fact, some characters are missing in their online reconstructions. For example, the smaller "e" loops (rows 3 and 4), the middle horizontal bar of "E" (row 4) and the apostrophe (row 2) are not covered. Figure 6a and 6b tend to over-segment the strokes, on the other hand, our approach predicts more accurate pen ups resulting in a far less number of strokes. Figure 6a often struggles with end-of-sequence predictions (first three rows).

Our method is observed to better capture the greater diversity in the stroke 2D ordering in Maths equations as illustrated by Fig. 7. For instance, the reconstruction in Fig. 7b shows greater variability compared to the strict X-Y ordering of Fig. 7a. Here the superscripts are predicted after the exponents and the operators. This is a less common way to write but is still plausible. As highlighted in Fig. 7b our network mistakenly re-crosses the first "+" sign (as highlighted in

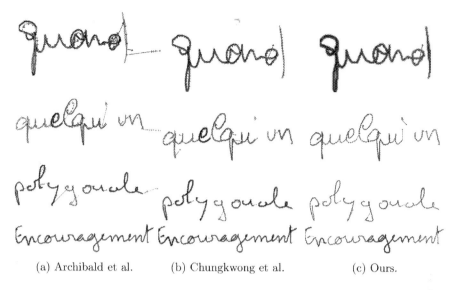

(a) Archibald et al.   (b) Chungkwong et al.   (c) Ours.

Fig. 6. Comparaison of our approach (c) to [2] (a) and [4](b) on IRONOFF samples. Each stroke is drawn with a distinct color. Blue arrows show the direction. The first and last stroke points are respectively yellow and red.

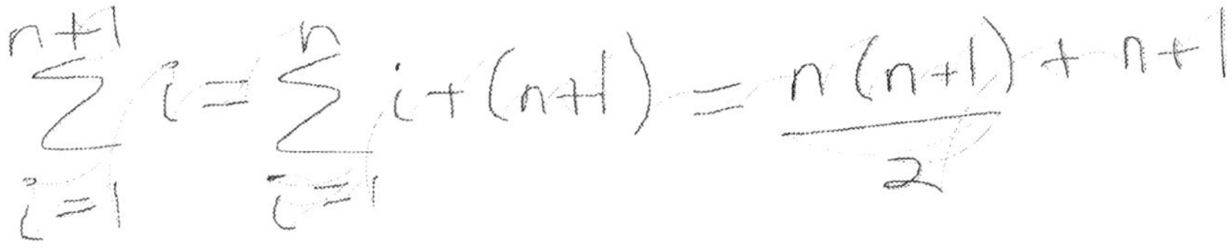

$$\sum_{i=1}^{n+1} i = \sum_{i=1}^{n} i + (n+1) = \frac{n(n+1)}{2} + n + 1$$

(a) Chungkwong et al. [4]

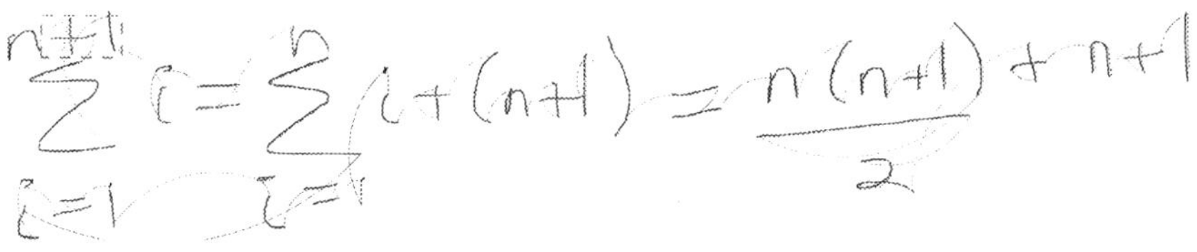

$$\sum_{i=1}^{n+1} i = \sum_{i=1}^{n} i + (n+1) = \frac{n(n+1)}{2} + n + 1$$

(b) Ours.

$$x = \frac{af(b) - bf(a)}{f(b) - f(a)} \qquad x = \frac{af(b) - bf(a)}{f(b) - f(a)}$$

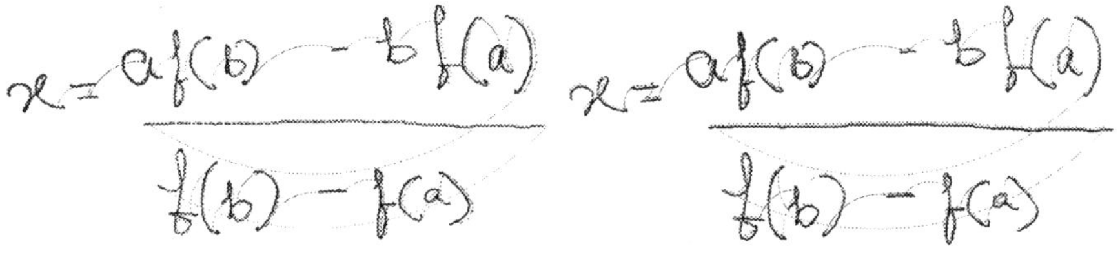

(c) Chungkwong et al.   (d) Ours.

Fig. 7. Inference results of [4] and our approach on CROHME datasets.

the red box), instead of drawing the "1", but is still able to recover the remaining strokes correctly. We hypothesize that it's due to the two strokes being of very similar shape and in close proximity to each other. We observe a few errors in Fig. 7a, the "*i*" dot is missing in one case. The ordering of the superscripts in the second "$\sum$" is far from ideal. In Fig. 7d, the parenthesis and their content are not always in the same order. Reflecting once again on the great diversity captured by our network. Thanks to our pen state prediction, we can more accurately segment the symbols Fig. 7d. In Fig. 7c the f in the term "$f(a)$" of the numerator is incorrectly segmented resulting in a bad recognition. Figure 8 shows the SORT prediction of the probability distribution of the next sub-stroke at every timestep. We observe that the network is very confident in its predictions and they are well centered around a small local region of the image. The network's reconstructed signal overall reflects the same temporal dynamics as the ground truth online signal. However, as depicted in Fig. 8b, in rare instances it locally drifts from the ground truth online.

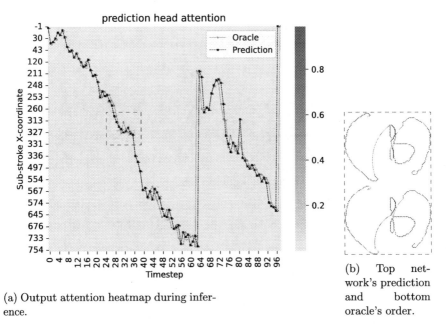

(a) Output attention heatmap during inference.

(b) Top network's prediction and bottom oracle's order.

Fig. 8. (a) Attention heatmap of the SORT decoder output layer for the Fig. 7d. The y-axis is the memory sub-strokes sorted from left to right (with the first point) for illustration purposes only. Network predictions (see Sect. 3.6) as well as the oracle answers are plotted on top of the heatmap. The *eos* sub-stroke is here indicated by a −1. (b) The divergence between the inferred sub-stroke order and the oracle's order, for timesteps from 24 to 40.

6 Discussion

Our approach is able to generalize to different handwriting domains. By transferring the learned knowledge from Latin words to Maths equations we are able to achieve better results compared to handcrafted rule-based systems. However, we need this transfer, in the form of a fine-tuning step, for new application domains. The existing online databases of handwriting in a multitude of languages and free-form charts can be exploited to train the system in order to generalize to all content types.

We focus our study application on offline images of at most 100 sub-strokes. Full offline documents can attain upwards of 6000 sub-strokes. Further research efforts are necessary to up-scale our sub-stroke ordering transformer to the document level. In fact, it presents two difficulties, firstly longer and more complex temporal dependencies to model. Secondly, the memory bottleneck of the quadratic multi-head attention needs to be addressed.

7 Conclusion

In this paper, we presented a novel sub-stroke level transformer approach to recover online from offline handwriting. Our approach consists of two steps: First, we embed the sub-strokes set, extracted from the inferred skeleton, using a sub-stroke encoding transformer (SET). The sub-strokes embeddings are ordered using a sub-strokes ordering Transformer (SORT) which also predicts the pen state. In contrast to other data-driven approaches, SORT is trained in a guided attention manner and is able to accurately string together the original sub-strokes rather than regressing a simplified approximation of the online. Our method's performance stands out when compared to the state-of-the-art on Latin words and Math equations. In future work, we would like to extend our system to full documents thus enabling a powerful combination of offline note-taking and seamless online editing.

Acknowledgements. We would like to express our gratitude to Robin Mélinand for his invaluable feedback and suggestions for this article.

References

1. Aksan, E., Deselaers, T., Tagliasacchi, A., Hilliges, O.: CoSE: compositional stroke embeddings. In: Proceedings of the 34th International Conference on Neural Information Processing Systems, NIPS 2020, pp. 10041–10052. Curran Associates Inc., Red Hook, December 2020. https://proceedings.neurips.cc/paper/2020/file/723e8f97fde15f7a8d5ff8d558ea3f16-Paper.pdf
2. Archibald, Taylor, Poggemann, Mason, Chan, Aaron, Martinez, Tony: TRACE: a differentiable approach to line-level stroke recovery for offline handwritten text. In: Lladós, Josep, Lopresti, Daniel, Uchida, Seiichi (eds.) ICDAR 2021. LNCS, vol. 12823, pp. 414–429. Springer, Cham (2021). https://doi.org/10.1007/978-3-030-86334-0_27

3. Bhunia, A.K., et al.: Handwriting trajectory recovery using end-to-end deep encoder-decoder network. In: 2018 24th International Conference on Pattern Recognition (ICPR), pp. 3639–3644, August 2018. https://doi.org/10.1109/ICPR.2018.8546093

4. Chan, C.: Stroke extraction for offline handwritten mathematical expression recognition. IEEE Access **8**, 61565–61575 (2020). https://doi.org/10.1109/ACCESS.2020.2984627

5. Doermann, D., Intrator, N., Rivin, E., Steinherz, T.: Hidden loop recovery for handwriting recognition. In: Proceedings Eighth International Workshop on Frontiers in Handwriting Recognition, pp. 375–380, August 2002. https://doi.org/10.1109/IWFHR.2002.1030939

6. Egiazarian, Vage, et al.: Deep vectorization of technical drawings. In: Vedaldi, Andrea, Bischof, Horst, Brox, Thomas, Frahm, Jan-Michael. (eds.) ECCV 2020. LNCS, vol. 12358, pp. 582–598. Springer, Cham (2020). https://doi.org/10.1007/978-3-030-58601-0_35

7. Guo, Y., Zhang, Z., Han, C., Hu, W., Li, C., Wong, T.T.: Deep line drawing vectorization via line subdivision and topology reconstruction. Comput. Graph. Forum **38**(7), 81–90 (2019). https://doi.org/10.1111/cgf.13818

8. Ha, D., Eck, D.: A neural representation of sketch drawings. In: ICLR 2018 (2018). https://openreview.net/pdf?id=Hy6GHpkCW

9. Holten, D., van Wijk, J.J.: A user study on visualizing directed edges in graphs. In: Proceedings of the SIGCHI Conference on Human Factors in Computing Systems, CHI 2009, pp. 2299–2308. Association for Computing Machinery, New York, April 2009. https://doi.org/10.1145/1518701.1519054

10. Jager, S.: Recovering writing traces in off-line handwriting recognition: using a global optimization technique. In: Proceedings of 13th International Conference on Pattern Recognition, vol. 3, pp. 150–154, August 1996. https://doi.org/10.1109/ICPR.1996.546812

11. Ji, B., Chen, T.: Generative Adversarial Network for Handwritten Text, February 2020

12. Mahdavi, M., Zanibbi, R., Mouchere, H., Viard-Gaudin, C., Garain, U.: ICDAR 2019 CROHME + TFD: competition on recognition of handwritten mathematical expressions and typeset formula detection. In: 2019 International Conference on Document Analysis and Recognition (ICDAR), pp. 1533–1538, September 2019. https://doi.org/10.1109/ICDAR.2019.00247

13. Marti, U.V., Bunke, H.: The IAM-database: an English sentence database for offline handwriting recognition. Int. J. Doc. Anal. Recogn. **5**(1), 39–46 (2002). https://doi.org/10.1007/s100320200071

14. Mohamed Moussa, Elmokhtar, Lelore, Thibault, Mouchère, Harold: Applying end-to-end trainable approach on stroke extraction in handwritten math expressions images. In: Lladós, Josep, Lopresti, Daniel, Uchida, Seiichi (eds.) ICDAR 2021. LNCS, vol. 12823, pp. 445–458. Springer, Cham (2021). https://doi.org/10.1007/978-3-030-86334-0_29

15. Mohamed Moussa, E., Lelore, T., Mouchère, H.: Point to Segment Distance DTW for online handwriting signals matching. In: Proceedings of the 12th International Conference on Pattern Recognition Applications and Methods, pp. 850–855. SCITEPRESS - Science and Technology Publications, Lisbon, Portugal (2023). https://doi.org/10.5220/0011672600003411

16. Nguyen, V., Blumenstein, M.: Techniques for static handwriting trajectory recovery: a survey. In: Proceedings of the 9th IAPR International Workshop on Doc-

ument Analysis Systems, DAS 2010, pp. 463–470. Association for Computing Machinery, New York, June 2010. https://doi.org/10.1145/1815330.1815390

17. Qiao, Y., Nishiara, M., Yasuhara, M.: A framework toward restoration of writing order from single-stroked handwriting image. IEEE Trans. Pattern Anal. Mach. Intell. **28**(11), 1724–1737 (2006). https://doi.org/10.1109/TPAMI.2006.216

18. Seki, Y.: Online and offline data collection of Japanese handwriting. In: 2019 International Conference on Document Analysis and Recognition Workshops (ICDARW), vol. 8, pp. 13–18, September 2019. https://doi.org/10.1109/ICDARW.2019.70135

19. Simo-Serra, E., Iizuka, S., Ishikawa, H.: Mastering Sketching: adversarial augmentation for structured prediction. ACM Trans. Graph. **37**(1), 11:1–11:13 (2018). https://doi.org/10.1145/3132703

20. Viard-Gaudin, C., Lallican, P.M., Knerr, S., Binter, P.: The IRESTE On/Off (IRONOFF) dual handwriting database. In: Proceedings of the Fifth International Conference on Document Analysis and Recognition, ICDAR '99 (Cat. No.PR00318), pp. 455–458, September 1999. https://doi.org/10.1109/ICDAR.1999.791823

21. Vinciarelli, A., Perone, M.: Combining online and offline handwriting recognition. In: Seventh International Conference on Document Analysis and Recognition, 2003. Proceedings, pp. 844–848, August 2003. https://doi.org/10.1109/ICDAR.2003.1227781

22. Zhang, J., Du, J., Dai, L.: Track, Attend, and Parse (TAP): an end-to-end framework for online handwritten mathematical expression recognition. IEEE Trans. Multimedia **21**(1), 221–233 (2019). https://doi.org/10.1109/TMM.2018.2844689

23. Zhang, T.Y., Suen, C.Y.: A fast parallel algorithm for thinning digital patterns. Commun. ACM **27**(3), 236–239 (1984). https://doi.org/10.1145/357994.358023

24. Zhao, B., Yang, M., Tao, J.: Pen tip motion prediction for handwriting drawing order recovery using deep neural network. In: 2018 24th International Conference on Pattern Recognition (ICPR), pp. 704–709, August 2018. https://doi.org/10.1109/ICPR.2018.8546086

25. Zhao, W., Gao, L.: CoMER: modeling coverage for transformer-based handwritten mathematical expression recognition. In: Avidan, S., Brostow, G., Cissé, M., Farinella, G.M., Hassner, T. (eds.) ECCV 2022. LNCS, pp. 392–408. Springer, Cham (2022). https://doi.org/10.1007/978-3-031-19815-1_23

Character Queries: A Transformer-Based Approach to On-line Handwritten Character Segmentation

Michael Jungo[1,2]([✉]), Beat Wolf[1], Andrii Maksai[3], Claudiu Musat[3], and Andreas Fischer[1,2]

[1] iCoSys Institute, University of Applied Sciences and Arts Western Switzerland, Fribourg, Switzerland
{michael.jungo,beat.wolf,andreas.fischer}@hefr.ch
[2] DIVA Group, University of Fribourg, Fribourg, Switzerland
[3] Google Research, Zurich, Switzerland
{amaksai,cmusat}@google.com

Abstract. On-line handwritten character segmentation is often associated with handwriting recognition and even though recognition models include mechanisms to locate relevant positions during the recognition process, it is typically insufficient to produce a precise segmentation. Decoupling the segmentation from the recognition unlocks the potential to further utilize the result of the recognition. We specifically focus on the scenario where the transcription is known beforehand, in which case the character segmentation becomes an assignment problem between sampling points of the stylus trajectory and characters in the text. Inspired by the k-means clustering algorithm, we view it from the perspective of cluster assignment and present a Transformer-based architecture where each cluster is formed based on a learned character query in the Transformer decoder block. In order to assess the quality of our approach, we create character segmentation ground truths for two popular on-line handwriting datasets, IAM-OnDB and HANDS-VNOnDB, and evaluate multiple methods on them, demonstrating that our approach achieves the overall best results.

Keywords: On-Line Handwriting · Digital Ink · Character Segmentation · Transformer

1 Introduction

Relevance. A significant advantage of using a stylus over a keyboard is its flexibility. As with pen and paper, users can draw, write, link objects and make gestures like circling or underlining with ease – all with a handful of strokes. For digital ink to have a compelling value proposition however, many features associated with all the use cases, that users have become accustomed to in an online environment, become relevant. They **go beyond the initial act of writing** and cover layout and ink generative models like autocompletion and spelling correction.

G. A. Fink et al. (Eds.): ICDAR 2023, LNCS 14187, pp. 98–114, 2023.
https://doi.org/10.1007/978-3-031-41676-7_6

Usefulness. On-line handwriting character segmentation has as a goal the understanding of which parts of the handwriting belong to which character. It complements handwriting recognition and enables functionalities like generative modeling, particularly **spellchecking and correction** [1], as well as ink-to-text conversion and layout handling [2,3]. What all these seemingly different tasks have in common is a need for character-level information.

Character-level knowledge opens up the possibility for layout-preserving processing. For instance, when converting the handwritten text into printed text, knowing the positions of the individual characters allow to generate printed text that is precisely superimposed on top of the printed text, retaining the feeling of agency over the document (e.g. in devices like Jamboard [2] and note-taking apps like FreeForm [3]). Moreover, for education and entertainment applications, knowing the positions of characters can unlock the capabilities such as animating individual characters (e.g. in the Living Jiagu project the symbols of the Oracle Bone Script were animated as the animals they represent [4]).

Individual character information is also important in handwriting generation models [1,5,6]. Examples include spellchecking and spelling correction. For spellchecking, knowledge of word-level segmentation helps to inform the user about the word that was misspelled, e.g. marking the word with a red underline, and the word-level segmentation is a natural byproduct of character-level segmentation. For spelling correction, users could strike out a particular character or add a new one, and the remaining characters could be edited such that the change is incorporated seamlessly, for example via handwriting generation models [1].

Difficulty. While the problem of character segmentation is fairly simple in case of printed text OCR, it is far from being solved for handwriting – both on-line and off-line. The problem is more difficult in settings like **highly cursive scripts** (e.g. Indic) or simply cursive writing in scripts like Latin. Difficult cases further include characters in Arabic script with ligatures, which vary in appearance depending on the surrounding of the character and position in which they appear, and characters differing only by diacritics [7].

In the academic on-line handwriting community, the progress on character segmentation is **limited by the absence of the datasets with character segmentation**. Two notable exceptions are Deepwriting [5] and BRUSH [8]. The authors of Deepwriting used a private tool to obtain a monotonic segmentation for the dataset. This is limited, as it cannot accommodate cursive writing. In the BRUSH dataset an image segmentation model was used to obtain the character segmentation.

For this reason, most of the works in the on-line handwriting community **rely on synthetic datasets**. These are created with either (1) segment-and-decode HMM-based approaches where character segmentation is a byproduct of recognition [9], or (2) hand-engineered script-specific features used in deep learning solutions, e.g. for Indic and Arabic script [10,11], as well as mathematical expressions [12].

For off-line handwriting, character segmentation is usually based on the position of skip-token class spikes in the Connectionist Temporal Classification (CTC) [13] logits – which works well for images as the segmentation is typically monotonic and separation of image by the spikes results in a reasonable segmentation (unlike on-line handwriting, where due to cursive writing the segmentation is not monotonic).

Another difficulty that is associated with character segmentation is the **annotation**. Individually annotating characters is hard and also time consuming. The most widely used handwriting datasets do not contain ground truth information on individual characters. We can, however, infer a high quality synthetic ground truth using a time consuming method that iteratively singles out the first character from the ink, based on temporal, spatial and stroke boundary information.

Methods. We compare multiple methods for character boundary prediction, with both a Long Short-Term Memory (LSTM) [14] and Transformer [15] backbone and further comparing them with a simple k-means baseline. A first classifier architecture, that accepts both an LSTM and a Transformer encoder, combines the individual point offsets with the CTC spikes to determine which points represent character boundaries. This initial approach has a clear limitation in that it is monotonic and cannot handle delayed strokes. We thus extend the Transformer classifier by including the character information, where each character in the text becomes a query in the Transformer decoder block. To show its efficacy we focus on the following approaches in an experimental evaluation on the publicly available IAM-OnDB and HANDS-VNOnDB datasets: k-means, LSTM, Transformer for character boundary prediction and Transformer with character queries.

The main **contributions** of this work can be summarized as follows:

- We obtain character segmentation ground truths for the publicly available datasets IAM-OnDB and HANDS-VNOnDB from a high-quality approximation.
- We present a Transformer-based approach to the on-line handwritten character segmentation, where each expected output character is represented as a learned query in the Transformer decoder block, which is responsible for forming a cluster of points that belong to said character.
- We compare our approach to other methods on the IAM-OnDB and HANDS-VNOnDB datasets thanks to the newly obtained ground truth and demonstrate that it achieves the overall best results.

The newly created ground truths and the source code of our methods are publicly available at https://github.com/jungomi/character-queries.

2   Datasets

2.1   IAM On-line Handwriting Database

IAM-OnDB [16] is a database of on-line handwritten English text, which has been acquired on a whiteboard. It contains unconstrained handwriting, meaning that it includes examples written in block letters as well as cursive writing, and any mixture of the two, because it is not uncommon that they are combined in a way that is most natural to the writer. With 221 different writers having contributed samples of their handwriting, the dataset contains a variety of different writing styles.

2.2 HANDS-VNOnDB

The HANDS-VNOnDB [17], or VNOnDB in short, is a database of Vietnamese handwritten words. A characteristic of Vietnamese writing, that is not found in English, is the use of diacritical marks, which can be placed above or below various characters and even stacked. This poses an additional challenge, as the diacritics are often written with delayed strokes, i.e. written after one or more characters have been written before finishing the initial character containing the diacritics. Most notably in cursive and hasty writing, it is very common that the diacritics are spatially displaced, for example hanging over the next character, which makes them disconnected in time as well as space and therefore much more difficult to assign to the correct character.

2.3 Ground Truth

Since both of the publicly available datasets we are using do not have the ground truth character segmentation, we resorted to obtaining a high-quality approximation of it (similar approach was applied, for example, by [18] where an image-based approach was used for obtaining the ground truth approximation). To obtain it, we repeatedly separated the first character from the rest of the ink, by performing an exhaustive search for the character boundary with potential cuts based on temporal information, spatial information, and stroke boundaries, and with the best candidate selected based on the likelihood that the first character indeed represents the first character of the label, and the rest matches the rest of the label, with likelihood provided by a state-of-the-art recognizer model [anonymized for review]. Such an approach is clearly not feasible in a practical setting due to the high computational cost, but allowed us to produce a high quality ground truth approximation, from which any of the models described below could be trained. Figure 1 illustrates some ground truth examples from the IAM-OnDB and the HANDS-VNOnDB. Despite the high quality of the ground truth it remains an approximation and therefore some small imperfections are present, as evidenced by some of the examples.

3   Methods

3.1   K-Means

For the initial baseline system, we chose to use a k-means [19] based approach. To segment the handwriting, the points are clustered into k different clusters,

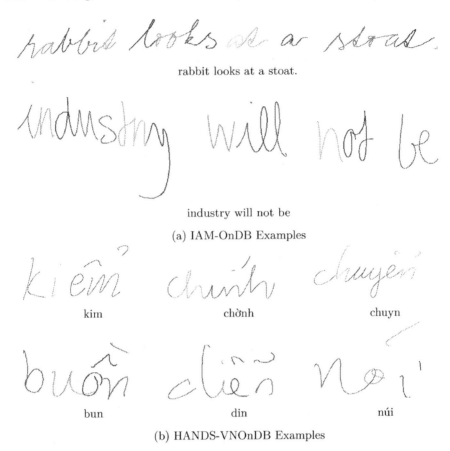

rabbit looks at a stoat.

industry will not be

(a) IAM-OnDB Examples

kim

chờnh

chuyn

bun

din

núi

(b) HANDS-VNOnDB Examples

Fig. 1. Examples of the ground truth character segmentations that were obtained by iteratively separating the first character through an exhaustive search in regards to the character boundary. Each color represents a different character. Some imperfections are to be expected as it remains an approximation.

where k is equal to the number of characters present in the already known text. Two methods have been implemented for the initialization of the centroids, the standard random implementation, and a second implementation which uses the points where the CTC spikes occurred as the initial centroids. Clustering is mainly based on the geometric locations of the points but the stroke information was also included, as it still adds value for points that are not clearly separable purely based on their position. Since the horizontal position is much more indicative of the character it might belong to, the x coordinate was weighted much stronger than the y coordinate. This heavily relies on the horizontal alignment of the writing and causes an inherent limitation for cases where the alignment deviates from the ideal representation, e.g. for strongly slanted handwriting.

3.2   Character Boundary Prediction with LSTM / Transformer

The input of the model for the character boundary prediction is a sequence of sampling points and the output is a classification of whether a point is a character boundary or not. Intuitively, an LSTM [14] can be employed for this, as it is particularly well suited to work with a sequence based representation. Given that the output remains a sequence but is not required to recognize which character it is, it is sufficient to have the tokens <start>, <char> and <none>, which signify the start of the character (boundary), being part of the current character and not belonging to a character at all, respectively. An important note about the <none> token is, that there is no point in the available ground truth that does not belong to any character, simply because of the exhaustive nature of the ground truth creation, as a consequence it is repurposed to indicate that the point does not belong to the current character, which primarily refers to delayed strokes. Due to the lack of back references in this approach, it will just be considered as not part of any character.

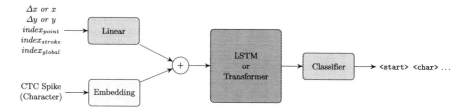

Fig. 2. Architecture of the boundary prediction model. A feature vector is created from the x, y-coordinates ($\Delta x, \Delta y$ for the LSTM and absolute coordinates for the Transformer) and the stroke information, where the pair of $index_{stroke}$ and $index_{point}$ indicate which stroke the point belongs to and which point it is within the stroke, as well as the global position with $index_{global}$. For points where a CTC spike occurred, an embedding vector of the identified character is added to the existing feature vector. The resulting feature vector is processed either by bidirectional LSTMs or a Transformer and followed by a linear classifier to produce the boundary prediction.

Figure 2 outlines the architecture of the boundary prediction model. A sequence of points is given as the input, where each point contains the x and y coordinates, as well as information at which part of the sequence it occurred. Handwriting is almost always performed in multiple strokes, which can be a helpful indicator of where a character might begin, therefore it is conveyed to the model with the pair of $index_{stroke}$ and $index_{point}$ to identify the stroke it belongs to and which point it is within the stroke, while $index_{global}$ is also provided in order facilitate locating the point globally. All this information is transformed by a linear layer to create a higher dimensionality feature vector that is more appropriate for the LSTM. Additionally, for each point where a CTC spike occurs, an embedding vector is created from the character it corresponds to, and added to the existing feature vector. Afterwards, the feature vector is processed by multiple bidirectional LSTMs followed by a linear classifier to produce the boundary predictions.

Transformers [15] are also widely used in sequence based task and are generally highly successful in many situations where LSTMs perform well, hence the same architecture can be used with a Transformer instead of an LSTM. Some minor changes to the input are required compared to the LSTM. The x, y-coordinates were given as deltas $(\Delta x, \Delta y)$ in regards to the previous point due to the recurrent nature of the LSTM, which turned out to work slightly better than the absolute coordinates. Since Transformers do not have any recurrence, there is no reference point for the deltas, therefore the absolute coordinates are the only viable option. Even though $index_{global}$ might be considered to be more important for Transformers, it is not used because the same effect is achieved by the positional encoding that is added to the Transformer to explicitly handle the positional information.

Post-processing. In order to assign the points to the respective characters, the sequence of tokens needs to be processed such that the point with the <start> token and all points marked as <char> up to the next <start> token are assigned to the expected output characters from left to right. Technically, the model is not limited to produce exactly number of expected characters, but is supposed to learn it. Unfortunately, it does occur that too many characters are predicted, hence we additionally restrict the output to the desired number of characters by removing the segments with the smallest number of points, as we are specifically interested in assigning the points to the expected characters. This is a limitation of this particular design for the character boundary.

3.3   Transformer with Character Queries

Given that the design of the character boundary prediction in Sect. 3.2 revolves around sequences of points, it is impossible to handle delayed strokes appropriately. Furthermore, the expected output characters are not integrated into the model, even though they are at least represented in the input features by the CTC spikes. This not only necessitates some post-processing, due to possible oversegmentation, but also eliminates any potential for the model to adapt to a specific character. To address these shortcomings, we design a Transformer-based architecture that integrates the expected output characters into the Transformer decoder block by using them as queries.

Related Work. In recent years, Transformers have been applied to many different tasks in various domains. With the necessity to adapt to domains not initially suited for Transformers, due to the inherently different structures compared to the familiar sequence based tasks, new Transformer-based approaches have been developed. In particular, in the domain of Computer Vision (CV), a lot of creative designs have emerged [20,21]. One of these novel approaches was pioneered by the DEtection TRansformer (DETR) [22], called object queries, where each learned query of the Transformer decoder block represents one object that has to be detected. Later on the query based approach has found its way to image segmentation tasks [23–25].

Only recently, the k-means Mask Transformer [26] introduced a Transformer-based approach that was inspired by the k-means clustering algorithm, where the authors discovered that updating the object queries in the cross-attention of the Transformer decoder block was strikingly similar to the k-means clustering procedure. While their approach was made for image based segmentation, it can easily be adapted to our task of on-line handwritten segmentation, which happens to remove a lot of the complexity that is only needed for images, mostly due to downsampling of the image and upsampling of the segmentation masks. Inspired by their findings and the fact that our baseline algorithm has been k-means, we design a Transformer-based architecture where the queries represent the characters that should be segmented.

Overview. For each character that needs to be segmented, a query in the Transformer decoder block is initialized with the embedding of that particular character and a positional encoding, which is necessary to distinguish two or more instances of the same character. In that regard, the character embedding provides the information to the model as to which particular patterns to pay attention to, while the positional encoding is primarily used to ensure that the order of the characters is respected. Having the available characters tightly integrated into the model, eliminates the post-processing completely, which is due to the fact that the characters were created from a sequence in the boundary prediction models, whereas now the points are simply assigned to the respective characters, reminiscent of clustering algorithms such as k-means. Additionally, it opens up the possibility to handle delayed strokes correctly without any modification as long as they are represented adequately in the training data.

Architecture. This model will subsequently be referred to as the Character Query Transformer and its architecture is outlined in Fig. 3. The input features remain the same as for the boundary prediction model, where the feature vector is created from the x, y-coordinates, the stroke information through the pair of $index_{stroke}$ and $index_{point}$, and the CTC spikes with an embedding of the identified character. A Transformer encoder is applied to the feature vector to create a new encoded vector, $E \in \mathbb{R}^{p \times d_h}$, that captures more pertinent information by virtue of the self-attention which incorporates the relation between the points. Afterwards, a Transformer decoder block takes the encoded vector in combination with the character queries, which are created from the expected output characters by applying a learned character embedding and positional embedding based on their position within the text.

The output of the decoder, $D \in \mathbb{R}^{c \times d_h}$, cannot be used directly to create a classification for each point, as it is merely a latent representation of the clusters, hence it has the dimensions $c \times d_h$, where c is the number of characters and d_h

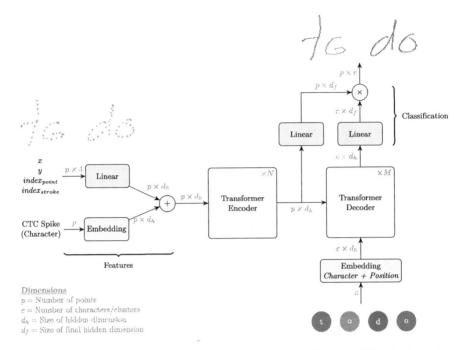

Fig. 3. Architecture of the Character Query Transformer. Like the boundary prediction models, the feature vector is created from the x, y-coordinates, the stroke information and the CTC spikes. A Transformer encoder takes the feature vector and creates an encoded vector, which is given to the Transformer decoder block in combination with an embedding of the desired characters to be segmented (including their position within the text). The classification is achieved by a matrix multiplication between the output of the encoder and the decoder, after a linear transformation of each of the respective outputs, in order to assign each point to one character.

the size of the hidden dimension. While the points have been assimilated into D through the cross-attention in the decoder, the exact association between points and characters must be done with an additional step. This can be achieved with $ED^T \in \mathbb{R}^{p \times c}$, a matrix multiplication between E and D, the outputs of the encoder and decoder respectively. Normally, a classifier would be applied afterwards, but because the dimensions of $p \times c$ are dynamic, since both p (number of points) and c (number of characters) vary depending on the input, that is not possible. As an alternative, a linear transformation is applied separately to E and D before the matrix multiplication.

Positional Encoding of the Character Queries. Transformers do not inherently have any sense of position of the inputs, as they do not contain any recurrence or convolutions, which implicitly take the order into account. To alleviate this issue, a separate positional encoding is added to the input to explicitly encode the positional information into the input. It is most commonly achieved with a sinusoidal positional encoding, where the sine and cosine functions are used with different frequencies:

$$PE_{(pos,2i)} = \sin(\frac{pos}{10000^{2i/d}})$$
$$PE_{(pos,2i+1)} = \cos(\frac{pos}{10000^{2i/d}})$$

$$(1)$$

Due to the characteristics of the sinusoidal functions the transition between the positions remains predictable and smooth, therefore the input features are not disrupted but rather slightly enhanced. Generally, this is a desirable property, but in the case of the character queries, a more recognizable distinction between the position is needed, because multiple instances of the same character need to be treated as completely separate. For this purpose, a learned positional encoding is used instead. Figure 4 depicts the normalized mean values of the vector at each position in the positional encoding for the sinusoidal (blue) and learned encodings (red) respectively. In the learned encoding it is clearly visible that there are a lot more large differences between two positions, indicating that a clear distinction between them does benefit the model and its capabilities to distinguish between multiple instances of the same character. On the other hand, the sinusoidal encoding keeps a smooth transition between the positions and therefore lacks the clear distinguishing aspect, and in our experience it was simply not enough to separate multiple instances of the same character.

4   Experiments

In this section we evaluate the four methods, namely the k-means, LSTM, Transformer (character boundary prediction) and Character Query Transformer on the IAM-OnDB and the HANDS-VNOnDB, as well as combining the two dataset to see whether more data with slightly different characteristics are beneficial to the overall results. And finally, an ablation study on the usefulness of the CTC spikes is conducted. All results are evaluated based on the mean Intersection over Union (mIoU) between the points in the segmented characters.

4.1   Setup

The experiments for the k-means are performed with Scikit-Learn [27] whereas all other methods are implemented in PyTorch [28]. For the k-means the weights of the input features are set to 1 for the x-coordinate, 0.04 for the y-coordinate and 224 for the stroke information. All PyTorch models use a dimension of 256 for all layers, i.e. embedding dimension, hidden dimension of LSTM/Transformer and the final hidden dimension, as well as a dropout probability of 0.2. The LSTM consist of 3 bidirectional layers with the Rectified Linear Unit (ReLU) [29]

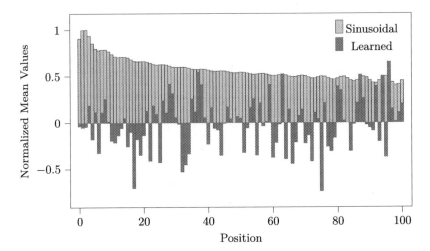

Fig. 4. Positional Encoding for Character Queries. For each position the normalized mean values of the vector in the positional encoding are displayed. The sinusoidal encoding (blue) follows a smooth trend with small variations between the positions, whereas the learned encoding (red) exhibits much larger differences between positions, which makes the distinction between multiple instances of the same character much more apparent to the model. (Color figure online)

as an activation function, whereas the Transformer uses 3 layers per encoder and decoder with 8 attention heads and Gaussian Error Linear Unit (GELU) [30] instead of ReLU. Label smoothing with $\epsilon = 0.1$ [31] is employed in the cross-entropy loss function. AdamW [32] is used as an optimizer with a weight decay of 10^{-4}, $\beta_1 = 0.9$ and $\beta_2 = 0.98$. The learning rate is warmed up over 4 000 steps by increasing it linearly to reach a peak learning rate of $3 \cdot 10^{-3}$ for the LSTM and 10^{-3} for the Transformer. Thereafter, it is decayed by the inverse square root of the number of steps, following the learning rate schedule proposed in [15]. Additionally, Exponential Moving Average (EMA) [33] is applied during the training to obtain the final weights of the model.

4.2   Results

IAM-OnDB. On the IAM-OnDB the k-means baseline already achieves very respectable results of up to 91.05% mIoU (Table 1). Considering that it is the simplest of the methods and does not need any training beforehand. The high mIoUs can be attributed to the fact that in the English language most characters do not have the potential of creating much overlap with the next one, unless the cursive writing is slanted excessively. As k-means relies heavily on the spatial position, it is capable of separating a majority of the cases, particularly on block letters. The LSTM is the strongest on this dataset with an mIoU of **93.72%** on the *Test Set F*, suggesting that the model is expressive enough to handle more difficult cases. A known limitation is that it cannot handle the delayed strokes but as only a small number of points are actually part of the delayed strokes, in addition to delayed strokes being fairly rare in the first place, the overall impact on the mean Intersection over Union (mIoU) is rather small. On the other hand, the Transformer does not reach the same level of accuracy, and is even slightly below the k-means. This is presumably due to limited amount of data, which does not satisfy the need of the generally data intensive Transformer models. Similarly, the Transformer with the character queries cannot establish the quality of results that is demonstrated in other experiments. Having an mIoU that is roughly 3% lower than the Transformer for the character boundary detection, is most likely because of the character queries being learned, and the same data limitation applies to it, hence it cannot reach its full potential.

Table 1. IAM-OnDB results. All models were trained using only the IAM-OnDB training set and the best model was determined by the mIoU on the validation set.

Model	Test Set T	Test Set F
K-Means	88.94	91.05
LSTM	**89.55**	**93.72**
Transformer	86.18	90.34
Character Query Transformer	83.53	87.48

mean Intersection over Union (mIoU)

HANDS-VNOnDB. The results for the HANDS-VNOnDB in Table 2 paint a very different picture from the IAM-OnDB results. The k-means does not reach quite the same level of accuracy on the Vietnamese characters as on English characters, which is mainly related to the additional complexity of Vietnamese characters, such as the use of diacritics, which can very easily shift in such a way that it might be considered as part of another character when focusing only on the spatial as well as temporal location of the points. This is a prime example, where additional language information is needed to accurately segment such characters. A much bigger difference to the previous results is observed in the LSTM, which is significantly worse than any other method with an mIoU of just under 50%. One expected reason for the degradation is the much more prominent use of delayed strokes. In this situation, the LSTM exhibits significant problems to accurately predict the character boundaries. The Transformer model performs better but only achieves a similar performance as the k-means baseline. By far the best results are achieved by the Character Query Transformer with a staggering **92.53%** mIoU on the *Test Set*, which is over 13% better than the next best method. This demonstrates that the approach is in fact working as the delayed strokes are no longer an inherent limitation.

Table 2. HANDS-VNOnDB results. All models were trained using only the HANDS-VNOnDB training set and the best model was determined by the mIoU on the validation set.

Model	Test Set
K-Means	79.78
LSTM	49.45
Transformer	78.22
Character Query Transformer	**92.53**

mean Intersection over Union (mIoU)

Combined. Finally, the models have been trained using the combined training sets, in order to see whether they are capable of scaling to multiple languages and improve the overall results by attaining additional information that can be found in the other dataset. It has to be noted that because these models use embeddings of the characters, the mutually exclusive characters are not directly benefiting from the combination of the dataset, in the sense of having more data points of the same character, but can still improve as the model's general segmentation capability improves. Even though k-means is not affected by changing the training data, it is still listed in Table 3 alongside the others for reference. The deterioration of the LSTM on the IAM-OnDB was foreseeable as it was not able to properly learn from the HANDS-VNOnDB. The drop of 6.48% (from 93.72% to 87.24%) on the *Test Set F* is significant but not to the point where the model fails completely. At the same time, its results on the HANDS-VNOnDB improved a little, from 49.45% to 53.66% on the *Test Set*, but less than the IAM-OnDB degraded. When it comes to the Transformer with the character boundary prediction, it is almost identical on the HANDS-VNOnDB as it was without using both datasets to train, but similarly to the LSTM the result on the IAM-OnDB *Test Set F* deteriorated from 90.34% to 86.18% (-4.16%), indicating that combining the two datasets has a negative effect on the models predicting character boundaries. The Character Query Transformer is the only model that benefited from training on both datasets. Even though the results on the HANDS-VNOnDB barely changed (-0.47%), the IAM-OnDb improved by almost 8% (from 87.48% to 95.11%). This demonstrates that the character queries are robust and that it is capable of scaling to multiple languages, especially as the additional data contributed to the large data requirements of the Transformer, even though it was not data from the same language.

4.3   Ablation Study: CTC Spikes

CTC spikes can be used as additional information whenever a CTC-based recognizer has been run beforehand, as it already broadly located the characters and therefore can serve as an initial guidance. There are other cases, where either the text was already known without having to run a recognizer, or when the recognizer does not utilize CTC. In this ablation study we remove the CTC

Table 3. Combined datasets results. All models were trained using the combined training sets of IAM-OnDB and HANDS-VNOnDB and the best model was determined by the mIoU across the validation sets.

| | IAM-OnDB | | HANDS-VNOnDB |
Model	Test Set T	Test Set F	Test Set
K-Means	88.94	91.05	79.78
LSTM	82.70	87.24	53.66
Transformer	80.93	86.18	78.25
Character Query Transformer	**92.28**	**95.11**	**92.06**

mean Intersection over Union (mIoU)

Table 4. CTC spikes ablation study. Comparison of results when using CTC spikes as a feature and without it. Including the CTC spikes improves the results significantly across all models.

| | | IAM-OnDB | | HANDS-VNOnDB |
Model	CTC Spikes	Test Set T	Test Set F	Test Set
K-Means		80.12	83.83	76.82
K-Means	✓	88.94	91.05	79.78
LSTM		74.60	80.59	42.24
LSTM	✓	82.70	87.24	53.66
Transformer		70.21	74.30	76.36
Transformer	✓	80.93	86.18	78.25
Character Query Transformer		86.03	90.78	87.58
Character Query Transformer	✓	**92.28**	**95.11**	**92.06**

mean Intersection over Union (mIoU)

spikes to see whether they are a meaningful addition to the models. In the case of k-means, the points where the CTC spikes occurred were used as the initial centroids, without the CTC spikes they are randomly initialized instead, as is common practice. All other models simply do not have the CTC spike information in the points. The ablation was conducted on the combined datasets.

Including the CTC spikes is a significant improvement across the board. The difference between using the CTC spikes and not, varies depending on the model, ranging from 1.89% for the Transformer on the HANDS-VNOnDB *Test Set* up to 11.88% for the character boundary predicting Transformer on the IAM-OnDB *Test Set F*. Generally, the CTC spikes are less impactful on the HANDS-VNOnDB, with the exception for the LSTM. The Character Query Transformer is the most consistent and hovers around a difference of 4.5% on all datasets, suggesting that it is very stable and is not tied to the CTC spikes but simply uses them to improve the results in a meaningful way. Even though the results without the CTC spikes are not quite as good, they can still be used in cases where no CTC spikes are available.

5   Conclusion

In this paper, we have introduced a novel Transformer-based approach to on-line handwritten character segmentation, which uses learned character queries in the Transformer decoder block to assign sampling points of stylus trajectories to the characters of a known transcription. In an experimental evaluation on two challenging datasets, IAM-OnDB and HANDS-VNOnDB, we compare the proposed method with k-means, LSTM, and a standard Transformer architecture. In comparing the four methods, we observe that approaches which rely on spatial information (k-means) perform reasonably well on non-monotonic handwriting but lack learned features to extract the exact character boundaries. The approaches that rely on temporal information (LSTM and standard Transformer) perform well on mostly-monotonic handwriting, but fail in highly non-monotonic cases. Using the Transformer decoder block in combination with character queries allows us to outperform all other approaches because it uses the strengths of the learned solutions, but does not have a strong inductive bias towards monotonic handwriting (Table 4).

We provide a character segmentation ground truth for the IAM-OnDB and HANDS-VNOnDB using a high-quality approximation. Producing a perfect ground truth for on-line handwritten character segmentation is impossible for cursive script, since even humans will not always agree on the exact start and end positions of the characters. Therefore, in future work, we aim to encode this uncertainty into the ground truth and into the evaluation measures. Another line of future research is to use the segmented characters for creating additional synthetic training material, which is expected to further improve the performance of the Character Query Transformer.

References

1. Maksai, A., et al.: Inkorrect: digital ink spelling correction. In: ICLR Workshop on Deep Generative Models for Highly Structured Data (2022)
2. Jamboard overview. URL: https://documentation.its.umich.edu/jamboard-overview. Accessed 20 Jan 2023
3. Freeform app overview. URL: https://www.apple.com/newsroom/2022/12/apple-launches-freeform-a-powerful-new-app-designed-for-creative-collaboration/. Accessed 20 Jan 2023
4. Living Jiagu. URL: https://www.adsoftheworld.com/campaigns/living-jiagu. Accessed 20 Jan 2023
5. Aksan, E., Pece, F., Hilliges, O.: DeepWriting: making digital ink editable via deep generative modeling. In: Proceedings of the 2018 CHI Conference on Human Factors in Computing Systems, pp. 1–14 (2018)
6. Haines, T.S.F., Mac Aodha, O., Brostow, G.J.: My text in your handwriting. ACM Trans. Graph. (TOG) **35**(3), 1–18 (2016)
7. Hassan, S., et al.: Cursive handwritten text recognition using bidirectional LSTMs: a case study on Urdu handwriting. In: 2019 International Conference on Deep Learning and Machine Learning in Emerging Applications (Deep-ML), pp. 67–72. IEEE (2019)

8. Kotani, A., Tellex, S., Tompkin, J.: Generating handwriting via decoupled style descriptors. In: Vedaldi, A., Bischof, H., Brox, T., Frahm, J.-M. (eds.) ECCV 2020. LNCS, vol. 12357, pp. 764–780. Springer, Cham (2020). https://doi.org/10.1007/978-3-030-58610-2_45

9. Keysers, D., et al.: Multi-language online handwriting recognition. IEEE Trans. Pattern Anal. Mach. Intell. **39**, 1180–1194 (2017)

10. Kour, H., Gondhi, N.K.: Machine Learning approaches for nastaliq style Urdu handwritten recognition: a survey. In: 2020 6th International Conference on Advanced Computing and Communication Systems (ICACCS), pp. 50–54. IEEE (2020)

11. Singh, H., Sharma, R.K., Singh, V.P.: Online handwriting recognition systems for Indic and non-Indic scripts: a review. Artif. Intell. Rev. **54**(2), 1525–1579 (2021)

12. Zhelezniakov, D., Zaytsev, V., Radyvonenko, O.: Online handwritten mathematical expression recognition and applications: A survey. IEEE Access **9**, 38352–38373 (2021)

13. Graves, A., et al.: Connectionist temporal classification: labelling unsegmented sequence data with recurrent neural networks. In: Proceedings of the 23rd International Conference on Machine Learning. ICML 2006, pp. 369–376. Association for Computing Machinery (2006). https://doi.org/10.1145/1143844.1143891

14. Hochreiter, S., Schmidhuber, J.: Long short-term memory. Neural Comput. **9**(8), 1735–1780 (1997). https://doi.org/10.1162/neco.1997.9.8.1735

15. Vaswani, A., et al.: Attention is all you need. Adv. Neural Inf. Process. Syst. **30** (2017)

16. Liwicki, M., Bunke, H.: IAM-OnDB - an on-line English sentence database acquired from handwritten text on a whiteboard. In: Eighth International Conference on Document Analysis and Recognition (ICDAR 2005), vol. 2, pp. 956–961. (2005) https://doi.org/10.1109/ICDAR.2005.132

17. Nguyen, H.T., et al.: A database of unconstrained Vietnamese online handwriting and recognition experiments by recurrent neural networks. Pattern Recogn. **78**, 291–306 (2018). https://doi.org/10.1016/j.patcog.2018.01.013

18. Kotani, A., Tellex, S., Tompkin, J.: Generating handwriting via decoupled style descriptors. In: Vedaldi, A., Bischof, H., Brox, T., Frahm, J.-M. (eds.) ECCV 2020. LNCS, vol. 12357, pp. 764–780. Springer, Cham (2020). https://doi.org/10.1007/978-3-030-58610-2_45

19. Lloyd, S.: Least squares quantization in PCM. IEEE Trans. Inf. Theory **28**(2), 129–137 (1982). https://doi.org/10.1109/TIT.1982.1056489

20. Liu, Y., et al.: A survey of visual transformers. arXiv: 2111.06091 (2021)

21. Khan, S., et al.: Transformers in vision: a survey. ACM comput. surv. (CSUR) **54**(10), 1–41 (2022)

22. Carion, N., Massa, F., Synnaeve, G., Usunier, N., Kirillov, A., Zagoruyko, S.: End-to-end object detection with transformers. In: Vedaldi, A., Bischof, H., Brox, T., Frahm, J.-M. (eds.) ECCV 2020. LNCS, vol. 12346, pp. 213–229. Springer, Cham (2020). https://doi.org/10.1007/978-3-030-58452-8_13

23. Wang, H., et al.: MaX-DeepLab: end-to-end panoptic segmentation with mask transformers. In: 2021 IEEE/CVF Conference on Computer Vision and Pattern Recognition (CVPR), pp. 5459–5470 (2020)

24. Fang, Y., et al.: Instances as queries. In: 2021 IEEE/CVF International Conference on Computer Vision (ICCV), pp. 6890–6899 (2021)

25. Cheng, B., et al.: Masked-attention mask transformer for universal image segmentation. In: 2022 IEEE/CVF Conference on Computer Vision and Pattern Recognition (CVPR), pp. 1280–1289 (2021)

26. Yu, Q., et al.: k-means Mask Transformer. In: Avidan, S., Brostow, G., Cissé, M., Farinella, G.M., Hassner, T. (eds.) Computer Vision – ECCV 2022. ECCV 2022. Lecture Notes in Computer Science, vol 13689. Springer, Cham (2022). https://doi.org/10.1007/978-3-031-19818-2_17

27. Pedregosa, F., et al.: Scikit-learn: machine learning in Python. J. Mach. Learn. Res. **12**, 2825–2830 (2011)

28. Paszke, A., et al.: PyTorch: an imperative style, high-performance deep learning library. In: Proceedings of the 33rd International Conference on Neural Information Processing Systems. Curran Associates Inc. (2019)

29. Nair, V., Hinton, G.E.: Rectified linear units improve restricted boltzmann machines. In: Proceedings of the 27th International Conference on International Conference on Machine Learning. ICML 2010, pp. 807–814. Omnipress, Haifa (2010)

30. Hendrycks, D., Gimpel, K.: Gaussian error linear units (GELUs). arXiv: 1606.08415 (2016)

31. Szegedy, C., et al.: Rethinking the inception architecture for computer vision. In: 2016 IEEE Conference on Computer Vision and Pattern Recognition (CVPR), pp. 2818–2826 (2016). https://doi.org/10.1109/CVPR.2016.308

32. Loshchilov, I., Hutter, F.: Decoupled weight decay regularization. arXiv: 1711.05101 (2017)

33. Polyak, B.T., Juditsky, A.B.: Acceleration of stochastic approximation by averaging. SIAM J. Control Optim. **30**(4), 838–855 (1992). https://doi.org/10.1137/0330046

Graphics Recognition 3: Math Recognition

Relative Position Embedding Asymmetric Siamese Network for Offline Handwritten Mathematical Expression recognition

Chunyi Wang[1], Hurunqi Luo[3], Xiaqing Rao[1], Wei Hu[1], Ning Bi[1,2], and Jun Tan[1,2(✉)]

[1] School of Mathematics, Sun Yat-Sen University, Guangzhou 510275, People's Republic of China
{wangchy53,raoxq5,huwei55}@mail2.sysu.edu.cn
[2] Guangdong Province Key Laboratory of Computational Science, Sun Yat-Sen University, Guangzhou 510275, People's Republic of China
{mcsbn,mcstj}@mail.sysu.edu.cn
[3] Meituan, Shanghai, China
luohurunqi@meituan.com

Abstract. Currently, Recurrent Neural Network(RNN)-based encoder-decoder models are widely used in handwritten mathematical expression recognition (HMER). Due to its recursive pattern, the problem of gradient disappearance or gradient explosion also exists for RNN, which makes them inefficient in processing long HME sequences. In order to solve above problems, this paper proposes a Transformer-based encoder-decoder model consisting of an asymmetric siamese network, relative position embedding Transformer (ASNRT). With the assistance of printed images, the asymmetric siamese network further narrows the difference betweeen feature maps of similar formula images and increases the encoding gap between dissimilar formula images. We insert coordinate attention into the encoder, additionally we replace RNN with Transformer as the decoder. Moreover, rotary position embedding is used, incorporating relative position information through absolute embedding ways. Given the symmetry of MEs, we adopt the bidirectional decoding strategy. Extensive experiments show that our model improves the ExpRate of state-of-the-art methods on CROHME 2014, CROHME 2016, and CROHME 2019 by 0.94%, 2.18% and 2.12%, respectively.

Keywords: Handwritten · Contrastive Learning · Encoder-Decoder

1 Introduction

The process of recognizing corresponding LaTeX sequences from pictures of handwritten mathematical expressions (MEs) is called Handwritten Mathematical Expression Recognition (HMER). With the rapid technological advancements in contemporary society, the accurate recognition of handwritten mathematical

G. A. Fink et al. (Eds.): ICDAR 2023, LNCS 14187, pp. 117–133, 2023.
https://doi.org/10.1007/978-3-031-41676-7_7

expressions (MEs) has emerged as a significant research challenge. Due to stylistic variability, writing irregularities, unique two-dimensional presentation structure, and logical structure [31], handwritten MEs are significantly different from traditional natural language, thus it's difficult to directly migrate traditional OCR techniques to the field of HMER.

The main methods of HMER for decades can be divided into two categories: traditional methods and deep learning-based methods. Traditional methods include sequential solving and global solving. The sequential solution consists of symbol segmentation, symbol recognition, and structure analysis, while the global solution attempts to place all three in a single pipeline. Some statistical learning methods such as Support Vector Machines (SVM) [12,19,20], Multilayer Perceptron (MLP) [2,6], and Hidden Markov Model (HMM) [33,34] have achieved high recognition accuracy in symbol recognition. The structural analysis focuses on determining the logical relationships between mathematical symbols by judging their positional relationships, which is essential for the correct recognition of mathematical expressions. Structural analysis models commonly use statistical methods, including SVM [26] and other machine learning classifiers. However, a sequential solution may lead to layer-by-layer transmission of errors, which can be minimized with the global solution method by [1,23] integrating and jointly optimizing symbol segmentation, symbol recognition, and structure analysis. In traditional HMER methods, there is an interaction between conformal recognition and structural analysis. Therefore, accurate recognition of mathematical expressions requires the correct execution of both symbol recognition and structural analysis, thereby imposing a higher demand on the methods employed.

The universal encoder-decoder [3] structure unifies symbol segmentation, symbol recognition, and structure analysis in a data-driven framework, which directly outputs sequences of characters in LaTeX format. Zhang et al. proposed a novel encoder-decoder method based on RNN, Watch, Attend, and Parse (WAP) [41], which has gained significant attention in the offline HMER field due to its excellent performance, thus resulting in numerous efforts to improve it by researchers [11,28,29].

Despite the fact that different writing habbits and writing patterns may lead to the diversity of handwritten MEs, printed mathematical symbols is undisputedly standard truth. Therefore, we believe that printed MEs will bring better results in handwritten recognition. Previously, some people used the idea of GAN [36,37] or established a new loss function [13] to use printed MEs in HMER, but still underutilized printing MEs. Combining the idea of contrastive learning, we design an asymmetric siamese network to assist image feature extraction. The proposed method incorporates two branches of image transformation, using character warping transformation and generation of printed mathematical expressions, and integrates projective MLP and predictive MLP. The multi-layer perception is used to generate the feature vector, the negative cosine similarity function is used as the loss function, and the stop gradient operation is used to make the model converge. The asymmetric siamese network improves encoding accuracy by encoding similar data similarly and encoding different classes of data as different as possible.

In the feature extraction module, we use DenseNet as the feature extractor. Some scholars [16] added attention to Densenet in the DWAP model [39]. Due to its ability to not only capture channel correlations but also take into account the position and orientation information of the image, Transformer exhibits improved performance in feature extraction on the MEs picture. The decoding part uses a Transformer structure. The rotary position embedding is computed as simply as the absolute position encoding but also contains relative position information. The model also adopts a bidirectional decoding strategy, which can better decode the symmetric information of the MEs. Our contributions are summarized as follows:

(1) We design an asymmetric siamese network for HMER to improve feature extraction.
(2) We incorporate coordinate attention into the DenseNet encoder.
(3) Introduce the rotary position embedding in the decoder, and the model adopts a bidirectional decoding strategy, which can better decode the symmetrical information of the MEs.
(4) Comprehensive experiments show that our proposed model outperforms state-of-the-art methods by a large margin on CROHME 2014, 2016, and 2019.

2 Related Work

2.1 Deep Learning Methods of HMER

Sequence learning applications are acquiring more popularity these days, encoder-decoder frameworks have been widely used to solve image-to-sequence tasks. In 2017, Deng et al. [7] introduced such a framework into HEMR for the first time. Since then there has been continuous improvements in the encoder-decoder model. For example, fully convolutional networks [32,41], DenseNet [14,28,39] and ResNet [15,38] are used as encoder, Transformer [42] and tree struct [40] are used as decoder. Overlay mechanisms [15,28,39,41] are also frequently used in decoders, which preserve history by tracking translation alignments information and guide the decoder to pay more attention to the part of the image that has not been translated, thus solving the problem of over-parsing and under-parsing to a certain extent. What's more, data augmentation [14,15] as well as printed expressions [13,36,37] are used to fully train the model because of the relatively small training dataset.

2.2 Contrastive Learning

Self-supervised learning is an approach of unsupervised learning that directly uses information from the data itself to supervise the learning of a model without artificial labels. In the representation learning of features, self-supervised learning has three main advantages.

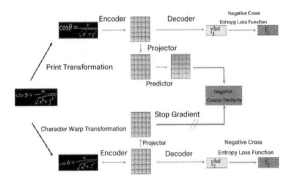

Fig. 1. The overall architecture of the model.

(1) It does not require a large amount of labeled data, which can reduce the high cost.
(2) It focuses more on the data itself, which contains more information than a single label.
(3) It can learn more general feature representations, and the learned features can be transferred and applied to other supervised learning tasks, such as image classification, image segmentation, etc.

Contrastive learning belongs to the category of self-supervised learning. It learns representations of features from labeled data, which are then used as input for supervised tasks. The idea of contrastive learning is simple: the model needs to identify which samples are similar and which are dissimilar. Similar or identical samples need to be as close as possible in the high-dimensional feature space, and dissimilar samples need to be as far apart as possible. The design idea of contrastive learning is to construct sample pairs and make them converge by maximizing the similarity of the two images enhanced by the same image data, avoiding the learning of collapsing solutions, which means that the learned feature is a constant and is not helpful for subsequent downstream tasks. The gradient stop operation proposed in the siamese network [5] has a good effect on avoiding learning collapsing solutions.

3   Method

In this paper, we use the encoder-decoder framework to improve the model, where the encoder is based on DenseNet and includes a coordinate attention module [9]. To enhance the encoder's feature extraction capabilities, inspired by siamese network [5], we design an asymmetric siamese network. Moreover, we replace RNN with Transformer as the decoder, incorporating rotary position embedding [27] and a bidirectional decoding strategy. The final prediction of LaTeX results is outputted using the beam search approach.

The overall structure of the model is shown in Fig. 1, where Y_1^{hat} and Y_2^{hat} represent predicted sequences, and Y_1 and Y_2 represent real sequences. The whole model consists of two parts, one part is supervised learning based on encoder-decoder, where the loss function is a cross-entropy function; the other part is self-supervised learning based on asymmetric siamese network, and the loss function is a negative cosine similarity function.

3.1   Coordinate Attention Module DenseNet Encode

The specific architecture of DenseNet in this paper is shown in Fig. 2. The model consists of four main modules, the first of which is used for initialization. The overall architecture of the second to fourth modules is the same, consisting of the dense block, transition layer, and coordinate attention module, except that the fourth module has fewer transition layers. Attention mechanisms can increase the amount of information obtained from neural networks, and most attention module designs focus on lightweight and portability, such as Squeeze-and-Excitation (SENet) [10], and Convolutional Block Attention Module (CBAM) [35]. However, SENet doesn't consider spatial information while CBAM only focuses on local information. This paper refers to the latest Coordinate Attention (CA) [9], which simultaneously models channel correlation and long-range dependence, taking account of both channel information and spatial information from local and long distances rather than SENet and CBAM. A comparison of SENet, CBAM, and CA is shown in Fig. 3.

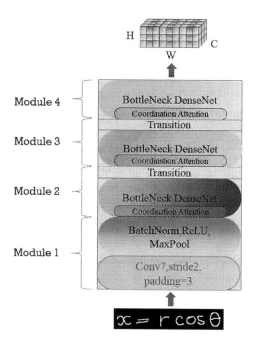

Fig. 2. The schematic diagram of the DenseNet architecture incorporating the coordinate attention module.

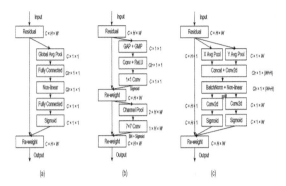

Fig. 3. Comparison of three attention modules (a)SENET (b)CBAM (c)CA

As you can see, CA takes two steps: the first is information embedding, and the second is attention generation.

Overall, it has two advantages:

(1) It can encode information between multiple channels and take into account the direction and location of the pixels.
(2) The module is flexible and lightweight, making it easy to plug into classical networks such as ResNet, DenseNet, and the inverted residual blocks in MobileNet v2 [24].

Therefore, adding CA to DenseNet is more conducive to extracting features.

3.2 Asymmetric Siamese Network

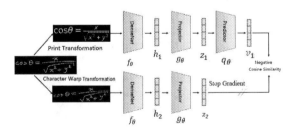

Fig. 4. The structure diagram of asymmetric siamese network.

As illustrated in Fig. 4, the process starts with data transformations performed twice on the original image, resulting in x_1 and x_2. The features are then extracted using the encoder f, with the two encoders sharing parameters. The output are feature vectors h_1 and h_2, which are further processed through a projected multilayer perceptron, represented by g_θ, to produce feature vectors z_1

and z_2. These projected multilayer perceptrons also share parameters. Finally, a predictive multilayer perception q_θ only works on z_1, leading v_1 to predict the feature vector z_2 of the other branch. The existence of the "Stop-Gradient" operation prevents the gradient update of the parameters of the lower half of the branch, which means that the feature vector z_2 of the lower half of the branch is regarded as a constant. The "Stop-Gradient" operation prevents the model from learning a "collapsing solution".

The above processes of feature extraction are summarized by following formulas:

$$v_1 = q_\theta(g_\theta (f_\theta (x_1))), \ z_2 = g_\theta(f_\theta (x_2)) \tag{1}$$

Then calculate negative cosine similarity for two features above:

$$\mathcal{D} (v_1, z_2) = - \frac{v_1}{||v_1||_2} * \frac{z_2}{||z_2||_2} \tag{2}$$

$|| \ ||_2$ represents the ℓ_2 norm. Drawing on the practice of Bootstrap Your Own Latent (BYOL) [8], the authors of SimSiam designed a loss function with a symmetric structure:

$$\mathcal{L} = \frac{1}{2} * \mathcal{D} (v_1, z_2) + \frac{1}{2} * \mathcal{D} (v_2, z_1) \tag{3}$$

Experiments show that the symmetry of the function has no effect on the convergence of the model, and the performance is slightly improved.

3.3   Relative Position Embedding

The addition of position embedding can provide more valid information to the model. Generally speaking, there are two types of position embedding: absolute position embedding and relative position embedding.

In the decoder model of this paper, a new embedding method called rotary position embedding (RoPE) [27] is adopted, which implements relative position embedding in an absolute position way, ensuring computational speed while also having the advantages of relative position embedding.

Suppose that the set of word vectors corresponding to each symbol is represented as $E_N = \{x_i\}_{i=1}^N$ where $x_i \in R^d$ is the word vector of the symbol w_i and does not contain position information. To achieve the above goal, it is assumed that absolute position information is added to x_m and x_n by the following operations:

$$\tilde{q}_m = f (x_m, m), \quad \tilde{k}_n = f (x_n, n) \tag{4}$$

That is to say, we design operations $f(, m)$ and $f(, n)$ for x_m and x_n respectively, so that $\tilde{q}_m$ and $\tilde{k}_n$ have absolute position information of positions m and n after such operations. Moreover, in order for the result of the inner product to contain relative position information, we might as well assume an identity relationship:

$$\langle f (x_m, m), f (x_n, n) \rangle = g (x_m, x_n, m - n) \tag{5}$$

124   C. Wang et al.

Where $g(\cdot)$ is an unknown function, also the initialization conditions is assumed $f(\boldsymbol{q},0) = \boldsymbol{q}$ and $f(\boldsymbol{k},0) = \boldsymbol{k}$. This gives the RoPE in complex numbers in two dimensions:

$$f(\boldsymbol{x}_m, m) = R_f(\boldsymbol{x}_m, m)\, e^{i\Theta_f(\boldsymbol{x}_m, m)} = \boldsymbol{x}_m e^{im\theta} \tag{6}$$

Here R_f represents the real part of the complex number, and $\theta \in \mathbb{R}$ is a non-zero constant. It can be seen from the geometric meaning of the complex vector that the transformation is a rotation operation on the vector, so it's called "rotary position embedding". Different from the position embedding in the original paper [30], the position embedding in HMER is slightly different as it takes a picture as input and output a sequence. For the position encoding of the picture, we only need to calculate the rotary position embedding $\boldsymbol{p}_{x,d/2}$ and $\boldsymbol{p}_{y,d/2}$ along the x-axis and y-axis of the two-dimensional points (x,y) on the image, and then stitch them together. Assuming that a two-dimensional image pixel point (x,y) is given, the same dimension d is encoded as the position of the sequence, then $\boldsymbol{p}_{(x,y),d}$ can be expressed as:

$$\bar{x} = \frac{x}{H}, \; \bar{y} = \frac{y}{W}, \; \boldsymbol{p}_{(x,y),d} = \left[\boldsymbol{p}_{\bar{x},\frac{d}{2}}, \boldsymbol{p}_{\bar{y},\frac{d}{2}}\right] \tag{7}$$

3.4 Bidirectional Decoding

Most mathematical expression recognition models only consider left-to-right information during the decoding process, which tends not to take full advantage of long-distance dependent information. In this case, the model in this paper adopts a bidirectional decoding strategy [42] with the addition of a right-to-left decoding direction. For the process of bidirectional decoding training, this paper uses ⟨sos⟩ and ⟨eos⟩ as the start and end symbols of LaTeX. For the target LaTeX sequence y = $\{y_1, \ldots, y_T\}$, this paper expresses the decoding direction from left to right as: y = $\{\langle sos\rangle, y_1, \ldots, y_T, \langle eos\rangle\}$, the decoding direction from right to left is expressed as: y = $\{\langle eos\rangle, y_T, \ldots, y_1, \langle sos\rangle\}$.

For a given input image x and model parameters, we use the autoregressive model to calculate the predicted probability value at each step:

$$p\left(\overrightarrow{y_j}\,\middle|\,\overrightarrow{y_{<j}}, x, \theta\right) \tag{8}$$

$$p\left(\overleftarrow{y_j}\,\middle|\,\overleftarrow{y_{<j}}, x, \theta\right) \tag{9}$$

where j is the index of the target sequence, represents the probability of the prediction at the j-th step when decoding from left to right, conditioned on the predicted value before the j-th step, the picture x and the model parameters θ.

3.5 Beam Search

Beam search is a heuristic graph search algorithm, usually used in cases where the solution space is relatively large. During the process of depth expansion, lower quality nodes are discarded, while higher quality nodes are retained, leading to a reduction in search space and time consumption. The cluster search is mainly used in the prediction phase and lets the value of beam width h be the parameter m, then in the first step m optimal starting characters are retained, and in the second step, the optimal m is searched for each character obtained in the first step. The second step searches for the optimal m characters for each of the characters obtained in the first step, and then selects the optimal m combination from the $m \times m$ character combinations to keep for the next step.

Greedy search occurs when $m = 1$. If the previous character is incorrectly predicted, then the whole sequence will also be incorrectly predicted, with no possibility of correction. Cluster search can overcome the disadvantages of greedy search, but it also has the problem of being computationally intensive.

3.6 Overall Loss Function

Given training samples $\{x^{(z)}, y^{(z)}\}$ the overall loss function of the model is as follows:

$$\overrightarrow{\mathcal{L}}_j^{(z)}(\theta) = -\log p\left(\overrightarrow{y_j}^{(z)} \mid \overrightarrow{y_{<j}}^{(z)}, x^{(z)}, \theta\right) \tag{10}$$

$$\overleftarrow{\mathcal{L}}_j^{(z)}(\theta) = -\log p\left(\overleftarrow{y_j}^{(z)} \mid \overleftarrow{y_{<j}}^{(z)}, x^{(z)}, \theta\right) \tag{11}$$

$$\mathcal{D}(v_1, z_2) = -\frac{v_1}{\|v_1\|_2} \cdot \frac{z_2}{\|z_2\|_2} \tag{12}$$

$$\mathcal{L}_{CL} = \frac{1}{2Z}\mathcal{D}(v_1, z_2) + \frac{1}{2Z}\mathcal{D}(v_2, z_1) \tag{13}$$

$$L(\theta) = \frac{1}{2ZL}\sum_{z=1}^{Z}\sum_{j=1}^{L}\left(\overrightarrow{\mathcal{L}}_j^{(z)}(\theta) + \overleftarrow{\mathcal{L}}_j^{(z)}(\theta)\right) \tag{14}$$

$$\mathcal{L} = L(\theta) + \lambda\mathcal{L}_{CL} \tag{15}$$

where Z is the number of training samples and L is the decoding step size. $\left(\overrightarrow{y}_j^{(z)} \mid \overrightarrow{y}_{<j}^{(z)}, x^{(z)}, \theta\right)$ represents decoding from left to right, under the condition of the predicted value before the j-th step, the picture x, the model parameters θ, and the probability of the prediction at the j-th step. Similarly, the meaning of $\left(\overleftarrow{y}_j^{(z)} \mid \overleftarrow{y}_{<j}^{(z)}, x^{(z)}, \theta\right)$ can be inferred. v_1 and z_2 in Eq. 1 are the eigenvectors of the upper and lower branches of the asymmetric siamese network. In addition, λ is a hyperparameter that measures the weight between the cross-entropy loss function and the negative cosine similarity function.

4   Experiments

4.1   Preprocess

Character Warp Transformation. Common image geometric transformations include rotation, scaling, and perspective transformations, which treat the image as a whole and enhance it globally, but these methods have certain disadvantages for handwritten character recognition. For example, if the image is rotated too much or scaled too much, characters can be lost. At the same time, the gain of these changes to the sample style diversity is limited.

This paper refers to the image transformation method proposed by Luo [17]. We divide the image into two parts by sampling three points along the upper and lower edges of the image. Based on these points, we scale and transform the image conforms to a specific distribution within a certain radius R. The principle of image transformation is based on moving least squares [25] for similarity deformation. Given a point u in the graph, the transformation of that point conforms to the following formula:

$$F(u) = (u - p^*) M + q^* \tag{16}$$

where $M \in R^{2 \times 2}$ is an affine transformation matrix that satisfies $M^T M = \lambda I$, and λ is the scaling coefficient. p^* is the weighted sum of the initialized datum points(p_i), q^* is the weighted sum of the datum points after the change(q_i), respectively satisfying:

$$p^* = \frac{\sum_{i=1}^{2(N+1)} w_i p_i}{\sum_{i=1}^{2(N+1)} w_i}, q^* = \frac{\sum_{i=1}^{2(N+1)} w_i q_i}{\sum_{i=1}^{2(N+1)} w_i} \tag{17}$$

where the weight w_i is calculated by the following formula:

$$w_i = \frac{1}{|p_i - u|^{2\alpha}}, u \neq p_i \tag{18}$$

It is easy to notice that the change of u is determined by the movement of the nearest reference point. Assuming that w_i is bounded, if $u = p_i$, then $F(u) = u$. Here the value of α is set to 1. The unique optimal solution $F(u)$ can be obtained by minimizing the following formula:

$$F(u) = \sum_{i=1}^{2(N+1)} w_i |F_u(p_i) - q_i|^2 \tag{19}$$

Through the above character distortion operation, we can increase the style variability of the characters in the formula image, which helps the asymmetric siamese network to learn more effective features.

Datasets and Metrics. CROHME [21] is an international handwritten mathematical expression recognition competition. Its corresponding CROHME dataset is the largest open dataset in the field of handwritten mathematical expression recognition. Our model uses the CROHME 2014 dataset containing 8,836 offline images as the training set. CROHME 2014 test set is used as the validation set, CROHME 2016 [22] test set, and CROHME 2019 [18] test set are used as the test sets. They contain 986, 1147, and 1199 images respectively.

The CROHME Handwritten Expression Recognition Competition uses two evaluation criteria, ExpRate and WER, to measure the effectiveness of the model. ExpRate is the strictest metric to measure the model at the expression level, requiring an edit distance of 0. In this paper, an edit distance of 1 is denoted as a ≤ 1 error, which allows for an editing error between the predicted value and the true label, and the ≤ 2 error is the same. This paper also uses the WER evaluation metric to measure the performance of the proposed model at the word level. WER is the edit distance divided by the total number of characters.

Implementation Details Setup. Our proposed method is optimized with Adadelta optimizer and adopts ReduceLROnPlateau as the learning rate adjuster. ReduceLROnPlateau is based on the error of the validation set (ExpRate) to reduce the learning rate. When it is detected that ExpRate no longer increases within a certain step size, the learning rate will be reduced according to a certain proportion. The reduction ratio set by the model is 0.1 and the tolerance is 10. All the models are trained/tested on four GeForce GTX 1080Ti 12.5G GPU.

4.2 Sensitivity Analysis

This paper evaluates the effect of the proposed model on the CROHME 2014, CROHME 2016, and CROHME 2019 test sets, and determines the optimal hyperparameters λ through multiple sets of experiments. It represents the weight ratio between the decoding loss function and the contrastive learning loss function in Eq. 15. The author did a series of experiments from $\lambda = 0.05$ to 1, according change in ExpRate about λ is shown in Fig. 5. When λ increases from 0, the ExpRate value gradually increases until the maximum value at $\lambda = 0.09$, and then begins to decrease with the increase of λ. Therefore, the model takes $\lambda = 0.09$ in the subsequent experiments.

4.3 Ablation Experiment

We evaluate our model on CROHME 2014 test dataset, CROHME 2016 test dataset, and CROHME 2019 test dataset. "Attention module" represents Coordinate Attention is added; "Positional Encoding" represents rotary Positional Encoding is used; "Bidirectional Decoding" represents the decoding strategy is

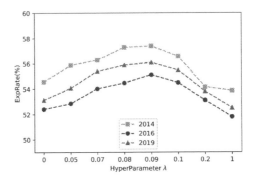

Fig. 5. Sensitivity analysis of λ.

Table 1. Ablation study.

Attention module	Positional Encoding	Bidirectional Decoding	Siamese Network	Integrated model	2014 ExpRate(%)	2016 ExpRate(%)	2019 ExpRate(%)
✗	✗	✗	✗	✗	47.49	45.90	44.62
✓	✗	✗	✗	✗	47.55	45.98	44.65
✓	✓	✗	✗	✗	50.01	48.37	49.24
✓	✓	✓	✗	✗	54.25	52.14	52.93
✓	✓	✓	✓	✗	57.79	55.10	56.08
✓	✓	✓	✓	✓	60.12	58.03	59.75

used; "Siamese Network" represents the asymmetric siamese network is used; "Integrated model" means integrating the results of different models.

To prove the effectiveness of each module in the model, a series of ablation experiments are designed in this paper. Table 1 shows that the above five modules improve the model during the encoding and decoding process, and finally improve the ExpRate.

4.4   Case Study

We use the transformer decoder containing single headed attention to draw the attention distribution diagram for the specific example in Fig. 6.

As shown in the schematic diagram, our proposed model focuses accurately on relevant regions during each decoding step, resulting in improvement on recognition accuracy.

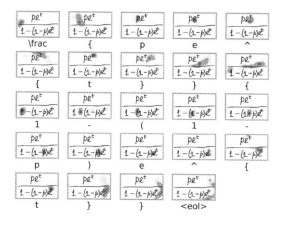

Fig. 6. attention distribution diagram for $\frac{pe^t}{1-(1-p)e^t}$

4.5 Comparison with the State-of-the-art Methods

We compare our model(ASNRT without "Integrated model") with the latest model of offline HMER. These models include PAL [36], PAL-v2 [37], WAP [41], DWAP [39] (WAP encoded by densenet), DWAP-TD [40] (DWAP encoded by tree decoder), DLA [13], WSWAP [28], BTTR [42], ABM [4]. As indicated in Tables 2, 3, and 4, the ExpRate results of our proposed model surpass those of the classic DWAP model by 6.79%, 7.67% and 8.38% on the three test sets respectively, and outperform the current optimal model ABM by 0.94%, 2.18% and 2.12%. This clearly demonstrates the superiority of the model proposed in this paper.

Table 2. Comparison with prior works on CROHME 2014 test dataset.

Methods	Exp-Rate(%)	≤ 1 error(%)	≤ 2 error (%)	WER (%)
PAL [36]	39.66	56.80	65.11	-
WAP [41]	46.55	61.16	65.21	17.73
PAL-v2 [37]	48.88	64.50	69.78	-
DWAP-TD [40]	49.10	64.20	67.80	-
DLA [13]	49.85	-	-	16.63
DWAP [39]	52.80	68.10	72.00	12.9
WSWAP [28]	53.65	-	-	11.48
BTTR [42]	53.96	66.02	70.28	-
ABM [4]	56.85	73.73	81.24	10.01
ASNRT	**57.79**	**75.01**	79.39	**9.65**

Table 3. Comparison with prior works on CROHME 2016 test dataset.

Methods	Exp-Rate(%)	≤ 1 error(%)	≤ 2 error (%)	WER (%)
WAP [41]	44.55	57.01	61.55	-
PAL-v2 [37]	49.61	64.08	70.27	-
DWAP-TD [40]	48.50	62.30	65.30	-
DLA [13]	47.34	-	-	15.82
DWAP [39]	50.10	63.80	67.40	13.7
WSWAP [28]	51.96	64.34	70.10	13.59
BTTR [42]	52.31	63.90	68.61	-
ABM [4]	52.92	69.66	78.73	-
ASNRT	**55.10**	**70.62**	78.12	**10.98**

Table 4. Comparison with prior works on CROHME 2019 test dataset.

Methods	Exp-Rate(%)	≤ 1 error(%)	≤ 2 error (%)	WER (%)
DWAP-TD [40]	51.40	66.10	69.10	-
BTTR [42]	52.96	65.97	69.14	-
ABM [4]	53.96	71.06	78.65	-
ASNRT	**56.08**	**71.48**	**79.07**	**10.37**

5 Conclusion

In this paper, based on the idea of contrastive learning, character warping transformation and printed ME transformation in the asymmetric siamese network are proposed for better image feature extraction. In addition, we not only embed a coordinate attention module into the DenseNet encoder, but also use the strategy of rotary position embedding and bidirectional decoding while adopting Transformer structure in the decoder part to improve the performance. Through detailed experimental analysis and comparison with state-of-the-art methods, we demonstrate that our proposed model has good generalization and superior performance for HMER.

Acknowledgments. This work was supported by Guangdong Province Key Laboratory of Computational Science at the Sun Yat-sen University (2020B1212060032), and the National Science Foundation of China (11971491, 12171490), and the Foundation of Guangdong-Hong Kong-Macao National Center for Applied Mathematics (2021B1515310002).

References

1. Álvaro, F., Sánchez, J.A., Benedí, J.M.: An integrated grammar-based approach for mathematical expression recognition. Pattern Recogn. **51**, 135–147 (2016)
2. Awal, A.M., Mouchère, H., Viard-Gaudin, C.: Towards handwritten mathematical expression recognition. In: 2009 10th International Conference on Document Analysis and Recognition, pp. 1046–1050. IEEE (2009)
3. Bahdanau, D., Cho, K., Bengio, Y.: Neural machine translation by jointly learning to align and translate. arXiv preprint arXiv:1409.0473 (2014)
4. Bian, X., Qin, B., Xin, X., Li, J., Su, X., Wang, Y.: Handwritten mathematical expression recognition via attention aggregation based bi-directional mutual learning. In: Proceedings of the AAAI Conference on Artificial Intelligence, vol. 36, pp. 113–121 (2022)
5. Chen, X., He, K.: Exploring simple siamese representation learning. In: Proceedings of the IEEE/CVF Conference on Computer Vision and Pattern Recognition, pp. 15750–15758 (2021)
6. Cheng, T., Khan, J., Liu, H., Yun, D.: A symbol recognition system. In: Proceedings of 2nd International Conference on Document Analysis and Recognition (ICDAR'93), pp. 918–921. IEEE (1993)
7. Deng, Y., Kanervisto, A., Ling, J., Rush, A.M.: Image-to-markup generation with coarse-to-fine attention. In: International Conference on Machine Learning, pp. 980–989. PMLR (2017)
8. Grill, J.B., et al.: Bootstrap your own latent-a new approach to self-supervised learning. Adv. Neural. Inf. Process. Syst. **33**, 21271–21284 (2020)
9. Hou, Q., Zhou, D., Feng, J.: Coordinate attention for efficient mobile network design. In: Proceedings of the IEEE/CVF Conference on Computer Vision and Pattern Recognition, pp. 13713–13722 (2021)
10. Hu, J., Shen, L., Sun, G.: Squeeze-and-excitation networks. In: Proceedings of the IEEE Conference on Computer Vision and Pattern Recognition, pp. 7132–7141 (2018)
11. Huang, G., Liu, Z., Van Der Maaten, L., Weinberger, K.Q.: Densely connected convolutional networks. In: Proceedings of the IEEE Conference on Computer Vision and Pattern Recognition, pp. 4700–4708 (2017)
12. Keshari, B., Watt, S.: Hybrid mathematical symbol recognition using support vector machines. In: Ninth International Conference on Document Analysis and Recognition (ICDAR 2007), vol. 2, pp. 859–863. IEEE (2007)
13. Le, A.D.: Recognizing handwritten mathematical expressions via paired dual loss attention network and printed mathematical expressions. In: Proceedings of the IEEE/CVF Conference on Computer Vision and Pattern Recognition Workshops, pp. 566–567 (2020)
14. Le, A.D., Indurkhya, B., Nakagawa, M.: Pattern generation strategies for improving recognition of handwritten mathematical expressions. Pattern Recogn. Lett. **128**, 255–262 (2019)
15. Li, Z., Jin, L., Lai, S., Zhu, Y.: Improving attention-based handwritten mathematical expression recognition with scale augmentation and drop attention. In: 2020 17th International Conference on Frontiers in Handwriting Recognition (ICFHR), pp. 175–180. IEEE (2020)
16. Lin, Q., Wang, C., Bi, N., Suen, C.Y., Tan, J.: An encoder-decoder approach to offline handwritten mathematical expression recognition with residual attention. In: International Conference on Pattern Recognition and Artificial Intelligence, pp. 335–345. Springer, Cham (2022). https://doi.org/10.1007/978-3-031-09037-0_28

17. Luo, C., Zhu, Y., Jin, L., Wang, Y.: Learn to augment: joint data augmentation and network optimization for text recognition. In: Proceedings of the IEEE/CVF Conference on Computer Vision and Pattern Recognition, pp. 13746–13755 (2020)

18. Mahdavi, M., Zanibbi, R., Mouchere, H., Viard-Gaudin, C., Garain, U.: Icdar 2019 crohme+ tfd: competition on recognition of handwritten mathematical expressions and typeset formula detection. In: 2019 International Conference on Document Analysis and Recognition (ICDAR), pp. 1533–1538. IEEE (2019)

19. Malon, C., Uchida, S., Suzuki, M.: Support vector machines for mathematical symbol recognition. In: Yeung, D.-Y., Kwok, J.T., Fred, A., Roli, F., de Ridder, D. (eds.) SSPR /SPR 2006. LNCS, vol. 4109, pp. 136–144. Springer, Heidelberg (2006). https://doi.org/10.1007/11815921_14

20. Malon, C., Uchida, S., Suzuki, M.: Mathematical symbol recognition with support vector machines. Pattern Recogn. Lett. **29**(9), 1326–1332 (2008)

21. Mouchere, H., Viard-Gaudin, C., Zanibbi, R., Garain, U.: Icfhr 2014 competition on recognition of on-line handwritten mathematical expressions (crohme 2014). In: 2014 14th International Conference on Frontiers in Handwriting Recognition, pp. 791–796. IEEE (2014)

22. Mouchère, H., Viard-Gaudin, C., Zanibbi, R., Garain, U.: Icfhr 2016 crohme: competition on recognition of online handwritten mathematical expressions. In: 2016 15th International Conference on Frontiers in Handwriting Recognition (ICFHR), pp. 607–612. IEEE (2016)

23. Rhee, T.H., Kim, J.H.: Efficient search strategy in structural analysis for handwritten mathematical expression recognition. Pattern Recogn. **42**(12), 3192–3201 (2009)

24. Sandler, M., Howard, A., Zhu, M., Zhmoginov, A., Chen, L.C.: Mobilenetv 2: Inverted residuals and linear bottlenecks. In: Proceedings of the IEEE Conference on Computer Vision and Pattern Recognition, pp. 4510–4520 (2018)

25. Schaefer, S., McPhail, T., Warren, J.: Image deformation using moving least squares. In: ACM SIGGRAPH 2006 Papers, pp. 533–540 (2006)

26. Simistira, F., Papavassiliou, V., Katsouros, V., Carayannis, G.: Structural analysis of online handwritten mathematical symbols based on support vector machines. In: Document Recognition and Retrieval XX, vol. 8658, pp. 332–339. SPIE (2013)

27. Su, J., Lu, Y., Pan, S., Wen, B., Liu, Y.: Roformer: enhanced transformer with rotary position embedding. arXiv preprint arXiv:2104.09864 (2021)

28. Truong, T.N., Nguyen, C.T., Phan, K.M., Nakagawa, M.: Improvement of end-to-end offline handwritten mathematical expression recognition by weakly supervised learning. In: 2020 17th International Conference on Frontiers in Handwriting Recognition (ICFHR), pp. 181–186. IEEE (2020)

29. Truong, T.-N., Ung, H.Q., Nguyen, H.T., Nguyen, C.T., Nakagawa, M.: Relation-based representation for handwritten mathematical expression recognition. In: Barney Smith, E.H., Pal, U. (eds.) ICDAR 2021. LNCS, vol. 12916, pp. 7–19. Springer, Cham (2021). https://doi.org/10.1007/978-3-030-86198-8_1

30. Vaswani, A., et al.: Attention is all you need. Advances in neural information processing systems 30 (2017)

31. Wang, H., Shan, G.: Recognizing handwritten mathematical expressions as latex sequences using a multiscale robust neural network. arXiv preprint arXiv:2003.00817 (2020)

32. Wang, J., Du, J., Zhang, J., Wang, Z.R.: Multi-modal attention network for handwritten mathematical expression recognition. In: 2019 International Conference on Document Analysis and Recognition (ICDAR), pp. 1181–1186. IEEE (2019)

33. Winkler, H.J.: Hmm-based handwritten symbol recognition using on-line and off-line features. In: 1996 IEEE International Conference on Acoustics, Speech, and Signal Processing Conference Proceedings. vol. 6, pp. 3438–3441. IEEE (1996)

34. Winkler, H.J., Lang, M.: Symbol segmentation and recognition for understanding handwritten mathematical expressions. In: Tagungs band 5th International Workshop on Frontiers in Handwriting Recognition IWFHR-5, pp. 407–412 (1997)

35. Woo, S., Park, J., Lee, J.Y., Kweon, I.S.: Cbam: convolutional block attention module. In: Proceedings of the European Conference on Computer Vision (ECCV), pp. 3–19 (2018)

36. Wu, J.-W., Yin, F., Zhang, Y.-M., Zhang, X.-Y., Liu, C.-L.: Image-to-Markup generation via paired adversarial learning. In: Berlingerio, M., Bonchi, F., Gärtner, T., Hurley, N., Ifrim, G. (eds.) ECML PKDD 2018. LNCS (LNAI), vol. 11051, pp. 18–34. Springer, Cham (2019). https://doi.org/10.1007/978-3-030-10925-7_2

37. Wu, J.W., Yin, F., Zhang, Y.M., Zhang, X.Y., Liu, C.L.: Handwritten mathematical expression recognition via paired adversarial learning. Int. J. Comput. Vision **128**(10), 2386–2401 (2020)

38. Yan, Z., Zhang, X., Gao, L., Yuan, K., Tang, Z.: Convmath: a convolutional sequence network for mathematical expression recognition. In: 2020 25th International Conference on Pattern Recognition (ICPR), pp. 4566–4572. IEEE (2021)

39. Zhang, J., Du, J., Dai, L.: Multi-scale attention with dense encoder for handwritten mathematical expression recognition. In: 2018 24th International Conference on Pattern Recognition (ICPR), pp. 2245–2250. IEEE (2018)

40. Zhang, J., Du, J., Yang, Y., Song, Y.Z., Wei, S., Dai, L.: A tree-structured decoder for image-to-markup generation. In: International Conference on Machine Learning, pp. 11076–11085. PMLR (2020)

41. Zhang, J., et al.: Watch, attend and parse: an end-to-end neural network based approach to handwritten mathematical expression recognition. Pattern Recogn. **71**, 196–206 (2017)

42. Zhao, W., Gao, L., Yan, Z., Peng, S., Du, L., Zhang, Z.: Handwritten mathematical expression recognition with bidirectionally trained transformer. In: Lladós, J., Lopresti, D., Uchida, S. (eds.) ICDAR 2021. LNCS, vol. 12822, pp. 570–584. Springer, Cham (2021). https://doi.org/10.1007/978-3-030-86331-9_37

EDSL: An Encoder-Decoder Architecture with Symbol-Level Features for Printed Mathematical Expression Recognition

Yingnan Fu[1], Tingting Liu[1], Ming Gao[1,2](✉), and Aoying Zhou[1]

[1] School of Data Science and Engineering, East China Normal University,
Shanghai, China
Yingnanfu@foxmal.com

[2] Shanghai Key Laboratory of Mental Health and Psychological Crisis Intervention,
School of Psychology and Cognitive Science, East China Normal University,
Shanghai, China

Abstract. Printed mathematical expression recognition (PMER) aims to transcribe a printed mathematical expression image into a structural expression. The task is useful in a wide spectrum of applications, including personalized question recommendation and automatic problem solving. In this paper, we propose a new method named EDSL, shorted for an Encoder-Decoder architecture with Symbol-Level features, to recognize printed mathematical expressions from input images. Its encoder consists of a segmentation module to identify all symbols and their spatial information from the image in an unsupervised manner, and a reconstruction module to recover symbol dependencies after symbol segmentation. Furthermore, we employ a position correction attention mechanism to capture the spatial relationship between symbols, and apply a transformer model to alleviate the negative impact from long output. We conduct extensive experiments on two real datasets to verify the effectiveness and rationality of our proposed EDSL model. The experimental results illustrated that EDSL outperformed state-of-the-art methods by an accuracy margin of 3.47% and 4.04% in the two datasets, respectively.

Keywords: printed mathematical expression recognition ·
encoder-decoder network · segmentation · position correction attention

1 Introduction

Mathematical expression understanding is a fundamental task that has been widely used in many intelligent education applications, including student performance evaluation [27], personalized exercise recommendation [11], and arithmetic problem solving [30,31]. As printed math expressions often exist in the form of images, it is crucial to convert images of printed math expressions into structural expressions, such as LaTeX or symbol layout trees [16]. The process is called printed mathematical expression recognition, or PMER for short. Compared to traditional Optical Character Recognition (OCR) problems, PMER is more challenging because it not only needs to identify all symbols from the images, but also captures their spatial relationships [36].

Traditional approaches decompose PMER into two tasks: symbol recognition and structural analysis, which require a lot of hand-crafted rules and may propagate recognition errors between tasks. Since PMER can be viewed as an image-to-text translation task, recent advances in PMER have focused on employing an encoder-decoder neural network to address the image captioning problem [13,21,22]. In particular, the encoder extracts semantic embeddings from an entire math expression image based on a convolutional neural network (CNN), and the decoder predicts LaTeX tokens using a recurrent neural network (RNN) [7,35]. Despite the performance improvements with traditional PMER approaches, we argue that these methods are suboptimal due to the following factors:

- **Output sequences of PMER are longer.** Math expressions in LaTeX format are normally much longer than image captions. For instance, in the MS COCO dataset, the average length of captions is only 10.47 [10]. In contrast, the average length of math expressions in academic papers is 62.78 [7]. Cho et al. demonstrate that the performance of the encoder-decoder network for text generation deteriorates rapidly with the increase of sentence length [6]. Thus, we argue that image captioning models cannot solve the PMER problem due to the design limitations of short outputs.
- **Spatial semantics of an expression can be complex and diverse.** In a math expression, a symbol could have different mathematical semantics with different position. As illustrated in Fig. 1, although there are six identical symbols(number '2'), they are in different positions with different semantics, such as subscripts, superscripts, above, and below, etc. In contrast, the image captioning may be independent on the position of an object. As demonstrated in Fig. 1(a), no matter where the tennis ball is, the semantics of this image will be the same. Image captioning models may degrade the performance when addressing the PMER problem since they only capture simple spatial information about objects.
- **PMER needs to provide a fine-grained description of a math expression.** Image captioning aims to provide a summarization, rather than a comprehensive, fine-grained description of the content in an image. As demonstrated in Fig. 1(b), although there are many objects in the image, the caption still only summaries the main content in the short sentence "*A group of people are shopping at the market.*" However, PMER not only identifies all Roman letters, Greek letters, and operator symbols, but also requires layout analysis of all symbols. If there is a token error or a lack of detail in the output sequence of PMER, the whole recognition process will fail. We point out that unlike image captioning, PMER requires fine-grained features to capture details of an image.

Recently, Deng et al. improved the image captioning model to solve the PMER problem and proposed the IM2Markup model [7]. IM2markup employs a small receptive field CNN encoder with a coarse-to-fine attention mechanism.

(a) A young girl is playing tennis.   (b) A group of people are shopping at the market

$$S\left(y\right) = \int \delta_1 x + \frac{1}{e^2}\left(\frac{2a}{a\partial + 4\alpha^2}\left(x - y\right) + \frac{\alpha}{2\left(b\partial + 4\alpha^2\right)}\delta_2\left(x - y\right)\right)dx$$

(c) S \left(y \right) = \int \delta _ 1 x + \frac { 1 } { e ^ { 2 } } \left(\frac { 2 a } { a \partial + 4 \alpha ^ { 2 } } \right) \left(x - y \right) + \frac { \alpha } { 2 \left(b \partial + 4 \alpha ^ { 2 } \right) } \delta _ 2 \left(x - y \right) \right) d x

Fig. 1. Comparison between image captioning and PMER. (a) image captioning only captures simple spatial information about objects. (b) summarizes an image, rather than a detailed description. (c) illustrates that the same symbol with different positions has different mathematical semantics.

A row encoder is also used to localize input symbols line by line from the feature map. We argue that this approach cannot capture fine-grained features and fails to extract cross-line spatial dependencies of symbols.

In this paper, we propose EDSL (shorted for an Encoder-Decoder architecture with Symbol-Level features), which addresses the above limitations of existing PMER methods. EDSL adopts a symbol-level image encoder that consists of a segmentation module and a reconstruction module. The segmentation module identifies both symbol features and their spatial information in a fine-grained manner. In the reconstruction module, we employ a position correction attention (PC-attention) to recover spatial dependencies of symbols in the encoder. To alleviate the negative impact from long output concerns, we apply a transformer model [29] to transcribe the encoded image into a sequential and structural output. The key contributions of this paper are summarized as follows:

(1) (1) We propose an encode-decoder framework with symbol-level features to address the PMER problem.
(2) Our encoder consists of a segmentation module to identify all symbols and their spatial information from images in an unsupervised manner, and a reconstruction module to recover symbol dependencies after symbol segmentation.
(3) We employ PC-attention to capture the spatial relationship between symbols, and apply a transformer model to alleviate the negative impact from long output.
(4) We conducted extensive experiments on two real datasets. The experimental results verify the superiority of our EDSL over state-of-the-art methods.

2   Related Work

Existing PMER methods can be categorized into two groups: traditional multi-stage methods, and end-to-end approaches.

2.1   Multi-stage Methods

A multi-stage PMER method can be simplified into two sub-tasks: symbol recognition [17] and layout analysis [3].

The main difficulty in symbol recognition is the problem caused by touching and over-segmented characters. Okamoto et al. [17] used a template matching method to recognize characters. Alternatively, the characters can be recognized using a supervised model. Malon et al. [15] and LaViola et al. [8] proposed an SVM and an ensemble boosting classifier to improve character recognition, respectively. In our proposed EDSL, we only employ an unsupervised method to segment symbols from images. Since EDSL only extracts symbol-level features from the segmented symbols and does not recognize the characters, the touching and over-segmented symbols do not affect the recognition accuracy.

The most common method used in symbol layout analysis is recursive decomposition [36]. Specifically, operator-driven decomposition recursively decomposes a math expression by using operator dominance to recursively identify operators [4]. Projection profile cutting recursively decomposes a typeset math expression using a method similar to X-Y cutting [20,24]. Baseline extraction decomposes a math expression by recursively identifying adjacent symbols from left to right on the main baseline of an expression [37,38]. In this paper, we propose an encoder-decoder framework with a PC-attention mechanism to preserve the spatial relationships, which has achieved the best performance compared to competitive baselines.

2.2   End-to-End Methods

Different from the multi-stage methods, PMER can also be addressed by an encoder-decoder network with an attention mechanism, where the encoder aims to understand mathematical expression images, and the decoder generates LaTeX text.

Zhang et al. [40] used a VGG network as the encoder to recognize handwritten formulas. To improve the accuracy of handwritten formula recognition, Zhang and Du proposed a multi-scale attention mechanism based on a DenseNet network [39]. Deng et al. [7] proposed the IM2Markup model based on the coarse-to-fine attention mechanism, which achieves state-of-the-art performance. Yin et al. [35] proposed a spotlight mechanism to recognize structural images, such as math formulas and music scores. Wang et al. leverage reinforcement learning to translate images of math formulas into LaTex markup sequences [32]. Yan et al. proposed ConvMath to solve the PMER problem, which is entirely based on convolution networks [34]. We argue that a CNN network is hard to directly apply for encoding math expression image features since the large receptive field cannot extract the fine-grained symbol features and the small receptive field is inevitable to increase the computational cost.

In addition, image captioning methods can also be applied to address the PMER problem [1,5,14]. However, we argue that image captioning methods are suboptimal due to improper design for text summarization.

3   Problem Formulation and Model Overview

3.1   Problem Formulation

For a printed mathematical expression $\mathbf{x}$, which is a grayscale and structural image. Let $\mathbf{y} = < y_1, y_2, \cdots, y_t >$ be a sequence of LaTeX text, where y_i is the i-th token in LaTeX sequence $\mathbf{y}$, t is the sequence length. The PMER task aims to transcribe a printed math expression into LaTeX text. Formally, the PMER problem can be defined as:

Definition 1 (PMER problem). *Given a printed math expression image $\mathbf{x}$, the goal of PMER is to learn a mapping function f, which can convert image $\mathbf{x}$ into a sequence of LaTeX text $\mathbf{y} = < y_1, y_2, \cdots, y_t >$, such that rendering $\mathbf{y}$ with a LaTeX compiler is the math expression in image $\mathbf{x}$.*

In the definition, PMER can be treated as a structural image transcription problem, where the structural content in an image is transcribed into a sequence of LaTeX text.

3.2   Model Overview

Figure 2 demonstrates the overall architecture of our proposed EDSL model, which consists of two main components: (1) **symbol-level image encoder**; (2) **transcribing decoder**. The encoder consists of a segmentation module and a reconstruction module, and is designed to capture fine-grained symbol features and their spatial information. The segmentation module divides an entire math expression image into symbol blocks in an unsupervised manner, such that each symbol block contains part of a symbol in the printed math expression. The reconstruction module is designed for recovering spatial relationships between symbols via employing PC-attention. To recover the expression, the transcribing decoder is designed to transcribe the encoded math expression image into a LaTeX sequence.

4   Details on EDSL Implementation

4.1   Segmentation Module

To get the symbol-level features of a math expression, we segment an input image into symbol blocks. EDSL applies a connected-component labeling algorithm to

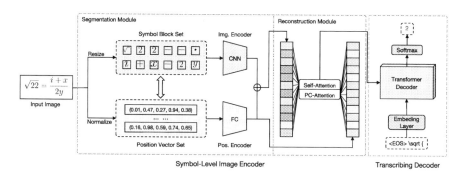

Fig. 2. The architecture of EDSL. EDSL consists of two main part: 1) a symbol-level image encoder with segmentation module and reconstruction module; 2) a transcribing decoder with transformer. FC is the fully-connected network.

find symbol blocks in a math expression image without any supervised information [23]. We then resize each symbol block to $b \times b$ pixels to extract visual features. Given an image $\mathbf{x}$ with height H and width W, let $S = \{\mathbf{s}_1, \mathbf{s}_2, \cdots, \mathbf{s}_n\}$ be a set of symbol blocks, where n is the total number of symbol blocks, and $\mathbf{s}_i \in \mathbb{R}^{b \times b}$.

It should be noted that, unlike the character segmentation used for addressing the traditional OCR problem, EDSL does not need to correctly and completely segment all symbols in the image. Each symbol block can be a complete symbol or part of a symbol, which is only utilized to extract features, rather than recognize symbols. As such, error propagation, which commonly arises in traditional OCR tasks, will not take place in our proposed EDSL. By calculating position vectors of all symbol blocks, the spatial information of each symbol block can be preserved.

As demonstrated in Fig. 3(a), we segment the two numbers into two components with a pre-defined threshold as shown in Fig. 3(b). Subsequently, we resize each component to $b \times b$ pixels as illustrated in Fig. 3 (c), where each $b \times b$ pixel is a symbol block.

Correspondingly, we calculate a set of position vectors $P = \{\mathbf{p}_1, \mathbf{p}_2, \cdots, \mathbf{p}_n\}$ associated with S, where $\mathbf{p}_i \in \mathbb{R}^5$ is the position vector of $\mathbf{s}_i$. The calculation of $\mathbf{p}_i$ is as follows:

$$\mathbf{p}_i = (\frac{t_i}{H}, \frac{d_i}{H}, \frac{l_i}{W}, \frac{r_i}{W}, \frac{H}{W}) \tag{1}$$

As illustrated in Fig. 3(d), t_i, d_i, l_i, and r_i are the distances from each edge (top, bottom, left, right) of the i−th symbol block to the upper and left of input image $\mathbf{x}$. For ease of training, we standardize each entry of the position vector into 0 to 1 with height H and width W of the input image. The last entry $\frac{H}{W}$ is the width/height ratio of input image $\mathbf{x}$, and will help us to reconstruct a symbol position when it is distorted after standardizing the first 4 entries.

For preserving symbol features and spatial information, we employ image and position encoders to map each symbol block and corresponding position vector

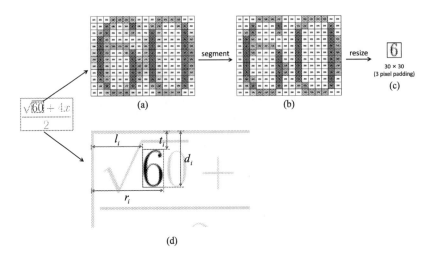

Fig. 3. A running example of segmentation. (a) is a grayscale image. A connected-component labeling algorithm is used to segment (a) into two components shown in (b). (c) is the symbol block by resizing the component of (b). (d) is a diagram of position vector features of a symbol block. t_i, d_i, l_i, and r_i are the distances from each edge (top, bottom, left, right) of the block to the upper and left of an entire image.

into low-dimensional spaces as demonstrated in Fig. 2. Specifically, we employ a six-layer CNN model with a fully connected layer [26] to encode all symbol blocks of S into an m-dimensional space, denoted by $S' = \{s'_1, s'_2, \cdots, s'_n\}$. Similarly, we employ a three-layer fully connected network(FC) to encode all position vectors, and also embed them into an m-dimensional space, denoted by $P' = \{p'_1, p'_2, \cdots, p'_n\}$.

$$s'_i = \text{CNN}(s_i) \quad p'_i = \text{FC}(p_i) \tag{2}$$

where $s'_i \in \mathbb{R}^m$, $p'_i \in \mathbb{R}^m$

Finally, we can get the symbol block embedding set

$$E = \{e_1, e_2, \cdots, e_n\} \tag{3}$$

where

$$e_i = s'_i + p'_i, \quad \text{for } i \in 1, \cdots, n; \; e_i \in \mathbb{R}^m \tag{4}$$

4.2 Reconstruction Module

Since the embedding vectors of all symbols are independent, it is necessary to reconstruct the spatial relationships between symbols. Although RNN is a commonly used approach to infer the dependencies between entries in a sequence, symbol blocks are in a two-dimensional space and cannot be modeled as a sequence. To reconstruct the spatial relationships between symbols, we employ a transformer model with a novel attention mechanism. In the architecture of a transformer, self-attention is a key concept. It refers to calculating the attention score for each pair of elements within a sequence, which can learn the dependencies between tokens. For each symbol block vector in E, we need to calculate the attention scores with all other symbols. As such, we can capture the internal spatial relationships between symbol blocks. The attention score between each pair of symbol block vectors is calculated by scaled dot-product attention [29]:

$$
\alpha_{ij} = \frac{(\mathbf{e}_i \mathbf{W}^Q)(\mathbf{e}_j \mathbf{W}^K)^T}{\sqrt{m}}
$$
$$
\text{attn}_{ij} = \frac{\exp{(\alpha_{ij})}}{\sum_{k=1}^{n} \exp{(\alpha_{ik})}}
$$
(5)

where m is the dimension of $\mathbf{e}_i$, $\mathbf{W}^Q \in \mathbb{R}^{m \times m}$ and $\mathbf{W}^K \in \mathbb{R}^{m \times m}$.

PC-Attention. Since self-attention is a global attention mechanism, it may be suboptimal to capture the spatial relationships in a long math expression since a symbol is not necessary to interact with the other symbols far away from it. For example, in the expression of Fig. 2, the first symbol '2' in the image only needs to calculate the dependencies with surrounding symbols. It does not need to interact with 'x' and 'y' because they are far away from it. Thus, we introduce the position correction attention (PC-attention), which utilizes the position vectors $\mathbf{p}'_i$ to calculate attention scores for a target symbol.

For each pair of symbol block vectors, PC-attention first calculates the attention score α^{pos} of their symbol block vectors followed by [14]. Then, we add it with α to obtain the new attention weight α'. Finally PC-attention score attn′ can be calculated by normalized α' with softmax function. The PC-attention score is calculated as follows:

$$
\alpha^{pos}_{ij} = \mathbf{v}_a^T \tanh(\mathbf{W}_a[\mathbf{p}'_i; \mathbf{p}'_j])
$$
$$
\alpha'_{ij} = \alpha_{ij} + \alpha^{pos}_{ij}
$$
$$
\text{attn}'_{ij} = \frac{\exp{(\alpha'_{ij})}}{\sum_{k=1}^{n} \exp{(\alpha'_{ik})}}
$$
(6)

where ';' is the concatenation operator, $\mathbf{W}_a \in \mathbb{R}^{2m \times m}$, $\mathbf{v}_a \in \mathbb{R}^m$. In practice, we utilize the multi-head variant to calculate attn′ [29].

The comparison of self-attention and PC-attention is illustrated in Fig. 4. Although PC-attention appears to be more complex, we share $\mathbf{W}_a$ in multiple transformer layers. As such, PC-attention is as efficient as self-attention since α^{pos} only need to be calculated once. PC-attention calculates the attention score via combining both symbol features and their spatial information. To avoid unnecessary long-distance dependencies, PC-attention focuses on the nearest symbols via adjusting self-attention with position information to better reconstruct the spatial relationships between symbols.

After Reconstruction Module, the symbol block embedding set E is encoded to embedding set $R = \{\mathbf{r}_1, \mathbf{r}_2, ..., \mathbf{r}_n\}$.

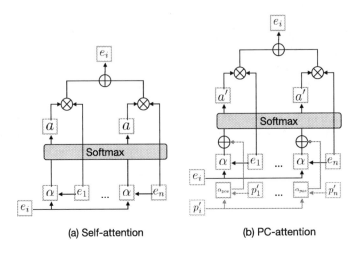

(a) Self-attention   (b) PC-attention

Fig. 4. Comparison of self-attention and PC-attention.

4.3   Transcribing Decoder

As with the general encoder-decoder architecture, given the symbol block embeddings set R, the transcribing decoder of EDSL generates one token of the sequence $o = (\mathbf{o}_1, ..., \mathbf{o}_t)$ at each time. When generating the next token, the previously generated tokens are used as additional input. We employ the transformer decoder proposed in [29] to transcribe the math expression since it is more conducive to generate a long LaTeX sequence compared with others. Finally, we used a softmax layer to predict the probability of the output token at time step t:

$$p(y_t|y_1, ..., y_{t-1}, \mathbf{r}_1, ..., \mathbf{r}_n) = \text{softmax}(\mathbf{W}_{out}\mathbf{o}_{t-1}) \tag{7}$$

where y_t is t−th token in the output LaTeX sequence, $\mathbf{o}_{t-1}$ is the output of transformer decoder in the $(t-1)$th step, $\mathbf{W}_{out} \in \mathbb{R}^{m \times |v|}$, and $|v|$ is the vocabulary size. The overall loss $\mathcal{L}$ is defined as the negative log-likelihood of the LaTeX token sequence:

$$\mathcal{L} = \sum_{t=1}^{T} -\log P(y_t|y_1, ..., y_{t-1}) \tag{8}$$

Since all calculations are deterministic and differentiable, the model can be optimized by standard back-propagation.

5   Experiment

To evaluate the performance of EDSL, we conduct extensive experiments on two real datasets.

Through empirical studies, we aim to answer the following research questions:

RQ1: How is the performance of EDSL when compared with state-of-the-art methods for PMER?

RQ2: How does the length of math expression affect the performance of EDSL?

RQ3: Is symbol-level image encoder helpful to improve the performance of EDSL?

In addition, we conduct a case study, which visualizes the role of different attention mechanisms.

5.1 Experimental Setup

Dataset. We evaluate the performance of EDSL on two public datasets, Formula [35] and IM2LATEX [7]. Before reporting the performance, we pre-process the two datasets as follows:

ME-20K: Dataset Formula collects printed math expression images and corresponding LaTeX representations from high school math exercises in Zhixue.com, which is an online education system. Due to many duplicates that existed in the dataset, we remove the duplicates and rename the new dataset as ME-20K.

ME-98K: Dataset IM2LATEX collects the printed formula and corresponding LaTeX representations from 60,000 research papers. As there are 4881 instances in the IM2LATEX dataset, which are tables or graphs, rather than math expressions, We remove these LaTeX strings and corresponding images from IM2LATEX, and get the dataset named ME-98K.

Baselines. We compare EDSL with two types of baselines:

PMER Method. Due to poor performance reported in [7], we do not report the performance of INFTY [28] and CTC [25]. We compare two methods with our proposed EDSL.

- STNR [35]: This method proposes a hierarchical spot-light transcribing network that consists of two stages. It designs a reinforcement learning method to refine the model.
- IM2Markup [7]: This method employs an encoder-decoder model with coarse-to-fine attention for recognizing math expressions.

Image Captioning Methods. We also compare our EDSL with several competitive image captioning methods, which are SAT [33], DA [14], TopDown [1], ARNet [5], LBPF [19] and CIC [2].

Evaluation Metrics. Our main evaluation method is to check the matching accuracy of the rendered prediction image compared to the ground-truth image. Followed by [7], we also employ Match-ws to check the exact match accuracy after eliminating white space columns. Besides, we also use standard text generation metrics, BLEU-1(abbr. as B@1), BLEU-4(abbr. as B@4) [18], ROUGE-1(abbr. as R@1) and ROUGE-4(abbr. as R@4) [9], to measure the precision and recall of the tokens in output sequences. All experiments are conducted three times and the average performance is reported.

Table 1. Performance comparison on ME-20K and ME-98K.

Dataset	Type	Method	Math-ws	Match	B@1	B@4	R@1	R@4
ME-20K	I.C.	CIC	70.91	70.56	84.15	79.27	90.25	83.37
		DA	77.31	76.92	94.01	87.27	96.45	89.08
		LBPF	80.88	80.46	94.98	88.82	97.32	90.57
		SAT	82.65	82.09	95.52	89.77	97.40	91.15
		TopDown	84.22	83.85	95.71	90.55	95.52	91.94
		ARNet	85.84	85.40	96.10	91.18	97.83	92.50
	PMER	STNR	78.13	77.62	94.67	89.31	95.81	90.39
		IM2Markup	89.63	89.23	97.26	92.83	98.27	93.74
		EDSL	**93.45**	**92.70**	**97.68**	**94.23**	**98.94**	**95.10**
ME-98K	I.C.	CIC	33.71	33.62	60.88	55.47	74.80	65.52
		DA	55.15	55.15	86.45	79.71	89.61	82.40
		LBPF	66.87	66.83	90.61	84.64	93.16	86.57
		SAT	71.04	70.85	92.34	84.64	93.16	86.57
		TopDown	72.85	72.65	93.27	87.56	95.45	89.32
		ARNet	68.98	68.55	92.05	86.04	94.66	88.27
	PMER	STNR	76.01	75.32	95.43	88.78	97.56	90.52
		IM2Markup	85.16	84.96	96.34	91.47	97.61	92.45
		EDSL	**89.34**	**89.00**	**97.39**	**92.93**	**97.93**	**93.30**

Implementation Details. As mentioned in [7], we group the images into similar sizes to facilitate batching for baselines. In EDSL, we employ two 8-layer transformer models with eight heads as reconstruction module and transcribing decoder. The embedding size m of EDSL is 256. The width b of symbol blocks is 30. We also use 160, 180, 200 as the segmentation thresholds on the training set and keep different symbol blocks of the same image as different training samples. In this way, the training samples roughly tripled. The default threshold is 160 for both validation set and test set. We make this approach as data augmentation for training the EDSL model. The effect of different thresholds for segmentation is further discussed in Sect. 5.5.

We train our models on the GTX 1080Ti GPU. The batch size of ME-20K and ME-98K are 32 and 16, respectively. We use Adam optimizer with an initial learning rate of 0.0003. Once the validation loss does not decrease in three epochs, we halve the learning rate. We stop training if it does not decrease in ten epochs.

5.2   Performance Comparison(RQ1)

Table 1 illustrates the performance of baselines and our proposed EDSL method, where we have the following key observations:

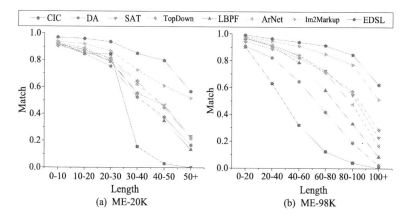

Fig. 5. Performance with different math expression lengths on ME-20K and ME-98K.

- PMER methods outperform the image captioning baselines. This is due to the factors that: (1) image captioning methods aim to summary an input image, rather than design for mining the fine-grained spatial relationships between symbols; (2) PMER methods, including STNR, IM2Markup, and EDSL are designed to reconstruct the spatial relationships between symbols in a fine-grained manner, which are more advantageous.
- EDSL is significantly better than STNR and IM2Markup. This improvement illustrates the effectiveness of EDSL, which employs the symbol-level image encoder to capture both symbol features and their spatial information, and preserves more details compared with STNR and IM2Markup.

5.3   Effect of Sequence Lengths (RQ2)

To demonstrate the effect of formula lengths, we vary the match expression lengths to evaluate the performances of baselines and our proposed EDSL method. As illustrated in Fig. 5, we have the following observations:

- The length of math expression affects the performances of all methods significantly. This is due to the factor that the neural encoder-decoder models will significantly decrease as the sequence length increases [6]. It indicates a negative impact on long math expressions.
- EDSL has achieved better performances when the math expression lengths vary. This sheds light on the benefit of preserving the fine-grained symbol-level features and their spatial information in the symbol-level image encoder. Although the performance of EDSL also decreases as the length of math expression increases, the performance declines are much smaller than the others. This indicates that EDSL is qualified to recognize the long math expressions.

Table 2. Comparison of the performances of EDSL, and its variant methods ED, ED + Seg and EDSL-S.

Dataset	Method	Match-ws	Match	B@1	B@4	R@1	R@4
ME-20K	ED	79.75	79.31	92.45	89.24	94.25	90.69
	ED+Seg	88.70	88.26	95.29	92.76	97.26	93.65
	EDSL-S	92.39	91.55	96.30	93.91	98.68	94.77
	EDSL	**93.45****	**92.70****	**97.68**	**94.23**	**98.94**	**95.10**
ME-98K	ED	68.71	68.54	90.13	86.15	91.38	87.31
	ED+Seg	81.15	80.57	95.29	91.58	96.21	91.97
	EDSL-S	88.02	87.50	97.24	92.65	97.46	93.08
	EDSL	**89.34****	**89.00****	**97.39**	**92.93**	**97.93**	**93.30**

Table 3. EDSL performance of varying segmentation thresholds on both dataset, where TH is the threshold used in the segmentation algorithm.

Dataset	TH	Match-ws	Match	B@1	B@4	R@1	R@4
ME-20K	160	92.56	91.82	97.24	93.97	98.13	94.85
	180	92.76	92.00	96.48	93.26	98.25	95.06
	200	91.82	91.77	97.02	93.66	98.80	94.52
	DA	**93.45**	**92.70**	**97.68**	**94.23**	**98.94**	**95.10**
ME-98K	160	87.35	87.06	97.18	92.75	97.36	93.11
	180	85.53	85.16	96.74	92.39	97.17	92.75
	200	85.72	85.38	97.13	92.35	97.37	92.72
	DA	**89.34**	**89.00**	**97.39**	**92.93**	**97.93**	**93.30**

5.4   Utility of Symbol-Level Image Encoder (RQ3)

To demonstrate the effectiveness of the symbol-level image encoder, we compare EDSL with their variants method ED, ED + Seg, and EDSL-S. ED only employs the CNN model as the encoder and a transformer model as the decoder, and takes the entire image of mathematical expression as input. ED + Seg removes the reconstruction module from the symbol-level image encoder of EDSL. EDSL-S employs the self-attention mechanism to capture the spatial relationships between symbols. From Table 2, we have the following key observations:

- Comparing ED with ED + Seg, the values of Match are improved by 8.95% and 12.03% on two datasets, respectively. This is due to the factor that ED + Seg encodes the fine-grained symbol features. These improvements prove the effectiveness of the fine-grained symbols features captured by the segmentation module.
- Comparing ED with ED + Seg on two datasets, the performance improvement on ME-98K is much higher. It reveals that our designed symbol-level

image encoder has a more obvious advantage in transcribing the longer math expression.

– EDSL outperforms the others significantly. This is due to the factor that PC-attention is designed for recovering the spatial relationships of symbols. This again points to the positive effect of employing PC-attention mechanism to reconstruct the spatial relationships between symbols in the reconstruction module of encoder.

5.5   Hyper-Parameter Studies

Different segmentation thresholds will produce different symbol blocks, which fundamentally affect the encoder to extract the symbol features and their spatial information. We therefore investigate the impact of threshold used for segmentation. As demonstrated in Table 3, we vary the threshold from 160 to 200, and observe that the different segmentation thresholds indeed influence the performance of EDSL. This is due to the factor that different segmentation thresholds will produce different symbol blocks, which affects the results of image feature extraction.

Inspired by the data augmentation, we retain segmented symbols given by different segmentation thresholds to increase the diversity of data for training EDSL, denoted as DA. We can observe that our EDSL method can be further improved after data augmentation. It indicates that we can use the diversity of segmentation results to improve the performance and avoid the difficulty of threshold selection.

5.6   Case Study

To better understand our proposed EDSL model, we visualize the attention scores for the tokens in the output LaTex text. We fetch the attention scores in the last layer of the transcribing decoder. Figure 6 demonstrates the predict tokens and the attention map. We can observe that: (1) For an output token, EDSL only focuses on the whole corresponding symbols, rather than a region given by image captioning methods [12,33]; (2) even if there are many identical symbols in a math expression image, EDSL can focus on the correct position. These shed light on the benefit of symbol-level image encoder, which is helpful to recognize all symbols and their spatial information.

As demonstrated in Fig. 7, we further visualized the differences between attention mechanisms used in the reconstruction module of the encoder, where the target symbol is in a red box. For each target symbol, we compare two attention mechanisms to address how they capture the spatial relationships between symbols in the symbol-level image encoder. From the visualization, we observe that PC-attention focuses on the nearest neighbors of the target symbol. For every target symbol, the found dependent symbols are reasonable in Fig. 7(b). However, it is hard to explain the self-attention mechanism, e.g., Columns 2–3 at Line 1, Columns 2, 4 at Line 2, and Columns 1, 3 at Line 3 in Fig. 7(a). Thus, we can conclude that PC-attention is more reasonable to recover the spatial relationships between symbols in the encoder.

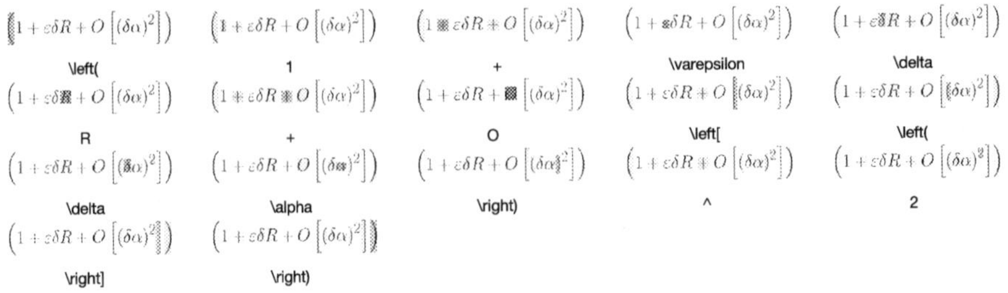

Fig. 6. Visualization of predicted tokens and attention maps.

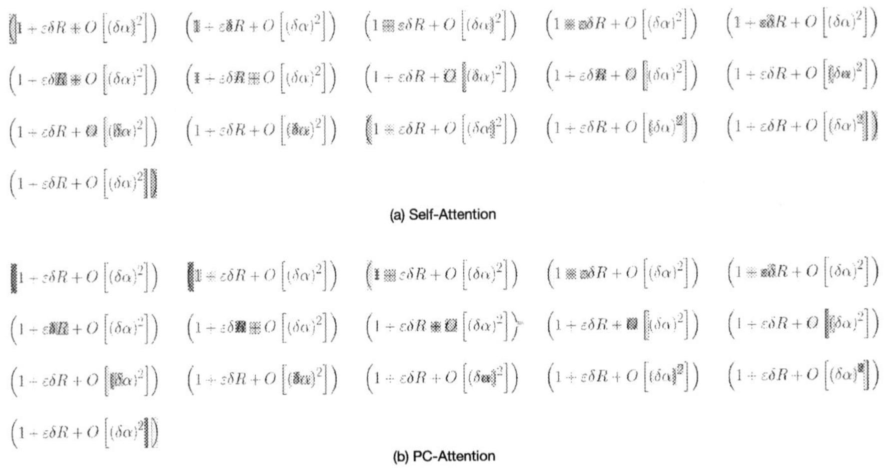

(a) Self-Attention

(b) PC-Attention

Fig. 7. Visualizing different attention mechanisms in reconstruction module.

6   Conclusion

In this paper, we propose an encoder-decoder framework with symbol-level features to address the PMER problem. Compared with the existing PMER method, the designed symbol-level image encoder aims to preserve the fine-grained symbol features and their spatial information. For recovering the spatial relationships between symbols, we propose the PC-attention mechanism to restore them in the reconstruction module of encoder. We have conducted extensive experiments on two real datasets to illustrate the effectiveness and rationality of our proposed EDSL method. In our future work, we plan to extend our proposed EDSL method to address more diversified applications of structural content recognition, including the recognition of handwritten math expression, music sheet chemical equations, and chemical molecular formulas.

References

1. Anderson, P., et al.: Bottom-up and top-down attention for image captioning and visual question answering. In: Proceedings of the IEEE Conference on Computer Vision and Pattern Recognition, pp. 6077–6086 (2018)
2. Aneja, J., Deshpande, A., Schwing, A.G.: Convolutional image captioning. In: Proceedings of the IEEE Conference on Computer Vision and Pattern Recognition, pp. 5561–5570 (2018)
3. Blostein, D., Grbavec, A.: Recognition of mathematical notation. In: Handbook of Character Recognition and Document Image Analysis, pp. 557–582. World Scientific (1997)
4. Chan, K.F., Yeung, D.Y.: Error detection, error correction and performance evaluation in on-line mathematical expression recognition. Pattern Recogn. **34**(8), 1671–1684 (2001)
5. Chen, X., Ma, L., Jiang, W., Yao, J., Liu, W.: Regularizing RNNs for caption generation by reconstructing the past with the present. In: Proceedings of the IEEE Conference on Computer Vision and Pattern Recognition, pp. 7995–8003 (2018)
6. Cho, K., Van Merriënboer, B., Bahdanau, D., Bengio, Y.: On the properties of neural machine translation: encoder-decoder approaches. arXiv preprint arXiv:1409.1259 (2014)
7. Deng, Y., Kanervisto, A., Ling, J., Rush, A.M.: Image-to-markup generation with coarse-to-fine attention. In: Proceedings of the 34th International Conference on Machine Learning-Volume 70, pp. 980–989. JMLR. org (2017)
8. LaViola, J.J., Zeleznik, R.C.: A practical approach for writer-dependent symbol recognition using a writer-independent symbol recognizer. IEEE Trans. Pattern Anal. Mach. Intell. **29**(11), 1917–1926 (2007)
9. Lin, C.Y.: Rouge: a package for automatic evaluation of summaries. In: Text Summarization Branches Out, pp. 74–81 (2004)
10. Lin, T.-Y., et al.: Microsoft COCO: common objects in context. In: Fleet, D., Pajdla, T., Schiele, B., Tuytelaars, T. (eds.) ECCV 2014. LNCS, vol. 8693, pp. 740–755. Springer, Cham (2014). https://doi.org/10.1007/978-3-319-10602-1_48
11. Liu, Q., et al.: Finding similar exercises in online education systems. In: Proceedings of the 24th ACM SIGKDD International Conference on Knowledge Discovery & Data Mining, pp. 1821–1830. ACM (2018)
12. Lu, J., Xiong, C., Parikh, D., Socher, R.: Knowing when to look: adaptive attention via a visual sentinel for image captioning. In: Proceedings of the IEEE Conference on Computer Vision and Pattern recognition, pp. 375–383 (2017)
13. Lu, J., Yang, J., Batra, D., Parikh, D.: Neural baby talk. In: Proceedings of the IEEE Conference on Computer Vision and Pattern Recognition, pp. 7219–7228 (2018)
14. Luong, M.T., Pham, H., Manning, C.D.: Effective approaches to attention-based neural machine translation. arXiv preprint arXiv:1508.04025 (2015)
15. Malon, C., Uchida, S., Suzuki, M.: Mathematical symbol recognition with support vector machines. Pattern Recogn. Lett. **29**(9), 1326–1332 (2008)
16. Mouchere, H., Zanibbi, R., Garain, U., Viard-Gaudin, C.: Advancing the state of the art for handwritten math recognition: the crohme competitions, 2011–2014. Int. J. Doc. Anal. Recogn. (IJDAR) **19**(2), 173–189 (2016)
17. Okamoto, M., Imai, H., Takagi, K.: Performance evaluation of a robust method for mathematical expression recognition. In: Proceedings of Sixth International Conference on Document Analysis and Recognition, pp. 121–128. IEEE (2001)

18. Papineni, K., Roukos, S., Ward, T., Zhu, W.J.: Bleu: a method for automatic evaluation of machine translation. In: Proceedings of the 40th Annual Meeting on Association for Computational Linguistics, pp. 311–318. Association for Computational Linguistics (2002)

19. Qin, Y., Du, J., Zhang, Y., Lu, H.: Look back and predict forward in image captioning. In: Proceedings of the IEEE Conference on Computer Vision and Pattern Recognition, pp. 8367–8375 (2019)

20. Raja, A., Rayner, M., Sexton, A., Sorge, V.: Towards a parser for mathematical formula recognition. In: Borwein, J.M., Farmer, W.M. (eds.) MKM 2006. LNCS (LNAI), vol. 4108, pp. 139–151. Springer, Heidelberg (2006). https://doi.org/10.1007/11812289_12

21. Redmon, J., Farhadi, A.: Yolov3: an incremental improvement. arXiv preprint arXiv:1804.02767 (2018)

22. Ren, S., He, K., Girshick, R., Sun, J.: Faster R-CNN: towards real-time object detection with region proposal networks. In: Advances in Neural Information Processing Systems, pp. 91–99 (2015)

23. Samet, H., Tamminen, M.: Efficient component labeling of images of arbitrary dimension represented by linear bintrees. IEEE Trans. Pattern Anal. Mach. Intell. **10**(4), 579–586 (1988)

24. Shafait, F., Keysers, D., Breuel, T.: Performance evaluation and benchmarking of six-page segmentation algorithms. IEEE Trans. Pattern Anal. Mach. Intell. **30**(6), 941–954 (2008)

25. Shi, B., Bai, X., Yao, C.: An end-to-end trainable neural network for image-based sequence recognition and its application to scene text recognition. IEEE Trans. Pattern Anal. Mach. Intell. **39**(11), 2298–2304 (2016)

26. Simonyan, K., Zisserman, A.: Very deep convolutional networks for large-scale image recognition. arXiv preprint arXiv:1409.1556 (2014)

27. Su, Y., et al.: Exercise-enhanced sequential modeling for student performance prediction. In: Thirty-Second AAAI Conference on Artificial Intelligence (2018)

28. Suzuki, M., Tamari, F., Fukuda, R., Uchida, S., Kanahori, T.: Infty: an integrated OCR system for mathematical documents. In: Proceedings of the 2003 ACM Symposium on Document Engineering, pp. 95–104. ACM (2003)

29. Vaswani, A., Shazeer, N., Parmar, N., Uszkoreit, J., Polosukhin, I.: Attention is all you need (2017)

30. Wang, L., Zhang, D., Gao, L., Song, J., Guo, L., Shen, H.T.: MathDQN: solving arithmetic word problems via deep reinforcement learning. In: Thirty-Second AAAI Conference on Artificial Intelligence (2018)

31. Wang, L., et al.: Template-based math word problem solvers with recursive neural networks (2019)

32. Wang, Z., Liu, J.C.: Translating math formula images to latex sequences using deep neural networks with sequence-level training. Int. J. Doc. Anal. Recogn. (IJDAR) **24**(1), 63–75 (2021)

33. Xu, K., et al.: Show, attend and tell: neural image caption generation with visual attention. In: International Conference on Machine Learning, pp. 2048–2057 (2015)

34. Yan, Z., Zhang, X., Gao, L., Yuan, K., Tang, Z.: ConvMath: a convolutional sequence network for mathematical expression recognition. In: 2020 25th International Conference on Pattern Recognition (ICPR), pp. 4566–4572. IEEE (2021)

35. Yin, Y., et al.: Transcribing content from structural images with spotlight mechanism. In: Proceedings of the 24th ACM SIGKDD International Conference on Knowledge Discovery & Data Mining, pp. 2643–2652. ACM (2018)

36. Zanibbi, R., Blostein, D.: Recognition and retrieval of mathematical expressions. Int. J. Doc. Anal. Recogn. (IJDAR) **15**(4), 331–357 (2012)
37. Zanibbi, R., Blostein, D., Cordy, J.R.: Baseline structure analysis of handwritten mathematics notation. In: Proceedings of Sixth International Conference on Document Analysis and Recognition, pp. 768–773. IEEE (2001)
38. Zanibbi, R., Blostein, D., Cordy, J.R.: Recognizing mathematical expressions using tree transformation. IEEE Trans. Pattern Anal. Mach. Intell. **24**(11), 1455–1467 (2002)
39. Zhang, J., Du, J., Dai, L.: Multi-scale attention with dense encoder for handwritten mathematical expression recognition. In: 2018 24th International Conference on Pattern Recognition (ICPR), pp. 2245–2250. IEEE (2018)
40. Zhang, J., et al.: Watch, attend and parse: an end-to-end neural network based approach to handwritten mathematical expression recognition. Pattern Recogn. **71**, 196–206 (2017)

Semantic Graph Representation Learning for Handwritten Mathematical Expression Recognition

Zhuang Liu[1] , Ye Yuan[1] , Zhilong Ji[1] , Jinfeng Bai[1] , and Xiang Bai[2( )]

[1] Tomorrow Advancing Life, Beijing, China
{liuzhuang7,jizhilong}@tal.com
[2] Huazhong University of Science and Technology, Wuhan, China
xbai@hust.edu.cn

Abstract. Handwritten mathematical expression recognition (HMER) has attracted extensive attention recently. However, current methods cannot explicitly study the interactions between different symbols, which may fail when faced similar symbols. To alleviate this issue, we propose a simple but efficient method to enhance semantic interaction learning (SIL). Specifically, we firstly construct a semantic graph based on the statistical symbol co-occurrence probabilities. Then we design a semantic aware module (SAM), which projects the visual and classification feature into semantic space. The cosine distance between different projected vectors indicates the correlation between symbols. And jointly optimizing HMER and SIL can explicitly enhances the model's understanding of symbol relationships. In addition, SAM can be easily plugged into existing attention-based models for HMER and consistently bring improvement. Extensive experiments on public benchmark datasets demonstrate that our proposed module can effectively enhance the recognition performance. Our method achieves better recognition performance than prior arts on both CROHME and HME100K datasets.

Keywords: Handwritten Mathematical Expression Recognition · Semantic Graph · Co-occurrence Probabilities

1 Introduction

Handwritten Mathematical Expression Recognition (HMER) is an important OCR task, which can be widely applied in question parsing and answer sheet correction. In recent years, with the rapid development of deep learning technology, scene text recognition approaches have achieved great progress [8,22,23,35]. However, due to the ambiguities brought by crabbed handwriting and the complicated structures of handwritten mathematical expressions, HMER is still a challenging task.

G. A. Fink et al. (Eds.): ICDAR 2023, LNCS 14187, pp. 152–166, 2023.
https://doi.org/10.1007/978-3-031-41676-7_9

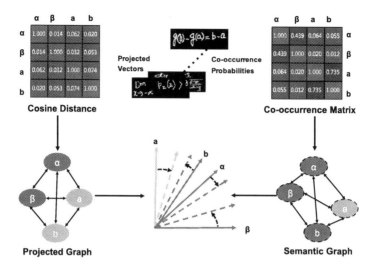

Fig. 1. Illustration of our method. The different colored graph nodes and arrows indicate different symbols.

Built upon the recent progress in sequence-to-sequence learning and neural networks [6,10,24], some studies have addressed HMER with end-to-end trained encoder-decoder models and showed significant improvement in performance. Nevertheless, the encoder-decoder framework do not fully explore the correlation between different symbols in the mathematical expression, which may be struggling when facing similar handwritten symbols or crabbed handwritings.

To address above issues, we argue that an effective HMER model should be improved from the following two aspects: (1) capturing semantic dependencies among different symbols in the mathematical expression; (2) integrating more semantic information to locate the regions of interest.

In this paper, we propose an simple but efficient method to improve the robustness of the model, which incorporate the learning of semantic relations among different symbols into the end-to-end training (Fig. 1). Firstly, we built a semantic graph rely on statistical co-occurrence probabilities, which can explicitly exhibit the dependencies among different symbols. Secondly, we propose a semantic aware module, which takes the visual and classification features as input and maps them into the semantic space. The cosine distance between different projected vectors suggests the correlation of symbols. Optimizing the distance to close to the corresponding graph value make the network capture the relationships between different symbols. Therefore, the search for regions of interest and the learning of symbols semantic dependencies are enhanced, which further improved the performance of the model.

The major contributions of this paper are briefly summarized as follows:

- To the best of our knowledge, we are the first to use co-occurrence to represent the relationship between symbols in mathematical expression and verify the effectiveness of enhancing semantic representation learning.
- We propose a semantic aware method that jointly optimizes the symbol relations learning and HMER, which can consistently improve the performance of the model for HMER.
- Our proposed semantic aware module can be easily plugged into attention based models for HMER and no extra computation during the inference stage.

To be specific about the performance, we adopt DWAP [40] as the baseline network. With the help of SAM, SAM-DWAP outperforms DWAP by 2.2%, 2.8% and 4.2% on CROHME 2014, 2016 and 2019, respectively. Moreover, with adopting the latest SOTA method CAN [16] as the baseline network, our method achieves new SOTA results (58.0% on CROHME 2014, 56.7% on CROHME 2016, 58.0% on CROHME 2019). This indicates that our method can be generalized to various existing encoder-decoder models for HMER and boost their performance.

2   Related Work

HMER is a fundamental OCR task, which has attracted research interests in the past several decades. In this section, we briefly introduce previous related works on HMER.

Traditional methods on HMER could be mainly separated into two steps: a symbol segmentation/recognition step and a grammar guided structure analysis step. In the first step, several classic classification techniques were studied, such as HMM [1,9,12,28], Elastic Matching [4,26], Support Vector Machines [11], etc. In the second step, formal grammars were designed to model the 2D and syntactic structures of expression. Lavirotte et al. [13] proposed to use graph grammar to recognize mathematical expression. Chan et al. [5] incorporated correction mechanism into parser based on definite clause grammar (DCG). Yamamoto et al. [32] modeled handwritten mathematical expressions with a stochastic context-free grammar and solved the recognition problem by using the CYK algorithm. In contrast to those traditional methods, our model incorporates grammatical structure and automatically learned encoder-decoder, therefore preventing from designing cumbersome rules.

Recently, deep learning techniques rapidly boosted the performance of HEMR. The mainstream framework was encoder-decoder networks [7,14,15, 20,25,27,29,37,38,40,42]. Deng et al. [7] firstly proposed an encoder-decoder framework to convert image to LATEX markup. A coarse-to-fine attention layer was used to reduce the attention complexity in their work. Zhang et al. [40] presented an encoder-decoder model, named WAP (Watch, Attend and Parse). In their model, the encoder is a FCN and a coverage vector is appended to the attention model. Wu et al. [30,31] focused on the pair-wise adversarial learning strategy to improve the recognition accuracy. To alleviate the challenge of lack of data, Le et al. [15] and Li et al. [17] employed distortion, decomposition and scale augmentation techniques, which achieved significant performance promotion. Le [14] proposed a dual loss attention model, which contains a new context match loss. Context matching loss is adapted to constrain the intra-class distance and enhance the discriminative power of model. Lately, Zhang et al. [39] devised a tree-based decoder to parse mathematical expression. At each step, a parent and child node pair was generated and the relation between parent node and child node reflects the structure type. Yuan et al. [34] firstly incorporate syntax information into the encoder-decoder, which achieved higer recognition accuracy while taking into account speed. Li et al. [16] design a weakly-supervised counting module and jointly optimizes HMER task and symbol counting task. With the help of integrated global information, it puts in a impressive performance.

3   Methodology

The overall framework of our approach is shown in Fig. 2. The pipeline includes several parts: densely connected convolutional network (DenseNet) [10] is applied as encoder to extract the features. The DenseNet takes a grayscale image X of size $H \times W \times 1$, where H and W are image height and image width, respectively, and returns a 2D feature map $\mathcal{F} \in \mathbb{R}^{H' \times W' \times 684}$, where $H/H' = W/W' = 16$. The decoder uses the feature map and gradually predicts the LaTeX markup. The Semantic Aware Module (SAM) comprises two branches with similar structure (visual branch and classification branch), which employ the visual and classification features, respectively. Visual and classification features are projected to semantic space to obtain projected visual and classification vectors, respectively. The cosine distance between projected vectors from different time steps indicates how related they are.

3.1   Semantic Graph

Capturing global context information has been proven to be an effective way to improve the robustness of recognition [21,33]. However, compared with words, the use of symbols in the mathematical expressions is relatively more casual. How to express the relationships among different symbols in the mathematical expressions is an open issue to be solved. Our intuition is that the magnitude of values in the co-occurrence graph reflects the relationship between different symbols, much like how different characters in text have different collocations. Making the distances close to the probabilities is aimed at enhancing the model's learning of the linguistic information in formulas.

Semantic graph is defined as $G = (S, E)$, where $S = \{s_1, s_2, ..., s_N\}$ represents the set of symbol nodes and E represents the edges, which suggest the dependence between any two symbols. The correlation matrix $R = \{r_{i,j}\}_{i,j=1}^N$ of graph G contains non-negative weights associated with each edge. The correlation matrix is a conditional probability matrix and the r_{ij} is set as $P(s_i/s_j)$, where P is calculated through training set. However, R is an asymmetric matrix,

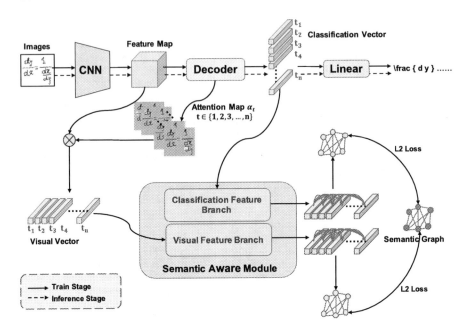

Fig. 2. The architecture of the proposed SAM-DWAP, which consists of a CNN, a decoder and a semantic aware module.

namely $r_{ij} \neq r_{ji}$. In order to facilitate the calculation, we turn the asymmetric matrix into a symmetric matrix following:

$$R' = \frac{1}{2}(R + R^T). \tag{1}$$

3.2 Semantic Aware Module

In this section, we present the detail of the proposed semantic aware module (SAM). As shown in Fig. 3 (a), SAM contains two branches, namely visual feature branch and classification feature branch. Each branch comprises two "LBR" block followed by a linear layer. A "LBR" block is built by stacking **L**inear layer, **B**atch Normalization and **R**eLU activation. We apply SAM to project the visual vectors (v_{vis}) and classification vector (v_{cls}) to semantic space to get projected visual vectors (v'_{vis}) and projected classification vectors (v'_{cls}):

$$v'_{vis} = W_3^{vis}(\sigma(\epsilon(W_2^{vis}(\sigma(\epsilon(W_1^{vis}v_{vis} + b_1^{vis}))) + b_2^{vis}))) + b_3^{vis} \tag{2}$$

$$v'_{cls} = W_3^{cls}(\sigma(\epsilon(W_2^{cls}(\sigma(\epsilon(W_1^{cls}v_{cls} + b_1^{cls}))) + b_2^{cls}))) + b_3^{cls} \tag{3}$$

where σ is the ReLU activation and ϵ refers to Batch Normalization. W_3^{vis}, W_2^{vis}, W_1^{vis}, W_3^{cls}, W_2^{cls} and W_1^{cls} are learnable parameters.

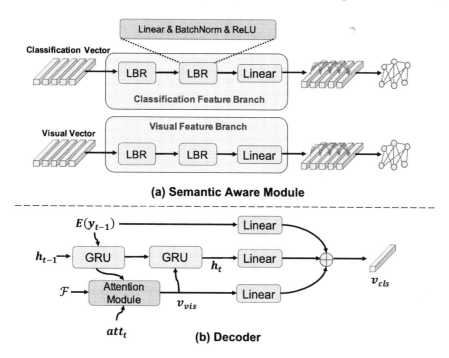

Fig. 3. The architecture of (a) semantic aware module (SAM) and (b) decoder.

Our goal is to optimize the projected visual vectors (v'_{vis}) and projected classification vectors (v'_{cls}). Such that $cos(v'_i, v'_j)$ is close to R_{ij} for all i, j, where $cos(v'_i, v'_j)$ denotes the cosine similarity between v'_i and v'_j:

$$cos(v'_i, v'_j) = \frac{v_i'^T v_j}{||v_i'^T|| \; ||v_j'^T||} \qquad (4)$$

3.3   Decoder

Figure 3 (b) shows the structure of decoder. The decoder mainly contains two Gated Recurrent Units (GRU) cells and an attention module. The first GRU takes the symbol embedding $(E(y_{t-1}))$ and historical state (h_{t-1}) predicted in the last step as input and output a new hidden state vector h'_t:

$$h'_t = GRU(E(y_{t-1}), h_{t-1}) \qquad (5)$$

Then the attention module calculates the attentional weights α_t through its attention mechanism:

$$e_t = W_\omega(tanh(W'_h h'_t + W_f \mathcal{F} + W_\alpha att_t)) \qquad (6)$$

$$\alpha_t = exp(e_t) / \sum exp(e_t) \qquad (7)$$

where W_ω, W_h', W_f and W_α are trainable parameters. $\mathcal{F}$ represents the feature map and att_t refers to coverage attention [40], which equals the sum of all past attention probabilities:

$$att_t = \sum_i \alpha_i, \quad i \in [0, t-1] \tag{8}$$

The α_t and $\mathcal{F}$ are multiplied to obtain visual features vectors v_{vis}:

$$v_{vis} = \alpha_t \otimes \mathcal{F} \tag{9}$$

The second GRU takes the v_{vis} and h_t' as input and returns the hidden state h_t:

$$h_t = GRU(v_{vis}, h_t') \tag{10}$$

Then we aggregate $E(y_{t-1})$, v_{vis} and h_t to obtain the classification feature vectors and symbol probabilities:

$$v_{cls} = W_e E(y_{t-1}) + W_h h_t + W_v v_{vis} \tag{11}$$

$$p_{symbol} = softmax(W_s v_{cls}) \tag{12}$$

where W_e, W_h, W_v and W_s are trainable parameters.

3.4 Loss Function

The overall function consists of three parts and is defined as follows:

$$\mathcal{L} = \mathcal{L}_{symbol} + \mathcal{L}_{vis} + \mathcal{L}_{cls} \tag{13}$$

where $\mathcal{L}_{symbol}$ is cross entropy classification loss of the predicted probability p_{symbol} with respect to its ground-truth. $\mathcal{L}_{vis}$ and $\mathcal{L}_{cls}$ are L2 regression loss defined as follows:

$$\mathcal{L}_{vis} = \sum_i^n \sum_j^n (cos(v_{vis,i}, v_{vis,j}), -R_{i,j})^2 \tag{14}$$

$$\mathcal{L}_{cls} = \sum_i^n \sum_j^n (cos(v_{cls,i}, v_{cls,j}), -R_{i,j})^2 \tag{15}$$

4 Experiments

We conduct experiments on three CROHME and HME100K benchmark datasets and compare the performance with previous state-of-the-art methods. In this section, we firstly specify the datasets, implementation details and evaluation protocol in Sect. 4.1, 4.2 and 4.3, respectively. Then, in Sect. 4.4 we evaluate our method on public datasets and compare it with other state-of-the-art methods. In Sect. 4.5, we exhibit the ablation studies and finally, in Sect. 4.6 we show few cases and discuss the effectiveness of our method.

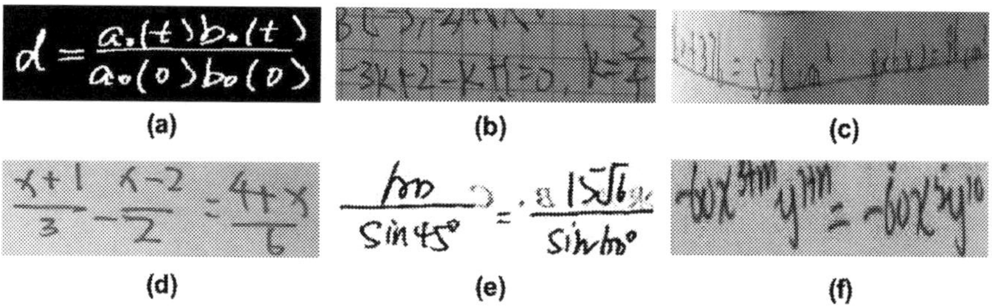

$$d = \frac{a_s(t)\,b_*(t)}{a_o(o)\,b_o(o)}$$

(a) (b) (c)

(d) (e) (f)

Fig. 4. Sample images from (a) CROHME dataset and (b-f) HME100K dataset.

4.1   Datasets

CROHME Dataset. CROHME dataset is from the competition on recognition of online handwritten mathematical expression, which is the most widely used public dataset. Images in CROHME dataset are synthesized from the handwritten stroke trajectory information in the InkML files. Therefore, the image background from CROHME dataset is clean (Fig. 4 (a)). The CROHME training set number is 8,836, while the test set contains 986, 1147 and 1199 images respectively due to different release years.

HME100K Dataset. HME100K dataset is a real scene dataset and consequently, HME100K dataset are varied in color, blur, complicated background, twist (Fig. 4 (b-f)). HME100K dataset contains 74,502 images for training and 24,607 images for testing. The data size of HME100K dataset is ten times larger than CROHME dataset. The number of math symbols included in the HME100K dataset is 245, which is two times larger than that of CROHME dataset.

4.2   Implementation Details

The proposed methods is implemented in PyTorch. A single Nvidia Tesla V100 with 32GB RAM is used to conduct experiment. The batch size is set at 8. Both the hidden state sizes of the two GRUs and dimension of word embedding are set at 256. The Adadelta optimizer [36] is used during the training process, in which ρ is set at 0.95 and ϵ is set at 10^{-6}. The learning rate starts from 0 and monotonously increases to 1 at the end of the first epoch. After that the learning rate decays to 0 following the cosine schedules [41]. For CROHME dataset, the total training epoch is set to 240 and for HME100K dataset, the training epoch is set to 40.

4.3   Evaluation Protocol

Recognition Protocol. We employ expression recognition rate (ExpRate) to evaluate the performance of different approaches. The definition of ExpRate is the percentage of predicted mathematical expressions that exactly match the ground truth.

Table 1. Expression Recognition Rate (ExpRate) performance of SAM-DWAP and SAM-CAN and other state-of-the-art methods on CROHME 2014, CROHME 2016 and CROHME 2019 test set. SAM-DWAP and SAM-CAN indicate adopting DWAP and CAN as the backbone, respectively. All results are reported as a percentage (%).

Method	CROHME2014	CROHME2016	CROHME2019	HME100K
UPV [18]	37.22	–	–	–
TOKYO [19]	–	43.94	–	–
PAL [30]	39.66	–	–	–
WAP [40]	46.55	44.55	–	–
PAL-v2 [31]	48.88	49.61	–	–
TAP [38]	48.47	44.81	–	–
DLA [14]	49.85	47.34	–	–
DWAP [37]	50.10	47.50	–	61.85
DWAP-TD [38]	49.10	48.50	51.40	62.60
DWAP-MSA [37]	52.80	50.10	47.70	–
WS-WAP [26]	53.65	51.96	–	–
MAN [27]	54.05	50.56	–	–
BTTR [42]	53.96	52.31	52.96	64.10
SAN [2]	56.20	53.60	53.50	67.10
ABM [3]	56.85	52.92	53.96	65.93
CAN [16]	57.00	56.06	54.88	67.31
SAM-DWAP (ours)	56.80	55.62	56.21	68.08
SAM-CAN (ours)	**58.01**	**56.67**	**57.96**	**68.81**

4.4 Comparison with State-of-the-Art

Results on the CROHME Datasets. Table 1 summaries the performance of our method and previous methods on the CROHME dataset. Since most of the previous work does not use data augmentation, we mainly discuss the results without data augmentation.

As shown in Table 1, using DWAP [37] as the backbone, SAM-DWAP achieves competitive results to the last SOTA method CAN [16] on CROHME 2014 and CROHME 2016. On CROHME 2019 dataset, our method ourperforms CAN by 1.33 %.

To further verify our proposed SAM is compatible with other models and can consistently bring performance improvements. We integrate SAM into CAN to construct SAM-CAN. As shown in Table 1, SAM-CAN achieves the best performance on all CROHME test set and outperforms CAN by 1.21 %, 0.61 % and 3.08 %, respectively. This result clearly demonstrates the effectiveness of our proposed module.

Results on the HME100K Dataset. As shown in Table 1 and 2, we compare our prosposed method with DWAP [37], DWAP-TD [39], BTTR [42], ABM [3], SAN [34] and CAN [16] on HME100K dataset. It is clear to notice that SAM-DWAP and SAM-CAN achieves the best performance. Specifically, as shown in

Table 2. Performance of SAM-DWAP and SAM-CAN versus DWAP, DWAP-TD, BTTR, ABM, SAN and CAN on the HME100K dataset on Easy (E.), Moderate (M.) and Hard (H.) HME100K test subsets. Our models achieve the best performance on the HME100K dataset.

HME100K	Easy	Moderate	Hard	Total
Image size	7721	10450	6436	24607
DWAP [37]	75.1	62.2	45.4	61.9
DWAP-TD [39]	76.2	63.2	45.4	62.6
BTTR [42]	77.6	65.3	46.0	64.1
ABM [3]	–	–	–	65.3
SAN [34]	79.2	67.6	51.5	67.1
CAN [16]	–	–	–	67.3
SAM-DWAP(ours)	79.3	68.4	54.0	68.1
SAM-CAN(ours)	**79.8**	**69.8**	**54.0**	**68.8**

Table 3. Ablation Studies on CROHME dataset. DWAP[†] and CAN[†] are our reproduced results. The effect of recognition performance with regard to the two components: visual feature branch and classification feature branch.

Method	CROHME		
	2014	2016	2019
DWAP[†]	54.6	52.8	52.0
Vis-DWAP	55.8	54.8	54.1
Cls-DWAP	55.6	55.2	54.7
SAM-DWAP	**56.8**	**55.6**	**56.2**
CAN[†]	57.1	55.3	54.9
Vis-CAN	57.5	56.3	56.6
Cls-CAN	57.4	56.5	55.8
SAM-CAN	**58.0**	**56.6**	**57.9**

Table 2, SAM-DWAP and SAM-CAN outperform SAN by 0.1 % and 0.6 % on easy subset, respectively. However, as the difficulty of the test subset increases, the leading margin of our method increases to 2.5 % and 2.5 % on the hard subset. This further proves the effectiveness of the proposed SAM.

4.5 Ablation Study

In this subsection, we evaluate the effectiveness of visual feature branch and classification feature branch. SAM-DWAP and SAM-CAN are the default models. Vis-DWAP and Vis-CAN have a visual feature branch but not a classification feature branch. Cls-DWAP and Cls-CAN have a classification feature branch but not a visual feature branch. DWAP† and CAN † are our reproduced results. The results are summarized in Table 3.

Impact of Visual Feature Branch. Table 4.5 shows adopting visual feature branch to DWAP improves the recognition performance ExpRate by 1.2 % on CROHME 2014, 2.0 % on CROHME 2016 and 2.1 % on CROHME 2019. Inserting visual feature branch into CAN also can enhance the performance by 0.4 % on CROHME 2014, 1.0 % on CROHME 2016 and 1.7 % on CROHME 2019. Hence integrating visual feature branch can effectively improve the performance.

Impact of Classification Feature Branch. Table 4.5 shows adopting classification feature branch to DWAP improves the recognition performance ExpRate

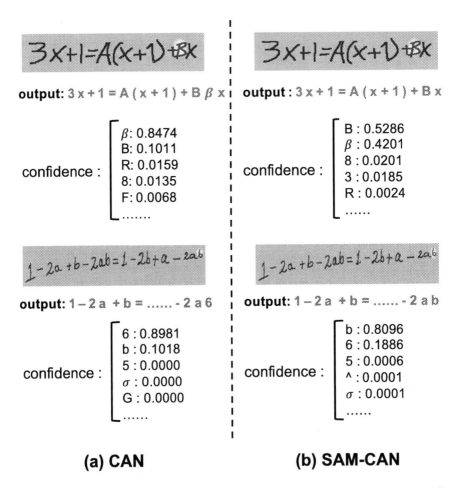

Fig. 5. Examples of (a) CAN and (b) SAM-CAN. Symbol in red color are mispredictions (Color figure online)

by 1.0 % on CROHME 2014, 2.4 % on CROHME 2016 and 2.7 % on CROHME 2019. Inserting visual feature branch into CAN also can enhance the performance by 0.3 % on CROHME 2014, 1.2 % on CROHME 2016 and 0.9 % on CROHME 2019. Hence integrating classification feature branch can effectively improve the performance.

4.6   Case Study

In this section, we show two examples to illustrate the effect of using SAM. As shown in Fig. 5 (a), although CAN correctly focuses on the region of interest, it misidentifies the symbol "B" as symbol "β" and misidentifies the symbol "b" as symbol "6". In contrast, the regions of interest of the SAM-CAN are similar to those of CAN, but SAM-CAN correctly predicts symbol "B" and "b". The confidences of symbol "B" and "b" also increase from 10.1 % to 52.9 % and increase from 10.2% to 81.0%, respectively. This phenomenon indicates that adopting SAM can improve the robustness of recognition especially the recognition performance of similar symbols.

5   Conclusion

This paper has presented a simple and efficient method for handwritten mathematical expression recognition by incorporate semantic graph representation learning into end-to-end training. To our best knowledge, the proposed method is the first to learn the correlation between different symbols through symbol co-occurrence probabilities. Experiments on the CROHME dataset and HME100K dataset have validated the effectiveness and efficiency of our method.

Acknowledgement. This work was supported by National Key R&D Program of China, under Grant No. 2020AAA0104500 and National Science Fund for Distinguished Young Scholars of China (Grant No.62225603).

References

1. Alvaro, F., Sánchez, J.A., Benedí, J.M.: Recognition of on-line handwritten mathematical expressions using 2D stochastic context-free grammars and hidden Markov models. Pattern Recogn. Lett. **35**, 58–67 (2014)
2. Anderson, R.H.: Syntax-directed recognition of hand-printed two-dimensional mathematics. In: Symposium on Interactive Systems for Experimental Applied Mathematics: Proceedings of the Association for Computing Machinery Inc., Symposium, pp. 436–459 (1967)
3. Bian, X., Qin, B., Xin, X., Li, J., Su, X., Wang, Y.: Handwritten mathematical expression recognition via attention aggregation based bi-directional mutual learning. In: Proceeding of the AAAI Conference on Artificial Intelligence, pp. 113–121 (2022)

4. Chan, K.F., Yeung, D.Y.: Elastic structural matching for online handwritten alphanumeric character recognition. In: Proceedings of the Fourteenth International Conference on Pattern Recognition (Cat. No. 98EX170), vol. 2, pp. 1508–1511. IEEE (1998)

5. Chan, K.F., Yeung, D.Y.: Error detection, error correction and performance evaluation in on-line mathematical expression recognition. Pattern Recogn. **34**(8), 1671–1684 (2001)

6. Chung, J., Gulcehre, C., Cho, K., Bengio, Y.: Empirical evaluation of gated recurrent neural networks on sequence modeling. arXiv preprint arXiv:1412.3555 (2014)

7. Deng, Y., Kanervisto, A., Ling, J., Rush, A.M.: Image-to-markup generation with coarse-to-fine attention. In: International Conference on Machine Learning, pp. 980–989. PMLR (2017)

8. Fang, S., Xie, H., Wang, Y., Mao, Z., Zhang, Y.: Read like humans: autonomous, bidirectional and iterative language modeling for scene text recognition. In: Proceedings of the IEEE/CVF Conference on Computer Vision and Pattern Recognition, pp. 7098–7107 (2021)

9. Hu, L., Zanibbi, R.: Hmm-based recognition of online handwritten mathematical symbols using segmental k-means initialization and a modified pen-up/down feature. In: 2011 International Conference on Document Analysis and Recognition, pp. 457–462. IEEE (2011)

10. Huang, G., Liu, Z., Van Der Maaten, L., Weinberger, K.Q.: Densely connected convolutional networks. In: Proceedings of the IEEE Conference on Computer Vision and Pattern Recognition, pp. 4700–4708 (2017)

11. Keshari, B., Watt, S.: Hybrid mathematical symbol recognition using support vector machines. In: Ninth International Conference on Document Analysis and Recognition (ICDAR 2007), vol. 2, pp. 859–863. IEEE (2007)

12. Kosmala, A., Rigoll, G., Lavirotte, S., Pottier, L.: On-line handwritten formula recognition using hidden Markov models and context dependent graph grammars. In: Proceedings of the Fifth International Conference on Document Analysis and Recognition. ICDAR'99 (Cat. No. PR00318), pp. 107–110. IEEE (1999)

13. Lavirotte, S., Pottier, L.: Mathematical formula recognition using graph grammar. In: Document Recognition, vol. 3305, pp. 44–52. International Society for Optics and Photonics (1998)

14. Le, A.D.: Recognizing handwritten mathematical expressions via paired dual loss attention network and printed mathematical expressions. In: Proceedings of the IEEE/CVF Conference on Computer Vision and Pattern Recognition Workshops, pp. 566–567 (2020)

15. Le, A.D., Indurkhya, B., Nakagawa, M.: Pattern generation strategies for improving recognition of handwritten mathematical expressions. Pattern Recogn. Lett. **128**, 255–262 (2019)

16. Li, B., et al.: When counting meets HMER: counting-aware network for handwritten mathematical expression recognition. In: Computer Vision-ECCV 2022: 17th European Conference, Tel Aviv, Israel, October 23–27, 2022, Proceedings, Part XXVIII, pp. 197–214. Springer, Cham (2022). https://doi.org/10.1007/978-3-031-19815-1_12

17. Li, Z., Jin, L., Lai, S., Zhu, Y.: Improving attention-based handwritten mathematical expression recognition with scale augmentation and drop attention. arXiv preprint arXiv:2007.10092 (2020)

18. Mouchere, H., Viard-Gaudin, C., Zanibbi, R., Garain, U.: ICFHR 2014 competition on recognition of on-line handwritten mathematical expressions (CROHME 2014). In: Proceeding of the International Conference on Frontiers in Handwriting Recognition, pp. 791–796 (2014)

19. Mouchère, H., Viard-Gaudin, C., Zanibbi, R., Garain, U.: ICFHR 2016 CROHME: competition on recognition of online handwritten mathematical expressions. In: Proceeding of the International Conference on Frontiers in Handwriting Recognition, pp. 607–612 (2016)

20. Nguyen, C.T., Nguyen, H.T., Morizumi, K., Nakagawa, M.: Temporal classification constraint for improving handwritten mathematical expression recognition. In: Barney Smith, E.H., Pal, U. (eds.) ICDAR 2021. LNCS, vol. 12917, pp. 113–125. Springer, Cham (2021). https://doi.org/10.1007/978-3-030-86159-9_8

21. Qiao, Z., Zhou, Y., Yang, D., Zhou, Y., Wang, W.: Seed: semantics enhanced encoder-decoder framework for scene text recognition. In: Proceedings of the IEEE/CVF Conference on Computer Vision and Pattern Recognition, pp. 13528–13537 (2020)

22. Shi, B., Bai, X., Yao, C.: An end-to-end trainable neural network for image-based sequence recognition and its application to scene text recognition. IEEE Trans. Pattern Anal. Mach. Intell. **39**(11), 2298–2304 (2016)

23. Shi, B., Yang, M., Wang, X., Lyu, P., Yao, C., Bai, X.: ASTER: an attentional scene text recognizer with flexible rectification. IEEE Trans. Pattern Anal. Mach. Intell. **41**(9), 2035–2048 (2018)

24. Sutskever, I., Vinyals, O., Le, Q.V.: Sequence to sequence learning with neural networks (2014)

25. Truong, T.N., Nguyen, C.T., Phan, K.M., Nakagawa, M.: Improvement of end-to-end offline handwritten mathematical expression recognition by weakly supervised learning. In: 2020 17th International Conference on Frontiers in Handwriting Recognition (ICFHR), pp. 181–186. IEEE (2020)

26. Vuong, B.Q., He, Y., Hui, S.C.: Towards a web-based progressive handwriting recognition environment for mathematical problem solving. Expert Syst. Appl. **37**(1), 886–893 (2010)

27. Wang, J., Du, J., Zhang, J., Wang, Z.R.: Multi-modal attention network for handwritten mathematical expression recognition. In: 2019 International Conference on Document Analysis and Recognition (ICDAR), pp. 1181–1186. IEEE (2019)

28. Winkler, H.J.: Hmm-based handwritten symbol recognition using on-line and off-line features. In: 1996 IEEE International Conference on Acoustics, Speech, and Signal Processing Conference Proceedings, vol. 6, pp. 3438–3441. IEEE (1996)

29. Wu, J.W., Yin, F., Zhang, Y., Zhang, X.Y., Liu, C.L.: Graph-to-Graph: towards accurate and interpretable online handwritten mathematical expression recognition. In: Proceedings of the AAAI Conference on Artificial Intelligence, vol. 35, pp. 2925–2933 (2021)

30. Wu, J.-W., Yin, F., Zhang, Y.-M., Zhang, X.-Y., Liu, C.-L.: Image-to-markup generation via paired adversarial learning. In: Berlingerio, M., Bonchi, F., Gärtner, T., Hurley, N., Ifrim, G. (eds.) ECML PKDD 2018. LNCS (LNAI), vol. 11051, pp. 18–34. Springer, Cham (2019). https://doi.org/10.1007/978-3-030-10925-7_2

31. Wu, J.W., Yin, F., Zhang, Y.M., Zhang, X.Y., Liu, C.L.: Handwritten mathematical expression recognition via paired adversarial learning. Int. J. Comput. Vis. **128**(10), 2386–2401 (2020)

32. Yamamoto, R., Sako, S., Nishimoto, T., Sagayama, S.: On-line recognition of hand-written mathematical expressions based on stroke-based stochastic context-free grammar. In: Tenth International Workshop on Frontiers in Handwriting Recognition. Suvisoft (2006)

33. Yu, D., et al.: Towards accurate scene text recognition with semantic reasoning networks. In: Proceedings of the IEEE/CVF Conference on Computer Vision and Pattern Recognition, pp. 12113–12122 (2020)

34. Yuan, Y., et al.: Syntax-aware network for handwritten mathematical expression recognition. In: Proceedings of the IEEE/CVF Conference on Computer Vision and Pattern Recognition, pp. 4553–4562 (2022)

35. Yue, X., Kuang, Z., Lin, C., Sun, H., Zhang, W.: RobustScanner: dynamically enhancing positional clues for robust text recognition. In: Vedaldi, A., Bischof, H., Brox, T., Frahm, J.-M. (eds.) ECCV 2020. LNCS, vol. 12364, pp. 135–151. Springer, Cham (2020). https://doi.org/10.1007/978-3-030-58529-7_9

36. Zeiler, M.D.: ADADELTA: an adaptive learning rate method. arXiv preprint arXiv:1212.5701 (2012)

37. Zhang, J., Du, J., Dai, L.: Multi-scale attention with dense encoder for handwritten mathematical expression recognition. In: 2018 24th International Conference on Pattern Recognition (ICPR), pp. 2245–2250. IEEE (2018)

38. Zhang, J., Du, J., Dai, L.: Track, Attend, and Parse (TAP): an end-to-end framework for online handwritten mathematical expression recognition. IEEE Trans. Multimedia **21**(1), 221–233 (2018)

39. Zhang, J., Du, J., Yang, Y., Song, Y.Z., Wei, S., Dai, L.: A tree-structured decoder for image-to-markup generation. In: International Conference on Machine Learning, pp. 11076–11085. PMLR (2020)

40. Zhang, J., et al.: Watch, attend and parse: an end-to-end neural network based approach to handwritten mathematical expression recognition. Pattern Recogn. **71**, 196–206 (2017)

41. Zhang, Z., He, T., Zhang, H., Zhang, Z., Xie, J., Li, M.: Bag of freebies for training object detection neural networks. arXiv preprint arXiv:1902.04103 (2019)

42. Zhao, W., Gao, L., Yan, Z., Peng, S., Du, L., Zhang, Z.: Handwritten mathematical expression recognition with bidirectionally trained transformer. In: Lladós, J., Lopresti, D., Uchida, S. (eds.) ICDAR 2021. LNCS, vol. 12822, pp. 570–584. Springer, Cham (2021). https://doi.org/10.1007/978-3-030-86331-9_37

An Encoder-Decoder Method with Position-Aware for Printed Mathematical Expression Recognition

Jun Long[2], Quan Hong[1(✉)], and Liu Yang[1(✉)]

[1] The School of Computer Science and Engineering, Central South University, Changsha 410075, Hunan, China
{hongquan,yangliu}@csu.edu.cn
[2] Big Data Institute, Central South University, Changsha 410083, Hunan, China
jlong@csu.edu.cn

Abstract. Printed mathematical expression recognition is to transform printed mathematical formula image into LaTeX sequence. Recently, many methods based on deep learning have been proposed to solve this task. However, the positional relationship between mathematical symbols is often ignored or represented insufficient, leading to the loss of structural features of mathematical formulas. To overcome this challenge, we propose a position-aware encoder-decoder model for printed mathematical expression recognition. We design a two-dimensional position encoding algorithm based on sin/cos function to capture positional relationship between mathematical symbols. Meanwhile, we adopt a more advanced image feature extraction network. In decoder component, we use Bi-GRU as the translator, and add attention mechanism to make decoder focus on the important local information. We conduct experiments on the public dataset IM2LaTeX-100K, and the results show that our proposed approach is more excellent than the majority of advanced methods.

Keywords: Deep learning · Mathematical expression recognition · Encoder-Decoder · Position encoding · Attention mechanism

1 Introduction

Mathematical formulas are widely used in online education system [9,14], scientific research literature, knowledge questions and answers [13] and other fields. Printed mathematical expression recognition(PMER) is to recognize printed mathematical expression image to the corresponding LaTeX sequences. So as to can provide more convenient services for above applications.

PMER is different from character recognition. The general steps of character recognition method first cuts a string into several independent characters, then extracts features of each character, and finally analyzes and classifies the extracted features to achieve character recognition. PMER is different, since mathematical formulas not only involve segmentation and recognition of mathematical characters, but also need to extract and analyze of two-dimensional structural features of mathematical formulas, it's crucial for the understand of formula semantic information. However, the complexity and diversity of the two-dimensional structure features contained in mathematical formulas are the biggest challenges for PMER.

The earliest solution for mathematical expression recognition can be traced back to 1967. Anderson et al. [1] proposed a formula structure analysis method based on syntax rules. With the development of this field, some other researchers propose to divide mathematical expression recognition into several sequential execution stages, among which the most famous multi-stage formula recognition system based on sequential execution is INFTY [20]. Recently, some researchers propose some deep learning methods to solve it. Deng et al. [4] of Harvard University proposed an end-to-end PEMR model called WYGIWYS, which is the beginning of implementing PEMR based on deep learning, afterwards some researchers put forward improved models based on WYGIWYS. However, most of these methods ignore positional relationship between mathematical symbols. For example, although these three mathematical formula xy, x_y and x^y are all composed of character x and y, the semantic information expressed by them is completely different due to the differently relative positions between x and y. Based on end-to-end neural network model structure, we propose a two-dimensional position-aware method for PMER task. The main contributions of this paper are as follows:

- We use a more advanced image feature extraction network, it can capture more abundant image feature information.
- We design a two-dimensional positional information awareness module based on sin/cos function for representing the positional relationship between mathematical symbols.
- We conducted comparison experiment, ablation experiment and hyper-parameter experiment on public dataset Im2LaTex-100K [4]. The experimental results show that our method is more excellent than most advanced models.

2   Related Work

In recent years, some researchers propose using deep learning algorithms to solve PMER, and it be regarded as an image to markup generation task [10,15]. Most of them use deep learning model which based on encoder-decoder structure. Starting with WYGIWYS [4] method, encoder component uses CNN as image feature extractor, and performs secondary processing on the extracted features, decoder component uses RNN to translate image features into LaTeX sequences. Wang et al. [23] proposed using DenseNet [8] as image feature extractor to improve the perception of formula image features. Yan et al. [27] proposed replacing RNN with CNN as decoder, and verified that this approach has a slight improvement. Peng et al. [18] proposed regarding mathematical formula as a graph structure, each mathematical symbol be regarded as a node in graph, and defined the corresponding rules for constructing edges between nodes. Some advanced neural network model has been proposed to solve handwritten mathematical expression recognition(HMER) [2,11,12,24,25,28]. These methods also bring some new solution ideas for PMER.

The positional relationships between symbols in formulas have a great significance for understanding the structural features of formulas. Most methods [4,5,29] choose to add a row encoder after image feature extractor based on RNN. Since RNN can extract order relationship between sequences, row encoder essentially incorporates the position information of symbols implicitly. Some methods [17,18] use sin/cos encoding function to encode position information of symbols, which is more accurate than RNN, so these methods can greatly improve the perception of positional information between symbols. But mathematical formula is a two-dimensional structure layout, it have horizontal and vertical positional relationships between mathematical symbols. However, few methods take into account this point.

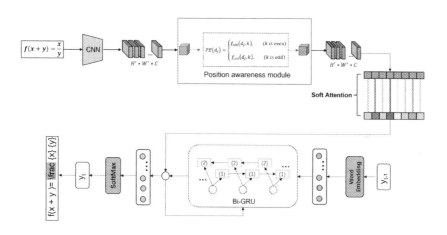

Fig. 1. The structure diagram of our PMER model with an encoder-decoder structure.

3 Proposed Method

We regard PMER task as an image to caption generation process, and we design the encoder-decoder structure as the overall PMER architecture. The input of our model is a gray image containing a mathematical formula, and the output is a LaTeX sequence corresponding to the formula. Figure 1 is the structure diagram of our PMER model.

3.1 Image Feature Extraction Network: ResNeXt

We use the recently proposed convolution neural network ResNeXt [26] based on ResNet [7] as our image feature extractor. Specially, this CNN structure retains residual block structure of ResNet, and introduces the idea of group convolution of Inception [21]. Compared the traditional CNN such as Vgg [19], ResNeXt can alleviate the problem of network degradation and gradient disappearance with the increase of network depth. Meanwhile, it also possess the ability of split-transform-merge to enhance the representation of image features. Figure 2 shows structural design of ResNeXt block.

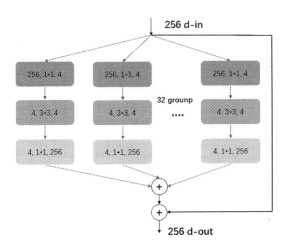

Fig. 2. The structure diagram of ResNeXt-block

We design a network structure based on ResNeXt with a depth of 26. The specific structure design and parameters are shown in Table 1.

Table 1. The structure of ResNeXt-26 with 32 groups for each ResNeXt block

layer-name	Input	Conv class	Conv kernel	output
conv-1	$H * W * 1$	Conv	k:7*7,s:2	$\frac{H}{2} * \frac{W}{2} * 64$
maxPool	$\frac{H}{2} * \frac{W}{2} * 64$	maxpool	k:3*3,s:2	$\frac{H}{4} * \frac{W}{4} * 64$
conv2-x	$\frac{H}{4} * \frac{W}{4} * 64$	Res-Block*2	k:3*3,s:2	$\frac{H}{8} * \frac{W}{8} * 128$
conv3-x	$\frac{H}{8} * \frac{W}{8} * 128$	Res-Block*2	k:3*3,s:2	$\frac{H}{16} * \frac{W}{16} * 256$
conv4-x	$\frac{H}{16} * \frac{W}{16} * 256$	Res-Block*2	k:3*3,s:2	$\frac{H}{32} * \frac{W}{32} * 512$
conv5-x	$\frac{H}{32} * \frac{W}{32} * 512$	Res-Block*2	k:3*3,s:1	$\frac{H}{32} * \frac{W}{32} * 512$

3.2   Position-Aware Module

The position information of formula symbols is one of a key point to understand formula structure. So, we design a position-aware module to capture the position information of each symbol in the mathematical formula. We use position encoding algorithm in Transformer [22] to represent the order and relative distance between words in a text sequence. Since text sequence is one-dimensional, it only contains positional relationship from left to right. But mathematical formula is two-dimensional, and the positional information of each mathematical symbol is composed of horizontal position and vertical position, as shown in Fig. 3. Therefore, we modify the position encoding algorithm in Transformer, and propose a two-dimensional position encoding algorithm.

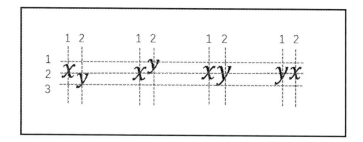

Fig. 3. Horizontal position and vertical position in mathematical formula.

The encoding object of position encoding algorithm is image feature matrix V_1(the size of V_1 is $H' * W' * C$). For each vector v_i in V_1, four kinds of values are calculated in the following order: the distance from left d_1, right d_2, top d_3, down d_4. And then a set of sin/cos coded values are generated for each above four values. If the subscript of current component in the vector is even, sin encoding function is used. Otherwise, cos encoding function is used. The specific calculation formula of $PE(v_i)$ is as follows: ($D = C$, $k \in [0, D/4]$.)

$$PE(d_j, 2k) = \sin \frac{d_j}{10000^{\frac{4k}{D}}} \tag{1}$$

$$PE(d_j, 2k + 1) = \cos \frac{d_j}{10000^{\frac{4k}{D}}} \tag{2}$$

$$PE(v_i) = Concat(PE(d_1), PE(d_2), PE(d_3), PE(d_4)) \tag{3}$$

We concatenate all the calculated position vectors to obtain a PE matrix V_{PE} with dimension $H' * W' * C$.

$$V_{PE} = Concat(PE(v_1), PE(v_2), ..., PE(v_{H'*W'})) \tag{4}$$

Finally, we add V_1 with position matrix V_{PE}, and obtain a new feature map V_2 with mathematical symbol position information.

$$V_2 = V_1 + V_{PE} \qquad (5)$$

Figure 4 is the algorithm process. In order to verify the gain effect of position-aware module to our model, we conduct some ablation experiments on it. See Sect. 4 for specific experimental results.

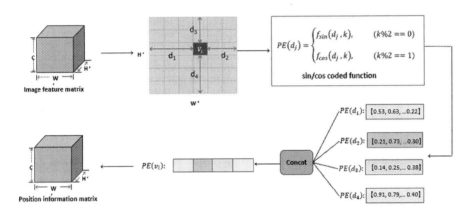

Fig. 4. The process of position encoding algorithm.

3.3   Decoder: Bi-GRU with Attention Mechanism

The decoder component mainly undertakes the translator task, which is used to translate context vector C generated by encoder into LaTeX sequences. Following the design idea of Deng [4], we use RNN as the core module of decoder component, to generate a LaTeX token at each stage. Then the LaTeX sequence generated after all prediction stages over is the translation result of decoder. Specially, we select bidirectional gate recurrent unit (Bi-GRU) as the encoder implementation instead of the standard RNN, since GRU can effectively compensate for the long term memory shortcomings of RNN model. Furthermore, Bi-GRU can work in a two-way parallel way that from front to back and from back to front at same time, it can better make full use of the context semantic information. Figure 5 is the structure diagram of our decoder.

In order to enhance translation performance, we learn from previous researchers [3,16] and introduce attention mechanism into decoder. Attention mechanism is inspired by human attention phenomenon. When analyzing and understanding information sources, human beings can always quickly focus on the important part of information, while ignoring or reducing the attention to unimportant information. Attention mechanism is widely used in the field of

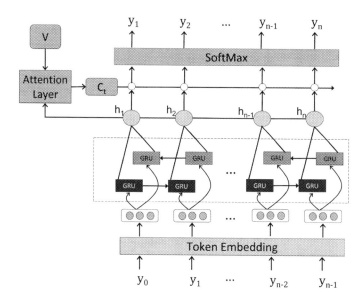

Fig. 5. The structure diagram of the decoder, It is composed of Bi-GRU, attention layer and token embedding layer.

visual processing. Therefore, we use attention mechanism to enhance the translation ability of our model. In each translation stage of decoder, the decoder only focus on the image features that are most closely related to the current stage when translating the context vector C. That is, the decoder can filter out a lot of useless redundant information without extracting every part of the image features in vector C, so the translation performance of the decoder is improved. Figure 6 shows the schematic diagram of attention phenomenon in PMER task.

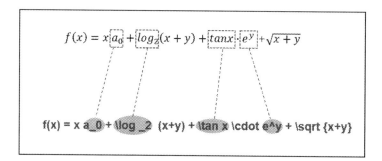

Fig. 6. Schematic Diagram of Attention Mechanism in PMER Task.

At each translation stage of decoder, it is necessary to pass in the context vector C_t containing image features provided by encoder. Before attention mechanism is introduced, C_t is a invariant matrix, while the situation changed after attention mechanism introduced. Decoder component dynamically calculates the context vector C_t for each stage, and in each translation stage, attention layer will calculate a corresponding weight value for each vector in V_2. The weight value represents attention value of decoder for the information represented by this vector. Dynamic weighting process is the embodiment of the attention mechanism implemented in our decoder. The calculation process of weight value $e_{t,i}$ is as follows:

$$\overline{h}_z = concat(\overrightarrow{h}_{t-1}, \overleftarrow{h}_{t-1}) \tag{6}$$

$$e_{t,i} = SoftMax(\beta^{\mathrm{T}} tanh(W_x \overline{h}_z + W_y v_i)) \tag{7}$$

After obtaining weight value $e_{t,i}$, the context vector C_t at stage t can be summed by multiplying e_{ti} with the corresponding vector v_i in V_2.

$$C_t = \sum_{i=1}^{H'*W'} e_{ti} v_i \tag{8}$$

Then a new vector K_t is gained by merging C_t with hidden layer vector h_t.

$$h_t = BiGRU(h_{t-1}, [w_{t-1}, K_{t-1}]) \tag{9}$$

$$K_t = tanh(W_c[h_t, C_t]) \tag{10}$$

Finally, the decoder output the probability distribution of LaTeX token y_t by:

$$y_t = SoftMax(W^k K_t) \tag{11}$$

3.4 Loss Function

We use cross entropy as our loss function, and y_i represents the ground truth of the i-th sample and $\hat{y}_i$ is the prediction result of i-th sample. M is the total number of samples in the dataset. The formula of Loss function as in Eq. (12):

$$Loss = -\frac{1}{M} (\sum_{i=1}^{M} p(y_i) \cdot log(p(\hat{y}_i))) \tag{12}$$

4 Experiments

4.1 Datasets

We conduct our experiments on the standard dataset IM2LaTeX-100K [4], which is widely used in printed mathematical expression recognition field. This dataset consists of about 100K Image-LaTeX matching pairs. The input of each sample is a gray-scale image containing a printed mathematical formula, and the output is a LaTeX string corresponding to the mathematical formula. We randomly divide the dataset into training set, test set and verification set according to the ratio of 8:1:1.

4.2 Parameter Setting and Experimental Environment

We use small batch random gradient descent algorithm to train our model. In the translation stage, in order to achieve global optimization of LaTex sequence generation, we adopted beam search strategy [6]. We also conduct many iterative experiments and based on the feedback of experimental results, constantly optimize the hyper-parameters values involved in the model, so as to achieve best recognition effect. See Table 2 for details of super parameter settings:

Table 2. Details of hyper-parameter setting in the experiment

Parameter name	description	Value
CNN_{layer}	Layer number of ResNeXt	26
$ResNeXt_{group}$	Group number of ResNeXt-block	32
v_d	Dimension of image feature matrix	512
h_d	Hidden layer dimension of Bi-GRU	512
L_{Bi-GRU}	Layer number of Bi-GRU	3
W_d	Dimension of LaTeX token embedding	100
B_{num}	Beam search scope	4
$Batch_{size}$	Batch size	6
I_{num}	Number of iterations of the experiment	40

At the same time, all of our experiments were conducted under the ubuntu-20.04 operating system, using Python as programming language and TensorFlow-1.12 as machine learning implementation framework. GPU used in the experiment is RTX-2080 12 GB.

4.3 Baseline

(1) **Infty**, **WYGIWYS**, **Coarse-to-fine** have been introduced in Sect. 2. (2) **DenseNet** [23]: This method uses DenseNet which with dense connection design as the image feature extractor (3) **ConvMath** [27]: In the design of this model, the author uses CNN to replace RNN as the translator of decoder components. (4) **Dual attention** [29]: An encoder-decoder neural network model with dual attention mechanism. (5) **Transformer** [17]: Based on the decoder-encoder network structure, this method introduces

Transformer to process the feature information extracted by the encoder. (6) **Im2Latex-GNN** [18]: This is a novel method, which proposes to convert mathematical formulas into graph structure. Each mathematical formula symbol is regarded as a vertex node of the graph, and the mathematical formula features are extracted based on graph structures. Then, the feature is fused with image feature extracted by the traditional encoder and transmitted to decoder for translation.

4.4 Experimental Results

We introduc four evaluation metrics during the experiment including BLEU-1, BLEU-4, edit distance, exact match. The first three metrics are often used in NLP tasks to judge the closeness of two sentences. We use them to compare the similarity between model translation results and ground truth. Exact match is used to compare the similarity of two images. It can be used to judge the matching degree between the formula image generated by the LaTeX sequence which output from the model with the formula image given by the dataset. The four evaluation metrics adopt a percentage system. The higher score represent the better result of the model.

We also reproduced other eight baseline methods. The specific effect is shown in Table 3. Among the nine methods, our method gains 94.39 in BLEU-1 and 92.31 in BLEU-4, and performs the best in these two metrics. Our method also gains 91.39 in edit dist and 82.07 in exact match, and performs the first best and second best respectively.

Table 3. Experimental results of different methods on Im2LaTeX-100K

Method	BLEU-1	BLEU-4	Edit dist	Exact Match
Infty	72.39	66.65	53.82	–
WYGIWYS	89.75	87.73	87.60	79.88
Coarse-to-Fine	90.33	87.07	87.32	78.10
DenseNet	91.75	88.25	**91.57**	79.10
ConvMath	91.94	88.33	90.80	**83.41**
Dual attention	92.15	88.42	88.57	79.81
Transformer	93.23	89.72	90.07	82.13
Im2Latex-GNN	93.88	90.19	–	81.82
Our Method	**94.39**	**92.31**	91.39	82.07

4.5 Ablation Study

In order to intuitively verify the effective of modules in our model, we conduct a series of ablation study. Our ablation experiments mainly consist of the following groups, and we consider each group as variant:

- **EDPA-pos:** This variant removes the position-aware module and directly transmits image feature matrix to decoder.
- **EDPA_{norm-pos}:** This variant uses a degenerate position encoding algorithm, with other parts of algorithm same as the original ones. The only difference is that it does not use sin/cos coding function, but directly uses the position value of each vector to fill each component in the vector.
- **EDPA-att:** This variant removes the attention module, and only use Bi-GRU for translation.
- **EDPA_{ResNet}:** This variant use ResNet as image feature extractor to replace ResNeXt.
- **EDPA_{Vgg}:** This variant uses Vgg network as image feature extractor to replace ResNeXt.
- **EDPA_{RNN}:** This variant uses standard RNN instead of Bi-GRU as the translator.
- **EDPA_{GRU}:** This variant uses standard GRU instead of Bi-GRU as the translator.

From the results in Table 4, we can learn that each module we designed has a certain impact on our model, of which attention mechanism module has the largest impact, followed by position-aware modules. These two modules improve performance of our model by about 3-4%. Compared with ResNet and Vgg, ResNeXt can improve the performance of our model by about 1%. Lastly, Bi-GRU is also better than standard GRU and standard RNN, and it raises the performance of our model about 1%.

4.6 Hyper-parameter Experiment

In order to make our model achieve the best recognition results, we conducted hyper-parameter experiments on the most important parameters in our model, and constantly adjust and optimiz the parameter values according to the feedback of the experimental results. We mainly carry out hyper-parameter experiments on the depth of ResNeXt and the layer number of Bi-GRU.

As for the depth of ResNeXt, we increased network depth with 9 as interval, and finally end with 98 layers. The experimental results are shown in Table 5. In the experiment, we find that the model performs the best when ResNeXt depth set to 26 and the performance decreases slightly as the depth exceed 44.

Table 4. Ablation study results of our model on IM2LaTex-100K

Method	BLEU-1	BLEU-4	Edit dist	Exact Match
EDPA	**94.39**	**92.31**	**91.39**	**82.07**
EDPA-pos	91.16	89.51	88.36	80.15
EDPA norm-pos	91.44	89.72	88.43	80.13
EDPA-att	90.87	88.92	87.75	78.89
EDPA ResNet	94.22	91.69	91.23	81.96
EDPA Vgg	93.67	91.56	90.89	81.77
EDPA RNN	93.51	91.34	90.33	80.54
EDPA GRU	93.94	91.42	90.55	80.96

Table 5. Experimental results of different depth of ResNeXt

depth	BLEU-1	BLEU-4	Edit dist	Exact Match
17-layer	93.79	91.32	90.43	81.56
26-layer	**94.39**	**92.31**	**91.39**	**82.07**
35-layer	93.33	90.52	89.89	80.91
44-layer	92.74	90.15	89.51	80.43

The layer number of Bi-GRU has a significant to the translation results of decoder. Thus, we design different Bi-GRU that stacked with 1, 2, 3 and 4 layers for comparative experiments. The experimental results are shown in Table 6. We find that the score of our model reaches the best when the layer of Bi-GRU is set to 3.

Table 6. Experimental results of different layers in Bi-GRU

Depth	BLEU-1	BLEU-4	Edit dist	Exact Match
1-layer	94.29	91.72	90.73	**82.53**
2-layer	93.88	91.51	90.89	81.78
3-layer	**94.39**	**92.31**	**91.39**	82.07
4-layer	93.74	91.43	90.81	81.63

5 Case Study

We show several test cases in the production environment to analyze the specific recognition effect of our model, as shown in Fig. 7. There are five cases in total, including two good cases and three bad cases. The error areas are marked in orange. We make a theoretical conjecture on the cause of these errors. It is found that our method mainly occurs recognition errors in the following two cases:

- The recognition errors caused by the confusion of Greek letters and English letters, such as "μ" and "u", "α" and "a".
- When containing multiple sets of brackets or complex structures, it may omit mathematical symbols or the right bracket early.

We analyze and concluded reasons for above problems may be caused by the following factors:

- The mathematical symbol features extracted by the image feature extractor are still not accurate, or the multiple iterations in the translation process lead to the loss of mathematical symbols features.
- Attention mechanism and position coding algorithm are not oriented at the level of mathematical symbols, and they cannot accurately divide which vectors correspond to which mathematical symbols, which may lead to confusion in the translation process.

Mathematical formula image	Recognize Result of Our Model
$I_{J-1} \equiv \int [\frac{du}{u}]^{J-1} 1 = i^{J-1} \int_{\theta}^{*} d^{J-1}\theta = \frac{(\pi i)^{J-1}}{(J-1)!}$	I_{J-1} \equiv \int [\frac {du} {u}]^{J-1} 1 = i^{J-1} \int_{0}^{\pi} d^{J-1} \theta = \frac {(\pi i)^{J-1}} {(J-1)!}
$K = \frac{3\alpha^2}{2} e^{-\alpha\phi_0} \sqrt{\frac{3\Lambda}{9\alpha^2 - 32\kappa_5^2}}$	K= \frac {3\alpha^{2}} {2} e^{-\alpha \phi_0} \sqrt {\frac {3 \Lambda} {9 \alpha^{2} - 32 \kappa_{5}^{2}}}
$-p_a = \frac{p_A^2}{4p_v} + \frac{1}{4\alpha'p_v}(\hat{N}_L + \hat{N}_R) + f(\hat{J}_L)$	-p_u=\frac {p_A^2} {4 p_v}+\frac {1} {4α^{\prime} p_v} \hat {N}_L+\hat {N}_R)+f(\hat {J}_L) \alpha
$\eta = \gamma(\mu^*, \theta) = \frac{\varepsilon\theta^2}{72}$	\eta =\gamma (u^{*}, \theta)=-\frac{\varepsilon \theta^{2}} {72} \mu
$ds^2 = V(r)(d\chi + nA)^2 + \frac{dr^2}{V(r)}$	d s^{2}=V(r)(d \chi+n A}^{2}+\frac{d r^{2}} {V(r)} }

Fig. 7. The case study results. We cover the area where recognition errors occur in orange, error correction in blue, and the added correction in dark gold. (Color figure online)

6   Conclusion

In this paper, we propose an improved end-to-end method for printed mathematical formulas image recognition. We use a more advanced image feature extractor to enrich the extracted image features. In order to better realize the representation of the positional relationship between mathematical symbols, we recast the position encoding algorithm proposed in Transformer and design a two-dimensional position coding algorithm. In the decoder part, we use Bi-GRU as core translation module, and we introduce attention mechanism to make translation to focus on the important local information of the feature map generated by encoder. We carry out experiments on the public dataset Im2LaTeX-100K, the experimental results show that our method is superior to most of the current advanced methods.

Acknowledgements. This work is being supported by the National Natural Science Foundation of China under the Grant No. U2003208 and No. 62172451, and supported by Open Research Projects of Zhejiang Lab under the Grant No. 2022KG0AB01.

References

1. Anderson, R.H.: Syntax-directed recognition of hand-printed two-dimensional mathematics. In: Symposium on Interactive Systems for Experimental Applied Mathematics: Proceedings of the Association for Computing Machinery Inc., Symposium, pp. 436–459 (1967)
2. Bian, X., Qin, B., Xin, X., Li, J., Su, X., Wang, Y.: Handwritten mathematical expression recognition via attention aggregation based bi-directional mutual learning. arXiv preprint arXiv:2112.03603 (2021)
3. Cho, K., et al.: Learning phrase representations using RNN encoder-decoder for statistical machine translation. arXiv preprint arXiv:1406.1078 (2014)
4. Deng, Y., Kanervisto, A., Rush, A.M.: What you get is what you see: a visual markup decompiler (2016)
5. Deng, Y., Kanervisto, A., Ling, J., Rush, A.M.: Image-to-markup generation with coarse-to-fine attention (2017)
6. Graves, A.: Sequence transduction with recurrent neural networks. arXiv preprint arXiv:1211.3711 (2012)
7. He, K., Zhang, X., Ren, S., Sun, J.: Deep residual learning for image recognition. IEEE (2016)
8. Huang, G., Liu, Z., Laurens, V., Weinberger, K.Q.: Densely connected convolutional networks. IEEE Computer Society (2016)
9. Huang, Z., et al.: Question difficulty prediction for reading problems in standard tests. In: Thirty-First AAAI Conference on Artificial Intelligence (2017)
10. Karpathy, A., Fei-Fei, L.: Deep visual-semantic alignments for generating image descriptions. In: Proceedings of the IEEE Conference on Computer Vision and Pattern Recognition, pp. 3128–3137 (2015)
11. Li, B., Yuan, Y., Liang, D., Liu, X., Ji, Z., Bai, J., Liu, W., Bai, X.: When counting meets hmer: Counting-aware network for handwritten mathematical expression recognition. In: European Conference on Computer Vision, pp. 197–214. Springer, Cham (2022). https://doi.org/10.1007/978-3-031-19815-1_12

12. Li, Z., Jin, L., Lai, S., Zhu, Y.: Improving attention-based handwritten mathematical expression recognition with scale augmentation and drop attention. In: 2020 17th International Conference on Frontiers in Handwriting Recognition (ICFHR), pp. 175–180. IEEE (2020)
13. Liu, Q., Huang, Z., Huang, Z., Liu, C., Chen, E., Su, Y., Hu, G.: Finding similar exercises in online education systems. In: Proceedings of the 24th ACM SIGKDD International Conference on Knowledge Discovery & Data Mining, pp. 1821–1830 (2018)
14. Liu, Q., et al.: Fuzzy cognitive diagnosis for modelling examinee performance. ACM Trans. Intell. Syst. Technol. (TIST) $9(4)$, 1–26 (2018)
15. Lu, J., Yang, J., Batra, D., Parikh, D.: Neural baby talk. In: Proceedings of the IEEE Conference on Computer Vision and Pattern Recognition, pp. 7219–7228 (2018)
16. Luong, M.T., Pham, H., Manning, C.D.: Effective approaches to attention-based neural machine translation. arXiv preprint arXiv:1508.04025 (2015)
17. Pang, N., Yang, C., Zhu, X., Li, J., Yin, X.C.: Global context-based network with transformer for image2latex. In: 2020 25th International Conference on Pattern Recognition (ICPR), pp. 4650–4656. IEEE (2021)
18. Peng, S., Gao, L., Yuan, K., Tang, Z.: Image to latex with graph neural network for mathematical formula recognition. In: International Conference on Document Analysis and Recognition, pp. 648–663. Springer (2021)
19. Simonyan, K., Zisserman, A.: Very deep convolutional networks for large-scale image recognition. arXiv preprint arXiv:1409.1556 (2014)
20. Suzuki, M., Tamari, F., Kanahori, T.: Infty – an integrated OCR system for (2003)
21. Szegedy, C., Vanhoucke, V., Ioffe, S., Shlens, J., Wojna, Z.: Rethinking the inception architecture for computer vision. In: Proceedings of the IEEE Conference on Computer Vision and Pattern Recognition, pp. 2818–2826 (2016)
22. Vaswani, A., et al.: Attention is all you need. Advances in neural information processing systems 30 (2017)
23. Wang, J., Sun, Y., Wang, S.: Image to latex with densenet encoder and joint attention. Procedia Comput. Sci. **147**, 374–380 (2019)
24. Wu, J.W., Yin, F., Zhang, Y.M., Zhang, X.Y., Liu, C.L.: Handwritten mathematical expression recognition via paired adversarial learning. Int. J. Comput. Vis., 1–16 (2020)
25. Wu, J.W., Yin, F., Zhang, Y.M., Zhang, X.Y., Liu, C.L.: Graph-to-graph: towards accurate and interpretable online handwritten mathematical expression recognition. In: Proceedings of the AAAI Conference on Artificial Intelligence, vol. 35, pp. 2925–2933 (2021)
26. Xie, S., Girshick, R., Dollár, P., Tu, Z., He, K.: Aggregated residual transformations for deep neural networks. In: Proceedings of the IEEE Conference on Computer Vision and Pattern Recognition, pp. 1492–1500 (2017)
27. Yan, Z., Zhang, X., Gao, L., Yuan, K., Tang, Z.: Convmath: a convolutional sequence network for mathematical expression recognition. In: 2020 25th International Conference on Pattern Recognition (ICPR), pp. 4566–4572. IEEE (2021)
28. Zhang, J., Du, J., Dai, L.: Track, attend, and parse (tap): an end-to-end framework for online handwritten mathematical expression recognition. IEEE Trans. Multimedia $21(1)$, 221–233 (2018)
29. Zhang, W., Bai, Z., Zhu, Y.: An improved approach based on CNN-RNNs for mathematical expression recognition. In: Proceedings of the 2019 4th International Conference on Multimedia Systems and Signal Processing, pp. 57–61 (2019)

Graphics Recognition 1: (Music, Chem, Charts, Figures)

A Holistic Approach for Aligned Music and Lyrics Transcription

Juan C. Martinez-Sevilla[✉], Antonio Rios-Vila[✉],
Francisco J. Castellanos[✉], and Jorge Calvo-Zaragoza[✉]

U. I. for Computing Research, University of Alicante, Alicante, Spain
{jcmartinez.sevilla,antonio.rios,francisco.castellanos,jorge.calvo}@ua.es

Abstract. In this paper, we present the Aligned Music Notation and Lyrics Transcription (AMNLT) challenge, whose goal is to retrieve the content from document images of vocal music. This new research area arises from the need to automatically transcribe notes and lyrics from music scores and align both sources of information conveniently. Although existing methods are able to deal with music notation and text, they work without providing their proper alignment, which is crucial to actually retrieve the content of the piece of vocal music. To overcome this challenge, we consider holistic neural approaches that transcribe music and text in one step, along with an encoding that implicitly aligns the sources of information. The methodology is evaluated on a benchmark specifically designed for AMNLT. The results report that existing methods can obtain high-quality text and music transcriptions, but posterior alignment errors are inevitably found. However, our formulation achieves relative improvements of over 80% in the metric that considers both transcription and alignment. We hope that this work will establish itself as a future reference for further research on AMNLT.

Keywords: Aligned Music Notation and Lyrics Transcription · Optical Music Recognition · Music Notation Recognition · Lyrics Recognition

1 Introduction

Music is an important vehicle for cultural transmission, which is a key element as regards understanding the social, cultural, and artistic trends of each period of history. For centuries, music has been shared and preserved by two traditions: aural transmission and written documents, usually referred to as music scores. Many of these works exist in the form of unpublished manuscripts and, therefore, are in danger of being lost through time.

A significant effort has been made in recent decades to digitize these documents by means of scanners for their storage and distribution [19]. In order to make the content depicted in the documents truly accessible, it is necessary for the images to be transcribed to a structured digital format that makes it possible to encode the information (notes, marks, tonality, etc.) of the score [12,15].

There exist several ways to encode music content such as the MusicXML or Music Encoding Initiative (MEI) formats [9,13]. Encoding music information in

G. A. Fink et al. (Eds.): ICDAR 2023, LNCS 14187, pp. 185–201, 2023.
https://doi.org/10.1007/978-3-031-41676-7_11

such structured formats not only allows the initial objective of preserving this heritage to be attained, but also makes it possible to perform other interesting tasks, such as large-scale computational music analysis, exact search, retrieval by content, or conversion between different music notations. However, the manual encoding of music is extremely time-consuming. The research field known as Optical Music Recognition (OMR), which involves automatically detecting and storing the content of a score from a scanned image [3], is postulated as an important resource to mitigate the disadvantages of manual transcription of music scores.

In addition to the classical multi-stage pipeline in OMR, extensively reviewed in the work of Rebelo et al. [18], modern advances in Deep Learning have diversified how OMR is approached. The *holistic* paradigm—also referred to as *end-to-end* formulation—consists of addressing the recognition problem in a single step, which helps to learn contextual relationships that eventually improve the quality of the result. This paradigm has been dominating the current state of the art in other applications such as text, speech, or mathematical formula recognition [6,7,28]. It has been also applied in the case of OMR, for which recent literature shows many efforts for music-notation recognition in an end-to-end way [1,2,5,22,24].

Thanks to the aforementioned advances, OMR is shifting towards new projects that were recently considered out of reach. Specifically, in this work, we pay special attention to documents of vocal music. These are of special interest since vocal music is the one with the longest tradition. In these documents, both the music and the text that accompanies it—commonly referred to as *lyrics*—are equally relevant. These sources of information can be considered as different and complementary modalities of the same piece of information. Nevertheless, OMR systems have traditionally ignored the recognition of the lyrics that may appear in the score, since this is a task that is commonly addressed by Optical Character Recognition (OCR) or Handwritten Text Recognition (HTR) systems.

Developing an automatic transcription system for vocal music documents opens up new scientific challenges. A vocal music score could be fully recognized using independent OMR algorithms for the music notation and OCR/HTR for the lyrics of the piece. However, this does not solve the underlying challenge: the alignment between the text, typically at the syllable level, and the music, typically at the note level. That is, retrieving the sequence of notes and the sequence of syllables independently does not solve the problem of retrieving which note or notes indicate how to sing each syllable. The proper alignment between the notes and the syllables is the key to further musicological analysis and search in vocal music heritage.

Here, we deal with the task of Aligned Music Notation and Lyrics Transcription (AMNLT) employing a holistic approach that retrieves both sources of information in a coordinated manner, forcing the alignment to be retrieved while notes and syllables are transcribed. We believe that addressing this challenge in a holistic way provides additional benefits such as learning the underlying model that relates notes and syllables which, despite being difficult to model entirely, must be able to guide the coordinated recognition process.

As a summary, we list the scientific contributions of this work in the following points:

- To formulate for the first time the task of automatic recognition of pieces of vocal music from written sources, which include both musical notes and text. As previously stated, the key is to produce not only the correct transcription of the musical notes and lyrics syllables but also to estimate their correct alignment (AMNLT), as this is essential for the correct preservation of the original source.
- To propose a first deep learning approach that directly transcribes and aligns the musical notes of a staff and the lyrics that accompany it. To do this, we resort to a reinterpretation of holistic formulations widely used in staff-level OMR and line-level HTR, and their recent adaptation to multi-line HTR. We also compare different alternatives (fully-convolutional, recurrent, and attention-based architectures) to implement the considered approach.
- To synthesize and release a dataset for the AMNLT task, in order to provide a benchmark for further research in vocal music transcription with enough data to train modern deep learning approaches.

Our experiments will empirically demonstrate that although AMNLT is related to widely studied graphic recognition endeavors (i.e., OCR and OMR), the real challenge is the precise alignment between the two information modalities. In this sense, our holistic approach stands out specifically in the metrics related to the alignment, in spite of losing some accuracy with respect to the case of the two modalities addressed independently.

The rest of the paper is structured as it follows: in Sect. 2 we first describe what a vocal music document looks like, how it can be modeled to be processed by deep learning algorithms, and the automatic generation mechanism that will be used as a benchmark in this work; in Sect. 3, we formulate the AMNLT problem and develop the solution proposed in this work based on an end-to-end approach with neural networks; the experiments conducted are presented in Sect. 4, along with the results and analysis that can be drawn from them; finally, we conclude the paper in Sect. 5, while pointing out some interesting avenues for future work.

2   Data

The main challenge in AMNLT, as introduced in Sect. 1, relies on retrieving multiple sequences, i.e. music and text, with their alignment information. To our best knowledge, neither OMR nor OCR/HTR fields have addressed similar problems. We, therefore, provide some fundamentals of the computational representations that are considered for the AMNLT task and describe the generated data to be used in the present work.

2.1   Score Representation

The key aspect of this work is to address the issue of aligning music and lyrics sequences while performing the transcription process. Figure 1 depicts some of the challenges that can be found in such a case. At first glance, we observe that lyrics are located before and after their associated note. For example, syllable *"Le-"* lasts more pixel frames than its related note D♯. Due to the spacing and size of some elements, it can also happen that characters from several syllables are also present in other music symbol areas, as it happens with syllable *"ter"* aligned with the note A but appearing in frames of prior note—i.e. C♯. These features make the transcription plus alignment process challenging even for a human expert.

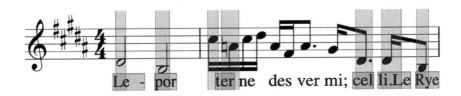

Fig. 1. Example of the alignment challenge between a music-symbol sequence and a text sequence in a chanted melody fragment. Red boxes refer in a pixel-wise viewpoint to the size inside the image of a music symbol, whereas blue boxes are the same for the syllables. (Color figure online)

For the sake of the success AMNLT task, we need an encoding format capable of staging the alignment aspects of the scores. In this work, we resort to the *Humdrum* KERN music encoding format [14]. This music notation format is one of the most frequently used representations in computational music analysis. Its features include a simple vocabulary and easy-to-parse file structure, which is very convenient for end-to-end OMR applications. Moreover, KERN files are compatible with dedicated music software [17,21] and can be automatically converted to other music encodings.

A KERN file is basically a sequence of lines. Each line is, in turn, another sequence of columns or *spines* that are separated by a *tab* character. Each column contains an instruction.[1] When interpreting a KERN file, all spines are simultaneously read, which is an advantageous feature for the goal of this work.

In our case, the KERN standard specifies a spine for each type of annotation. In our case, the ***kern* and the ***text* spines are referred to music and lyrics, respectively. Figure 2 depicts an example of how this information is encoded.

---

[1] The KERN standard defines an instruction as anything that belongs to the music score, such as the creation or ending of spines, or the encoding of musical symbols such as clefs, key signatures, meter, bar lines, notes, or lyrics, to name a few.

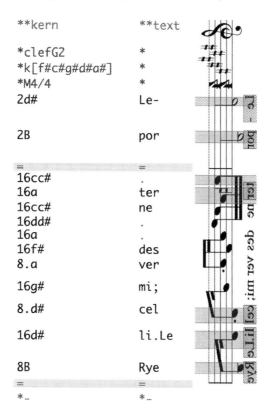

Fig. 2. Monophonic chanted melody fragment with its alignment information in Humdrum KERN format.

2.2 Score Generator

Finding fully annotated-musical excerpts to assess the performance of holistic AMNLT models remains a difficult task, as there are no well-known datasets. On the one hand, fully-annotated datasets, such as DeepScores [25] and the MUSCIMA dataset [11] are not conceived for end-to-end OMR, and do not provide ground-truth in a digital music standard format. On the other hand, well-known datasets conceived for end-to-end OMR [4,5,16,20] do not provide lyrics or alignment information.

The current situation—along with the requirement of large datasets to train, validate and test deep learning-based models—demanded the design of a fully-annotated vocal music generator.[2]

---

[2] The implementation of this generator is provided in the repository given in Sect. 4.

A score generator can be simplified as a concatenation of rhythms that fulfill a specific space or measure—constant along the piece—having all of them a pitch associated. The presented strategy divides the generator into four tasks, all of them crucial to come about with a realistic sample:

- **Measure, key, and tesiture selection.**[3] from dictionaries designed by a musical expert—given the musical significance between key and tesiture—it selects a triplet with a measure (e.g., 4/4, 3/2, 7/8), a key (e.g., C Major, F Major, E♭ Major) and a tesiture (from A– to ccc and others, in KERN format).
- **Rhythm cells generation**: a dictionary of rhythmic cells associated with the measure chosen and its space, in conjunction with random patterns, brings about a concatenation of rhythms. This step is key in rule-based algorithms given that previous ones failed in generating "good" or "credible" rhythmic patterns bringing unreliable music as result.
- **Pitches generation**: having a key (e.g. C Major), the generator randomly assigns pitches, from an associated key note list, to the already existing rhythms. To deliver a genuine melody we narrow the possible pitches from the last one created, avoiding excessive *jumps* in tesiture.
- **Lyrics generation**: using a sentence generator and a hyphenation tool, syllables are aligned with some musical notes. With this approach, we ensure realistic syllables and an actual text sequence. To improve performance in future scenarios, multiple languages have been added to the tool, having in mind the existing undocumented repertoire (e.g., Latin, Italian, German, Spanish, and English). Most of them share their character list considering numbers, punctuation marks, and accents.

In addition to this generator pipeline, several filters are applied (e.g., rotations, chops. swirls, noise, or waving) to create a version of the corpus that resembles a real optical capturing process. An example is depicted in Fig. 3.

(a) Ideal score   (b) Camera-based score

Fig. 3. Example of a generated input for AMNLT. The first example depicts ideal conditions, while the second one resembles the conditions of an imperfect capture process.

---

[3] Understanding tesiture as the range of notes in a score. In instruments, from the lowest to the highest note possible to play.

3 Holistic Aligned Music and Lyrics Transcription

In this section, we describe how state-of-the-art OMR methods deal with transcription. Then, we analyze why this formulation cannot be directly applied to AMNLT and propose an adaptation to this challenge while preserving the holistic nature of these methods.

3.1 Single-Task End-to-end Transcription

State-of-the-art OMR seeks the most probable symbolic representation $\hat{\mathbf{s}}$ for each staff-section image x:

$$\hat{\mathbf{s}} = \arg \max_{\mathbf{s} \in \Sigma_a} P(\mathbf{s} \mid x). \tag{1}$$

Neural networks approximate this probability by training with the Connectionist Temporal Classification (CTC) loss [10]. This alignment-free expectation-maximization method forces the network to maximize the sum of the probability of all the possible alignments between a ground-truth sequence s and the input source x. Since our input is an image, we treat x as a sequence of frame columns.

The output of the network consists of a *posteriorgram*, which contains the probabilities of all the tokens within the vocabulary Σ_a, along with an additional "blank" label (ϵ) that indicates time-step separations. The output vocabulary of the network becomes $\Sigma'_a = \Sigma_a \cup \{\epsilon\}$. At prediction, a *greedy* decoding is usually considered, for which the most probable token per frame is chosen. Then, consecutive frames with the same token are merged and, finally, the "blank" label is removed.

The presented formulation assumes the transcription task as a sequence retrieval problem, and the output of the network is, therefore, always a token (i.e., character or music-notation symbol) sequence. A sequence of this nature is obtained from an image by converting the image domain $\mathcal{R}^{h \times w \times c}$—which is defined by its width w, height h, and the number of channels c—into a sequence domain $\mathcal{R}^{l, \Sigma'_a}$, where l stands for the output sequence length and Σ'_a is the aforementioned music notation vocabulary. CTC-based methods typically define a reshape function $h : \mathcal{R}^{h \times w \times c} \rightarrow \mathcal{R}^{l, \Sigma'_a}$ that vertically collapses the feature map.[4] Thus, symbols can be read from left to right, and frame columns always contain information about a single symbol, the one that is currently being read.

3.2 The Challenge of AMNLT

The presented methodology in Sect. 3.1 works for single-staff transcription (or analogously, single-line OCR/HTR), as the premise of having a single symbol information per frame holds. However, when dealing with transcription tasks that require multiple pieces of information to be given at once, this formulation becomes inadequate.

---

[4] The specific implementation depends on the method itself.

In the case of AMNLT, the issue is observed through the example given in Fig. 1: multiple notes and lyrics are found in the same columns, unsatisfying the mentioned premise. This becomes challenging for state-of-the-art technologies, as the number of available frames to predict each piece of information becomes tight. In fact, this issue becomes relevant when approaches aim to perform aligned transcription, as both elements—music and lyrics—have to be predicted in very close *timesteps*. This fact hinders the performance of current systems in the form that the correlation between the representation of the image and the ground-truth information becomes challenging.

To handle this issue, a *divide and conquer* approach may be taken, where separate OMR and OCR/HTR systems handle their respective content, and alignment is retrieved through postprocessing operations. Although this approach is feasible—since state-of-the-art staff-level and line-level OMR and OCR/HTR report good accuracy—the alignment of sequences is non-trivial, as will be demonstrated in our experiments. In contrast to adopting a heuristic method to combine the results of note and text transcription, in this work we resort to a method that deals with AMNLT in one single step.

Upon closely studying the KERN format, it can be noted that each text line represents a specific *timestep* in the music score. That is, all the information present in a KERN line happens at the same time, as they belong to different spines. The reading order—from a graphical perspective—of these documents is from top to bottom and left to right, which matches the left-to-right reading of the music score. It is, therefore, possible to obtain a graphic alignment between them by rotating the source image 90° clockwise and flipping it horizontally. When applying this transformation—as illustrated in Fig. 2—it is observed that the same reading order is performed for both sources.

By following this reinterpretation, we obtain both a document and a ground-truth text representation that are read like a text paragraph. This consequently makes it possible to propose solutions based on segmentation-free multi-line transcription approaches.

3.3   Simultaneous Transcription via Score Unfolding

Segmentation-free multi-line document transcription is a text methodology whose objective is to transcribe document images that contain more than one line without the need to perform any previous line detection processes. We have taken inspiration from the document unfolding methods of this field, where the model learns to unfold text lines in order for them to be read sequentially [8,27].

A graphic visualization of our methodology is depicted in Fig. 4. Here, rather than concatenating frame-wise elements along the height axis (h) during the vertical collapse, we reshape the feature map by concatenating all of its rows (w)—in the same way as [8]—to subsequently obtain a $(c, h \times w)$ sequence, in which c is the number of filters used by the convolutional layers of the model. The input image of the system—which is a music staff (with lyrics) rotated 90° and flipped horizontally—is therefore reshaped as a sequence where music notes can be read. As AMNLT can be seen as a multi-task challenge, the network output vocabulary is a combination of both the KERN music notation Σ_k, the union of all the languages covered by our generator character sets Σ_t and the blank CTC token ϵ, forming $\Sigma' = \Sigma_k \cup \Sigma_t \cup \{\epsilon\}$.

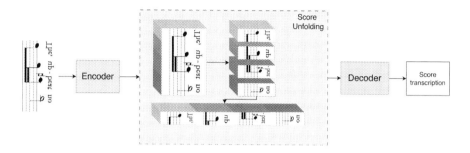

Fig. 4. Graphical scheme of the considered neural method to address AMNLT.

3.4   Considered Implementations

In this paper, we are adapting state-of-the-art OMR models to perform AMNLT. All the implemented solutions contain a fully convolutional block, which acts as an encoder of the input image features. This network is composed of stacked convolutional layers, which end up producing a feature map of size $(h/32, w/4, c, b)$, h and w being the height and the width of the input image, c the filters in the last convolutional layer, and b the batch size. Then, the following decoding architectures are proposed:

Recurrent Neural Network. We follow the implementation of the original CRNN-CTC staff transcription model from [5], where the reshaped feature map is fed into a Bidirectional LSTM (BLSTM) and linearly projected onto the notation dictionary. Specifically, we implemented a BLSTM with 512 units.

The Transformer. As observed in the reshaping step, the model has to process long sequences in one step, which can have a negative impact on the performance of RNNs. For this reason, we consider an alternative encoder based on the Transformer [26] (referred to as CNNT). In particular, we implemented one encoder layer with an embedding size of 512, a feed-forward dimension of 1 024, and 8 attention heads. We implemented two versions of this model, one with the standard one-dimensional Positional Encoding and another with 2D Positional Encoding, which has proven to retrieve good results in multi-line transcription tasks [23].

Sequence-Processing-Free Module. As mentioned previously, the proposed methodology with which to transcribe music and lyrics is based on analogous works for multi-line transcription in the HTR field [8, 27]. These works are based on convolutional-only architectures—in which no sequence processing decoders are implemented, as the solution lies in preserving the prediction space in two dimensions, and applying backpropagation directly to the feature map retrieved before being reshaped. In order to carry out our study on the architecture, we implemented an encoder-only network. As it is based only on fully convolutional layers, it will be referred to as FCN in the results section.

Baseline Implementations. Since there is no AMNLT state-of-the-art reference method to compare our results, we resorted to implementing a baseline approach based on the *divide and conquer* strategy mentioned in Sect. 3.2. Specifically, we implemented two CRNN models, one for OMR and the other for OCR, to deal with the music-notation and text modalities separately. We consider two experiments for this baseline:

- *Plain*: in this case, both OMR and OCR methods are provided with cropped samples of the music staff and lyrics line, respectively. Then, the model must predict the raw music and text sequences, which are eventually merged. Since there is no alignment information in this scenario, the elements of both hypotheses are paired one to one according to the order in which they were predicted. To balance the size of the two sequences, we use the null KERN token "." to pad the shortest sequence.
- *With alignment information*: in an alignment information-based scenario, both models are fed with the complete staff image—including both music and lyrics—and predict the sequences with alignment information, i.e. the null KERN token ".". As in the previous scenario, both retrieved music and text sequences are merged to create the output KERN file.

4 Experiments

In this section, we present the experimental setup, by including the considered metrics and a description of the experiments involved in this work.[5]

4.1 Metrics

A sequence-based metric—henceforth Music Error Rate (MER)—is typically considered for assessing the performance of the OMR transcription. This metric, as well as its analog versions in text recognition such as Character Error Rate

---

[5] The data and source code required to replicate the experiments can be found in https://github.com/antoniorv6/icdar-2023-amnlt.

(CER) and Word Error Rate (WER), is based on the normalized mean edit distance between a hypothesis sequence $\hat{s}$ and a reference one s in the form of:

$$\mathcal{E}(\hat{S}, S) = \frac{\sum_{i=0}^{n} d(\mathbf{s}_i, \hat{\mathbf{s}}_i)}{\sum_{i=0}^{n} |\mathbf{s}_i|} \tag{2}$$

where $\hat{S}$ is the hypotheses set, S stands for the ground-truth set, $d(\cdot, \cdot)$ represents the edit distance between the tokens of each paired hypothesis and ground-truth sequences $(\mathbf{s}_i, \hat{\mathbf{s}}_i)$ and $|\mathbf{s}_i|$ is the number of tokens of the reference sequence.

Since MER is widely used for music-transcription evaluations, we considered this metric to determine the quality of the retrieved music notation. In text recognition, CER and WER represent the standard evaluation metrics; however, it should be noted that lyrics do not follow the word structure that can be found in common lines and paragraphs. Lyrics are typically divided into syllables and aligned with the music notes with which they have to be sung. Therefore, we consider the Syllable Error Rate (SER), which performs the evaluation at the syllable level, instead of using WER.

Another consideration is that our proposal provides KERN-format information with data of different natures: music and text. Moreover, these sequences are aligned with each other, in the way that each music symbol is paired to a syllable, even if it were empty—represented in KERN as ".". In other words, we need to consider a metric to assess the alignment capability of the considered approach. To evaluate this, we make use of the Kern Line Error Rate (KLER), which measures the error at the combined note-syllable level (which is encoded in a single line in the KERN format). As shown in the example of Fig. 2, KERN files contain the music and text information organized in columns built by means of indents, whose lines represent sequential time steps. Given the lines from a two-column KERN file with music and text information in the form <music><tab><text>, KLER computes the number of structural errors of the hypothesis with respect to the expected lines by means of $\mathcal{E}(\hat{S}, S)$. In short, the idea behind this metric is to check whether the predicted KERN data follows the expected structure and whether the note-syllable pairing is accurate.

4.2 Data Partitions

Data generated for the proposal is separated into two different scenarios: "Ideal" images and "Camera-based" images. The former refers to synthetically rendered scores without any distortion, while the latter refers to its distorted version, as aforementioned in Sect. 2.2. Each of these scenarios contains three partitions at a file level with sizes of 10 000, 1 000, and 5 000 samples for the train, validation, and test sets, respectively.

4.3 Results

Table 1 presents the results over the test partition obtained with the proposed experimental scheme, in terms of MER, CER, SER, and KLER.

Attending to the baseline reported values, similar recognition rates are observed in the considered scenarios. More precisely, the MER column reports 0.4% and 1.9% in the scenarios with ideal images and camera-based ones, respectively. Concerning the lyrics, CER and SER report the same results for both scenarios (1.2% and 5.2% for each metric, respectively). This fact demonstrates the adequacy of using a single-task end-to-end transcription method for the independent modalities. However, according to the KLER metric, this formulation is not proper for alignment purposes, since they yield values of 42.0% for ideal images and 42.9% for the camera-based scenario.

Focusing on our holistic single-step procedures, MER, CER and SER metrics degrade compared with the baseline methods, having the best results in ideal images with the CNNT PE 2D and the CRNN architectures in the camera scenario. However, KLER values are drastically improved when they are set side by side with the baseline ones, decreasing from 42% to 6.7% in the non-distorted images and from 42.9% to 8.6% in Camera. Such a fact proves the goodness of the considered holistic approaches in order to deal with AMNLT. Despite losing precision in the independent sequence transcriptions (i.e. MER, CER and SER metrics), the alignment process is more correctly performed. Specifically, the best holistic approach reports a 84% of relative KLER improvement with respect to the best baseline case in the ideal scenario, while increasing up to 80% for the camera one.

In addition to the previous analysis, while all the presented approaches prove to be successful—except for the CNN—for the posed task, it is observed that CNNT with 2D positional encoding, due to Transformer characteristics, yields more competitive results when graphic consistency is present, with a KLER result of 6.7%. The results also suggest that CNNT PE 1D works better than the 2D version for distorted or inconsistent inputs according to the KLER metric, but the MER, CER and SER are detrimental. In the camera scenario, the CRNN yields the best KLER result, with a value of 8.6%.

4.4   Discussion

From the experiments reported above, we observed certain phenomena that deserve further discussion. Some of these are depicted in Fig. 5, where representations of ground-truth and prediction KERN files with their rendered content are provided. Annotations representing different types of errors (i.e., the most common ones) can be found: red circles (key and accidentals), pink boxes (alignment), and blue boxes (text transcription). Figures refer to samples 4909.krn—MER 5.9%, CER 4.6%, SER 20%, KLER 11.3%—and 748.krn—MER 11.1%, CER 13.9%, SER 45.5%, KLER 17.8%—from the test partition.

In Fig. 5a, a two-measure fragment in **C**—equivalent to 4/4—is displayed. As described, the blue box present represents a text issue when trying to predict word *"beat"*, obtaining *"beal"*. Similar errors appear in Fig. 5b, where regardless of the perfect alignment with the music, some characters are transcribed incorrectly.

Table 1. Results for two scenarios: *Ideal images* stands for the case with perfect image conditions; *Camera-based images* represents the case in which the images have been distorted to simulate a camera-capturing process. The figures have been measured in terms of MER (%) for music, CER (%) and SER (%) for text, and KLER (%) for assessing the accuracy of the alignment. Bold figures represent the best values for each metric per scenario, while underlined figures point out the best results among our proposals.

Scenario	Music	Lyrics		Alignment
Method	MER	CER	SER	KLER
Scenario I: Ideal images				
Baseline				
Plain	**0.4**	**1.2**	**5.2**	42.0
With alignment information	15.8	49.1	81.2	54.8
Ours				
CNN	15.6	19.5	53.6	21.4
CRNN	5.2	10.8	30.9	9.7
CNNT PE 1D	5.2	11.6	33.9	10.7
CNNT PE 2D	<u>3.1</u>	<u>7.8</u>	<u>22.4</u>	**<u>6.7</u>**
Scenario II: Camera-based images				
Baseline				
Plain	**1.9**	**1.2**	**5.2**	42.9
With alignment information	29.1	19.0	57.1	67.9
Ours				
CNN	13.0	31.7	90.5	25.5
CRNN	<u>4.8</u>	<u>8.1</u>	25.1	**<u>8.6</u>**
CNNT PE 1D	5.5	17.2	44.6	13.5
CNNT PE 2D	5.5	13.6	<u>24.6</u>	16.2

Attending to red circles in Fig. 5b, it is necessary to clarify that the key—represented as *k[b-e-a-d-g-]* in this example—prevents the need for writing every single note of the piece with its accidental. As a result, notes—graphically identical—are annotated differently (i.e., when it renders, it omits the ♭ or ♯ inherited from the key). Red circles indicate this context knowledge problem.

Finally, pink boxes specify where there have been alignment problems and text transcription errors. In Fig. 5a, a syllable—"ver"—was predicted one *timestep* earlier than expected. This can also be noticed in Fig. 5b. The underlying problems when transcribing simultaneously lyrics and music can be visually seen in these examples.

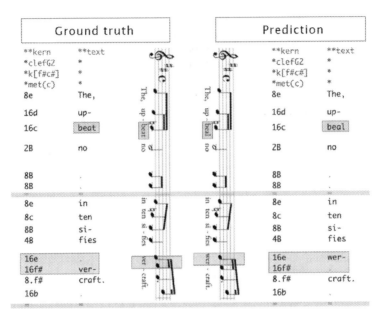

(a) Prediction result of 4909.krn from the test partition.

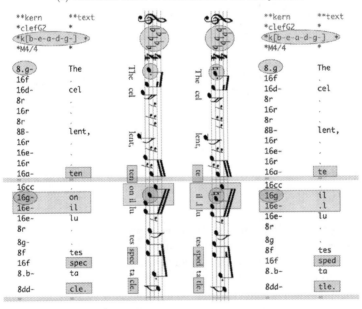

(b) Prediction result of 748.krn from the test partition.

Fig. 5. Representative errors in AMNLT in ideal conditions. Predictions are obtained from the CNNT PE 2D model. Errors are highlighted in different colors according to their nature. (Color figure online)

It is also well noticed that even with an evaluation process done with what we consider pessimistic metrics (i.e., CER and SER), to our best knowledge predictions obtained as showed and described in this section are rather reliable for automatic score transcription.

5   Conclusions

In this work, we define a novel research line derived from the Optical Music Recognition (OMR) field, denoted as Aligned Music Notation and Lyrics Transcription (AMNLT): the challenge of automatically obtaining the aligned music and lyrics transcription from music score images. In addition, a generated dataset is provided for future benchmarking on AMNLT.

This work also presents the first neural formulation to simultaneously obtain music and text transcriptions and their respective alignments directly in one step. Our proposal adapts holistic approaches widely used in staff-level OMR and line-level and multi-line HTR to perform this process. For the specific implementation of this approach, we considered fully-convolutional, recurrent, and Transformer architectures, in order to provide a robust baseline for AMNLT.

Our experiments were divided into two scenarios: one using ideal images without any distortion and the other using distorted images to simulate a camera-capturing process. The results demonstrate that our AMNLT proposal drastically reduces the alignment errors compared to an OMR+OCR baseline. improvement of 84% and 80% in the metric that considers both transcription and alignment (KLER), for the scenarios with ideal and camera-based images, respectively.

In future work, we plan to extend this approach to other handwritten corpora, where alignment is likely to be even more complex. In addition, according to the results obtained in our experiments, a promising approach for the near future would be to leverage the quality of the independent transcriptions (music and text), which are more accurate if we ignore the alignment, as an auxiliary input of the holistic AMNLT model.

Acknowledgments. This paper is part of REPERTORIUM project, funded by the European Union's Horizon Europe programme under grant agreement No 101095065. The second author is supported by grant ACIF/2021/356 from the "Programa I+D+i de la Generalitat Valenciana".

References

1. Alfaro-Contreras, M., Ríos-Vila, A., Valero-Mas, J.J., Iñesta, J.M., Calvo-Zaragoza, J.: Decoupling music notation to improve end-to-end optical music recognition. Pattern Recogn. Lett. **158**, 157–163 (2022)
2. Baró, A., Riba, P., Calvo-Zaragoza, J., Fornés, A.: From optical music recognition to handwritten music recognition: a baseline. Pattern Recogn. Lett. **123**, 1–8 (2019)
3. Calvo-Zaragoza, J., Hajič Jr., J., Pacha, A.: Understanding optical music recognition. ACM Comput. Surv. **53**(4) (2020)

4. Calvo-Zaragoza, J., Rizo, D.: Camera-primus: neural end-to-end optical music recognition on realistic monophonic scores. In: Proceedings of the 19th International Society for Music Information Retrieval Conference, ISMIR 2018, Paris, France, 23–27 September 2018, pp. 248–255 (2018)
5. Calvo-Zaragoza, J., Toselli, A.H., Vidal, E.: Handwritten music recognition for mensural notation with convolutional recurrent neural networks. Pattern Recogn. Lett. **128**, 115–121 (2019)
6. Chiu, C.C., et al.: State-of-the-art speech recognition with sequence-to-sequence models. In: 2018 IEEE International Conference on Acoustics, Speech and Signal Processing (ICASSP), pp. 4774–4778 (2018)
7. Chowdhury, A., Vig, L.: An efficient end-to-end neural model for handwritten text recognition. In: British Machine Vision Conference 2018, BMVC 2018, Newcastle, UK, 3–6 September 2018. p. 202. BMVA Press (2018)
8. Coquenet, D., Chatelain, C., Paquet, T.: SPAN: a simple predict & align network for handwritten paragraph recognition. In: Lladós, J., Lopresti, D., Uchida, S. (eds.) ICDAR 2021. LNCS, vol. 12823, pp. 70–84. Springer, Cham (2021). https://doi.org/10.1007/978-3-030-86334-0_5
9. Good, M., et al.: MusicXML: an internet-friendly format for sheet music. In: Xml Conference and Expo, pp. 03–04. Citeseer (2001)
10. Graves, A., Fernández, S., Gomez, F.J., Schmidhuber, J.: Connectionist temporal classification: labelling unsegmented sequence data with recurrent neural networks. In: Proceedings of the Twenty-Third International Conference on Machine Learning, (ICML 2006), Pittsburgh, Pennsylvania, USA, 25–29 June 2006, pp. 369–376 (2006)
11. Jan Hajič, J., Pecina, P.: The MUSCIMA++ dataset for handwritten optical music recognition. In: 14th International Conference on Document Analysis and Recognition, ICDAR 2017, Kyoto, Japan, 13–15 November 2017, pp. 39–46. Department of Computer Science and Intelligent Systems, Graduate School of Engineering, Osaka Prefecture University. IEEE Computer Society, New York, USA (2017)
12. Hankinson, A., Burgoyne, J.A., Vigliensoni, G., Fujinaga, I.: Creating a large-scale searchable digital collection from printed music materials. In: Proceedings of the 21st International Conference on World Wide Web, pp. 903–908. ACM (2012)
13. Hankinson, A., Roland, P., Fujinaga, I.: The music encoding initiative as a document-encoding framework. In: Klapuri, A., Leider, C. (eds.) Proceedings of the 12th International Society for Music Information Retrieval Conference, ISMIR 2011, Miami, Florida, USA, 24–28 October 2011, pp. 293–298. University of Miami (2011)
14. Huron, D.: Humdrum and kern: selective feature encoding BT - beyond MIDI: the handbook of musical codes. In: Beyond MIDI: The Handbook of Musical Codes, pp. 375–401. MIT Press, Cambridge, January 1997
15. Meredith, D. (ed.): Computational Music Analysis. Springer, Cham (2016). https://doi.org/10.1007/978-3-319-25931-4
16. Parada-Cabaleiro, E., Batliner, A., Schuller, B.W.: A diplomatic edition of Il Lauro Secco: ground truth for OMR of white mensural notation. In: Proceedings of the 20th Interntional Society for Music Information Retrieval Conference, pp. 557–564 (2019)
17. Pugin, L., Zitellini, R., Roland, P.: Verovio - a library for engraving MEI music notation into SVG. In: International Society for Music Information Retrieval, January 2014

18. Rebelo, A., Fujinaga, I., Paszkiewicz, F., Marcal, A.R., Guedes, C., Cardoso, J.S.: Optical music recognition: state-of-the-art and open issues. Int. J. of Multimedia Inf. Retrieval **1**(3), 173–190 (2012)

19. Riley, J., Fujinaga, I.: Recommended best practices for digital image capture of musical scores. OCLC Syst. Serv. **19**(2), 62–69 (2003)

20. Ros-Fábregas, E.: Codified Spanish music heritage through Verovio: the online platforms Fondo de Música Tradicional IMF-CSIC and books of hispanic polyphony IMF-CSIC. In: Proceedings of Music Encoding Conference (MEC), Alicante, Spain. Alicante, Spain, July 2021

21. Sapp, C.S.: Verovio humdrum viewer. In: Proceedings of Music Encoding Conference (MEC), Tours, France (2017)

22. Shi, B., Bai, X., Yao, C.: An end-to-end trainable neural network for image-based sequence recognition and its application to scene text recognition. IEEE Trans. Pattern Anal. Mach. Intell. **39**(11), 2298–2304 (2017)

23. Singh, S.S., Karayev, S.: Full page handwriting recognition via image to sequence extraction. In: Lladós, J., Lopresti, D., Uchida, S. (eds.) ICDAR 2021. LNCS, vol. 12823, pp. 55–69. Springer, Cham (2021). https://doi.org/10.1007/978-3-030-86334-0_4

24. Torras, P., Baró, A., Kang, L., Fornés, A.: On the integration of language models into sequence to sequence architectures for handwritten music recognition. In: Proceedings of the 22nd International Society for Music Information Retrieval Conference, ISMIR 2021, Online, 7–12 November 2021

25. Tuggener, L., Elezi, I., Schmidhuber, J., Pelillo, M., Stadelmann, T.: DeepScores-a dataset for segmentation, detection and classification of tiny objects. In: Proceedings of the 24th International Conference on Pattern Recognition, pp. 3704–3709 (2018)

26. Vaswani, A., et al.: Attention is all you need. In: Advances in Neural Information Processing Systems, vol. 30 (2017)

27. Yousef, M., Bishop, T.E.: Origaminet: weakly-supervised, segmentation-free, one-step, full page textrecognition by learning to unfold. In: The IEEE Conference on Computer Vision and Pattern Recognition (CVPR), June 2020

28. Zhang, J., et al.: Watch, attend and parse: an end-to-end neural network based approach to handwritten mathematical expression recognition. Pattern Recogn. **71**, 196–206 (2017)

A Multi-level Synthesis Strategy for Online Handwritten Chemical Equation Recognition

Haoyang Shen[1], Jinrong Li[2(✉)], Jianmin Lin[2], and Wei Wu[1(✉)]

[1] Xidian University, Xi'an, China
shenhaoyang@stu.xidian.edu.cn, wwu@xidian.edu.cn
[2] CVTE Research, Guangzhou, China
{lijingrong,linjianmin}@cvte.com

Abstract. Handwritten chemical equation recognition is an appealing task, but its development is hampered by the lack of publicly available datasets. To this end, we propose a multi-level synthesis strategy to synthesize the corresponding handwritten equations from LaTeX expressions and regard the chemical equation recognition as an image-to-markup task. In particular, our approach first decomposes the LaTeX expression into a symbol layout tree (SLT) and obtains different multi-level components in stages by traversing the SLT. Then, online isolated symbols are placed in appropriate locations consistent with handwritten habits through a baseline-based layout strategy. Furthermore, expression patterns are enhanced at the local, component, and global levels to increase the diversity of synthesized data. It is worth noting that our synthesis strategy is theoretically applicable to any LaTeX-based expression. We also collected a real dataset containing 1595 handwritten chemical equations, and the experimental results confirm that our proposed method can effectively improve the performance of handwritten chemical equation recognition systems. The dataset we generated will be released.

Keywords: Handwritten chemical equations · Synthesis strategy · LaTeX decomposition · Symbol layout tree

1 Introduction

As a form of representing the reaction relationships between chemical substances, chemical equations play an integral role in academic and educational scenarios. Traditional human-computer interaction when inputting chemical equations relies on the click-and-drag style [15], which has severely limited the possibility of developing the digital level of chemical equations. Therefore, as a more natural and efficient interaction method, handwritten chemical equation recognition (HCER) technology has attracted the attention of many researchers.

Chemical equations include both inorganic and organic chemical equations. Because of the complexity of the ring structure, the recognition of organic chemical equations usually requires compound partitioning and classification first

G. A. Fink et al. (Eds.): ICDAR 2023, LNCS 14187, pp. 202–217, 2023.
https://doi.org/10.1007/978-3-031-41676-7_12

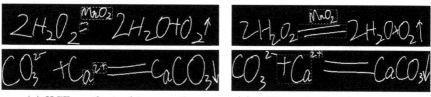

(a) HCE synthesized incorrectly.   (b) HCE synthesized by our strategy

Fig. 1. Deficiencies in directly applying synthesis methods of HMEs to generate hand-written chemical equations.

[6,21]. In this paper, we will focus on end-to-end handwritten inorganic chemical equation recognition.

Inorganic chemical equations consist mainly of numbers, letters, operators, and graphical symbols, which cause many challenges in the recognition of hand-written chemical equations, such as special 2-D structures, complex reaction condition symbols, variable handwritten styles, and so on [2,4,17]. Due to the strong structural similarity with mathematical expressions (MEs), some work [9,15] uses the seq2seq approach, which has proven effective in HMER on inorganic chemical equation recognition. For these methods based on deep learning, a massive and diverse dataset is necessary. However, to our knowledge, there is no publicly available dataset of online handwritten chemical equation (OHCE), which is an obstacle to the development of corresponding recognition tasks.

Given the high cost and inefficiency of data input and subsequent proofreading, manually synthesizing large quantities of HCEs is considered a better solution. However, in chemistry, there are few synthesis strategies for single molecular formulas and no mature work on handwritten chemical equations. Although there has been a lot of work on synthesizing HMEs with impressive results, they are difficult to use directly for generating handwritten chemical equations.

In most HME synthesis work, expressions are synthesized with symbols as the base unit. The position of the last symbol is only relevant to the previous one, which may lead to ambiguities in the layout of some chemical equations due to the writing habits of upper and lower case letters. On the other hand, in the writing habits of CEs, the length of "=" usually changes dynamically depending on the content of the reaction terms above or below it, which cannot be achieved in the synthesis strategy of HMEs. These situations arise because only the relative position of the previous symbol is considered in the layout, rather than considering the previous component as a unit. Figure 1(a) shows these possible layout deficiencies when synthesizing handwritten chemical equations only on the symbol level.

To solve the above problems, we consider the recognition task as an image2markup task and propose a multi-level synthesis strategy that can synthesize arbitrary inorganic HCEs from LaTeX expressions. In this work, we first transform the collected LaTeX expression into a symbol layout tree (SLT) based on positional relationships, in which the depth-first search (DFS) result of each node is a sub-string of the related LaTeX expression. Then, the concept of component is introduced to allow for multi-level trajectory synthesis, which can be seen as a group of connected nodes of the SLT. For example, as shown in Fig. 1(b), the sub-tree representing "2+" is an ionic component, and "+Ca" is a chemical component, so the ionic component can be placed correctly on the top right of the chemical component. Based on the SLT, the trajectory of one or more symbols is merged into different single-branch components with different positional relationships. To ensure that the non-horizontal components can be placed in the right place, we start by merging the non-horizontal components into multi-branch components and then placing all the multi-branch components horizontally to obtain the final handwritten chemical equation. In this process, we also design a baseline-based layout strategy based on chemical equation writing habits to reduce the impact caused by the irregular symbol position. Specifically, we divide the vertical space into three grids with four lines according to English writing habits. Then we place the upper and lower case symbols by aligning them between different baselines instead of simply aligning them centrally. In addition, to further enhance the expression patterns, we also distort each symbol obtained during the traversal process by random translation, scaling, and rotation models at three levels to generate more diverse samples. The experiment shows that the synthesized handwritten chemical equations can significantly improve the accuracy of the recognition system.

The main contributions of our work are summarized as follows:

- To the best of our knowledge, this is the first attempt to solve the HCER task by obtaining the corresponding LaTeX expressions end-to-end.
- A multi-level synthetic strategy is proposed to obtain HCEs from arbitrary LaTeX expressions. Specifically, benefiting from SLT, we use multi-level components as the basic unit for synthesizing HCEs, which can effectively avoid spatial location errors. And a new baseline-based layout strategy is adopted to make synthesized data more compatible with handwritten habits.
- Compared with only real data for training, the Expression Recognition Rates increased by 43.70% and the Word Error Rates decreased by 17.76% after using extra data synthesized by our proposed method.

The rest of this paper is organized as follows: Sect. 2 reviews the advanced methods of handwritten expressions synthesis. Section 3 describes our proposed synthesis strategy and the recognition method we used. The experimental results are given in Sect. 4, and the conclusions are given in Sect. 5.

2   Related Work

2.1   Data Generation in HCEs

As an effective method to improve the accuracy of model recognition, data generation and augmentation have been widely used in the field of optical character recognition (OCR) [1,8]. However, relatively little work has been done in the field of chemistry in terms of data synthesis, except for a small number of synthetic strategies for individual molecular formulas. Liu et al. [15] collected corpus data of 97 different chemical equations at the middle school and high school levels, and called about 200 teachers and students to write the individual molecular formulas in the equations. The formulas collected were then randomly rotated to obtain a more diverse dataset of handwritten molecular formulas, which is the earliest publicly available handwritten chemical formula data. However, this dataset is less diverse due to the lack of complete chemical equations and the lack of graphical symbols commonly used in chemical equations.

The earliest dataset we are aware of for chemical equations is from Microsoft Research Asia [6], a dataset of 35,932 manually written chemical equations containing 2,000 chemical expressions, 25% of which are organic expressions, but this dataset is not publicly available. Hagag et al. [9] extracted the corresponding chemical equations and individual molecular formulas from chemical equations commonly used in educational institutions, but the final sample images were also completely written manually, which is inefficient and costly to obtain, resulting in a small overall data volume. To the best of our knowledge, there is no well-formed work on the synthesis of handwritten chemical equations.

2.2   Data Generation in HMEs

In contrast to the generation of chemical equations, there have been many excellent works in the field of mathematical expression synthesis, most of which are based on the CROHME dataset [18]. Le et al. [11] decomposed the original HMEs into multiple sub-HMEs by grammar rules. However, each sub-HME has the same trajectory and writing style as the original HME, which may lead to overfitting due to the lack of local diversity [10]. Truong et al. [20] first decomposed LaTeX expressions in CROHME into a syntactic parse tree and then randomly exchanged the sub-components with approximate aspect ratios to generate new syntactically valid HMEs. However, this method of scaling and replacing sub-HMEs may lead to severe symbol distortion during subsequent rendering, and such substitutions are only based on finite expressions in CROHME.

Synthesizing HMEs from arbitrary LaTeX expressions can improve the spatial diversity of the data. MacLean et al. [16] first generated the random HME templates through a grammar model, then transcribed and annotated all the templates to obtain a large database of HMEs. However, manual transcription of expression templates in this work is labor- and material-intensive. Awal et al. [3] first obtained a corpus of LaTeX expression counterparts and then generated a database of handwritten mathematical expressions from the collected isolated symbols, but they only use one handwritten sample template for each symbol. Deng et al. [7] extracted a large corpus of LaTeX expressions from academic papers and then rendered the expressions using individual handwritten symbols in Detexify. Similar to this work is the HME synthesis method proposed by Khuong et al. [10]. However, both synthesis methods are based on the symbolic level in the layout process, which is different from our habit of writing expressions in blocks according to their spatial locations, so the synthesized handwritten expressions have a large gap with the real writing trajectory.

In addition to this, there are many other efforts in data augmentation. Li et al. [13] increased the training samples by first randomly scaling the formula to a scale while ensuring the aspect ratio, and then zero-filling it to a fixed size. However, this approach only provides diversity on a global scale. Le et al. [11] proposed a pattern generation strategy to increase the training samples by distorting the shape of HMEs through operations such as scaling, perspective, and rotation, but this kind of augmentation cannot produce various spatial relationships. Although the current HME synthesis strategies have a good effect on HMER [7,10], they do not consider the spatial positional relationships between English symbols, so they are difficult to extend to the field of chemistry.

3   The Proposed Approach

The task of synthesizing handwritten chemical equations is divided into three main stages: single symbol trajectory sampling, LaTeX expressions decomposing, and symbol layout. There are already many publicly available datasets that provide handwritten symbol trajectory data. In this section, we focus on how to better decompose the 2-D structure of expressions and a more rational layout strategy. The specific process of our synthesis strategy is illustrated in Fig. 2. In addition to this, we also introduce the end-to-end recognition model we used in the experimental phase in this section.

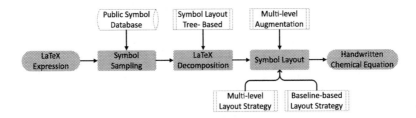

Fig. 2. Pipeline of the proposed synthesis strategy.

3.1   LaTeX Decomposition

LaTeX expression trajectories can be described at stroke level, symbol level, and expression level, respectively, most commonly based on expression level. However, handwritten expressions have a special 2-D structure and such positional relationships between symbols cannot be explicitly expressed in 1-D LaTeX expressions. Considering that subsequent layout operations require accurate spatial structure information, we chose to decompose LaTeX into a symbol layout tree (SLT) instead of a symbol operator tree before synthesis [25], which represents the symbol position on the baseline.

We decompose chemical equations according to symbol operators by the method proposed by Zanibbi et al. [22], and the symbol layout tree can be obtained through the official toolkit provided by CROHME. Firstly, the positional relationships present in chemical equations are classified into the following five types: above, below, sup, sub, and right. Then, in the process of decomposition, the first symbol in the LaTeX expression is regarded as the root node of this symbol layout tree, and adjacent symbols are decomposed into different branches of the parent node based on their positional relationships. Each node in the SLT stores the ground truth of the symbol and its spatial location relative to the parent node, with the leaf node representing the end of the decomposition at the current spatial position.

As an example, for the chemical equation "CO_{3}^{2-}+Ca^{2+} = CaCO_{ 3}\downarrow", its symbol layout tree obtained by the above decomposition strategy is shown in Fig. 3, in where each node represents the ground truth of the corresponding symbol, and the labels between the nodes represent the position of the child nodes relative to their parent node. Specifically, The first symbol "C" on the baseline is the root node of the SLT, and "Right" represents the branch of "O" is on the same baseline as "C". For the symbol "O", "Sub" represents the spatial location of the branch where "3" belongs is a subscript, same as "Sup" represents the spatial location of the branch where "2" belongs is a superscript, and "Right" represents the symbol "+" is on the same baseline as the component composed of "C" and "O". By traversing the symbol layout tree by Depth First Search (DFS), we can obtain the complete chemical equation $CO_3^{2-} + Ca^{2+} = CaCO_3 \downarrow$. In addition, we can also obtain the sub-strings of the chemical equation based on the tree structure by traversing any non-leaf node.

3.2   Symbol Layout

After collecting a sufficient corpus of chemical expressions, the generation of the corresponding HCEs requires us to scale and move each symbol in expressions to its corresponding position using a proper strategy, which is called the symbol layout of HCEs, and the trajectory coordinates of each symbol are randomly extracted from three different public datasets. The overall layout of symbols is based on Cartesian coordinates, where the upper left corner is defined as the origin of the coordinate system, the horizontal coordinate is oriented to the right, and the vertical coordinate is oriented downward.

Multi-level Layout Strategy. In the field of HMEs synthesis, most layout strategies are performed with the symbol as the basic unit, in which the scale size and layout position of the present symbol are determined according to its previous symbol. However, in chemical equations, such a strategy may cause ambiguities in spatial positions due to the writing habits of upper and lower case letters. As shown

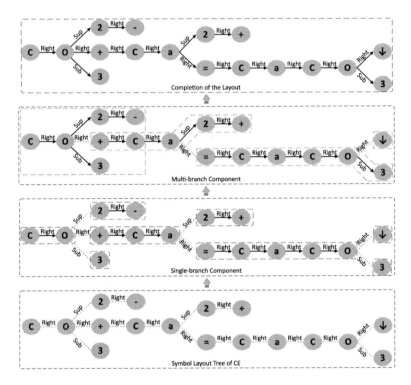

Fig. 3. SLT of chemical equations and multi-level layout steps based on the SLT. The different colored boxes represent the different components at each stage, and the branches with only leaf nodes represent components with an isolated symbol. The dashed grey line indicates that the two branches have not been merged. (Color figure online)

in Fig. 4(a), for the ionic symbol such as "Ca^{2+}", if only use the previous symbol "a" to determine the layout of "2", the final position of "2+" will be very close to the writing baseline, which may lead the system to recognize the positional relationship as "Right". Therefore, we need to consider the preceding "+Ca" as a unit to determine the layout of "2+" (see Fig. 4(b)). Also, for some of the more complex spatial structures, the size of the symbol needs to be determined according to its multiple symbols in different positions (as the symbol "=" shown in Fig. 1(b)), so it is essential to consider all symbols within the same positional relationship as a component to improve the regularity of synthesis.

Using the decomposition method in Sect. 3.1, we can decompose an arbitrary LaTeX expression into a symbol layout tree, whose nodes store the ground truth and relative positional relationships of individual symbols. Since SLT is obtained by parsing the positional relationships between symbols, if the number of children in any parent node is greater than one, it means that there is a different spatial layout of subsequent symbols, and this kind of node is defined as multi-branch nodes. Based on this characteristic, instead of using only the symbol-level

(a) HCE synthesized with symbols as the basic unit.

(b) HCE synthesized with components as the basic unit.

Fig. 4. Different handwritten chemical equations generated from the symbol-level and the multi-level respectively, and the different colored boxes represent different levels of symbols or components (Color figure online)

layout as in previous work, we propose a multi-level strategy for synthesizing handwritten chemical equations by first merging symbols into components at different scales and then using components as the basic unit for synthesis.

As shown in Fig. 3, starting with the first symbol of each branch, all symbols on each branch are merged into a component until reaching a multi-branch node or leaf node, and such a component is called a single-branch component (green boxes). Considering that components in the horizontal direction are the last to be written during the writing process, then we merge all the non-horizontal sub-trees into a multi-branch component (orange boxes), according to the relationships of the components to their parent components. Subsequently, during the layout process, we place all the multi-branch components horizontally to obtain the final handwritten chemical equation. This layout method takes more account of real handwriting habits, paying more attention to the position of the different components, effectively avoiding potential ambiguities in the synthesis process and making the synthesized data more standardized.

Baseline-Based Layout Strategy. The baseline in the symbol layout tree represents the horizontal writing line in the writing process. Symbols on the baseline are usually of the same height and size. However, chemical equations are mainly composed of English symbols, so even if the upper and lower case letters are in a horizontal relationship, there will still be variations in size and relative position (see Fig. 5(a)). To avoid ambiguities caused by such writing habits and to reflect the position of symbols more accurately, we have redefined the baseline in SLT based on the 4-line-3-grid layout in English writing and proposed a new baseline-based layout strategy that divides the horizontal space position into three grids through four lines. All uppercase letters and numbers occupy the top two grids, while lowercase letters and some other special symbols occupy different grids according to their shape (see Fig. 5(b)). In the layout process, symbols obtained by the traversal are first normalized to the appropriate size according to the size of the grids they occupy, and divided into components according to the SLT. The components are then scaled and shifted according to the multi-level layout strategy. The baseline-based layout strategy makes the spatial relationships between symbols in synthetic HCE more obvious, and because this constraint is more compatible with the writing conventions of English symbols, it works better for chemical equations.

(a) HCE obtained by only using the central writing line for alignment in the layout process.

(b) HCE obtained by using four parallel baselines for alignment in the layout process.

Fig. 5. The different handwritten molecular formulas obtained by our proposed baseline-based approach and using only the center line for alignment.

3.3   Data Augmentation

Because of the limited amount of trajectory data for individual symbols, in addition to enhancing the 2-D structure by defining arbitrary LaTeX expressions, another point to consider is how to increase the shape diversity of the symbols. At the layout stage of generating a new handwritten expression, we first adopt a certain range of random scaling, translation, and rotation as local distortions at the symbol level and component level, respectively, and then adopt random rotation and scaling as global distortions at the whole expression level. The overall process of our data augmentation strategy is shown in Fig. 6.

We use both equal scaling and random stretching when scaling at the symbol level to enrich the symbol shape as much as possible. At the same time, because adjacent symbols are generally tilted in the same direction in real writing, we obtain random rotation angles in the same rotation direction when rotating symbols in the same component. All of the above distortion operations are performed based on matrix operations on trajectory coordinates. For offline handwritten chemical equations, they can be obtained by rendering the corresponding trajectory data.

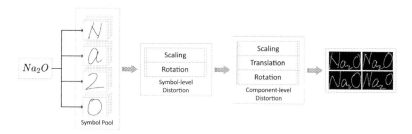

Fig. 6. The overall process of our data augmentation strategy, the symbol pool is composed of different trajectories for each symbol, from which it is randomly extracted during the synthesis process.

3.4 HCE Recognition Method

A typical end-to-end recognition network system consists of an encoder and a decoder. The encoder is usually a convolutional neural network, which is used to extract features from the input image and send them to the decoder, and the decoder is usually a recurrent neural network with GRU or LSTM. At moment t, the decoder computes the next output y_t by using the context vector c_t, the hidden state h_{t-1} of the previous moment, and the current embedding x_t, and finally obtains the recognition results in sequence, i.e. the LaTeX expression corresponding to the formula.

We chose the Counting-Aware Network (CAN) [12] for our experiments, which obtained an accuracy of 57% on the CROHME dataset in HMER. CAN is in a similar structure to the traditional end-to-end recognition model, with the addition of a weakly supervised counting module, which is used to alleviate overfitting and underfitting of symbols at the global level. Its overall structure is shown in Fig. 7, which is composed of three modules: backbone, multi-scale counting module (MSCM), and counting-combined attentional decoder (CCAD). The backbone of the CAN is densenet [23], which extracts the corresponding features and sends the feature map to MSCM and CCAD. The MSCM consists of two parallel branches with two different sizes of convolutional kernels to efficiently handle symbol changes at various scales and finally obtain the counting vector $V \in R^{1*N}$, where N is the number of symbol classes.

In most end-to-end HMER methods, the output is calculated using the context vector c_t, hidden state h_{t-1}, and embedding x_t, all of which contain only local information from the beginning to the current moment t [5,24,27]. Since v_t is obtained from the global level of the input image, we regard it as additional global information to make the prediction more accurate. At moment t, the hidden state h_t is calculated by the GRU using h_{t-1} and x_t, then attention a_t is obtained jointly by h_t, M', and the cover attention, where M' is obtained by spatial location encoding [19] based on the feature map M. After that, a_t is multiplied with the M to get the context vector c_t, which is weighted to guide

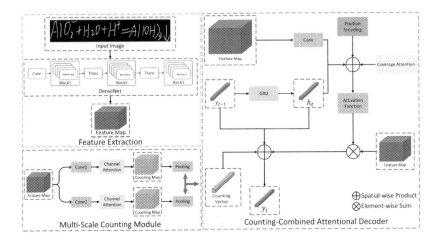

Fig. 7. The overall structure of the Counting-Aware Network [12].

the model to focus on the important positions in the current moment. Finally, the prediction result y_t is calculated sequentially by c_t, v_t, h_t and x_t together.

4   Experiments

4.1   Datasets

We collected 1182 common inorganic chemical equations from high school and college and then gathered volunteers to write 1,595 real handwritten chemical equations based on them, which were stored as trajectory coordinates. These equations contain a total of 133 symbols including letters, numbers, operators, e.g., "=", "+", "−", and other graphical symbols such as "↑", "↓" "△" etc. We obtained the images of the handwritten chemical equations by rendering the trajectory coordinates of the symbols, of which 650 images were randomly selected as a test set and the remaining 945 as a training set to provide a reference baseline for the experiment.

In contrast to the wealth of LaTeX expressions available in the mathematical field, there is only a very limited corpus in the field of chemistry. To further enrich the corpus information, we use the collected chemical equations as a basis from which we extract all the individual chemical equations and special symbols to synthesize a batch of nonsense LaTeX expressions for synthesis. To make our work reproducible, the individual symbol trajectories needed to synthesize HCEs were extracted from currently publicly available datasets, we parsed and extracted the required parts from each of the three publicly available datasets, Detexify[1], OLHWDB [14] and HIT-OR3C [28].

---

[1] http://detexify.kirelabs.org/classify.html.

Based on these chemical expressions of LaTeX, we generated 7092 handwritten equations using the proposed synthesis method mentioned in Sect. 3 and added them to the initial training set to obtain a new dataset with 8037 data. Then, the data augmentation operation and the improved baseline-based layout strategy were added in turn to generate different datasets with the same amount of data for comparison. The random rotation angle is set to $\theta \in (-10, 10)$, and the scaling ratio is limited to $\alpha \in (0.8, 1.2)$. Considering the actual writing habits, we do not change the direction of rotation of symbols in the same component. Figure 8 shows some samples of handwritten chemical equations synthesized by our proposed strategy.

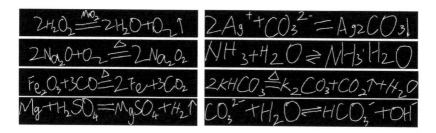

Fig. 8. The different chemical equations synthesized by our proposed strategy.

4.2 Experimental Details and Metrics

We used the PyTorch framework to train the recognition model, which was trained on four 1080Ti with 12G of memory per GPU. The total number of model iterations was 120, and the batch size was set to 16. To prevent overfitting of the training process, we set the dropout to 0.6 and the weight decay in Adelta optimizer to 1e−6, and the initial learning rate grows from 0 to 1 in the warm-up phase and then decays to 0 by cosine schedules [26]. Since some symbols in LaTeX are not visible in the actual formula [12], we ignore the "sos", "eos", "{", "}", "^", "_" in the counting module.

Referring to the ranking rules in the CROHME competition, we measure the performance of the model based on the Expression Recognition Rate (ExpRate), which represents the proportion of correctly predicted formulas out of all formulas. We denote the total number of CEs in the test set as E_N, where the number of symbols contained in each LaTeX is $\{n_1, n_2, ..., n_N\}$. In the recognition process, we represent the number of expressions with incorrect symbols as E_w, and the number of incorrectly recognized symbols in each formula as $\{e_1, e_2, ...e_i, i \leq N\}$. ExpRate is obtained as shown in Eq. 1.

$$ExpRate = \frac{E_N - E_w}{E_N} \tag{1}$$

We use "e1" and "e2" to represent the percentage of formulas where the number of incorrectly identified symbols is $\leq 1 (or\ 2)$, which can be a good metric for the global performance metric of the recognition model. To more accurately evaluate the improvement due to the synthetic dataset, we also use the Word Error Rate (WER) at the symbol level to measure the model as shown in Eq. 2, WER represents the symbol level recognition accuracy, which is a more accurate measure of model performance.

$$WER = \frac{\sum_{i=1}^{N} e_i}{\sum_{i=1}^{N} n_i} \qquad (2)$$

4.3 Effect of Data Generation

The model was trained using the different datasets we synthesized in Sect. 4.1 as a way to evaluate the performance of our synthesis strategy for handwritten chemical equations. We first trained the model using the 945 real datasets collected and then tested it on the remaining 650 real datasets as a benchmark. On this basis, we separately added data obtained by different synthesis strategies to the training set to obtain different training sets, where n = 1182 is the number of chemical equations we collected, and the amount of synthesized data is 6n. The experimental results are shown in Table 1, from which it can be seen that our synthetic data without the modified layout strategy and data augmentation could lead to a 40.31% improvement in the recognition accuracy of the model. The addition of the new layout strategy and data augmentation to the synthesis method further improves the accuracy by 1.08% and 1.23% respectively. In addition, the model achieved 89.95% accuracy when both strategies were combined to generate handwritten chemical equations, and the WER decreased by 17.76% compared to training with only real data.

Table 1. Effect of different synthetic strategies on model performance.

Synthesis strategy			Result			
Generation	Layout	Augmentation	ExpRate	≤ 1 error	≤ 2 error	WER
✗	✗	✗	46.15	50.77	54.31	19.85
✓	✗	✗	86.46	88.77	90.15	3.44
✓	✗	✓	87.54	86.62	88.31	4.38
✓	✓	✗	87.69	90.62	92.77	2.57
✓	✓	✓	**89.85**	**92.00**	**93.54**	**2.09**

To analyze the effect of the data volume of the synthesis HCEs on training, we used the optimal synthesis strategy discussed above to generate $1182 \times T$ synthetic data by making T=2, 4, 6, 8, and 10, respectively. This data was then mixed with 945 real data as the training set, and the remaining 650 real data were also used as the test set. The final test results are shown in Table 2, which shows that relatively good results can be achieved when T=6. Thereafter, as n grows, the effect of synthetic chemical equation data on model performance is no longer apparent.

Table 2. Effect of synthetic data volume on model performance.

Synthetic data volume $n = 1182$	Result			
	ExpRate	≤ 1 error	≤ 2 error	WER
0	46.15	50.77	54.31	19.85
2n	88.00	90.46	92.15	2.75
4n	88.92	91.54	93.54	1.95
6n	**89.85**	**92.00**	**93.54**	**2.09**
8n	89.23	91.38	93.08	2.26
10n	89.38	92.77	93.69	1.99

5   Conclusions

In this paper, we propose a multi-level synthesis strategy based on the symbol layout tree, which can synthesize handwritten chemical equations by arbitrary LaTeX expressions to improve the accuracy of recognition models. We first decompose LaTeX expressions into SLTs, then extract individual symbol trajectories from the three publicly available datasets and place them in appropriate locations according to the layout strategy. To further improve the diversity of synthesis, we also explore the impact of different data augmentation strategies such as symbol shape, writing style, and synthetic data volume.

In our future work, we will focus on organic chemical equations with more complex structures and variability, which will be more challenging. On the other hand, considering that the adjustment of parameters such as symbol scaling and rotation angle is an important but laborious task in the synthesis process, it will be a valuable research direction to automatically adjust the relevant parameters through model recognition results.

References

1. Alonso, E., Moysset, B., Messina, R.: Adversarial generation of handwritten text images conditioned on sequences. In: 2019 International Conference on Document Analysis and Recognition (ICDAR), pp. 481–486. IEEE (2019)
2. Anderson, R.H.: Syntax-directed recognition of hand-printed two-dimensional mathematics. In: Symposium on Interactive Systems for Experimental Applied Mathematics: Proceedings of the Association for Computing Machinery Inc., Symposium, pp. 436–459 (1967)

3. Awal, A.-M., Mouchère, H., Viard-Gaudin, C.: Towards handwritten mathematical expression recognition. In: 2009 10th International Conference on Document Analysis and Recognition, pp. 1046–1050. IEEE (2009)
4. Belaid, A., Haton, J.-P.: A syntactic approach for handwritten mathematical formula recognition. IEEE Trans. Pattern Anal. Mach. Intell. **PAMI-6**(1), 105–111 (1984)
5. Bian, X., Qin, B., Xin, X., Li, J., Xuefeng, S., Wang, Y.: Handwritten mathematical expression recognition via attention aggregation based bi-directional mutual learning. In: Proceedings of the AAAI Conference on Artificial Intelligence, vol. 36, pp. 113–121 (2022)
6. Chang, M., Han, S., Zhang, D.: A unified framework for recognizing handwritten chemical expressions. In: 2009 10th International Conference on Document Analysis and Recognition, pp. 1345–1349. IEEE (2009)
7. Deng, Y., Kanervisto, A., Ling, J., Rush, A.M.: Image-to-markup generation with coarse-to-fine attention. In: International Conference on Machine Learning, pp. 980–989. PMLR (2017)
8. Graves, A.: Generating sequences with recurrent neural networks. arXiv preprint arXiv:1308.0850 (2013)
9. Hagag, A., Omara, I., Alfarra, A.N.K., Mekawy, F.: Handwritten chemical formulas classification model using deep transfer convolutional neural networks. In: 2021 International Conference on Electronic Engineering (ICEEM), pp. 1–6. IEEE (2021)
10. Khuong, V.T.M., Huy, U.Q., Masaki, N., Phan, M.K.: Generating synthetic handwritten mathematical expressions from a latex sequence or a mathml script. In: 2019 International Conference on Document Analysis and Recognition (ICDAR), pp. 922–927. IEEE (2019)
11. Le, A.D., Indurkhya, B., Nakagawa, M.: Pattern generation strategies for improving recognition of handwritten mathematical expressions. Pattern Recogn. Lett. **128**, 255–262 (2019)
12. Li, B., et al.: When counting meets HMER: counting-aware network for handwritten mathematical expression recognition. In: Avidan, S., Brostow, G., Cissé, M., Farinella, G.M., Hassner, T. (eds.) ECCV 2022. LNCS, vol. 13688, pp. 197–214. Springer, Cham (2022). https://doi.org/10.1007/978-3-031-19815-1_12
13. Li, Z., Jin, L., Lai, S., Zhu, Y.: Improving attention-based handwritten mathematical expression recognition with scale augmentation and drop attention. In: 2020 17th International Conference on Frontiers in Handwriting Recognition (ICFHR), pp. 175–180. IEEE (2020)
14. Liu, C.-L., Yin, F., Wang, D.-H., Wang, Q.-F.: Casia online and offline Chinese handwriting databases. In: 2011 International Conference on Document Analysis and Recognition, pp. 37–41. IEEE (2011)
15. Liu, X., Zhang, T., Yu, X.: An end-to-end trainable system for offline handwritten chemical formulae recognition. In: 2019 International Conference on Document Analysis and Recognition (ICDAR), pp. 577–582. IEEE (2019)
16. MacLean, S., Labahn, G., Lank, E., Marzouk, M., Tausky, D.: Grammar-based techniques for creating ground-truthed sketch corpora. Int. J. Doc. Anal. Recogn. (IJDAR) **14**, 65–74 (2011)
17. Miller, E.G., Viola, P.A.: Ambiguity and constraint in mathematical expression recognition. In: AAAI/IAAI, pp. 784–791 (1998)

18. Mouchere, H., Viard-Gaudin, C., Zanibbi, R., Garain, U., Kim, D.H., Kim, J.H.: ICDAR 2013 crohme: third international competition on recognition of online handwritten mathematical expressions. In: 2013 12th International Conference on Document Analysis and Recognition, pp. 1428–1432. IEEE (2013)
19. Parmar, N., et al.: Image transformer. In: International Conference on Machine Learning, pp. 4055–4064. PMLR (2018)
20. Truong, T.-N., Nguyen, C.T., Nakagawa, M.: Syntactic data generation for handwritten mathematical expression recognition. Pattern Recogn. Lett. **153**, 83–91 (2022)
21. Wang, Y., Zhang, T., Yu, X.: A component-detection-based approach for interpreting off-line handwritten chemical cyclic compound structures. In: 2021 IEEE International Conference on Engineering, Technology and Education (TALE), pp. 785–791. IEEE (2021)
22. Zanibbi, R., Blostein, D.: Recognition and retrieval of mathematical expressions. Int. J. Doc. Anal. Recogn. (IJDAR) **15**, 331–357 (2012)
23. Zhang, J., Du, J., Dai, L.: Multi-scale attention with dense encoder for handwritten mathematical expression recognition. In: 2018 24th International Conference on Pattern Recognition (ICPR), pp. 2245–2250. IEEE (2018)
24. Zhang, J., et al.: Watch, attend and parse: an end-to-end neural network based approach to handwritten mathematical expression recognition. Pattern Recogn. **71**, 196–206 (2017)
25. Zhang, T., Mouchère, H., Viard-Gaudin, C.: A tree-BLSTM-based recognition system for online handwritten mathematical expressions. Neural Comput. Appl. **32**, 4689–4708 (2020)
26. Zhang, Z., He, T., Zhang, H., Zhang, Z., Xie, J., Li, M.: Bag of freebies for training object detection neural networks. arXiv preprint arXiv:1902.04103 (2019)
27. Zhao, W., Gao, L., Yan, Z., Peng, S., Du, L., Zhang, Z.: Handwritten mathematical expression recognition with bidirectionally trained transformer. In: Lladós, J., Lopresti, D., Uchida, S. (eds.) ICDAR 2021. LNCS, vol. 12822, pp. 570–584. Springer, Cham (2021). https://doi.org/10.1007/978-3-030-86331-9_37
28. Zhou, S., Chen, Q., Wang, X.: Hit-OR3C: an opening recognition corpus for Chinese characters. In: Proceedings of the 9th IAPR International Workshop on document analysis systems, pp. 223–230 (2010)

Context-Aware Chart Element Detection

Pengyu Yan$^{(\boxtimes)}$ ⓘ, Saleem Ahmed$^{(\boxtimes)}$ ⓘ, and David Doermann ⓘ

Department of Computer Science and Engineering, University at Buffalo,
Buffalo, NY, USA
{pyan4,sahmed9,doermann}@buffalo.edu

Abstract. As a prerequisite of chart data extraction, the accurate detection of chart basic elements is essential and mandatory. In contrast to object detection in the general image domain, chart element detection relies heavily on context information as charts are highly structured data visualization formats. To address this, we propose a novel method **CACHED**, which stands for **C**ontext-**A**ware **Ch**art **E**lement **D**etection, by integrating a local-global context fusion module consisting of visual context enhancement and positional context encoding with the Cascade R-CNN framework. To improve the generalization of our method for broader applicability, we refine the existing chart element categorization and standardized 18 classes for chart basic elements, excluding plot elements. Our CACHED method, with the updated category of chart elements, achieves state-of-the-art performance in our experiments, underscoring the importance of context in chart element detection. Extending our method to the bar plot detection task, we obtain the best result on the PMC test dataset. Our code and model are available at https://github.com/pengyu965/ChartDete.

Keywords: Chart Detection · Chart Data Extraction · Chart Understanding · Document Analysis

1 Introduction

Charts are highly abstract data visualization formats, which are convenient for readers to obtain trends or comparisons between different entities but hard to extract the exact data value from them. Therefore, automated chart data extraction could reduce the human effort to summarize data from scientific, financial analysis, marketing charts, etc. The chart data extraction typically involves but is not limited to, element detection, text OCR, data interpretation, and semantic data conversion. Basic element detection is the most fundamental part of chart data extraction and would affect all downstream tasks. Thus, accurately detecting and recognizing the basic elements of the chart outside the plot area (see Fig. 1) is the first critical step. Basic element detection would be challenging due to the highly diverse chart design.

Several methods [24,25] have been proposed for data extraction but only focused on data plot detection, such as bar plot and line key points, while neglecting most of the fundamental element detection. Meanwhile, some works [1,12,26] use often-seen standard two-stage detectors to detect chart elements. These two-stage detectors [2,9,10,13,30] have a limited ability to utilize the context in images, where context is vitally important for accurate detection in the chart images domain. Unlike common

G. A. Fink et al. (Eds.): ICDAR 2023, LNCS 14187, pp. 218–233, 2023.
https://doi.org/10.1007/978-3-031-41676-7_13

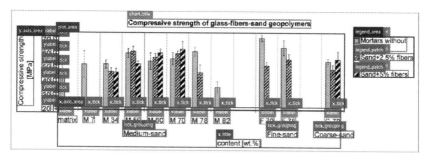

(a) Chart Element Detection Objective

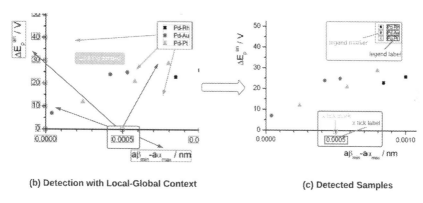

(b) Detection with Local-Global Context   **(c) Detected Samples**

Fig. 1. Context-aware Chart Element Detection. (a) is a sample of chart element detection objective, (b) and (c) illustrate the detection in chart images relies on context.

objects in general images, many elements in chart images share a similar visual appearance but different roles. It could only be distinguished by referring to the context, *e.g.*, legend label and tick labels are text blocks with different roles according to the functioning position and relationship to the other elements in the chart images (as shown in Fig. 1(b)(c)). Therefore, a detector that uses local-global context features is needed to tackle the chart element detection task. Additionally, a comprehensive and reasonable categorization of the chart elements would help improve generalization on various charts.

In this paper, we propose a context-aware chart element detection method by integrating the local-global context fusion module between the cascaded RoI head in Cascade R-CNN to draw the importance of context in charts. As Fig. 2 shows, the local-global context fusion module contains two parts–visual context enhancement (VCE) and positional context encoding (PCE). The VCE enables the model to gain better visual context information by incorporating the global feature map into the local feature map. The PCE allows the model to learn the particular object distribution pattern from the bbox coordinates. Besides, we look into multiple datasets for analyzing and refining chart element categorization. A total of 18 classes of chart elements are summarized, excluding any plot elements. The dataset is updated accordingly for model training and testing. The quantitative evaluation and qualitative analysis of samples show that our method achieves accurate chart element detection. **Overall, our contributions can be summarized as follows:**

- A detector with the local-global context fusion module is proposed for accurate chart element detection. The module emphasizes the context from two aspects - visual and positional features. Our method achieves state-of-the-art results on the chart element detection task.
- Refine chart element categorization and generate additional structural-area objects to assist the detector in better chart understanding. A total of 18 classes are summarized, and the accordingly updated PMC dataset can be accessed at https://github.com/pengyu965/ChartDete
- Our method and several common two-stage detectors, including those used in existing related works, are trained and evaluated on the updated datasets. This can offer an overview of the performance of these methods on such tasks.

2   Related Work

2.1   Object Detection

Since 2014, many well-designed object detectors have been invented, and most of them could be divided into two kinds - one-stage detector [22,27–29] and two-stage detector [2,9,10,13,30]. They have a similar first part – a convolutional neural network (CNN) based backbone like [8,14–16,31,32,35,36] to extract visual features. The difference is two-stage detector involves a region proposal module to generate category-independent region proposals and then classify these objects and refine their localization in the second stage, while one-stage detectors directly predict the object position and category from input images and their feature maps. Intuitively second-stage detectors like Faster R-CNN [30] and Cascade R-CNN [2] achieve higher accuracy in object localization and recognition but sacrifice the inference speed. The chart element detection task demands high accuracy on object localization and classification than inference speed, so we stay with the two-stage detectors for the task.

2.2   Transformers

With the success of Transformer-based methods [7,34] in natural language processing, many works [17,19,33] put effort into adopting Transformer in vision-related tasks. Transformer could extract strong feature representations from both visual and text inputs. In 2020, Carion et al. [3] proposed the first Transformer-based end-to-end object detector. However, DETR lacks training stability on small-scale datasets and cannot accurately localize and recognize small or overlapping objects. In a more recent work, [23] introduced a hierarchical Transformer that leverages shifted windows to extract features from input images, and the resulting framework serves as a new backbone for

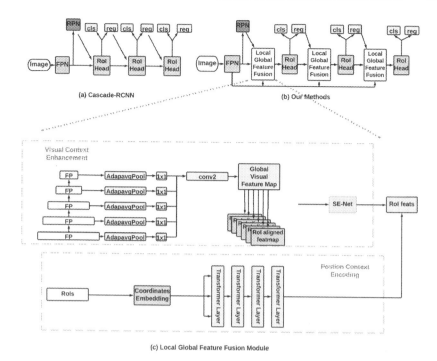

Fig. 2. Framework of our Method. (a) is the standard Cascade R-CNN model, (b) is the overall framework of our method, and (c) shows the details of the local-global feature fusion module in our method.

two-stage detectors. This new backbone has been shown to be effective in improving the performance of two-stage detectors. Due to the attention mechanism in Transformer architecture, the Swin Transformer can draw spatial attention and obtain context-aware image features. The Swin-Transformer backbone is used exclusively in our model and experiments.

2.3 Chart Data Extraction

In 2017, Jung et al. [26] proposed a semi-interactive system to extract underlying data from charts. However, it uses human interaction as the first step to set the starting and ending points. Then rule-based methods are utilized to interpret the data value. This method heavily relies on human interaction for data extraction and would fail in complicated cases. From 2019–2022, Davila et al. [4–6] organized the chart harvesting competition and offered two valuable datasets: Human-annotated real-world chart images from PubMed Central documents (PMC dataset) and Adobe synthetic chart dataset. As a participating group in the 2020 chart harvesting competition, Ma et al. [25] use Cascade R-CNN and a heatmap-based keypoint detector to detect bar box and line key points, respectively. It focused on data plot detection within the plot area, while in data interpretation, they used the elements' ground truth with the predicted data plots for semantic conversion. Later, another EXCEL400K dataset is proposed in [24] by utilizing the Microsoft Excel API. Similarly to CornerNet [18], an hourglass network backbone with key-point heatmap generation is used to predict and group the sets of corner points on objects. Using the key point detection method, the author generalized their detection on different charts. However, the dataset proposed in this work only has data plot annotations within the plot area, and most essential basic elements are left without annotations.

3   Our Method

In this section, we explain our CACHED method in the following aspects: the local-global context fusion module, which consists of visual context enhancement and positional context encoder; loss function for class-wise objects imbalance; the standardized categorization for broader applicability and generalization.

3.1   Local-Global Context Fusion Module

As Fig. 2 shows, local-global context fusion modules are designed and integrated between each RoI head for context extraction and fusion. It brings local-global visual features and relative positional features towards each region of interest before sending them to the RoI head for regression and classification. The three modules share the same architecture but are trained with unshared weights, and each module consists of the following two parts.

Visual Context Enhancement (VCE).   Although the field of view of each anchor would increase with stacking of the convolutional neural network (CNN) layer, it is still limited to the local field, which has fragile context awareness. However, accurate element detection and labeling in chart images require a much larger field of context.(see Fig. 1). For instance, a text block can be classified as a legend label by the context that nearly comes from the whole image, where the legend label is beside some (legend) markers that share the same color with plot elements in the plot area. To address this issue, similar to SCNet [11], we introduce visual context enhancement (VCE) by incorporating the feature maps from the backbone feature pyramid as global visual features and bringing them into each region of interest. These global visual features are combined with each RoI-aligned local feature map to amplify local-global visual context. As shown in Fig. 2(c), we first average pool the feature maps from all stages of the feature pyramid to the same size $[N, 256, 7, 7]$, then concatenate them together, where the shape becomes $[N, 256*5, 7, 7]$. A convolutional neural network further abstracts these global visual features and reduces the number of channels. Then the global feature maps are attached to each RoI-aligned feature map. The feature map for each region of interest with size $[N, 512, 7, 7]$ now consists of local visual features in the first half channels and global visual features in the rest of the channels. A SE-Net and a 1×1 convolutional layer are followed to fuse the local-global visual features channel-wise and reduce the number of channels, respectively. The final feature maps with size $[N, 256, 7, 7]$, which align with the original RoI feature map dimension but contain rich local-global visual context, are used as visual feature representations for each region of interest.

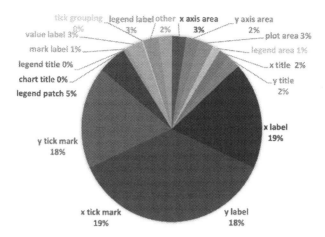

Fig. 3. The class-wise chart object number distribution in PMC dataset. Chart images have naturally unbalanced objects class-wisely. The number of x/y tick labels and x/y tick marks is inevitably larger than others. Some elements are extremely small, like tick groupings and legend titles.

Positional Context Encoding (PCE). In the general image domain, the detection of common objects such as people, vehicles, and animals depends mainly on visual features, and the positions of the objects in the image have less impact on accurate localization and classification. However, since chart images are highly structured data visualization formats, where object rendering always follows some specific patterns, *e.g.*, x-tick labels are always below or beside the x-tick marks and chart titles typically appear on the edges of the image while rarely appearing in the center part, the relative positional context among the chart elements is very trivial. We propose a Transformer-based positional context encoder to obtain the positional context feature. As Fig. 2 shows, the Transformer-based architecture draws attention among objects' bounding-box coordinates. Firstly, we normalize the 4-dim coordinates to $[0, 1]$ by the height and width of the input images. Then a linear layer takes the normalized coordinates and embeds them into a 512-dim vector. Zero mask paddings are filled in the embedded bbox vector sequence, where the length of the final embedding bbox sequence is fixed to the max block size of 1024. The block size of 1024 is enough to cover the maximum sampling bbox number from the RPN results in training and testing, which is set to a maximum of 1000. We use each head's output to represent the relative positional context information from the corresponding bbox to all other bboxes. Then, the encoding vector for each bbox is concatenated to its RoI visual feature vectors.

3.2 Loss for Class-Wise Objects Imbalance

Unlike the general image domain data imbalance can be reduced by augmentation, the unbalanced number of objects among different categories is the natural effect in chart images and is hard to undermine, *e.g.*, the number of tick marks and tick labels would always be multi-times larger than the number of chart titles and legend labels (see Fig. 3). We use the Focal Loss [20] on the classification to undermine the object imbalance in chart images:

$$L_{cls} = -\sum_{i=1}^{i=n}(i - p_i)^\gamma \log_b(p_i) \tag{1}$$

Then, smooth L1 loss is used for bounding box regression and balancing it with classification:

$$loss_\mu = L_{cls}(\mu) + \lambda[\mu \geq 1]L_{loc}(t^\mu, v) \tag{2}$$

where L_{loc} is the localization loss between the regression results for class μ and the regression targets. λ is used to tune the loss for outliers where their loss is greater than 1.

3.3 Categorization Refinement

The categorization of chart elements in the dataset is crucial for detection performance. We analyze the datasets with the annotations of the chart elements and comprehend the categorization of the elements to generalize the various chart situations (see Table 1).

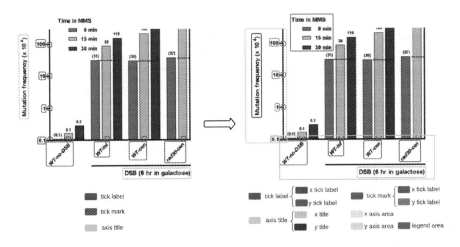

Fig. 4. Categorization Refinement. It only shows the updated categories for illustration. Some elements are not visualized.

From the chart competition [5,6], the Adobe Synthetic and PMC datasets offer the most valuable annotations for chart element detection tasks (a detailed introduction of the datasets can be found in the experiment section). We refer to these two datasets for element categorization. In Table 1, the second column is the category in the Adobe

Synthetic dataset, and the third column is the category annotated in the PMC dataset. In the PMC dataset, all objects related to the axes, such as the axis title, the tick label, and the tick mark, are jointly labeled without separation into the x and y axes. Early experiments showed that if we jointly treat the items based on the x- and y-axis as the same category, it would be easier to separate these labels later by post-processing. This is caused by the definition of the x-axis and y-axis, in which the x-axis is defined as the axis with independent value/label, while the y-axis has dependent values. Such definitions are more based on the content's semantic meaning than on the visual or positional information. Therefore, separating these elements by axis in advance would help avoid this issue when using the trained model for the application. Meanwhile, these separated labels also assist our method in understanding the chart's content and potentially obtaining better detection results.

Table 1. Categorization Refinement. We look into the Adobe Synthetic Dataset and PMC Dataset from Chart Competition [5, 6] and comprehend the element categories shown in the third column.

	Synthetic	**PMC**	**Refined Categories**
Chart Basic Skeleton Elements	x-axis title	axis title	x-axis title
	y-axis title		y-axis title
	x tick label	tick label	x tick label
	y tick label		y tick label
	x tick mark	tick mark	x tick mark
	y tick mark		y tick mark
	chart title	chart title	chart title
	legend patch	legend marker	legend marker
	legend label	legend label	legend label
	-	legend title	legend title
	plot text	value label	value label
		mark label	mark label
	-	tick grouping	tick grouping
	-	others	others
Chart Structural Area	plot area	plot area	plot area
			x-axis area
	-	-	y-axis area
			legend area

The categories in the second part of Table 1 are four additional labels of structural objects that cover the specific area in the chart images. We break down the chart images into the four most crucial structures—the x-axis area, the y-axis area, the plot area, and the legend area. The definitions of these four areas are as follows:

– **Plot area**: The plotting area is formed by the x and y axes.
– **X/Y axis area**: Cover all x/y ticks, x/y labels, and x/y-axis titles.
– **Legend area**: The area covers all items related to the legend, including legend labels, legend markers, and legend titles.

A sample of category refinement is shown in Fig. 4. We can see that each axis-related element is separated and structural-area objects cover the specific area in the chart image. After refinement, 18 categories for the chart elements are summarized. Based on the new categories, we update the PMC dataset accordingly by employing rule-based methods to separate the axis-related elements and generate three additional structure-area elements than the originally offered plot area. Then the annotation format is converted to the standard COCO object detection format for convenience. The updated PMC dataset with conversion tools can be accessed from the link in the contribution summary from Sect. 1.

4   Datasets and Experiments

In this section, we introduce the existing datasets used in our experiments and perform quantitative evaluation and qualitative analyzes.

4.1   Dataset

Adobe Synthetic Dataset. The adobe synthetic dataset was first proposed in 2019 [4] and refined in 2020 Chart-infographic Competition [5]. This dataset is synthesized using Matplotlib and contains 14400 images for 12 types of charts. The annotations include the chart data information and all elements' label and location. Such annotations are valuable for chart classification, element detection, and data extraction tasks. However, the diversity of chart samples in this dataset is severely limited. The best results showed in ICPR 2020 Chart Competition [5] in chart classification, text role classification, tick mark label association, and legend marker detection are close to 100%, further indicating the limited variance of the data. Low data variance typically causes unstable performance and low generalization from the trained deep-learning model on samples outside the dataset. We only refer to the annotation standards in this dataset for element categorization refinement.

PubMed Central (PMC) Chart Dataset. This dataset was released and updated with the Chart Competition in ICDAR 2019 [5], ICPR 2020 [6] and ICPR 2022 [6]. Unlike the Adobe Synthetic dataset, the PMC dataset is a real-world dataset collected from PubMed Central Documents and manually annotated. Taking into account the much more diverse and high-fidelity samples in this dataset, the PMC dataset became the primary dataset in most recent chart competitions. The most up-to-date PMC dataset is released in the ICPR 2022 CHART-Infographic competition [6], which contains 5614 images for chart element detection, 4293 images for final plot detection and data extraction, and 22924 images for chart classification. Although slightly limited by the number of available training samples due to the time-consuming human annotation process, this real-world dataset is most valuable, and much more challenging than most synthetic datasets.

Table 2. Comparison of Methods in ICPR 2022 Chart Competition. Evaluation on task 2 and task 3 in chart competition [6] is based on the original PMC categorization. We train our method with the updated categories but rewind the predicted results backward to be compatible with the chart competition evaluation metrics.

Team Methods	Task 2 Text Detection	Task 3 Text Role Classification		
	Average IoU	Recall	Precision	F-measure
six_seven_four	0.435	-	-	-
IIIT_CVIT_Chart_Understanding	0.790	-	-	**0.821**
Ystar	0.810	-	-	-
UB-ChartAnalysis	0.820	-	-	0.736
Ours	**0.869**	**0.735**	**0.846**	**0.787**

Excel400K. In [24], the author proposed the ExcelChart400k dataset, which contains a total of 360k training samples. This dataset is generated using the Excel data sheet and was used to train the plot element detection model in [24]. However, the dataset is focused only on annotating data plots inside the plot area, and most basic elements are left without annotation. We use this additional dataset to qualitatively evaluate the results of our method.

After conducting early experiments, we observed that including the Adobe synthetic training dataset decreased the detection performance on real-world chart images. This was attributed to the uniform chart rendering patterns and limited sample variance in the synthetic dataset. Taking into account the high diversity and real-world chart data distribution, we use the PMC dataset as our primary dataset for training and quantitative evaluation purposes.

4.2 Quantitative Evaluation

To quantitatively evaluate our method, we have undertaken three experiments: (i) ICPR 2022 Chart Competition Evaluation [6], wherein we employed backward adapted results from our method; (ii) COCO object detection evaluation with the refined PMC dataset; and (iii) an extended experiment on detecting bar plots.

ICPR 2022 Chart Competition Evaluation. We compare with the results from the ICPR 2022 Chart Competition [6]. However, our method shares different detection routines, where the chart competition treats the text block detection task as a 1-class detection task (task 2) and leaves the recognition as an additional task (task 3). In the chart competition, teams could use the ground truth of previous tasks for the current task. Our method follows the regular detection routine (localization + recognition). After refining the categorization, the overall categories (see Table 1) expand from the original PMC dataset used in the competition. Taking these two facts, we compare by adapting our

Table 3. COCO Evaluation on Refined PMC Dataset. First 4 rows show our trained public detector models, and the last 3 rows are our method ablation study.

Model	Context Module	Backbone	AP	AP_{50}	AP_{75}	AP_S	AP_M	AP_L
DETR	-	-	0.536	0.762	0.572	0.384	0.555	0.809
Faster R-CNN	-	ResNeXt101-FPN	0.665	0.815	0.737	0.557	0.697	0.827
Cascade R-CNN	-	ResNeXt101-FPN	0.696	0.825	0.759	0.579	0.725	0.899
Cascade R-CNN	-	SwinT-FPN	0.699	0.838	0.772	0.589	0.732	0.885
Cascade R-CNN	PCE	SwinT-FPN	0.708	0.842	0.775	0.591	0.742	0.903
Cascade R-CNN	VCE+PCE	SwinT-FPN	0.713	**0.851**	0.786	0.597	0.741	0.909
Cascade R-CNN^L	VCE+PCE	SwinT-FPN	**0.729**	0.845	**0.790**	**0.602**	**0.763**	**0.939**

prediction results backwards to be compatible for evaluation and splitting the prediction result into task 2 and task 3. Table 2 shows the result of the official 2022 chart competition. Our method in task 2 outperforms the best result from 'UB-ChartAnalysis'. In task 3, our method detects and recognizes all elements once, without taking the element localization ground truth as prior knowledge, and this may result in a lower recall for task 3 due to imperfect detection. Overall, we achieve the best detection result on task 2 and the comparable F-measure score on task 3.

COCO Evaluation on Refined PMC Datasets. To provide a better general overview of the performance of our method, we use the COCO object detection evaluation metric [21] to evaluate our method with several often used two-stage detectors (shown in Table 3). To our best knowledge, there isn't any trained chart element detection model conducted on PMC datasets publicly available for inference and fine-tuning. We trained these two-stage detectors on the refined PMC dataset at our end. In Table 3, the first part includes results from popular two-stage detectors. DETR converges slowly in training due to the scale of our dataset and the Transformer architecture. The accuracy of small object detection from DETR also lags behind. An enormous amount of small elements in chart images result in limited overall accuracy. The Cascade R-CNN with Swin-Transformer performs best in the standard Cascade R-CNN model zoo. The second part shows our methods with several ablated setups on the local-global context module, which consists of visual context enhancement (VCE) and positional context encoding (PCE). Since the Swin-Transformer backbone can draw spatial attention and potentially obtain context-aware visual feature representations, we made an ablation by only integrating the positional context encoding (PCE) into the Swin-Transformer-based Cascade R-CNN. As expected, the Swin-Transformer-based Cascade R-CNN with the PCE performs better than the standard Swin-Transformer Cascade R-CNN. Although the Swin-Transformer can obtain context-aware visual features, adding VCE could further enhance the context extraction capability, giving better results. The focal loss with balanced smooth L1 loss can help with sample imbalance problems, and our method with this loss setup achieves state-of-the-art performance (see the last row in Table 3).

Table 4. Bar detection evaluation on PMC test dataset. The comparison results are from [25], and we evaluate our method with the same evaluation metric. The first three columns are F-measure with different IoU thresholds, and the last column is the calculated score using Chart Competition [5,6] evaluation criteria.

Model	IoU=0.5	IoU=0.7	IoU=0.9	Score_a
SSD	43.65	26.28	2.67	25.83
YOLO-v3	58.84	36.14	4.14	60.97
Faster R-CNN	66.37	60.88	29.13	70.03
Faster R-CNN+FPN	85.81	78.05	31.30	89.65
Cascade R-CNN+FPN	86.92	83.53	55.32	91.76
Our Methods	**89.30**	**88.73**	**76.94**	**93.75**

Extended Experiment on Bar Detection. Although our goal is to offer a robust basic element detection method, we extend our experiment to bar plot detection. We fine-tune our method using the PMC bar chart subset and test on the PMC test set (see Table 4). The results of the first five models were from [25], and we evaluate our results with the same criteria. The first three columns are F-measure scores with different IoU thresholds - 0.5, 0.7, and 0.9. The last column, 'Score_a', is calculated using the ICPR 2022 chart competition [5] evaluation metric for task 6a. Our method achieves state-of-the-art performance on bar chart detection on the PMC dataset.

4.3   Qualitative Evaluation on Element Detection

Although the Excel400K dataset does not have the ground truth for chart basic elements, we visualize the prediction results from our method (see Fig. 5) for qualitative evaluation. Our method is able to locate each element accurately on most samples in Excel400K datasets. However, our method struggles with the first sample in the third row of Fig. 5 as the table attached at the bottom confuses the detector due to the lack of similar samples in the PMC training dataset. Our method generally delivers accurate localization and classification of the basic elements in charts.

230   P. Yan et al.

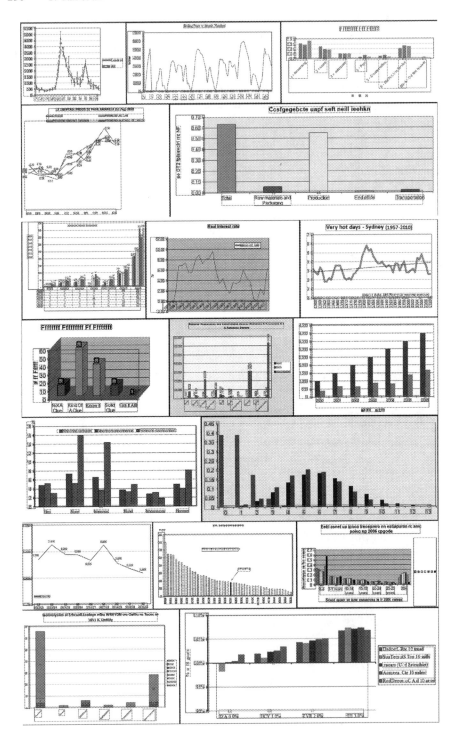

Fig. 5. Visualization of Our methods on Excel400K.

5 Conclusion and Future Work

In this work, we propose a method that focuses on the importance of visual and positional context in chart images for accurate chart element detection. The categories of chart elements are analyzed and refined to provide a better generalization of various chart designs, which could benefit data interpretation-related downstream tasks. Our method trained on the refined PMC dataset achieves state-of-the-art performance on the chart element detection task.

Considering the context information, the text contains rich information. Due to the challenging OCR task on chart images, where many symbols are easily confused with characters or numbers, and the rotation is hard to detect when the text is short, we don't include the text embedding into the context extraction. In the future, robust OCR for extracting chart text and adding chart text embedding as additional context information to each region of interest may improve performance.

References

1. Balaji, A., Ramanathan, T., Sonathi, V.: Chart-text: a fully automated chart image descriptor. arXiv preprint arXiv:1812.10636 (2018)
2. Cai, Z., Vasconcelos, N.: Cascade R-CNN: delving into high quality object detection. In: Proceedings of the IEEE Conference on Computer Vision and Pattern Recognition, pp. 6154–6162 (2018)
3. Carion, N., Massa, F., Synnaeve, G., Usunier, N., Kirillov, A., Zagoruyko, S.: End-to-end object detection with transformers. In: Vedaldi, A., Bischof, H., Brox, T., Frahm, J.-M. (eds.) ECCV 2020. LNCS, vol. 12346, pp. 213–229. Springer, Cham (2020). https://doi.org/10.1007/978-3-030-58452-8_13
4. Davila, K., et al.: ICDAR 2019 competition on harvesting raw tables from infographics (chart-infographics). In: 2019 International Conference on Document Analysis and Recognition (ICDAR), pp. 1594–1599 (2019). https://doi.org/10.1109/ICDAR.2019.00203
5. Davila, K., Tensmeyer, C., Shekhar, S., Singh, H., Setlur, S., Govindaraju, V.: ICPR 2020 - competition on harvesting raw tables from infographics. In: Del Bimbo, A., et al. (eds.) ICPR 2021. LNCS, vol. 12668, pp. 361–380. Springer, Cham (2021). https://doi.org/10.1007/978-3-030-68793-9_27
6. Davila, K., Xu, F., Ahmed, S., Mendoza, D.A., Setlur, S., Govindaraju, V.: ICPR 2022: challenge on harvesting raw tables from infographics (chart-infographics). In: 2022 26th International Conference on Pattern Recognition (ICPR), pp. 4995–5001. IEEE (2022)
7. Devlin, J., Chang, M.W., Lee, K., Toutanova, K.: Bert: Pre-training of deep bidirectional transformers for language understanding. arXiv preprint arXiv:1810.04805 (2018)
8. Ghiasi, G., Lin, T.Y., Le, Q.V.: NAS-FPN: learning scalable feature pyramid architecture for object detection. In: Proceedings of the IEEE/CVF Conference on Computer Vision and Pattern Recognition, pp. 7036–7045 (2019)
9. Girshick, R.: Fast R-CNN. In: Proceedings of the IEEE International Conference on Computer Vision, pp. 1440–1448 (2015)
10. Girshick, R., Donahue, J., Darrell, T., Malik, J.: Rich feature hierarchies for accurate object detection and semantic segmentation. In: Proceedings of the IEEE Conference on Computer Vision and Pattern Recognition, pp. 580–587 (2014)
11. Han, K., et al.: ScNet: learning semantic correspondence. In: Proceedings of the IEEE International Conference on Computer Vision, pp. 1831–1840 (2017)

12. Hassan, M.Y., Singh, M., et al.: Lineex: data extraction from scientific line charts. In: Proceedings of the IEEE/CVF Winter Conference on Applications of Computer Vision, pp. 6213–6221 (2023)
13. He, K., Gkioxari, G., Dollár, P., Girshick, R.: Mask R-CNN. In: Proceedings of the IEEE International Conference on Computer Vision, pp. 2961–2969 (2017)
14. He, K., Zhang, X., Ren, S., Sun, J.: Deep residual learning for image recognition. In: Proceedings of the IEEE Conference on Computer Vision and Pattern Recognition, pp. 770–778 (2016)
15. Howard, A.G., et al.: Mobilenets: efficient convolutional neural networks for mobile vision applications. arXiv preprint arXiv:1704.04861 (2017)
16. Iandola, F.N., Han, S., Moskewicz, M.W., Ashraf, K., Dally, W.J., Keutzer, K.: Squeezenet: alexnet-level accuracy with 50x fewer parameters and 0.5 mb model size. arXiv preprint arXiv:1602.07360 (2016)
17. Kafle, K., Price, B., Cohen, S., Kanan, C.: DVQA: understanding data visualizations via question answering. In: Proceedings of the IEEE Conference on Computer Vision and Pattern Recognition, pp. 5648–5656 (2018)
18. Law, H., Deng, J.: CornerNet: detecting objects as paired keypoints. In: Ferrari, V., Hebert, M., Sminchisescu, C., Weiss, Y. (eds.) Computer Vision – ECCV 2018. LNCS, vol. 11218, pp. 765–781. Springer, Cham (2018). https://doi.org/10.1007/978-3-030-01264-9_45
19. Li, L.H., Yatskar, M., Yin, D., Hsieh, C.J., Chang, K.W.: VisualBERT: a simple and performant baseline for vision and language. arXiv preprint arXiv:1908.03557 (2019)
20. Lin, T.Y., Goyal, P., Girshick, R., He, K., Dollár, P.: Focal loss for dense object detection. In: Proceedings of the IEEE International Conference on Computer Vision, pp. 2980–2988 (2017)
21. Lin, T.-Y., et al.: Microsoft COCO: common objects in context. In: Fleet, D., Pajdla, T., Schiele, B., Tuytelaars, T. (eds.) ECCV 2014. LNCS, vol. 8693, pp. 740–755. Springer, Cham (2014). https://doi.org/10.1007/978-3-319-10602-1_48
22. Liu, W., et al.: SSD: single shot multibox detector. In: Leibe, B., Matas, J., Sebe, N., Welling, M. (eds.) ECCV 2016. LNCS, vol. 9905, pp. 21–37. Springer, Cham (2016). https://doi.org/10.1007/978-3-319-46448-0_2
23. Liu, Z., et al.: Swin transformer: hierarchical vision transformer using shifted windows. arXiv preprint arXiv:2103.14030 (2021)
24. Luo, J., Li, Z., Wang, J., Lin, C.Y.: ChartOCR: data extraction from charts images via a deep hybrid framework. In: Proceedings of the IEEE/CVF Winter Conference on Applications of Computer Vision, pp. 1917–1925 (2021)
25. Ma, W., et al.: Towards an efficient framework for data extraction from chart images. arXiv preprint arXiv:2105.02039 (2021)
26. Oglan, V.A.: Chart sense: common sense charts to teach 3–8 informational text and literature. Lang. Arts **92**(5), 368 (2015)
27. Redmon, J., Divvala, S., Girshick, R., Farhadi, A.: You only look once: unified, real-time object detection. In: Proceedings of the IEEE Conference on Computer Vision and Pattern Recognition, pp. 779–788 (2016)
28. Redmon, J., Farhadi, A.: Yolo9000: better, faster, stronger. In: Proceedings of the IEEE Conference on Computer Vision and Pattern Recognition, pp. 7263–7271 (2017)
29. Redmon, J., Farhadi, A.: Yolov3: an incremental improvement. arXiv preprint arXiv:1804.02767 (2018)
30. Ren, S., He, K., Girshick, R., Sun, J.: Faster R-CNN: towards real-time object detection with region proposal networks. Adv. Neural. Inf. Process. Syst. **28**, 91–99 (2015)
31. Sandler, M., Howard, A., Zhu, M., Zhmoginov, A., Chen, L.C.: Mobilenetv 2: inverted residuals and linear bottlenecks. In: Proceedings of the IEEE Conference on Computer Vision and Pattern Recognition, pp. 4510–4520 (2018)

32. Simonyan, K., Zisserman, A.: Very deep convolutional networks for large-scale image recognition. arXiv preprint arXiv:1409.1556 (2014)
33. Tan, H., Bansal, M.: LxMERT: learning cross-modality encoder representations from transformers. arXiv preprint arXiv:1908.07490 (2019)
34. Vaswani, A., et al.: Attention is all you need. In: Advances in Neural Information Processing Systems, pp. 5998–6008 (2017)
35. Xie, S., Girshick, R., Dollár, P., Tu, Z., He, K.: Aggregated residual transformations for deep neural networks. In: Proceedings of the IEEE Conference on Computer Vision and Pattern Recognition, pp. 1492–1500 (2017)
36. Zhang, X., Zhou, X., Lin, M., Sun, J.: Shufflenet: an extremely efficient convolutional neural network for mobile devices. In: Proceedings of the IEEE Conference on Computer Vision and Pattern Recognition, pp. 6848–6856 (2018)

SCI-3000: A Dataset for Figure, Table and Caption Extraction from Scientific PDFs

Filip Darmanović[(✉)] [iD], Allan Hanbury [iD], and Markus Zlabinger [iD]

TU Wien, Vienna, Austria
filip.darmanovic.96@gmail.com,
{allan.hanbury,markus.zlabinger}@tuwien.ac.at

Abstract. Extracting figures and similar visual elements from PDFs of scientific publications is important but non-trivial, and progress is impeded by a lack of datasets for evaluation and machine learning. In this work, we describe and publish the *SCI-3000 dataset*, containing 3 000 PDFs of scientific publications (34 791 pages) with annotations of figures, tables, and corresponding captions, from the fields of *computer science*, *biomedicine*, *chemistry*, *physics*, and *technology*. We demonstrate the use of the dataset to benchmark two figure, table, and caption extraction approaches from recent literature: one rule-based and one deep learning-based.

Keywords: Figure Extraction · Table Extraction · Caption Extraction

1 Introduction

Scientific papers are generally published electronically in PDF format. Extracting information from PDFs and making it machine-actionable has proven to be a challenge, sought to be addressed by research in the field of *Page Object Detection* (POD), also referred to as *Semantic Document Segmentation*. The reason behind this challenge is that vector graphics, various symbols, tables, and other miscellaneous elements like page decorations get represented by rudimentary vector drawing commands in a PDF. This makes it difficult to extract individual blocks of text, figures, tables, etc.

Use cases for extracting these elements are numerous, especially in academia. In fields like *biomedicine* and *computer science*, interest in mining figures from previous publications is notably high [19,30]. Examples range from various figure search engines [16,18], to extracting semantic information from graphs [29], to using curated databases for training machine learning models [1,2]. Going beyond the benefits to the respective research communities, having textual descriptions of figures in the form of captions provides input data for training cross-media machine learning systems, which use different forms of the same data to extract deeper semantic meaning, for example, neural networks which

G. A. Fink et al. (Eds.): ICDAR 2023, LNCS 14187, pp. 234–251, 2023.
https://doi.org/10.1007/978-3-031-41676-7_14

learn to describe images with natural language [23]. Furthermore, Clark and Divvala [5] show that the number of figures per paper page and the average caption length have been rising steadily over the past few decades, suggesting that the amount of information presented visually has been on the increase relative to plain text.

Although there exist several tools that can take apart a PDF file with varying degrees of success [24], the task of figure, table, and caption extraction is an area with much potential for improvement. For the sake of brevity, we refer to this task simply as *figure extraction* unless explicitly noted otherwise, as in [5,30]. The previously-mentioned internal structure of the PDF poses a challenge for most tools available today, as they are usually not able to discern an entire graphical element, but instead output its individual pieces, like the background, text and so on. Extracted separately, these elements are far less valuable than the entire semantic unit they belong to. Including the corresponding caption further increases the difficulty of this task. While captions contain essential information for understanding figures and tables they describe, the number of document layouts and designs possible makes their extraction difficult. The few tools that extract both captions and graphical elements from scientific publications are, in most cases, usable only on works from specific research fields [19]. This has created the need for more advanced approaches, motivating the recent increase of research in the field of POD [20]. Nonetheless, researchers have been vocal regarding the lack of standardized metrics and datasets for evaluation and machine learning. Most extraction tools from the literature have either been tested on unpublished validation sets, or datasets that are not specifically tailored for the discipline, for example, by including non-scientific publications. Other validation datasets currently available have different issues, e.g., only containing images of pages instead of full PDFs, requiring the user to piece together the dataset from multiple sources, or focusing on only one scientific field or element type. Therefore, addressing the lack of standardized metrics and datasets is a critical research topic.

With that in mind, this paper has three main contributions:

1. A novel dataset, SCI-3000, built by annotating figures, tables, and captions in 3000 documents (34,791 pages) from the fields of *computer science, biomedicine, chemistry, physics,* and *technology*.
2. A suite of tools for evaluating figure, table, and caption detection, as well as annotation of such elements.
3. A SCI-3000-based evaluation of two figure-extraction methods from recent literature; a rule-based approach (PDFFigures2 [5]), and a deep learning-based approach (DeepFigures [30]).

While previous research efforts have predominantly focused on *computer science* and *biomedicine*, they have produced methods that do not perform well on other fields [25]. By including five research disciplines, we make our dataset more general. SCI-3000 also includes the original PDFs of publications. This is in contrast to many other datasets [30] [5] that require the user to manually acquire them because the underlying licenses prevent redistribution. We publish SCI-3000[1] under CC-BY 4.0.

Finally, our evaluation of existing figure-extraction approaches demonstrates their acceptable effectiveness for tasks where perfect recall or precision is not required. Still, it shows that there is room for improvement, especially regarding caption extraction.

2   Available Annotated Datasets

One characteristic of previous work on figure extraction and on POD in general, is the focus on single scientific disciplines. Arguably, the most focused-on domains are *computer science* (CS) and *biomedicine.*

The former was the most represented discipline in our literature research, with 12 papers either using a predominantly CS-based dataset in their evaluation phase, or focusing on building one. The two most prominent open datasets in this field are the CS-150 [6], containing 150 papers sampled from three CS conferences, and CS-Large [5], with 350 CS papers published after 1999. The ICDAR2013 [8] dataset facilitated multiple challenges regarding table detection and interpretation in PDFs. It was later extended with data on graphs in [14]. A well-used pair of datasets from this group are the ICDAR2016 and ICDAR2017 [7] challenge validation sets, sampled from the CS-focused repository CiteSeer. These datasets were used by Saha et al. [27] and Li et al. [20], before Younas et al. [36] pointed out a lack of quality in the annotations and posted an amended version. These datasets have the disadvantage of only containing rasterized versions (i.e., each page is available as an image) of papers, which means that approaches taking advantage of PDF structure cannot use them. Younas et al. [36] also noted that a dataset with more types of page objects, e.g., captions, is needed to push the field forward. CiteSeer appears to be a popular choice for source material, as many other publications sourced their datasets from it [4,29,35], even though these were never made public. Three papers from Kuzi et al. [16–18] sourced their dataset from the ACL Anthology, which is a repository consisting of work from the areas of *Natural Language Processing* and *Computational Linguistics.* Finally, Chiu et al. [3] sampled 30 papers from two CS conferences: ACM UIST and IEEE ICME. Their test dataset was also not made public.

The second research field in terms of representation was *biomedicine.* PubMed and repositories like Biomedcentral are the main sources for building PDF extraction datasets [22,28,32,34]. One popular dataset that came up during our literature review was the ImageCLEF 2016 Medical dataset [11], used by Tsutsui and Crandall [33] and Yu et al. [37]; however, this is a collection of already extracted images from medical publications. The most recent, and largest dataset in this category is PubLayNet [38] [13], which includes figures and tables along with other typical document elements. It does not, however, include relationships between those elements.

---

[1]   DOI: 10.5281/zenodo.6564971

We found two papers focusing on both *computer science* and *biomedicine.* One is [19], using the previously mentioned CS-150 dataset, and the other is [30]. For the latter dataset, several aspects are missing in terms of usability. While around 5 million annotations were released publicly on Github[2], the licenses of the underlying publications hosted on arXiv do not allow for redistribution without explicit permission from each individual author. PubMed, which they used to source PDFs from the biomedical domain, does have an open publishing arrangement with the authors[3], but the scope of this repository is focused only on biomedicine and other sub-fields of life science. This limitation means that the dataset must be pieced together from three sources. While there is no doubt that these challenges can be overcome, they definitely present hurdles for re-use in future efforts.

3 Evaluation Methodology

Correctly assessing if and how two sets of annotations differ is an essential part of our work. When evaluating existing figure extraction approaches, we need to analyze if their output matches the ground truth. In the crowd-sourced annotation stage (Sect. 4), we need to know if two people agree in their annotation of the same page. Both of these use cases can be served by a single automated annotation assessment system. Furthermore, we argue that using the same methodology in both stages is a requirement for the consistency of our work.

To make sure we implement the correct evaluation strategy, we examined related research on figure extraction from scientific publications. Most researchers described the performance of their extraction approaches through metrics like *accuracy, recall, F1 measure,* and *precision.* To apply these metrics to bounding boxes, an adaptation of the Jaccard index to 2D space was often used, called Intersection Over Union (IOU) [30]. The IOU is computed by dividing the intersection surface of two bounding boxes by the surface area of their union. Authors of [5,6,20,30] used an IOU of 0.8 as the minimum threshold when deciding if a predicted bounding box matched the ground truth.

Going beyond these similarities however, the information is so scarce, that recreating evaluation setups from most papers becomes impossible. This lack of clarity and standardized evaluation sets was also observed by Choudhury et al. [4]. Even the most influential papers in the broader field of *object detection* like [26] vaguely reference other work instead of giving a detailed description of their evaluation setup. This makes it hard to know how exactly that referenced work was applied when benchmarking new systems. Essentially, we had to resolve three main ambiguities during the implementation of our automated annotation evaluation system:

1. Mapping of bounding boxes between annotation sets.
2. Handling of misclassifications between Figures, Tables and Captions.

---

[2] https://github.com/allenai/deepfigures-open, accessed on 15.09.2021
[3] https://www.ncbi.nlm.nih.gov/pmc/tools/openftlist, accessed on 15.09.2021

3. Evaluation of relations between captions and elements they describe (refer to).

The first ambiguity was addressed by Liu and Haralick [21] by modeling the task as an *optimal assignment problem*. This formulation entails two distinct finite sets, K and L. These represent the two sets of bounding boxes we are comparing. Let there also be a cost function for associating a $k \in K$ with an $l \in L$ denoted with $q(k, l)$ (Euclidean distance between the centers of two bounding boxes in our case). The goal is to find an optimal assignment $a : K \to L$, such that the sum of costs for all one-to-one mappings is the smallest possible. If the cost is a rational valued function, like in our case, the optimal solution(s) can be found in $O(N^3)$ by applying the Hungarian algorithm [15]. The only change that has to be made to the original problem formulation is to allow K and L to have different sizes, since the prediction and ground truth sets do not necessarily have the same cardinality.

Where our approach differs from [21], and by extension from [12], is the way we handle classes and misclassification errors. In these two papers, detection (localization) and classification errors are handled separately. For example, if a predicted bounding box matches the ground truth in the IOU metric, but its class is wrong, some points are still given. In our case, however, we run the Hungarian algorithm for each class separately, meaning that a correctly detected but misclassified element would incur both a false positive for the predicted class and a false negative for the ground truth class. While our approach makes the evaluation more strict, it simplifies the result, as each prediction can either be entirely correct or incorrect.

In contrast, a more lenient approach is taken when assessing the correctness of relation assignments between captions and tables or figures. More specifically, we run the Hungarian algorithm for all classes together and match each bounding box in one annotation set to its closest corresponding annotation in the other set (if one exists). For every caption-figure/table pair, we then check if the reference relation exists between their respective closest elements in the other annotation set. A true positive is recorded if the corresponding pair of elements is linked in the same manner. When assigning the closest corresponding element, misclassification or IOU do not play a role. Only the proximity between the center points of bounding boxes is considered (Fig. 1). This design decision was made in order to make evaluating the assignment of relations between elements less dependent on the precision of drawn bounding boxes and their predicted classes.

We have made our implementation of the entire evaluation pipeline available as a python package[4] We encourage other researchers to contribute parsers and exporters for a variety of tool outputs to it.

---

[4] https://pypi.org/project/sci-annot-eval

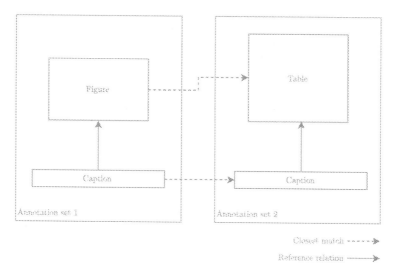

Fig. 1. Example of a matching reference relation across annotation sets. Even though the shape and class of the referenced element is not the same on both sides, we match both bounding boxes by proximity and conclude that the reference relation has been assigned in the same manner.

4   Building the SCI-3000 Dataset

Before starting with the acquisition of PDFs and the crowd-sourced annotation, we set four properties that the dataset must fulfill:

1. The included scientific works must be published with a license that allows redistribution.
2. The dataset should be relevant and useful as a training and benchmarking aid to other researchers in the area of figure extraction.
3. The dataset should facilitate focus on scientific fields that have previously been under-represented as targets of figure extraction research but have a need for such systems.
4. Each scientific field considered in the dataset should have sufficient documents to act as an individual validation dataset, able to produce performance metrics comparable based on statistical significance.

4.1   Data Source and Sampling

To obtain PDFs eligible for redistribution, we turned to the DOAJ[5], a meta-repository hosting millions of open-access publications under the CC BY-SA 4.0 license. DOAJ gives access to the metadata of all indexed work, including journal language, year of publication, and field of research according to the Library of

---

[5] https://doaj.org/

Congress Classification Scheme (LCC)[6]. The following decisions were made when downloading the papers: (i) to ensure that all papers are in English, we omitted papers from journals that are indicated as publishing papers in languages other than English; (ii) for papers that list multiple fields of research, we took the first field listed as the main one; and (iii) we used the fields at the second level of the LCC hierarchy and mapped all classifications to this level (except for papers that only provided classifications at the top level).

We included the research fields of *computer science* and *biomedicine* due to the existing extensive related work on POD in these areas.

For the further fields, the results of a meta-study [31] on the number and sizes of figures and captions in scientific publications across different research areas helped us identify research fields with an above-average number of per-page figures and caption lengths. We also identified fields in which at least some initial work on POD has been done.

Praczyk et al. [25] focused on the automatic extraction of figures from the field of *high-energy physics*. When referencing *physics* in the previously-mentioned meta-study [31], the authors found that the field has an above-average number of figures (0.8 compared to 0.7), charts (5.7 versus 3.6), as well as caption length (468 characters versus 411), which further reinforces the field as a relevant target of figure-caption extraction approaches.

Choudhury et al. [4] describe an end-to-end figure-caption extraction and search engine system for chemistry. In the meta-study [31], the authors found a slightly above-average number of graphs per paper (3.7 compared to 3.6), as well as caption length (416 versus 411 characters), though the number of images is significantly lower than the mean (0.3 compared to 0.7).

Kuzi et al. [18] explored the use of their system FigExplorer in supporting mechanical failure diagnosis. Referencing the meta-study [31], the field of mechanical engineering has more than double the average number of charts per paper and almost five times more images than the mean; however, the average caption length lies significantly under the mean (119.8 characters compared to 411). Looking at other similar fields, we noticed the same trend, even more pronounced. Therefore, we decided to generalize by including the entire first-level classification of *technology* (T) from the LCC in our dataset.

To summarize, we have identified five research fields for which figure extraction is critical: *computer science*, *medicine*, *physics*, *chemistry*, and *technology*. We equally split the entire corpus into these five research fields and sample 3000 documents, containing 34,791 pages in total. Rasterized versions of these pages were created using version 22.02.0 of Poppler[7], using the default media box cropping. We limited the maximum number of pages per paper to 20 to prevent a sampling bias towards longer publications, which would reduce the variety of visual styles in the dataset. Note that some of the sampled papers are cross-discipline, belonging to more than just one of the five selected research fields.

---

[6] https://www.loc.gov/catdir/cpso/lcco/
[7] https://poppler.freedesktop.org/, accessed on 24.04.2023

4.2 Data Annotation

We go into detail on how we used Amazon Mechanical Turk (AMT) to crowd-source the annotations of the 34,791 pages. The annotation tool used in this process is available on GitHub[8]

Task Specification. Our task is defined as follows: Bounding boxes have to be drawn around *figures*, *tables* and their corresponding *captions* in rasterized document pages. In addition to determining their location and size, the element to which each caption refers needs to be established. A reference relation has the cardinality of 1:1, meaning that captions refer to a single element and vice-versa, although ones without a reference relation are also allowed. While this should not come up in the context of an entire document, focusing on one page at a time makes elements without a reference possible if they refer to each other across pages [19], like a figure whose caption is on the following page. Another case where this might happen is if a table or figure is broken into multiple parts to fit on one page. Although the problem could be solved by assigning multiple bounding boxes to one element, it makes the system needlessly complicated, so in our formulation of the task, the caption always references its closest part of figure or table it describes, while all others are considered separate elements without a reference.

Submission Review Policy. To ensure that submissions are accepted and rejected consistently and to asses the quality of our dataset, we designed a clear and transparent submission review process. We describe this process and explain how it was used to build a pool of workers for our task.

Our previous experience with AMT has shown that picking a few top workers to annotate the entire corpus is more efficient than opening the task for everyone and manually reviewing erroneous submissions. To rank workers, we built a scoring system by giving a worker one point each time a submission was manually verified as correct by us and subtracting five if it was rejected. This grading disparity is motivated by the fact that around half of the pages have no elements to annotate. Making the reward and penalty equal would make the score a less meaningful indicator of the quality of work. In terms of review criteria, we aim to be fair and only reject submissions that are intentionally wrong. For example, assigning random bounding boxes, skipping clearly visible elements, and submissions that violate our instructions. In cases where less severe mistakes are made, like imprecise bounding boxes, we simply correct the submission. In such cases, the worker is still compensated after 72 h without rewarding or discounting points.

---

[8] DOI: 10.5281/zenodo.7878627

To build a pool of qualified workers, we submit pages to AMT in batches of 100 and wait until they are all worked through. At this stage, each page is annotated by only one person and subsequently reviewed by us. The batching is done to avoid a small number of workers speeding through all of the possible assignments hoping they would get the payment without review. Since it would be infeasible for us to manually review all thirty-four thousand pages, once the qualified worker pool reaches around 15 members, we start the main annotation phase by posting more tasks and letting two annotators label each page. We set a threshold using AMT's *Qualification* feature so that only workers with over a certain number of points could see and work on them.

Manual disagreement resolution would only be needed when two submissions for the same page have not passed the automated evaluation procedure. It works by first cropping the whitespace in every bounding box and then applying the evaluation framework described in Sect. 3, with an IOU threshold set to 95%. We have released our system for administrating annotation by AMT on GitHub[9]

Task Pricing. To help us determine a fair compensation amount, we turned to observational studies of the crowdsourcing marketplace. Two in-depth studies by Hara et al. [9,10] used an opt-in browser plugin to collect metadata for 3.8 million task instances from AMT, including the compensation. They found that, once unpaid work like searching for tasks was accounted for, the mean and median hourly wages were \$3.31/h and \$1.77/h, respectively. With this way of calculating wages, only 4% of workers earned more than the U.S. federal minimum wage of \$7.25/h. Ignoring the unpaid work, the median and mean wages rise to \$3.18/h and \$6.19/h, respectively. Splitting the earnings by task type, the authors found that the task of image transcription, which is closely related to our work, is by far the lowest-paid task type on the platform, with a median wage of \$1.13/h, while at the same time having the most instances compared to other types.

With the above-mentioned findings in mind, we settle on a price around the U.S. federal minimum wage of \$7.25/h, which we believe is fair considering that the task does not require any special qualifications. The workers are paid per annotated page.

After settling on an hourly wage, we measure the median time needed to annotate a single page and use the result to infer the final compensation amount per page. We enlisted three volunteers that have never done this task or used our tool to annotate 40 pages each. The average annotation time measured in this experiment was between 20 and 35 s. Therefore, we set the payment per page to \$0.04.

Annotation Results. The crowd-sourced annotation process was started by building a pool of qualified workers. The threshold for the worker score was set to 25 and stayed in that range during the entire run. This distilled about

---

[9] DOI: 10.5281/zenodo.7878638

Table 1. Annotation statistics by research field.

Research Field	Page Count	Figures	Tables	Captions	Empty Pages
Chemistry	7,664	3,880	1,226	5,092	3,981
Computer Science	7,796	4,277	1,784	5,999	4,244
Medicine	6,144	1,752	1,431	3,147	3,872
Physics	5,520	4,025	683	4,703	2,576
Technology	7,667	4,448	1,613	6,027	3,793
Total	**34,791**	**18,382**	**6,737**	**24,968**	**18,466**

15 workers out of the 164 from the phase where each submission was manually reviewed from our side. As the second stage, where each page was annotated by two workers, took us three weeks, a few more runs of single-worker annotations were performed to increase the size of the worker pool. At the end of the process, we had 241 workers and 62,100 submissions but only 20 workers were responsible for more than 90% of them.

77.8% of all annotations in our dataset were the case where a page was annotated by two workers and their submissions passed our automatic evaluation procedure (IOU between the annotations greater than 95%). Since those submissions were nearly identical, a random one was picked as the final annotation in our dataset. The second group of annotations, at 16.4%, resulted from either in-house annotation or corrections to submissions from the initial annotation phase (one worker per page). The final 5.7% are disagreements between workers that we manually resolved by either picking the correct submission or amending annotation mistakes. When taking into consideration only pages that were annotated by at least two workers, we derive an inter-annotator agreement of 93.1%.

Throughout the experiment, workers would send requests to overturn our rejections of their submissions. We handled each of these on a case-by-case basis and always made sure to explain what was wrong with the submission and why we rejected it. In total, our rejection rate was less than 1%, most of which were submissions from the initial pool-building stage.

When analyzing the working time, our initial estimates were correct, as the average time per task was just over 22 s.

A per-field breakdown of the annotated objects is shown in Table 1. The entire annotated dataset contains 18,382 figures, 6,737 tables, and 24,968 captions. Roughly every second page has contains one annotation.

All annotations in the published dataset have had white space surroundings cropped to make them as precise as possible. Full details on all aspects of the dataset are available in the thesis[10] on which this paper is based.

---

[10] https://doi.org/10.34726/hss.2022.94800

5   Evaluation

We test the performance of two existing approaches for figure, table, and caption extraction from scientific publications on the SCI-3000 dataset.

5.1   Evaluated Approaches

Approaches in this research field of POD can be classified on a spectrum between rule-based systems on one end and machine-learning-based ones on the other. Between them are systems using both approaches in various proportions. The best representative of the rule-based group in terms of impact is PDFFigures 2.0 [5]. This system by Clark and Divvala (including the first version [6]) was referenced by a significant number of papers in the field [4, 16, 18, 19, 29–31, 33, 36].

As a representative approach for figure extraction using machine learning, we selected DeepFigures [30]. This system was trained on the largest dataset for our task currently available, containing over a million papers and over 5.5 million labels. Additionally, the dataset contains works from several fields, including *biomedicine, computer science, biology*, and *physics*. This should make the model more robust than PDFFigures 2.0, which was fine-tuned only on *computer science* papers. A possible drawback of DeepFigures in the context of our comparison is that it uses PDFFigures 2.0 for detecting and assigning captions to graphical elements, meaning that both systems share the same approach for this sub-task. The authors justify this decision by the way of reduced performance when the model is trained to also identify captions, although they describe a different design that could produce a neural network capable of performing both sub-tasks equally well.

5.2   Experiment Setup

To compute predictions from the selected systems, their respective source codes were downloaded from GitHub[11][12].

During the benchmarks, both systems had difficulties with some documents because of special PDF features or encodings. We skipped thirteen PDFs for Deepfigures and four for PDFFigures 2.0. A lack of output was considered as an empty prediction, meaning that a false negative prediction is counted for each ground truth annotation in skipped documents. We ran both systems on a machine with 8 AMD EPYC 7542 cores, 8 GB of memory, and around 300 GB of storage. PDFFigures 2.0 took three hours to complete, and Deepfigures needed more than 72. However, our aim is not to compare runtimes, and therefore, we have not used any optimizations that could improve these results.

For the actual evaluation process, we use the strategy described in Sect. 3 to get True Positive (TP), False Negative (FN), and False Positive (FP) per-page

---

[11] https://github.com/allenai/pdffigures2, accessed on 15.05.2022

[12] https://github.com/allenai/deepfigures-open, accessed on 15.05.2022

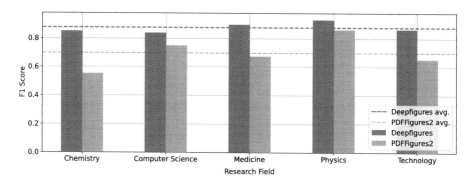

Fig. 2. F1 Score comparison between PDFigures 2.0 and DeepFigures for the task of figure detection. The bars represent average F1 scores grouped by research fields, while the dotted lines represent overall averages.

counts for four element types: *figures, tables, captions* and *references between captions and their corresponding elements* at an IOU of 0.8. From these counts, we derive the key evaluation metrics: *Precision, Recall*, and the *F1* score.

The metrics are calculated per *element type* and *research field*. We use the macro-averaging strategy for any averages displayed in the next section by first computing the mean inside each element group and then using the results to derive the overall average. Because there is a caption for almost every graphical element and a reference relation between each caption and a figure/table, computing averages over all element types in one step (micro-averaging) would skew the performance metrics towards these two larger groups. The same reasoning is applied to research fields, as the amount of graphical elements varies between them.

5.3 Results

An in-depth breakdown of performance metrics is provided in Table 2. We guide the reader through these results in a visual manner, starting with figure-specific performances.

As shown in Fig. 2, there is a substantial performance difference between PDFFigures 2.0 and DeepFigures regarding figure detection. The former reaches an F1 score of 0.68, while the latter does better, with a score of 0.79. The hand-tuned nature of PDFFigures 2.0 can also be seen in the difference in its performance across different research fields. The system seems to struggle with publications in the fields of *chemistry* and *technology*, while *physics* publications seem to be a better extraction target than *computer science*: the field for which the system was optimized. On the other hand, DeepFigures shows a similar F1 score across all research fields.

Moving to table detection (Fig. 3), a similar discrepancy can be seen between the two systems, as both reach almost the same scores for extracting figures, demonstrating the similarities between those two tasks. This time, however,

Table 2. Evaluation results for PDFFigures2 and DeepFigures

Research Field	El. Type	PDFFigures2			DeepFigures		
		F1	Prec	Rec	F1	Prec	Rec
Average	Average	0.68	0.75	0.62	0.79	0.84	0.76
	Caption	0.52	0.51	0.53	0.52	0.56	0.48
	Figure	0.70	0.83	0.60	0.88	0.95	0.81
	References	0.81	0.88	0.75	0.95	0.95	0.95
	Table	0.69	0.79	0.61	0.83	0.89	0.77
Chemistry	Average	0.60	0.68	0.55	0.80	0.85	0.76
	Caption	0.50	0.46	0.54	0.51	0.57	0.47
	Figure	0.55	0.74	0.44	0.85	0.97	0.76
	References	0.69	0.75	0.64	0.94	0.92	0.96
	Table	0.67	0.77	0.60	0.88	0.93	0.84
Computer Science	Average	0.71	0.79	0.65	0.75	0.82	0.70
	Caption	0.54	0.55	0.53	0.52	0.58	0.47
	Figure	0.75	0.87	0.66	0.84	0.93	0.76
	References	0.87	0.94	0.81	0.94	0.95	0.92
	Table	0.70	0.81	0.61	0.72	0.81	0.65
Medicine	Average	0.67	0.77	0.61	0.82	0.86	0.78
	Caption	0.53	0.53	0.53	0.54	0.58	0.51
	Figure	0.68	0.83	0.57	0.90	0.95	0.85
	References	0.81	0.90	0.73	0.97	0.98	0.97
	Table	0.68	0.81	0.59	0.85	0.93	0.79
Physics	Average	0.76	0.81	0.72	0.84	0.87	0.81
	Caption	0.60	0.59	0.60	0.60	0.63	0.58
	Figure	0.86	0.93	0.80	0.93	0.97	0.90
	References	0.92	0.96	0.88	0.98	0.98	0.97
	Table	0.67	0.77	0.59	0.85	0.91	0.80
Technology	Average	0.64	0.71	0.59	0.76	0.80	0.73
	Caption	0.43	0.40	0.46	0.42	0.46	0.39
	Figure	0.65	0.80	0.55	0.86	0.95	0.79
	References	0.77	0.84	0.72	0.94	0.93	0.94
	Table	0.71	0.79	0.64	0.83	0.87	0.79

PDFFigures 2.0 reaches consistent results across research fields. On the other hand, DeepFigures underperforms on publications from *computer science*.

For the task of detecting correct references between captions and tables/figures, both systems performed better than in the previous two (Fig. 4). DeepFigures achieves a precision and recall of 0.95, indicating that even when the underlying bounding boxes do not perfectly overlap with the ground truth,

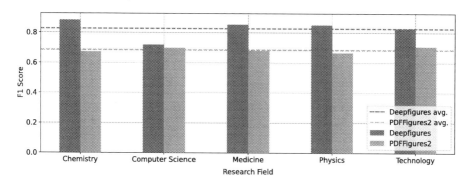

Fig. 3. F1 Score comparison between PDFigures 2.0 and DeepFigures in the task of table detection. The bars represent average F1 scores grouped by research fields, while the dotted lines represent overall averages.

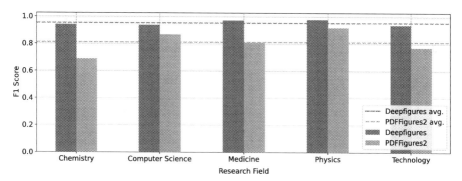

Fig. 4. F1 Score comparison between PDFigures 2.0 and DeepFigures for the task of reference assignment between captions and tables or figures. The bars represent average F1 scores grouped by research fields, while the dotted lines represent overall averages.

the system has a good idea of which elements reference each other. PDFFigures 2.0 achieves an average precision of 0.87 and an average recall of 0.75. The system shows similar performance across research fields for the figure extraction task, suggesting that its reduced performance in reference matching arises from its inability to consistently detect figures.

For caption detection, PDFFigures 2.0 and DeepFigures reach an F1 score of around 0.5. We skip a direct comparison between the systems for this sub-task, as DeepFigures relies on the output of PDFFigures 2.0, making their performance nearly identical. The reason for this performance drop compared to other sub-tasks is that PDFFigures 2.0 often produces caption bounding boxes that do not enclose the entire caption (see example in Fig. 5). This difference is small when considering the absolute area of the boxes; however, the small size of captions makes the relative difference significant enough, that the prediction does not pass an IOU threshold of 0.8. This problem was cited by the authors of DeepFigures [30] as one of the main hurdles in the neural network training process and was

F. Darmanović et al.

Table 1. Stations information and rainfall missing values

(a) Imprecise bounding box - Produced by PDFFigures 2.0.

Table 1. Stations information and rainfall missing values

(b) Correct bounding box.

Fig. 5. Example of an imprecise caption bounding box produced by PDFFigures 2.0, compared with our annotation.

the reason why they decided to use PDFFigures 2.0 as the underlying caption extraction mechanism. The authors of PDFFigures 2.0 even reported this as an issue in the evaluation phase and used OCR-ed text as a fallback matching technique [5]. The problem could be fixed by snapping the bounding boxes to a grid in order to make them less sensitive to changes, but that introduces another variable to the evaluation process. In our case, the F1 metric for the reference detection task shows that PDFFigures 2.0 is effective at detecting captions but ineffective at precisely defining their bounding boxes.

6 Conclusion

We addressed one of the most prevalent problems currently plaguing research on figure, table, and caption extraction from scientific PDFs: the lack of a large, cross-discipline, and easily-accessible dataset. We published SCI-3000: a novel dataset of annotated scientific publications from five research areas: *computer science, biomedicine, chemistry, physics*, and *technology*. Two state-of-the-art figure, table, and caption extraction methods were evaluated on our dataset, using an evaluation protocol we made publicly available as a python library.

The SCI-3000 dataset not only surpasses most of its predecessors in size and scope by incorporating new scientific fields, but also provides source publications in PDF format, made possible by the permissive licensing of the sourced PDF articles. This characteristic makes the dataset viable for extension and re-publication, for example, by adding new annotations for elements like equations and paragraphs. An alternative future research path would be to make the available annotations more specific, for example, by classifying figures into different types such as graphs, light-photography, or biomedical images.

Our evaluation of state-of-the-art methods showed that there is still room for improvement, especially for the task of caption detection. Therefore, developing more effective extraction and caption detection methodologies is another viable path for future research.

References

1. Ahmed, Z., Zeeshan, S., Dandekar, T.: Mining biomedical images towards valuable information retrieval in biomedical and life sciences. Database **2016**, baw118 (2016). https://doi.org/10.1093/database/baw118

2. Asgari Taghanaki, S., Abhishek, K., Cohen, J.P., Cohen-Adad, J., Hamarneh, G.: Deep semantic segmentation of natural and medical images: a review. Artif. Intell. Rev. **54**(1), 137–178 (2021)

3. Chiu, P., Chen, F., Denoue, L.: Picture detection in document page images. In: Proceedings of the 10th ACM Symposium on Document Engineering, pp. 211–214. DocEng 2010, Association for Computing Machinery, New York (2010). https://doi.org/10.1145/1860559.1860605

4. Choudhury, S.R., et al.: A figure search engine architecture for a chemistry digital library. In: Proceedings of the 13th ACM/IEEE-CS joint Conference on Digital libraries, pp. 369–370. JCDL 2013, Association for Computing Machinery, New York (2013). https://doi.org/10.1145/2467696.2467757

5. Clark, C., Divvala, S.: PDFFigures 2.0: Mining figures from research papers. In: 2016 IEEE/ACM Joint Conference on Digital Libraries (JCDL), pp. 143–152 (2016)

6. Clark, C.A., Divvala, S.: Looking beyond text: extracting figures, tables and captions from computer science papers. In: Workshops at the Twenty-Ninth AAAI Conference on Artificial Intelligence (2015), https://www.aaai.org/ocs/index.php/WS/AAAIW15/paper/view/10092

7. Gao, L., Yi, X., Jiang, Z., Hao, L., Tang, Z.: ICDAR2017 competition on page object detection. In: 2017 14th IAPR International Conference on Document Analysis and Recognition (ICDAR), vol. 01, pp. 1417–1422 (2017). https://doi.org/10.1109/ICDAR.2017.231, ISSN: 2379-2140

8. Göbel, M., Hassan, T., Oro, E., Orsi, G.: ICDAR 2013 table competition. In: 2013 12th International Conference on Document Analysis and Recognition, pp. 1449–1453 (2013). https://doi.org/10.1109/ICDAR.2013.292, ISSN: 2379-2140

9. Hara, K., Adams, A., Milland, K., Savage, S., Callison-Burch, C., Bigham, J.P.: A data-driven analysis of workers' earnings on amazon mechanical turk. In: Proceedings of the 2018 CHI Conference on Human Factors in Computing Systems, pp. 1–14. Association for Computing Machinery, New York (2018), https://doi.org/10.1145/3173574.3174023

10. Hara, K., et al.: Worker demographics and earnings on amazon mechanical turk: an exploratory analysis. In: Extended Abstracts of the 2019 CHI Conference on Human Factors in Computing Systems, pp. 1–6. CHI EA 2019, Association for Computing Machinery, New York (2019). https://doi.org/10.1145/3290607.3312970

11. García Seco de Herrera, A., Schaer, R., Bromuri, S., Müller, H.: Overview of the ImageCLEF 2016 medical task. In: Working Notes of CLEF 2016 - Conference and Labs of the Evaluation forum (2016)

12. Hoiem, D., Chodpathumwan, Y., Dai, Q.: Diagnosing error in object detectors. In: Fitzgibbon, A., Lazebnik, S., Perona, P., Sato, Y., Schmid, C. (eds.) ECCV 2012. LNCS, vol. 7574, pp. 340–353. Springer, Heidelberg (2012). https://doi.org/10.1007/978-3-642-33712-3_25

13. Jimeno Yepes, A., Zhong, P., Burdick, D.: ICDAR 2021 competition on scientific literature parsing. In: Lladós, J., Lopresti, D., Uchida, S. (eds.) ICDAR 2021. LNCS, vol. 12824, pp. 605–617. Springer, Cham (2021). https://doi.org/10.1007/978-3-030-86337-1_40

14. Kavasidis, I., et al.: A saliency-based convolutional neural network for table and chart detection in digitized documents. In: Ricci, E., Rota Bulò, S., Snoek, C., Lanz, O., Messelodi, S., Sebe, N. (eds.) ICIAP 2019. LNCS, vol. 11752, pp. 292–302. Springer, Cham (2019). https://doi.org/10.1007/978-3-030-30645-8_27
15. Kuhn, H.W.: The Hungarian method for the assignment problem. Naval Res. Logistics Q. **2**(1–2), 83–97 (1955)
16. Kuzi, S., Zhai, C.X.: Figure retrieval from collections of research articles. In: Azzopardi, L., Stein, B., Fuhr, N., Mayr, P., Hauff, C., Hiemstra, D. (eds.) ECIR 2019. LNCS, vol. 11437, pp. 696–710. Springer, Cham (2019). https://doi.org/10.1007/978-3-030-15712-8_45
17. Kuzi, S., Zhai, C.X.: A study of distributed representations for figures of research articles. In: Hiemstra, D., Moens, M.-F., Mothe, J., Perego, R., Potthast, M., Sebastiani, F. (eds.) ECIR 2021. LNCS, vol. 12656, pp. 284–297. Springer, Cham (2021). https://doi.org/10.1007/978-3-030-72113-8_19
18. Kuzi, S., Zhai, C., Tian, Y., Tang, H.: FigExplorer: a system for retrieval and exploration of figures from collections of research articles. In: Proceedings of the 43rd International ACM SIGIR Conference on Research and Development in Information Retrieval, pp. 2133–2136. SIGIR 2020, Association for Computing Machinery, New York (2020). https://doi.org/10.1145/3397271.3401400
19. Li, P., Jiang, X., Shatkay, H.: Figure and caption extraction from biomedical documents. Bioinformatics **35**(21), 4381–4388 (2019)
20. Li, X.H., Yin, F., Liu, C.L.: Page object detection from PDF document images by deep structured prediction and supervised clustering. In: 2018 24th International Conference on Pattern Recognition (ICPR), pp. 3627–3632 (2018). https://doi.org/10.1109/ICPR.2018.8546073, ISSN: 1051-4651
21. Liu, G., Haralick, R.M.: Optimal matching problem in detection and recognition performance evaluation. Pattern Recogn. **35**(10), 2125–2139 (2002)
22. Lopez, L.D., Yu, J., Arighi, C.N., Huang, H., Shatkay, H., Wu, C.: An automatic system for extracting figures and captions in biomedical PDF documents. In: 2011 IEEE International Conference on Bioinformatics and Biomedicine, pp. 578–581 (2011). https://doi.org/10.1109/BIBM.2011.26
23. Peng, Y.X., et al.: Cross-media analysis and reasoning: advances and directions. Front. Inf. Technol. Electron. Eng. **18**(1), 44–57 (2017). https://doi.org/10.1631/FITEE.1601787
24. Pitale, S., Sharma, T.: Information extraction tools for portable document format. Int. J. Comput. Technol. Appl. **2**, 2047–2051 (2012)
25. Praczyk, P.A., Nogueras-Iso, J.: Automatic extraction of figures from scientific publications in high-energy physics. Inf. Technol. Libr. **32**(4), 25–52 (2013)
26. Redmon, J., Divvala, S., Girshick, R., Farhadi, A.: You only look once: unified, real-time object detection. In: 2016 IEEE Conference on Computer Vision and Pattern Recognition (CVPR), pp. 779–788 (2016). https://doi.org/10.1109/CVPR.2016.91, ISSN: 1063-6919
27. Saha, R., Mondal, A., Jawahar, C.V.: Graphical object detection in document images. In: 2019 International Conference on Document Analysis and Recognition (ICDAR), pp. 51–58 (2019). https://doi.org/10.1109/ICDAR.2019.00018, ISSN: 2379-2140
28. Shao, M., Futrelle, R.P.: Recognition and classification of figures in PDF documents. In: Liu, W., Lladós, J. (eds.) GREC 2005. LNCS, vol. 3926, pp. 231–242. Springer, Heidelberg (2006). https://doi.org/10.1007/11767978_21

29. Siegel, N., Horvitz, Z., Levin, R., Divvala, S., Farhadi, A.: FigureSeer: parsing result-figures in research papers. In: Leibe, B., Matas, J., Sebe, N., Welling, M. (eds.) ECCV 2016. LNCS, vol. 9911, pp. 664–680. Springer, Cham (2016). https://doi.org/10.1007/978-3-319-46478-7_41

30. Siegel, N., Lourie, N., Power, R., Ammar, W.: Extracting scientific figures with distantly supervised neural networks. In: Proceedings of the 18th ACM/IEEE on Joint Conference on Digital Libraries, pp. 223–232. JCDL 2018, Association for Computing Machinery, New York (2018). https://doi.org/10.1145/3197026.3197040

31. Sohmen, L., Charbonnier, J., Blümel, I., Wartena, Ch., Heller, L.: Figures in scientific open access publications. In: Méndez, E., Crestani, F., Ribeiro, C., David, G., Lopes, J. (eds.) TPDL 2018. LNCS, vol. 11057, pp. 220–226. Springer, Cham (2018). https://doi.org/10.1007/978-3-030-00066-0_19

32. Stahl, C.G., Young, S.R., Herrmannova, D., Patton, R.M., Wells, J.C.: DeepPDF: a deep learning approach to extracting text from PDFs. In: Proceedings of the 7th International Workshop on Mining Scientific Publications (2018), https://www.osti.gov/biblio/1460210

33. Tsutsui, S., Crandall, D.J.: A data driven approach for compound figure separation using convolutional neural networks. In: 2017 14th IAPR International Conference on Document Analysis and Recognition (ICDAR), vol. 01, pp. 533–540 (2017). https://doi.org/10.1109/ICDAR.2017.93, ISSN: 2379-2140

34. Yang, S.T., et al.: Identifying the central figure of a scientific paper. In: 2019 International Conference on Document Analysis and Recognition (ICDAR), pp. 1063–1070 (2019). https://doi.org/10.1109/ICDAR.2019.00173, ISSN: 2379-2140

35. Yi, X., Gao, L., Liao, Y., Zhang, X., Liu, R., Jiang, Z.: CNN based page object detection in document images. In: 2017 14th IAPR International Conference on Document Analysis and Recognition (ICDAR), vol. 01, pp. 230–235 (2017). https://doi.org/10.1109/ICDAR.2017.46, ISSN: 2379-2140

36. Younas, J., et al.: Fi-Fo detector: figure and formula detection using deformable networks. Appl. Sci. **10**(18), 6460 (2020)

37. Yu, Y., Lin, H., Meng, J., Wei, X., Zhao, Z.: Assembling deep neural networks for medical compound figure detection. Information **8**(2), 48 (2017)

38. Zhong, X., Tang, J., Jimeno Yepes, A.: PubLayNet: largest dataset ever for document layout analysis. In: 2019 International Conference on Document Analysis and Recognition (ICDAR), pp. 1015–1022 (2019). https://doi.org/10.1109/ICDAR.2019.00166, ISSN: 2379-2140

Document NLP

Consistent Nested Named Entity Recognition in Handwritten Documents via Lattice Rescoring

David Villanova-Aparisi[1(✉)] , Carlos-D. Martínez-Hinarejos[1] ,
Verónica Romero[2] , and Moisés Pastor-Gadea[1]

[1] PRHLT Research Center, Universitat Politècnica de València, Camí de Vera, s/n,
46021 València, Spain
`davilap@inf.ups.es`
[2] Departament d'Informática, Universitat de València, 46010 València, Spain

Abstract. Named Entity Recognition (NER) on handwritten documents is often approached as a decoupled process where Handwritten Text Recognition (HTR) is used to obtain a transcription and classic NER techniques are applied to this transcription. This approach is error-prone due to the noisy nature of the HTR output. On the other hand, coupled approaches, where HTR and NER are simultaneously performed, have been proposed. These coupled approaches still have difficulties to obtain consistent tagging in certain tasks because of long-term dependencies between opening and closing tags. In this paper we propose the usage of a Finite State Automata (FSA), which acts as a tagging template during the decoding stage of a coupled HTR-NER system to syntactically constrain the output. This ensures consistent tagging and improves the performance of the base system with a substantially smaller computational cost compared to other output-constraining methods.

Keywords: Named Entity Recognition · Historical documents · Consistent tagging · Computational cost

1 Introduction

Historical archives are vulnerable to degradation as time goes on. Consequently, digitization of their contents is a powerful tool to keep their data available for humankind and allow their research. However, having raw data (digitized images) could be insufficient to work with. Therefore, additional processes, such as transcription of their documents or detection of relevant entities in them, are usually performed.

In this context, the objective of historical Handwritten Text Recognition (HTR) [20] is to implement systems to obtain accurate transcriptions from

This work was supported by Grant PID2020-116813RB-I00 funded by MCIN/AEI/ 10.13039/501100011033, by Grant ACIF/2021/436 funded by Generalitat Valenciana and by Grant PID2021-124719OB-I00 funded by MCIN/AEI/10.13039/5011 00011033 and by ERDF, EU A way of making Europe.

G. A. Fink et al. (Eds.): ICDAR 2023, LNCS 14187, pp. 255–268, 2023.
https://doi.org/10.1007/978-3-031-41676-7_15

scanned pages of historical documents. This technology can help preserve ancient documents and make them more accessible for future research. Historical HTR is a mature but open area of study, as the present technology can give good results but still has to deal with the challenges associated with each corpus [2].

However, there are some tasks in which obtaining a perfect transcription is not necessarily the objective. In these cases, the goal is to perform some kind of information retrieval from the images, targeting specific fields within the records [17]. A way to solve this problem is to rely on Named Entity Recognition (NER) [13], which is a process that allows the identification of parts of text based on their semantic meaning, such as proper names or dates.

The majority of NER technology employs Natural Language Processing (NLP) models [11,30]. These models are usually trained from clean text, which makes them susceptible to input errors. This is especially significant for hand-written documents since automatic transcription by using HTR systems is not free of errors. However, there are works in which NER is performed directly over sequences of text images [19] without text recognition.

The alternative approach is to perform both tasks with the same model, known as the coupled approach. The output of this coupled approach is the transcription of the text along with the tags that help identify and categorize the Named Entities. While most of these proposals achieve a better transcription and NER due to the avoidance of error propagation [4,5], they sometimes provide inconsistent NER tagging, i.e., sequences of tags that do not fit the correct tag syntax. Some recent research has dealt with the production of consistently tagged hypotheses [24] by performing additional decoding steps, although with high computational overhead.

This work aims to obtain precise transcriptions and to efficiently perform consistent NER tagging in the same process. The main contribution of our work is the usage of a Finite State Automata (FSA) to syntactically restrict the system's output. The proposed decoding process ensures consistent tagging and, as a result, improves the model's performance. This decoding process introduces substantially less overhead compared to other output-constraining processes [24].

Our work approaches the task of historical HTR and NER over the HOME corpus [4], a multilingual and multi-author historical document database. The employed corpus poses additional challenges due to its unique tagging of Named Entities, in which nested entities appear frequently.

The rest of the paper is structured as follows. Section 2 reviews works related to this one. Section 3 overviews the employed architecture and describes the error avoidance strategies that have been considered. Section 4 describes the experimental methodology followed and discusses the obtained results. Lastly, Sect. 5 concludes the paper by remarking the key takeaways.

2   Related Work

When approximating the task of obtaining transcription and tagging entities in scanned documents, one initial idea could be to use HTR to obtain the transcription of the document and then employ NER techniques on the HTR output. This is known as the decoupled approximation [1,14]. In contrast, the coupled approximation tries to obtain the transcription and tagging of the document in a single step.

As presented above, the decoupled approximation has an error propagation problem caused by the noisy nature of the HTR output and the input requirements of the standard NER techniques, which expect clean text. Consequently, recent approximations [4,5,22] have shown that the coupled approach improves transcription and tagging performance by avoiding this mismatch between HTR output and NER input. This is well-known in other tasks such as NER from speech recognition [8].

Nevertheless, it is necessary to point out that recent decoupled approximations, such as those presented in [1,14,23], still obtain good results by using pre-trained word embeddings (such as those obtained from BERT [6]) on the HTR output. These sequences of embeddings are used for tagging, generally by using a Conditional Random Field model [9]. Other NER techniques are presented in [26,29].

In contrast, the coupled approximation [4] employs an HTR architecture adapted to the NER task. This adaptation consists of adding the tagging symbols to the alphabet of the model, which is included in the Convolutional Recurrent Neural Network (CRNN) responsible for the HTR transcription. This way, the HTR model can deal with Named Entity tagging during the decoding process. Although CRNN is one of the most popular models for this process, recent approaches have proposed the Transformer architecture [27] as well for HTR decoding.

However, current HTR models have difficulties in learning the long-term dependencies between opening and closing tags imposed by the parenthesized tagging syntax, especially for complex tasks with many categories and where nested tags may appear. This is the example of the HOME corpus and why consistency techniques are necessary to provide a consistent tag output [24].

3   Framework

3.1   Characteristics of the Task

As said above, the task we are dealing with is HTR and NER on a multilingual historical corpus [4], the HOME dataset. In this dataset, the inputs are colored images, and the expected output is the transcription of the contents of the handwritten text in the image, along with their corresponding Named Entity tags.

HOME is annotated using parenthesized notation for the Named Entities [4,5]. Therefore, opening and closing tags are required for each Named Entity.

This notation, in turn, increases the tagging process's complexity, particularly by enabling the appearance of nested Named Entities. For example, inside a proper noun entity, the name of a location may appear, causing the words of the place to pertain to the proper name as well. When decoding these nested tag structures automatically, the model must consider syntactic constraints on tag order (i.e., an opening and closing tag must match adequately). To the best of our knowledge, HOME is the only corpus reported in the literature where this phenomenon appears. Figure 1 contains an example of this type of productions.

> Ja **\<persName\>** Oldrzych z **\<placeName\>** Hradczie
> **\</placeName\>** **\</persName\>** wyznawam tiem to listem wssyem,
> ktoz gey uzrzye neb cztucz usslyssye, tak jakoss sem prodal

Fig. 1. An example of a location Named Entity nested inside a proper name Named Entity in a Czech line transcription.

Another problem comes from Named Entities that span over several lines. To solve this problem, keeping the context of previous lines to the one currently being processed would be necessary. In our case, since we are dealing with HTR and NER at line level, we assume that Named Entities that would remain open at the end of one line are automatically closed at the end of that line. This simplifies the problem and allows us to focus our efforts on the nested Named Entities challenge.

3.2   A Coupled Approach for HTR and NER

The formal definition of the coupled HTR and NER decoding starts from the classic probabilistic definition of the HTR problem. In HTR, given a text line image as input, the most likely word sequence, $\hat{w} = \hat{w}_1\hat{w}_2...\hat{w}_l$, is searched given the feature vector input sequence, $x = x_1x_2...x_m$, obtained from the input line image. Thus, it is a search problem in the probability distribution $p(w \mid x)$:

$$\hat{w} = \operatorname*{argmax}_{w} p(w \mid x) \tag{1}$$

If we consider the tagging sequence $t = t_1t_2...t_l$, related with the word sequence w, as a hidden variable in Eq. 1, we obtain the following equation:

$$\hat{w} = \operatorname*{argmax}_{w} \sum_{t} p(w, t \mid x) \tag{2}$$

If we follow the derivation presented in [4] and search for the most probable tagging sequence $\hat{t}$ during the decoding process, the resulting equation is:

$$(\hat{w}, \hat{t}) \approx \operatorname*{argmax}_{w,t} p(x \mid w, t) \cdot p(w, t) \tag{3}$$

If the information present in t is merged with the transcription w we obtain the tagged transcription h, resulting in the following equation:

$$\hat{h} \approx \underset{h}{\operatorname{argmax}}\, p(x \mid h) \cdot p(h) \tag{4}$$

Equation 4 is similar to that of the original HTR problem. The main difference is that the hypothesis $\hat{h}$ to be generated contains the most likely transcription and tagging. Therefore, we estimate both the optical probability, $p(x \mid h)$, and the syntactical probability, $p(h)$, to perform the search for the best hypothesis. We decided to implement a decoding architecture based on a Convolutional Recurrent Neural Network (CRNN) [28] to estimate the optical probability and on a character n-gram to estimate the syntactical one. Both models are combined following the approach introduced in [3].

3.3   Ensuring Consistent Tagging

Syntactical Errors at Named Entity Level. When decoding HTR and NER, the ideal system should be able to correctly tag Named Entities by enclosing them between the corresponding opening and closing tags, apart from obtaining a perfect transcription of their words. Nested entities should be opened and closed in the correct order (closing the last pending opening tag), and no nested entities of the same category may appear.

Thus, given these conditions, some errors may happen, which consist of syntactically wrong sequences at the Named Entity level. Those errors include incorrect closing of nested tags, tags that remain open at the end of the line, closing tags without a matching opening tag, and sequences of nested Named Entities of the same type. Consequently, the proposed error-correcting strategies try to discard these syntactic errors.

Exploration of the n-Best Hypotheses. In order to avoid syntactical errors, it is possible to improve the decoding process by restricting the output. Instead of obtaining the best hypothesis during the decoding phase, the authors of [24] obtain the lattice, a graph containing a sizeable number of hypotheses, using Kaldi [15]. This graph is then pruned to obtain a smaller version containing only the n-best outputs, n being a parameter that can be adjusted. The sequence of the n-best hypotheses for an input line can be obtained from this reduced lattice.

The decoding process for a line is an exploration from the 1-best hypothesis to the n-best. In such exploration, each hypothesis is reviewed by performing a left-to-right analysis to detect syntactical errors. If the current i-best hypothesis contains a syntactical error, it is rejected, and the exploration follows with the hypothesis $i+1$. This search for the best valid hypothesis finishes when a hypothesis without syntactical errors is found. For some lines, the correct output may be deep in the lattice; consequently, a large enough number of hypotheses n should be considered to obtain such output. This results in a significant overhead during the decoding process.

It may also happen that no error-free hypothesis is found among the n-best hypotheses. This exception should happen less often as the parameter n increases, but there is no possibility of knowing it beforehand, so it must always be considered as possible. In those cases, the policy chosen by the authors in [24] is to keep the 1-best hypothesis as the selected output. Therefore, the result of this decoding process will be the best valid hypothesis for each line or, in case of not finding any correct output, the one with the highest probability.

Pruning the Lattice with a Tagging Template. In this work, we propose a more refined approach that includes a prior probability to indicate the correctness of the tagged hypothesis during the decoding process. That is possible if, starting from Eq. 4, we model the probability $p(h)$ as the product of the probability given by the character language model, $p_L(h)$, and the probability given by a tagging template, $p_T(h)$. The resulting equation is:

$$\hat{h} \approx \underset{h}{\operatorname{argmax}} \, p(x \mid h) \cdot p_L(h) \cdot p_T(h) \tag{5}$$

In order to model $p_T(h)$, we analyzed the syntax of the tagged data. Our study revealed that the set of all possible outputs is a regular language, since no example of two nested tags of the same category was found within the tagged samples. Therefore, we could represent said language with a Finite State Automata (FSA). Even though the output language is complex, we can give some insight into how it is defined and, therefore, into how to build the FSA.

First, let us assume the alphabet Σ and three regular languages P, D, and O that represent different types of Named Entities.

$$\Sigma = \{a, .., Z, 0, .., 9\}$$
$$P = \{\text{<persName>}s\text{</persName>} \mid s \in \Sigma^*\}$$
$$D = \{\text{<date>}s\text{</date>} \mid s \in \Sigma^*\}$$
$$O = \{\text{<orgName>}s\text{</orgName>} \mid s \in \Sigma^*\}$$

We can define the operation $\odot$ between two different languages, e.g. P and D, describing Named Entities such that the productions of the second may be nested in the transcription of the first. This allows the production of more than one entity of the nested category inside the root entity. This behavior is intended since we found multiple appearances of this phenomenon in the chosen task. An example of this operation would be:

$$P \odot D = \{\text{<persName>}s\text{</persName>} \mid s \in (\Sigma^* \cup D)^*\}$$

The resulting FSA for this case can be seen in Fig. 2.

As we can see, the resulting language is regular. The operation is right associative and it can be applied to languages resulting from the union of languages

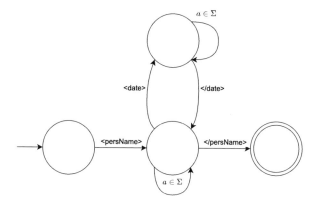

Fig. 2. FSA resulting from the operation $P \odot D$.

that describe Named Entities. For example:

$$(P \odot D \odot O) \cup (P \odot O \odot D) = (P \odot (D \odot O)) \cup (P \odot (O \odot D)) =$$
$$P \odot ((D \odot O) \cup (O \odot D)) =$$
$$\{<\text{persName}>s</\text{persName}> \mid s \in (\Sigma^* \cup ((D \odot O) \cup (O \odot D)))^*\}$$

Using the nesting operation and the union, we can define the output language as the union between Σ^* and every permutation without repetition of nested Named Entities. There are five types of Named Entities in the chosen corpus, leading us to 325 possible permutations of different lengths: 5 for one type, 20 for two types, 60 for three types, 120 for four types, and 120 for five types. Figure 2 shows a two-type permutation example. The output language has been modelled with a FSA with 205 nodes and 17715 edges.

In order to use the FSA that describes the output language as a tagging template, we set the weight of each node edge to 1. With this, the score for sequences accepted by the FSA will remain unchanged. On the other hand, sequences not belonging to the output language will have their score set to 0.

If we rescore the lattice with this model, we successfully discard every syntactically wrong sequence. From this pruned lattice, the 1-best hypothesis is the best syntactically correct hypothesis. It may happen, however, that the lattice does not contain any correct path. In those exceptional cases, we automatically correct the 1-best hypothesis from the original lattice by applying some heuristics, guaranteeing consistent tagging. The considered heuristics are: ignoring closing symbols of unopened Named Entities, closing Named Entities before reading an opening symbol of the same category or upon line ending, and using a stack to ensure that nested Named Entities are closed in the inverse order in which they were opened.

3.4   Evaluation Metrics

Several metrics have been used to assess the performance of our proposal: Character Error Rate (CER), Word Error Rate (WER), Precision, Recall, F1 Score, Entity Character Error Rate (ECER), and Entity Word Error Rate (EWER). ECER and EWER, presented in [24], take into account the Named Entity tagging error along with the inner words transcription quality. Consequently, our focus will mainly be the improvement of the ECER and EWER scores computed, for each line, as the edit distance between the sequence of recognized and transcribed Named Entities and the sequence of Named Entities in the ground truth.

We will also evaluate the system's performance in terms of computational cost. As such, we will compare the overhead introduced by each decoding strategy without considering the time it takes to generate the original lattice, which contains a large number of hypotheses generated by the base model. This step is not accounted for since it is necessary to apply the n-gram character language model, so it can be considered as a part of the baseline decoding strategy.

4   Experimental Method

4.1   HOME Dataset

Experimentation has been carried out with the HOME corpus. This corpus is composed of handwritten medieval charters, including 499 letters written by several authors in three different languages: Latin, Czech, and German. Despite its ancient nature (letters date from 1145 to 1491), the contents are well preserved. A more detailed description of the dataset can be found in [4].

We followed the same experimental scheme as in [24], which is similar to that presented in [4]. Therefore, the available data is split into three parts: a training set with 80% of the charters, a validation set with 10% of the letters, and a testing set with the remaining 10%. Table 1 presents the details on how data is partitioned.

Throughout the corpus, we can find five types of Named Entities: the name of a person, the name of a place, a date, the name of an organization, and an extra miscellaneous category for Entities that do not match any of those types. Therefore, the model must be able to generate the corresponding opening and closing tags for each type of Named Entity, which is a total of 10 additional symbols. However, the number of different syntactical structures that can be generated is significant due to the appearance of nested Named Entities.

During previous experimentation, other authors [4,14] assumed different simplifications over the original problem. In [4], nested tagging sequences were removed and Named Entities spanning over different lines (Continued Entities) were split so that their occurrences were always contained within single lines. In [14], the authors kept the Named Entities that spanned over different lines but assumed a simplification over nested Named Entities. In the case of nesting,

Table 1. HOME dataset split, equal to that proposed in [4].

	Number	**Czech**	**German**	**Latin**	**All**
Train	pages	161	138	99	398
	lines	2,905	2,556	1,585	7,046
	tokens	52,708	60,427	28,815	141,950
	N. Entities	4,973	6,024	2,809	13,806
Validation	pages	21	18	12	51
	lines	300	252	150	702
	tokens	5,997	5,841	2,467	14,305
	N. Entities	440	461	188	1,089
Test	pages	20	17	13	50
	lines	381	388	229	998
	tokens	6,891	9,843	3,995	20,729
	N. Entities	467	744	295	1,506

the authors "flattened" the Entities, splitting the parent and the nested tagged sequences into two separate Named Entities.

Our solution employs a coupled model that works at line level. Therefore, we will maintain the simplification of splitting the Continued Entities, as not doing so would lead to a noticeable performance decrease. However, we forgo the other simplification and guarantee syntactical consistency with nested Named Entities.

4.2 Implementation Details

The architecture of the system is based on the coupled approach presented in Sect. 3.2. The only preprocessing steps applied to line images are scaling to 64 pixels of height, contrast enhancement, and noise removal, as described in [25].

The optical model is composed of a CRNN with four convolutional layers, where the n-th layer has $16n$ 3×3 filters, and a Bidirectional Long Short-Term Memory (BLSTM) unit of three layers of size 256 plus the final layer with a Softmax activation function. The rest of the hyperparameters have the same values that were used in [16]. This model is implemented and trained with the PyLaia toolkit [12], which was regarded as one of the best HTR toolkits in [10].

The employed language model is a character 8-gram with Kneser-Ney back-off smoothing. The estimation of its probabilities was done with the SRILM toolkit [21] by considering the tagged transcriptions in the training partition.

The optical model and the language model are combined into a Stochastic Finite State Automata (SFSA) with the Kaldi toolkit [15]. Some additional parameters (such as Optical Scale Factor and Word Insertion Penalty) were tuned over the validation partition using the Simplex algorithm provided by

SciPy[1]. The final SFSA is used to obtain the hypothesized transcription for each test line. Our code is available at https://github.com/DVillanova/HTR-NER.

4.3 Experimental Protocol

We have replicated the experiments performed by [24] and obtained similar results with both the baseline model and the n-best decoding strategy. In order to compare the performance of the FSA tagging template to the n-best decoding, we have evaluated both strategies using the same hardware and considering the time it takes to generate the 998 test hypotheses from the same lattice generated by the baseline model. We do not consider the time it takes to generate and compile the FSA tagging template, as it can be done prior to the decoding process and is not dependent on the size of the dataset.

4.4 Obtained Results

Table 2 shows the best results obtained with each approach. The first column reports the results obtained with the baseline version of the coupled approach. The second column shows the best results obtained with the n-best decoding strategy. Lastly, the third column presents the results obtained with the usage of the proposed tagging template.

Table 2. Reference results and evaluation scores of the best proposed models, with 95% confidence intervals. Consistency stands for the percentage of syntactically correct line hypotheses within the 998 test samples.

Metric	Coupled model (baseline) [24]	Coupled model + 2500-best decoding [24]	Coupled model + tagging template
CER (%)	**9.23 ±1.80**	9.24 ±1.80	9.53 ±1.82
WER (%)	28.20 ±2.79	**28.14 ±2.79**	28.86 ±2.81
Precision (%)	43.14 ±3.07	40.05 ±3.04	**51.57 ±3.10**
Recall (%)	37.58 ±3.00	**39.97 ±3.04**	39.11 ±3.03
F1 (%)	40.17 ±3.04	40.01 ±3.04	**44.49 ±3.08**
ECER (%)	31.94 ±2.89	**28.69 ±2.81**	29.31 ±2.82
EWER (%)	46.62 ±3.10	**44.42 ±3.08**	45.62 ±3.09
Consistency (%)	75.15	97.29	**100.00**

As we can see, using a tagging template provides very similar results to those obtained with the n-best decoding strategy. There is a slight drop in transcription quality, as we can observe from the CER and WER metrics, which happens because we are selecting hypotheses with lower likelihood (deeper in the lattice). This is similar to what would happen if we used a big number of hypotheses n in the n-best decoding strategy. The difference, however, is not large enough to be considered statistically significant.

---

[1] The documentation for the employed Simplex implementation is available at: https://docs.scipy.org/doc/scipy/reference/optimize.linprog-simplex.html.

Recall decreases because the number of True Positives is slightly lower than with the n-best decoding. However, there is a statistically significant improvement in terms of Precision, as the system produces around half the amount of False Positives. As a result, there is an increase in the F1 score, although not a statistically significant one.

Compromising transcription quality, together with the lower amount of tagged Named Entities, results in a subtle decrease in ECER and EWER performance. However, even with these results, the usage of the proposed tagging template improves the base coupled model.

The main advantage of using a tagging template instead of the n-best decoding is that consistency is guaranteed without tuning the parameter n. Even in the exceptional case where no correct hypothesis is found, which happened three times in the 998 test lines, the considered heuristics are enough to produce a syntactically sound transcription. This, in turn, makes the overhead imposed by the method only depend on the number of samples to decode. In contrast, it would be necessary to employ a larger n with the n-best decoding in more complex tasks, resulting in longer decoding times. Furthermore, the temporal cost of the FSA rescoring is considerably smaller than that of the n-best decoding when n grows past a certain threshold, as we show in Fig. 3.

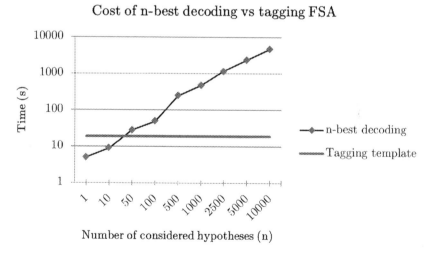

Fig. 3. Comparison between the time to decode the 998 test lines with n-best decoding, for different values of n, and with FSA rescoring. Note that the time to generate the original lattice is not considered, as it is part of the baseline decoding process.

5 Conclusions

We have presented a technique to improve the performance of the coupled model on a NER task with nested tagging. Using the proposed FSA tagging template ensures syntactical correctness while not adding as much overhead as the previously explored techniques. Therefore, we consider the proposed method a more viable solution when dealing with syntactical constraints in more extensive databases. Even if we have achieved consistent nested tagging, the challenge of dealing with Named Entities spanning over several lines remains. As future work, it would be interesting to find a way to deal with the complete task, possibly employing paragraph level recognition [7,18].

References

1. Abadie, N., Carlinet, E., Chazalon, J., Duménieu, B.: A benchmark of named entity recognition approaches in historical documents application to 19th century French directories. In: Uchida, S., Barney, E., Eglin, V. (eds.) Document Analysis Systems, pp. 445–460. Springer, Cham (2022). https://doi.org/10.1007/978-3-031-06555-2_30
2. Babu, N., Soumya, A.: Character recognition in historical handwritten documents-a survey. In: 2019 International Conference on Communication and Signal Processing (ICCSP), pp. 0299–0304. IEEE (2019)
3. Bluche, T.: Deep Neural Networks for Large Vocabulary Handwritten Text Recognition. Ph.D. thesis, Université Paris Sud-Paris XI (2015)
4. Boroş, E., et al.: A comparison of sequential and combined approaches for named entity recognition in a corpus of handwritten medieval charters. In: 2020 17th International Conference on Frontiers in Handwriting Recognition (ICFHR), pp. 79–84. IEEE (2020)
5. Carbonell, M., Villegas, M., Fornés, A., Lladós, J.: Joint recognition of handwritten text and named entities with a neural end-to-end model. In: 2018 13th IAPR International Workshop on Document Analysis Systems (DAS), pp. 399–404. IEEE (2018)
6. Catelli, R., Casola, V., De Pietro, G., Fujita, H., Esposito, M.: Combining contextualized word representation and sub-document level analysis through BI-LSTM+ CRF architecture for clinical de-identification. Knowl.-Based Syst. **213**, 106649 (2021)
7. Coquenet, D., Chatelain, C., Paquet, T.: End-to-end handwritten paragraph text recognition using a vertical attention network. IEEE Trans. Pattern Anal. Mach. Intell. (TPAMI) (2022). https://doi.org/10.1109/TPAMI.2022.3144899
8. Ghannay, S., et al.: End-to-end named entity and semantic concept extraction from speech. In: 2018 IEEE Spoken Language Technology Workshop (SLT), pp. 692–699 (2018). https://doi.org/10.1109/SLT.2018.8639513
9. Lafferty, J.D., McCallum, A., Pereira, F.C.N.: Conditional random fields: probabilistic models for segmenting and labeling sequence data. In: Proceedings of the Eighteenth International Conference on Machine Learning, ICML 2001, pp. 282–289. Morgan Kaufmann Publishers Inc., San Francisco (2001)
10. Maarand, M., Beyer, Y., Kåsen, A., Fosseide, K.T., Kermorvant, C.: A comprehensive comparison of open-source libraries for handwritten text recognition in norwegian. In: Uchida, S., Barney, E., Eglin, V. (eds.) Document Analysis Systems, pp. 399–413. Springer, Cham (2022). https://doi.org/10.1007/978-3-031-06555-2_27
11. Mikheev, A., Moens, M., Grover, C.: Named entity recognition without gazetteers. In: Ninth Conference of the European Chapter of the Association for Computational Linguistics, pp. 1–8. Association for Computational Linguistics, Bergen, Norway, June 1999. https://aclanthology.org/E99-1001

12. Mocholí Calvo, C.: Development and experimentation of a deep learning system for convolutional and recurrent neural networks. Degree's thesis, Universitat Politècnica de València (2018)
13. Mohit, B.: Named entity recognition. In: Zitouni, I. (ed.) Natural Language Processing of Semitic Languages. TANLP, pp. 221–245. Springer, Heidelberg (2014). https://doi.org/10.1007/978-3-642-45358-8_7
14. Monroc, C.B., Miret, B., Bonhomme, M.L., Kermorvant, C.: A comprehensive study of open-source libraries for named entity recognition on handwritten historical documents. In: Uchida, S., Barney, E., Eglin, V. (eds.) Document Analysis Systems, pp. 429–444. Springer, Cham (2022). https://doi.org/10.1007/978-3-031-06555-2_29
15. Povey, D., et al.: The kaldi speech recognition toolkit. In: IEEE 2011 Workshop on Automatic Speech Recognition and Understanding. No. CFP11SRW-USB, IEEE Signal Processing Society (2011)
16. Puigcerver, J.: Are multidimensional recurrent layers really necessary for handwritten text recognition? In: 2017 14th IAPR International Conference on Document Analysis and Recognition (ICDAR), vol. 1, pp. 67–72. IEEE (2017)
17. Romero, V., Fornés, A., Serrano, N., Sánchez, J.A., Toselli, A.H., Frinken, V., Vidal, E., Lladós, J.: The esposalles database: an ancient marriage license corpus for off-line handwriting recognition. Pattern Recogn. 46(6), 1658–1669 (2013). https://doi.org/10.1016/j.patcog.2012.11.024
18. Rouhou, A.C., Dhiaf, M., Kessentini, Y., Salem, S.B.: Transformer-based approach for joint handwriting and named entity recognition in historical document. Pattern Recogn. Lett. 155, 128–134 (2022). https://doi.org/10.1016/j.patrec.2021.11.010
19. Rowtula, V., Krishnan, P., Jawahar, C.: Pos tagging and named entity recognition on handwritten documents. In: Proceedings of the 15th International Conference on Natural Language Processing, pp. 87–91 (2018)
20. Sánchez, J.A., Bosch, V., Romero, V., Depuydt, K., De Does, J.: Handwritten text recognition for historical documents in the transcriptorium project. In: Proceedings of the First International Conference on Digital Access to Textual Cultural Heritage, pp. 111–117 (2014)
21. Stolcke, A.: Srilm - an extensible language modeling toolkit. In: Proc. 7th International Conference on Spoken Language Processing (ICSLP 2002), pp. 901–904 (2002)
22. Tarride, S., Lemaitre, A., Coüasnon, B., Tardivel, S.: A comparative study of information extraction strategies using an attention-based neural network. In: International Workshop on Document Analysis Systems, pp. 644–658. Springer, Cham (2022). https://doi.org/10.1007/978-3-031-06555-2_43
23. Tüselmann, O., Wolf, F., Fink, G.A.: Are end-to-end systems really necessary for NER on handwritten document images? In: Lladós, J., Lopresti, D., Uchida, S. (eds.) ICDAR 2021. LNCS, vol. 12822, pp. 808–822. Springer, Cham (2021). https://doi.org/10.1007/978-3-030-86331-9_52
24. Villanova-Aparisi, D., Martínez-Hinarejos, C.D., Romero, V., Pastor-Gadea, M.: Evaluation of named entity recognition in handwritten documents. In: International Workshop on Document Analysis Systems, pp. 568–582. Springer, Cham (2022). https://doi.org/10.1007/978-3-031-06555-2_30
25. Villegas, M., Romero, V., Sánchez, J.A.: On the modification of binarization algorithms to retain grayscale information for handwritten text recognition. In: Paredes, R., Cardoso, J.S., Pardo, X.M. (eds.) IbPRIA 2015. LNCS, vol. 9117, pp. 208–215. Springer, Cham (2015). https://doi.org/10.1007/978-3-319-19390-8_24

26. Wen, Y., Fan, C., Chen, G., Chen, X., Chen, M.: A survey on named entity recognition. In: Liang, Q., Wang, W., Liu, X., Na, Z., Jia, M., Zhang, B. (eds.) CSPS 2019. LNEE, vol. 571, pp. 1803–1810. Springer, Singapore (2020). https://doi.org/10.1007/978-981-13-9409-6_218

27. Wick, C., Zöllner, J., Grüning, T.: Transformer for handwritten text recognition using bidirectional post-decoding. In: Lladós, J., Lopresti, D., Uchida, S. (eds.) ICDAR 2021. LNCS, vol. 12823, pp. 112–126. Springer, Cham (2021). https://doi.org/10.1007/978-3-030-86334-0_8

28. Xingjian, S., Chen, Z., Wang, H., Yeung, D.Y., Wong, W.K., Woo, W.C.: Convolutional LSTM network: a machine learning approach for precipitation nowcasting. In: Advances in Neural Information Processing Systems, pp. 802–810 (2015)

29. Yadav, V., Bethard, S.: A survey on recent advances in named entity recognition from deep learning models. In: Proceedings of the 27th International Conference on Computational Linguistics, pp. 2145–2158. Association for Computational Linguistics, Santa Fe, August 2018. https://aclanthology.org/C18-1182

30. Yang, Z., Ma, J., Chen, H., Zhang, J., Chang, Y.: Context-aware attentive multilevel feature fusion for named entity recognition. IEEE Trans. Neural Networks Learn. Syst., 1–12 (2022). https://doi.org/10.1109/TNNLS.2022.3178522

Search for Hyphenated Words in Probabilistic Indices: A Machine Learning Approach

José Andrés[1,2,3(✉)] , Alejandro H. Toselli[1,2] , and Enrique Vidal[1,2,3]

[1] PRHLT Research Center, Universitat Politècnica de València, Valencia, Spain
{joanmo2,ahector,evidal}@prhlt.upv.es
[2] tranSkriptorium AI, Valencia, Spain
{joanmo2,ahector,evidal}@transkriptorium.com
[3] Valencian Graduate School and Research Network of Artificial Intelligence,
Camí de Vera s/n, 46022 Valencia, Spain

Abstract. Hyphenated words are one of the most common challenges in historical handwritten documents. For information retrieval, users issue an entire-word query and expect to retrieve all occurrences of this word, including the hyphenated ones. Thus, methods for predicting hyphenated word fragments and joining them must be developed. In this paper, we build upon and extend the work of Vidal and Toselli (2021) based on probabilistic indexing. We propose a new probabilistic framework to merge prefix/suffix word fragments into "combined spots", searchable through entire-word queries, and assess different techniques to estimate the corresponding relevance (or "spotting") probabilities. Additionally, we also consider the use of a hyphenation tool to join these text fragments at query time. We discuss the obtained retrieval results and storage cost using either probabilistic indices or plain automatic 1-best transcripts. The results show that it is possible to train a machine-learning system to join prefix/suffix word fragments automatically, with good information retrieval performance and reasonable storage usage.

Keywords: Hyphenated words · Probabilistic Indexing · Document Retrieval · Handwritten Text Recognition

1 Introduction

One major problem when transcribing and/or indexing historical manuscripts is the presence of hyphenated words. A hyphenated word (HypWrd) can be defined as a word that, due to a line break, has been split into two text fragments (HwF). These fragments account for the prefix and the suffix of the HypWrd respectively. Some examples of entire and hyphenated words can be seen in Fig. 1.

Hyphenated words present a challenge due to a variety of reasons: first, diverse hyphenation markup styles can be found in historical manuscripts, making their recognition more difficult. Typically, hyphenation is denoted using a

G. A. Fink et al. (Eds.): ICDAR 2023, LNCS 14187, pp. 269–285, 2023.
https://doi.org/10.1007/978-3-031-41676-7_16

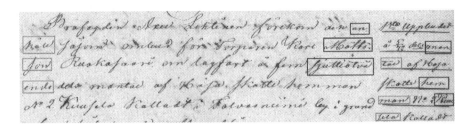

Fig. 1. Image region from a document of the *Finnish Court Records* collection. Prefix and suffix HwF's are marked in blue and orange, respectively. All the text tokens not marked with a BB are entire words.

special symbol (such as "-") at the end of the prefix. Nevertheless, this convention is not necessarily followed in historical documents: there are documents where this mark is found at the beginning of the suffix, others where it is both at the end of the prefix and the beginning of the suffix and even documents where there is no hyphenation mark at all. Furthermore, there are also other possible hyphenation symbols such as "=", ".", "∼", etc. Examples of some of the different hyphenation symbols that are used in the FCR collection can be found in Fig. 2. In addition, the position (or positions) where the hyphenation symbol is placed and the symbol itself might vary, even on the same page!

Another difficulty of this task is that historical documents might not follow the syllable-based hyphenation rules used nowadays in many modern languages. Depending on the language and time when the text was written, it is possible that hyphenation rules were different or even that they did not use rules for splitting words at all. Some examples of hyphenated words that do not follow modern hyphenation rules can be seen in Fig. 2.

Fig. 2. Examples of hyphenated words where different hyphenation symbols have been employed: "." in a), "∼" in b), "=" in c), "-" in d) and no hyphenation symbol in e). Moreover, the first three HypWrd's do not follow nowadays hyphenation rules. The word written in a) is "agande", b) is "berorde", c) is "Februari" and d) and e) are "hemman".

Last but not least, since our objective is information retrieval, it is not admissible that users have to type all the possible prefix/suffix combinations of the entire word they want to query. Therefore, it is not enough to recognize the HwF's: we need to join the HwF's to assemble the corresponding HypWrd. Moreover, note that while the correspondence between a prefix HwF and a suffix HwF might seem trivial in a relatively modern and simple document, finding this correspondence becomes challenging for documents with non-uniform and complex layouts.

In the Handwritten Text Recognition (HTR) literature, the most common approach to deal with hyphenated words is to ignore their hyphenated nature [2, 16–18, 21], considering each HwF just as another token written in the image. This scenario presents the inconvenience of increasing the lexicon with tokens that the users will not search for. Some of these approaches, [2, 16, 18] acknowledge the presence of hyphenated words and try to improve the quality of the obtained transcripts relying on the usage of hyphenation tools to artificially augment the vocabulary and enhance the language model used in the HTR decoding process. Nonetheless, tempting as it might seem, the usage of these tools present various downsides: first, as previously discussed, hyphenation rules were inconsistently used in ancient manuscripts. Second, hyphenation rules are different depending on the language and there are documents which contain multiple languages. Finally, these tools are only provided for a few modern languages, making it impossible for minority languages that lack these tools.

Other approaches [24, 25] rely on the accurate detection of hyphenated symbols to spot hyphenated words. However, as previously discussed, the usage of hyphenation symbols is not consistent in historical documents, varying the positions where the symbols might appear and the employed symbols (if any).

The next approach that we consider is the use of systems that make end-to-end recognition of paragraphs [4] or pages [5]. One of the advantages of these systems is that, as they have the context of all the lines belonging to the same paragraph or page respectively, they can naturally deal with hyphenated word fragments by joining them. Nevertheless, despite they have shown promising results when the aim is to obtain the transcription of a document, they do not match the levels of precision and recall that can be achieved employing the probabilistic indexing technology [23] when there is poor preservation of the documents, complex and non-uniform layouts, variable and erratic writing styles and other inherent difficulties of ancient manuscripts that make the recognition process more difficult. This fact makes probabilistic indexing a more appropriate technology to allow information retrieval.

Finally, in [22] a method is proposed to optically predict hyphenated word fragments and then use hand-crafted geometrical rules to join *probabilistic indexing spots* which contain likely matching HwF's. However, despite the good information retrieval results achieved, an obvious drawback is to have to develop the appropriate rules for each corpus.

In this paper, we continue and extend the work of [22] with the following contributions: first, we provide a new probabilistic formulation which better allows to rely on machine learning methods, rather than on handcrafted rules. Second, we have developed different techniques to estimate the required probabilities. Third, we have also considered an on-line technique that relies on a hyphenation tool to decompose entire query words into their possible HwF's. Next, we have also measured the storage space needed by each technique when assessing the information retrieval performance. Additionally, we also report results obtained using pruned probabilistic indices and even plain 1-best transcripts as input. Finally, we show examples of the most common errors produced by each technique and discuss probable causes of these errors.

2   Probabilistic Indexing and Search

In previous works, a Probabilistic Indexing (PrIx) framework was conceived to deal with the intrinsic word-level uncertainty exhibited by handwritten text in images in general and, in particular, images of historical manuscripts [15,23]. It stems from previous developments for keyword spotting (KWS), both in speech signals and text images. Nonetheless, rather than searching for "key" words, any element in an image which is likely enough to be interpreted as a word is spotted and stored, along with its *relevance probability* (RP) and location in the image. These text elements are referred to as *"pseudo-word spots"*.

KWS can be interpreted as the binary classification problem about whether a particular image region x is *relevant* or not for a given query word v, i.e. try to answer the question: "Is v actually written in x?". As in [15,21], we denote this image-region word RP as $P(R=1 \mid X=x, V=v)$, but for the sake of conciseness, we will omit the random variable names, and for $R=1$, we will simply write R. As discussed in [15,23], this RP can be approximated as:

$$P(R \mid x, v) \approx \max_{b \sqsubseteq x} P(R \mid x, b, v) \approx \max_{b \sqsubseteq x} P(v \mid x, b) \tag{1}$$

where b is any small, word-sized image sub-region or bounding box (BB), and with $b \sqsubseteq x$ we mean the set of all BBs contained in x. $P(v \mid x, b)$ is just the posterior probability needed to "recognize" the BB image (x, b). Therefore, assuming the computational complexity entailed by the maximization in (1) is algorithmically managed, any sufficiently accurate isolated word classifier can be used to obtain $P(R \mid x, v)$ [15].

An alternative to Eq. (1) to compute $P(R \mid x, v)$ is to use a suitable segmentation-free word-sequence recognizer [15,21,23]:

$$P(R \mid x, v) = \sum_{w} P(R, w \mid x, v) = \sum_{w:v \in w} P(w \mid x) \tag{2}$$

where w is the sequence of words of any possible transcript of x and with $v \in w$ we mean that v is one of the words of w. So the RP can be computed using state-of-the-art optical and language models and processing steps similar to those employed in handwritten text recognition, even though no actual text transcripts are explicitly produced in PrIx.

In any case, image region RPs do not explicitly take into account where the considered words may appear in the region x, but the precise positions of the words within x are easily obtained as a by-product. See details in [15,22]. All in all, the PrIx of an image x consists of a list of *"spots"* of the form:

$$[x, v, p, b] \, , \quad p \stackrel{\text{def}}{=} P(R \mid x, v, b) \tag{3}$$

where v is a "pseudo-word", b is a BB for v and x here is understood as an identifier of the corresponding page in the image collection. See details in [22].

Equation (1) will be used in Sect. 3.2 to explain our proposal to build entire-word spots for HypWrd's on the base of conventional PrIx's which contain HwF's. On the other hand, Eq. (2) will be used in Sect. 4.3 to explain how the conventional PrIx's are obtained using HTR models.

3   Searching for Hyphenated Words

In this section, we describe our proposal to allow searching for hyphenated words using PrIx's. It is based on two main components: first, prefix and suffix HwF's are optically predicted as described in Sect. 3.1. Then, those HwF's which are likely to correspond to actual HypWrd's are joined to build the corresponding entire word spots. The statistical framework to deal with such an "off-line" scenario is explained in Sect. 3.2.

We also consider an alternative "on-line" scenario where a hyphenation tool is used at query time to convert query words into boolean-AND/OR combinations of HwF's, as detailed in Sect. 3.3.

3.1   Optical and Language Modeling for Dealing with HypWrd's

Special optical and language models have been used to predict whether the last or first token of a given text line is likely to be a prefix or suffix HwF of a HypWrd, or rather a normal entire word. For this purpose, prefix and suffix HwF's of the training transcripts were tagged by appending and prepending respectively the special symbol ">". Note that if a specific hyphen character was annotated, it was removed. So a HypWrd such as Ma– ria is converted into Ma> >ria.

For *Optical Modeling*, a standard character optical model was trained using the HypWrd-tagged training transcripts. It is expected that the (often blank) right and left context of HwF's help the optical (and language) models learn to distinguish HwF's from entire words.

For *Language Modeling*, a special character N-gram is used both to probabilistic model usual character concatenation regularities, and to enforce two deterministic hyphenation-derived constraints: a HypWrd prefix can only appear at the end of a line, while a HypWrd suffix can only appear at the beginning of a line. Therefore, no HypWrd tags are allowed in the middle of a line. Once a conventional N-gram character language model has been trained, these constraints can be easily enforced by manually editing the trained back-off parameters.

3.2   Off-Line Merging of Prefix/Suffix HwF Pairs

In this approach we aim to join PrIx spots containing HwF's into new entire-word spots. To this end, the following probabilistic framework is proposed to estimate the RP $P(R \mid x, v)$ of these HwF-combined spots, where x is an image and v a HypWrd in x. Let r and s be a possible prefix and suffix of v (i.e., $v = rs$) and let b_r and b_s be possible bounding boxes (BB) of r and s respectively. From Eq. (1), $P(R \mid x, v)$ can be approximated in terms of r, s, b_r and b_s as:

$$P(R \mid x, v) \approx \max_{b \sqsubseteq x} P(R \mid x, v, b) \approx \max_{\substack{b_r, b_s, r, s: \\ rs=v, b_r, b_s \sqsubseteq x}} P(R \mid x, b_r, b_s, r, s) \quad (4)$$

Note that this expression is still valid for entire words by just considering that s is the empty string and b_s a null BB, while r is the entire word and b_r its BB.

The boolean variable R can be considered the conjunction of three boolean random variables, $R = R_r \wedge R_s \wedge R_h$, where R_r and R_s denote whether r and s are actually written in b_r and b_s, respectively, and R_h, denotes whether b_r is the BB which b_s is referencing to. Clearly R is true *iff* the three R_r, R_s and R_t are true, so we can write $P(R \mid x, b_r, b_s, r, s) \equiv P(R_r, R_s, R_h \mid x, b_r, b_s, r, s)$. Now, according to [3,6,7], as in [20], this joint probability can be adequately approximated as:

$$
\begin{aligned}
&P(R_r, R_s, R_h \mid x, b_r, b_s, r, s) \\
&\approx \min \left(P(R_r \mid x, b_r, b_s, r, s), P(R_s \mid x, b_r, b_s, r, s), P(R_h \mid x, b_r, b_s, r, s) \right) \\
&\approx \min \left(P(R_r \mid x, b_r, r), P(R_s \mid x, b_s, s), P(R_h \mid x, b_r, b_s) \right) \\
&\approx \min \left(P(r \mid x, b_r), P(s \mid x, b_s), P(R_h \mid x, b_r, b_s) \right)
\end{aligned} \quad (5)
$$

In the second step, conditional independence has been applied to simplify dependencies and in the last step we just proceed as in Eq. (1) of Sect. 2 for the RPs $P(R_r \mid x, b_r, r)$ and $P(R_s \mid x, b_s, s)$. Finally, from Eqs. (4) and (5):

$$P(R \mid x, v) \approx \max_{\substack{r, b_r, s, b_s: \\ rs=v, b_r, b_s \sqsubseteq x}} \min \left(P(r \mid x, b_r), P(s \mid x, b_s), P(R_h \mid x, b_r, b_s) \right) \quad (6)$$

The first terms, $P(r \mid x, b_r)$ and $P(s \mid x, b_s)$, are the probabilities that r and s are written in b_r and b_s respectively, as provided by the original PrIx's. The last term accounts for the (pure geometrical) probability that b_r is paired with b_s.

For the maximization (6), the *min* term is computed for each pair (r, s) found in the original PrIx of x such that $P(R_h \mid x, b_r, b_s) \geq \gamma$. Then a entire-word spot $[x, v, p, b]$ (see Eq. (3)) is built, where $v = rs$ and b is a special BB, representing the "union" of b_r and b_s. So, as byproduct of Eq. (6), all the relevant HypWrd spots of x are computed and stored among the other, regular spots.

$P(R_h \mid x, b_r, b_s)$ can be estimated in different ways, as discussed below:

- *Plain:* HwF's are not united at all. In Eq. (6) this corresponds to setting $P(R_h \mid x, b_r, b_s)$ to 0 (except for entire words). This "method" just ignores the hyphenated search problem altogether and is considered as a baseline.
- *All combinations:* All the possible pairings of prefix/suffix HwF's that are found in an image x are allowed, disregarding geometric constraints. This corresponds to unconditionally setting $P(R_h \mid x, b_r, b_s) = 1$.

- *Heuristic:* $P(R_h \mid x, b_r, b_s)$ is set to 1 *iff* some hand-crafted geometric constraints are fulfilled. We use the same rules proposed in [22]: b_s is on the left
  and a little (no more than 200 pixels) below of b_r in x.
- *MLP:* $P(R_h \mid x, b_r, b_s)$ is estimated using a Multilayer Perceptron (MLP),
  where the input features are the center, width and height of b_r and b_s.
- *Oracle:* This last method amounts to just setting $P(R_h \mid x, b_r, b_s)$ to 1 *iff*,
  according to the GT, there are two consecutive textlines beginning and ending
  with prefix and suffix HwF's, respectively, to which b_r and b_s belong. This
  method obviously yields the maximum performance that could be achieved if
  the proposed approaches to merge prefix/suffix pairs never failed.

3.3   Using Hyphenation Software to Generate HwF Queries On-line

Here we consider the idea of using a hyphenation tool at query time to decompose
a query (entire) word into its prefix/suffix HwF's and build an adequate query
with the resulting HwF's. Specifically, a boolean AND/OR query is constructed
as illustrated in the following example, where the user is assumed to search
for pages where the word "Katarina" appears. Then, this single-word query is
transformed into the following AND/OR boolean query:

$$\text{Katarina} \vee (\text{Ka}{>} \wedge {>}\text{tarina}) \vee (\text{Kata}{>} \wedge {>}\text{rina}) \vee (\text{Katari}{>} \wedge {>}\text{na})$$

   This is a rather naive approach, prone to diverse types of errors. As discussed
in Sect. 1, any non-regularly hyphenated instance of this word, such as Katar–
ina, will obviously be a *miss*. And the above AND/OR query will yield *false
positives* in images containing pairs of hyphenated words such as Kata–ria and
Seve–rina.
   Another significant downside of this idea is the larger complexities it entails
at query time. First, an adequate hyphenation tool and the corresponding word-
to-boolean-query conversion code need to be embedded into the search interface;
and then, each query becomes much more computationally complex than if only
entire words are assumed.

4   Dataset, Assessment and Empirical Settings

To evaluate empirically the performance of the proposed approaches for hyphen-
ated-word PrIx and search, the following sections describe the dataset, query
sets, evaluation measures and experimental setup adopted.

4.1   Dataset and Query Sets

The FCR-HYP [22] dataset, freely available at zenodo,[1] has been selected for
the present experiments because of its large proportion of hyphenated words.
It is composed of 600 pages images from the 18th-century manuscripts of the

---

[1] https://doi.org/10.5281/zenodo.4767732.

Finnish Court Records collection held by the National Archives of Finland. These manuscripts were written mostly in Swedish by many hands and consist mainly of records of deeds, mortgages, traditional life-annuity, among others. All the dataset images were manually transcribed, including annotation of hyphenated words, according to what was explained in Sect. 3.1.

Following a common practice in search engines and KWS systems alike, our system retrieves all instances of a given query word without matter of unessential spelling nuances. To this end, ground-truth transcripts were transliterated. As in [10], case and diacritic folding was applied, and some symbols were mapped onto ASCII equivalences to facilitate their typing on standard keyboards.

Basic statistics of the FCR-HYP dataset are reported in Table 1 for the training and test partitions, including the proportions of HwF's. More details about how this dataset was compiled and annotated can be consulted in [22].

Table 1. Basic statistics of the FCR-HYP dataset and their hyphenated word fragments (HwF's). All the text has been transliterated and the punctuation marks ignored.

Dataset partition:	Dataset		HwF's		
	Train-Val	Test	Train-Val	Test	Overall %
Images	400	200	–	–	–
Lines	25 989	13 341	10 973	5 609	42%
Running words	147 118	73 849	13 081	6 589	9%
Lexicon size	20 710	13 955	4 091	2 677	20%
ALLWORDS query set	–	10 416	–	–	–
MAYBEHYPH query set	–	1 972	–	–	–

Regarding query sets, we adopt the same ones of [22]: one named MAYBE-HYPH, with 1 972 keywords for which at least one instance in the test set is hyphenated; and other, ALLWORDS, with 10 416 keywords, which also includes the MAYBEHYPH set (see Table 1).

4.2   Evaluation Metrics

Since HypWrd's are finally indexed in the same way as entire words, search performance can be assessed as usual for a standard PrIx; namely, using the standard *recall* and *interpolated precision* [11] measures. Results are then reported in terms of both global and mean *average precision* (AP and mAP, respectively) [12], with confidence intervals at 95% ($\alpha = 0.025$) calculated employing the bootstrap method with 10 000 repetitions [1]. Furthermore, to indirectly evaluate the quality of optical and language modeling, we conduct simple handwritten text recognition experiments using these models and report word error rate to assess the 1st best recognition hypotheses (see Sect. 5.1), along with its confidence intervals at 95%.

In addition, to measure how much does the size of the PrIx increase when using the proposed methods, the *density* [19] of the obtained PrIx has been calculated as the average number of spots per running word. Therefore, the higher the density, the larger the storage required for the PrIx.

4.3   Optical and Language Model Settings

PrIx's were obtained according to Eq. (2). The required word-sequence posterior $P(w \mid x)$ is computed using character-level optical and language models. As described in [22], rather than a 1-best decoding as for plain HTR, a *character lattice* (CL) is obtained for each line-shaped image region x.

To this end, an optical model based on Convolutional-Recurrent Neural Networks (CRNN) was trained using the PyLaia Toolkit [14][2]. The same CRNN architecture of [22] was adopted, with the last output layer in charge to compute the probabilities of each of the 59 characters appearing in the training alphabet plus a non-character symbol. On the other hand, for language modelling, a 8-gram character language model (with Kneser-Ney back-off smoothing [9]) was estimated from the training transcripts. Both optical and language models were trained on transliterated transcripts, with the HypWrd prefix/suffix tagged with the ">" symbol, and used later by the Kaldi decoder [13] to produce CLs for all the test-set lines images.

Finally, a large set of the most probable word-like subpaths are extracted from the CL of each line image following the indexing methods developed in [15]. Such subpaths define the character sequences referred to as "pseudo-words" which, along with their geometric locations and the corresponding relevance probabilities, constitute the spots of the resulting PrIx.

For hyphenated words, their RPs are computed based on the PrIx entries of possible HwF's, as explained in Sect. 3, and further detailed in the next section (Sect. 3.2).

In addition, for testing the quality of produced optical and character language models, the first-best character sequences obtained from the CLs were used in the HTR evaluation described in Sect. 5.1.

4.4   Settings for Off-Line Merging Prefix/Suffix HwF's

First, in the case of the *heuristic* method for uniting prefix/suffix HwF's, the same restrictions employed in [22] are used. These are: b_s must be to the left of b_r and also a little below (no more than 200 pixels).

Second, in the case of the *MLP* approach, the network is formed by two hidden layers of 512 neurons each plus the softmax classification layer. Each hidden layer is followed by a batch normalization layer and a ReLU activation function. The Adam solver [8] was used for training with a learning rate of 10^{-3} and a batch size of 256. Next, as input features for the MLP we have considered the center, width and height of b_r and b_s, conveniently normalized by the page

---

[2] https://github.com/jpuigcerver/PyLaia.

width and/or height. As training samples, we have employed all the possible combinations of prefix/suffix HwF's that are annotated in the GT.

Finally, different values for γ in the range (0,1) have been considered. Due to this range, note that the only technique that will be affected by γ will be the *MLP* approach, since in the rest of techniques $P(R_h \mid x, b_r, b_s)$ is either 0 or 1.

4.5   Hyphenation Software Settings

To determine the different ways a word can be hyphenated at query time, we have employed the Pyphen[3] library, utilizing Swedish as the chosen language for this tool.

5   Experiments and Results

The following sections present information retrieval and storage performance results. The plain HTR transcription word error rate (WER) is also reported to more directly assess the quality of optical and language models. Moreover, a set of illustrative examples is also shown to provide a better understanding of the virtues and limitations of each approach.

5.1   Basic HTR Transcription Performance

For a direct evaluation of the quality of the trained optical and language models, the test-set images were automatically transcribed, as discussed in Sect. 4.3. Using only the optical model a 31.1% WER was achieved, while using also an 8-gram character language model the WER went down to 23.0%. In both cases, the 95% confidence interval was ±0.3%. Under the same conditions, but considering only HwF's, the respective WER values were 44.0% and 39.3%, with a 95% confidence interval of ±1.2%.

5.2   Retrieval and Storage Performance

Table 2 reports mAP, AP and density, for the MAYBEHYP query set detailed in Sect. 4.1, using the different techniques explained in Sects. 3.2 and 3.3.

The densities shown in Table 2 for not pruned PrIx correspond to a storage usage which is clearly unsuitable for large-scale applications, even without allowing HypWrd search (*plain*). These results are only aimed to show the maximum information retrieval performance that could be achieved. To obtain practical results, suitable for large-scale applications, we have considered pruning all PrIx spots with RP$< 10^{-5}$, as well as 1-best HTR transcripts (which is roughly equivalent to drastically pruning the PrIx down to one hypothesis per running word).

---

[3] https://pyphen.org/.

Table 2. mAP, AP and density for each method with the MAYBEHYP queryset and three levels of PrIx pruning: non-pruned, pruned by $RP > 10^{-5}$ and 1-best HTR transcription. The *MLP* threshold, γ, is indicated in parentheses. 95% confidence intervals are never larger than 0.02 for mAP and 0.01 for AP.

Metric	Non-Pruned PrIx			PrIx Pruned by 10^{-5}			1-best HTR		
	mAP	AP	density	mAP	AP	density	mAP	AP	density
Plain	0.44	0.81	109	0.43	0.80	10	0.35	0.72	1
Pyphen	0.70	0.88	109	0.65	0.87	10	0.43	0.74	1
All combin.	0.74	0.89	37 948	0.68	0.88	271	0.44	0.75	2
Heuristic	**0.79**	**0.91**	1 463	**0.71**	**0.89**	21	0.45	0.76	1
MLP (10^{-4})	**0.79**	0.90	3 218	**0.71**	**0.89**	33	**0.46**	**0.77**	1
MLP (0.04)	0.77	0.90	1 852	0.70	0.88	24	0.45	**0.77**	1
MLP (0.35)	0.76	0.90	1 245	0.69	0.88	19	0.45	0.76	1
Oracle	0.80	0.91	345	0.71	0.89	12	0.46	0.77	1

The density of the *all combinations* method also entails huge storage costs, even with pruned PrIx's. So these results are also aimed only at setting a simple baseline. The huge density comes from joining the HwF pairs of all the PrIx spots of each page image. Clearly, many (most) of these pairings are wrong, and do not lead to real entire words; nevertheless, since such non-sense words will never be queried, they do not hinder retrieval performance significantly.

As expected, the retrieval performance of 1-best transcripts is much worse than that of PrIx. Also as expected, the mAP and AP of all the HypWrd-aware approaches are significantly better than those of the HypWrd-agnostic *plain* "method". This clearly puts forward the need of methods which allow HypWrd searching, since failure to do so entails important losses of retrieval performance.

On the other hand, the on-line *pyphen* method also achieves a much better retrieval performance than the *plain* "method", without any density increase. Nonetheless, it does entail a significant increase in retrieval computing time and overall system complication, as discussed in Sects. 1 and 3.3. Moreover, the retrieval performance of this technique falls short with respect to all the other HypWrd-aware off-line methods, because of the inherent downsides of hyphenation tools also discussed previously.

Now, let us analyze the results of the *heuristic* and *MLP* approaches. When a small threshold γ such as 10^{-4}, is applied to the *MLP* outputs, the retrieval performance is similar to that of the *heuristic* approach, but with a significant increase of density (and storage usage). However, for larger γ values such as 0.04 or 0.35, a better density is achieved, even better than with the *heuristic* method, at the cost of a slightly worse retrieval performance. This trade-off can be seen in more detail in Fig. 3 where we observe that, as γ is increased, both the mAP and the density of the *MLP* method fall smoothly.

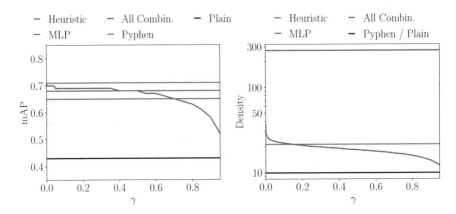

Fig. 3. MAYBEHYP mAP (left) and density (right), as a function of the MLP threshold γ, for the different techniques to unite prefix/suffix HwF's explained in Sects. 3.2 and 3.3, using PrIx's pruned by 10^{-5}.

Finally, the *oracle* results clearly show that the *heuristic* and *MLP* methods provide information retrieval performance values which are very close to the maximum achievable with the current PrIx's (and 1-best transcripts). Nonetheless, there is still room for improvement in storage usage.

Let us now consider the results for the ALLWORDS query set, reported in Table 3. The best performing approaches are also *MLP* and the *heuristic* method. Moreover, the *MLP* with $\gamma = 0.35$ provides almost identical performance as the *heuristic* approach, but at lower density. Finally, note that the difference in retrieval performance between the *plain* "method" and the other approaches is smaller than for the MAYBEHYP query set. The obvious reason is the hyphenation rate, which is 4.5% for ALLWORDS, but 13% for MAYBEHYP.

Table 4 summarizes the most relevant results. Only pruned PrIx's are considered since they have provided the best balance between storage usage and retrieval performance. All the results are very similar and very close to the maximum performance that can be achieved with the indexing accuracy currently provided by PrIx. MLP with $\gamma = 0.35$ yields the best compromise between retrieval accuracy and density. It hits the most important goal of the present work; namely, to achieve essentially the same retrieval performance and density as the *heuristic* approach, using *machine learning* methods to learn how to merge HwF's, rather than hand-crafted geometric rules which would have to be manually devised for each manuscript collection. However, these results are achieved with a density that is larger than the minimum achievable (19 versus 12 spots per running word). Therefore, future developments should focus on reducing storage usage (in addition to improve the quality of the PrIx in general).

Table 3. AP, mAP and density for each method with the ALLWORDS queryset and three levels of PrIx pruning: non-pruned, pruned by $RP > 10^{-5}$ and 1-best HTR transcription. The *MLP* threshold, γ, is indicated in parentheses. 95% confidence intervals are never larger than 0.02 for mAP and 0.01 for AP.

Metric	Non-Pruned PrIx			PrIx Pruned by 10^{-5}			1-best HTR		
	mAP	AP	density	mAP	AP	density	mAP	AP	density
Plain	0.78	0.84	109	0.72	0.83	10	0.46	0.69	1
Pyphen	0.81	0.86	109	0.75	0.85	10	0.47	0.69	1
All combin.	0.81	0.86	37 948	0.75	0.85	271	0.47	0.69	2
Heuristic	**0.84**	**0.87**	1 463	**0.77**	**0.86**	21	**0.48**	**0.71**	1
MLP (10^{-4})	**0.84**	**0.87**	3 218	**0.77**	**0.86**	33	**0.48**	**0.71**	1
MLP (0.04)	**0.84**	**0.87**	1 852	0.76	**0.86**	24	**0.48**	**0.71**	1
MLP (0.35)	**0.84**	**0.87**	1 245	0.76	**0.86**	19	**0.48**	**0.71**	1
Oracle	0.85	0.88	345	0.77	0.86	12	0.48	0.71	1

Table 4. Summary of results. mAP, AP and density for each method with the ALL-WORDS and MAYBEHYP query sets when utilizing PrIx pruned by 10^{-5}.

Query set	ALLWORDS		MAYBEHYP		
Method/Metric	mAP	AP	mAP	AP	density
Heuristic	**0.77**	**0.86**	**0.71**	**0.89**	21
MLP ($\gamma = 10^{-4}$)	**0.77**	**0.86**	**0.71**	**0.89**	33
MLP ($\gamma =0.35$)	0.76	**0.86**	0.69	0.88	19
Oracle	0.77	0.86	0.71	0.89	12

5.3   Illustrative Examples

To better understand the main virtues and most common errors of the proposed approaches, in this section we show and discuss some illustrative examples.

Figure 4 shows six pairs of HwF's accounting for prefixes (denoted in blue) and suffixes (denoted in orange). The correct prefix/suffix pairing is indicated with a letter next to the BB. In this example, all the off-line techniques have succeeded in correctly joining the six prefix/suffix pairs (except the *plain* "method", which does not merge HwF's at all). On the other hand, *pyphen* has failed to retrieve the pair of HwF's labeled with "E" because this hyphenation does not follow the rules supported by the tool.

Now, let us discuss some of the most common failures made by the different off-line approaches.

First, we consider the *all combinations* method. While this approach correctly joins the six HwF pairs of Fig. 4, it also creates thirty HypWrd spots with wrong prefix/suffix pairs, thereby increasing the density and storage requirements with useless spots.

Fig. 4. Example text region of the FCR collection. There are six prefix BBs (denoted in blue) and six suffix BBs (denoted in orange). The correct pairing between prefixes and suffixes, according to the GT, is denoted employing letters. In the text below it is detailed which techniques successfully join which pair. (Color figure online)

Second, we analyze some errors made by the *heuristic* approach. One of the most common errors is merging a prefix that is found in a marginalia with a suffix found in another text region. Clearly, the simple hand-crafted geometric rules of this method can not cope with the variabilities which appear in the many collection images. While this method does succeed in joining all the correct pairs of Fig. 4, it also adds wrong, useless HypWrd spots. Specifically, in this case the *heuristic* approach has joined the prefix B with the suffixes D and E, and the prefix C with the suffixes E and F. These errors do not occur in the *Pyphen* method, since these prefix/suffix pairs do not yield any real word that a user would query. The *MLP* approach does not make these errors either, thanks to a correctly learnt geometry of the HwF BBs b_r and b_s.

Another common source of errors made by the *heuristic* approach is sloped text lines: in Fig. 5 a), the suffix BB is slightly higher than the prefix BB and the strict geometric rules of the method do not allow joining that prefix/suffix pair. Similarly, in Fig. 5 b), the suffix BB is further down than allowed by the geometric constraints and the prefix/suffix pair has also failed to be merged. These errors do not happen in either the *Pyphen* approach, since these pairs of HwF's do follow modern hyphenation rules, or *all combinations*, which due to its nature joins all the prefix/suffix HwF pairs in a page, or the *MLP* approach, which is more robust to slope variability than the *heuristic* method.

Third, we analyze the MLP method. Figure 6 show two examples of failures made by this approach, due to very low estimates of $P(R_h \mid x, b_r, b_s)$ in both cases. These errors do also happen with the *Pyphen* method, given that these HwF's do not follow modern hyphenation rules. Nevertheless, they are correctly joined by the *all combinations* "method" (at the expense of creating many wrong, useless spots), and in the *heuristic* approach, as the BBs happen to met the hand-crafted geometrical constraints.

Finally, we would like to offer a public demonstrator which allows the search for hyphenated words using each of the off-line techniques studied in this paper. It is available at the following url: http://prhlt-carabela.prhlt.upv.es/fcr-hyp-icdar23/.

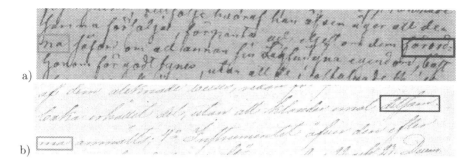

Fig. 5. Examples of errors made by the *heuristic* method due to sloped textlines. In a) the prefix/suffix pair has not been merged given that the suffix BB is above the prefix. In b) the system has not joined them because the suffix BB is farther below than allowed by the geometric restrictions. These errors are not made by any of the other HypWord-aware approaches.

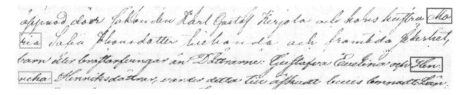

Fig. 6. Examples of prefix/suffix pairs that have not been joined by the MLP method. These errors also happen in the on-line approach, given that these HwF pairs do not match the allowed HwF pairs provided by the hyphenation tool. Nonetheless, note that these failures do not happen in the all combinations and *heuristic* methods.

6   Conclusion

To sum up, in this paper we have developed methods to allow the search of hyphenated words in untranscribed text images. To this end, we rely on probabilistic indexing (PrIx) with optical prediction of hyphenated word fragments (HwF's). In the present work we have proposed, developed and assessed a probabilistic framework for off-line merging HwF's of PrIx spots. We have compared this approach with other methods, including an on-line approach where a hyphenation tool is used at query time to convert entire words into AND/OR boolean combinations of HwF's.

The obtained results show that the off-line methods outperform the hyphenation tool approach. Both the *heuristic* method and the here proposed *MLP* approach achieve the best performances in terms of retrieval performance and storage usage. But *MLP* has the advantage of using *machine learning* to estimate a probabilistic model of HwF geometry from simple training data, rather than depending on hand-crafted geometric restrictions. This makes it more robust to diverse layout difficulties such as marginalias and sloped text lines. Moreover, it also presents the advantage of allowing adequate trade-offs between storage usage and information retrieval performance by simply choosing wisely a probability threshold.

In future works we will mainly focus on improving the storage usage of the proposed approaches. To this end, we plan to go beyond plain geometry and further take into account the lexical contents of the HwF pairs to better estimate $P(R_h \mid x, b_r, b_s, r, s)$ used in Eq. 5. Moreover, we also aim to assess these techniques employing automatically detected lines.

Acknowledgment. Work partially supported by: the SimancasSearch project as Grant PID2020-116813RB-I00a funded by MCIN/AEI/ 10.13039/501100011033, grant DIN2021-011820 funded by MCIN/AEI/ 10.13039/501100011033, the valgrAI - Valencian Graduate School and Research Network of Artificial Intelligence and the Generalitat Valenciana, and co-funded by the European Union. The second author is supported by a María Zambrano grant from the Spanish Ministerio de Universidades and the European Union NextGenerationEU/PRTR.

References

1. Bisani, M., Ney, H.: Bootstrap estimates for confidence intervals in ASR performance evaluation. In: 2004 IEEE International Conference on Acoustics, Speech, and Signal Processing, vol. 1, pp. I-409. IEEE (2004)
2. Bluche, T., Ney, H., Kermorvant, C.: The LIMSI handwriting recognition system for the HTRtS 2014 contest. In: 2015 13th International Conference on Document Analysis and Recognition (ICDAR), pp. 86–90. IEEE (2015)
3. Boole, G.: An investigation of the laws of thought: on which are founded the mathematical theories of logic and probabilities. Dover (1854)
4. Coquenet, D., Chatelain, C., Paquet, T.: SPAN: a simple predict & align network for handwritten paragraph recognition. In: Lladós, J., Lopresti, D., Uchida, S. (eds.) ICDAR 2021. LNCS, vol. 12823, pp. 70–84. Springer, Cham (2021). https://doi.org/10.1007/978-3-030-86334-0_5
5. Coquenet, D., Chatelain, C., Paquet, T.: DAN: a segmentation-free document attention network for handwritten document recognition. IEEE Trans. Patt. Anal. Mach. Intell. **PP**(99), 1–17 (2023)
6. Fréchet, M.: Généralisation du théoreme des probabilités totales. Fundam. Math. **1**(25), 379–387 (1935)
7. Fréchet, M.: Sur les tableaux de corrélation dont les marges sont données. Ann. Univ. Lyon, 3 serie, Sciences, Sect. A **14**, 53–77 (1951)
8. Kingma, D.P., Ba, J.L.: Adam: a method for stochastic optimization. In: 3rd International Conference on Learning Representations, ICLR 2015 - Conference Track Proceedings (2015)
9. Kneser, R., Ney, H.: Improved backing-off for N-gram language modeling. In: International Conference on Acoustics, Speech and Signal Processing, vol. 1, pp. 181–184 (1995)
10. Lang, E., Puigcerver, J., Toselli, A.H., Vidal, E.: Probabilistic indexing and search for information extraction on handwritten German parish records. In: 2018 16th International Conference on Frontiers in Handwriting Recognition (ICFHR), pp. 44–49 (2018)
11. Manning, C.D., Raghavan, P., Schtze, H.: Introduction to Information Retrieval. Cambridge University Press, New York, NY, USA (2008)

12. Perronnin, F., Liu, Y., Renders, J.M.: A family of contextual measures of similarity between distributions with application to image retrieval. In: 2009 IEEE Conference on Computer Vision and Pattern Recognition, pp. 2358–2365 (2009). https://doi.org/10.1109/CVPR.2009.5206505

13. Povey, D., et al.: The Kaldi speech recognition toolkit. In: IEEE 2011 Workshop on Automatic Speech Recognition and Understanding. IEEE Signal Processing Society (2011)

14. Puigcerver, J.: Are multidimensional recurrent layers really necessary for handwritten text recognition? In: 14th International Conference on Document Analysis and Recognition (ICDAR), vol. 01, pp. 67–72 (2017)

15. Puigcerver, J.: A probabilistic formulation of keyword spotting, Ph.D. thesis, Univ. Politècnica de València (2018)

16. Sánchez, J.A., Romero, V., Toselli, A.H., Vidal, E.: ICFHR2014 competition on handwritten text recognition on transcriptorium datasets (htrts). In: 2014 14th International Conference on Frontiers in Handwriting Recognition, pp. 785–790. IEEE (2014)

17. Sánchez, J.A., Romero, V., Toselli, A.H., Villegas, M., Vidal, E.: A set of benchmarks for handwritten text recognition on historical documents. Pattern Recogn. **94**, 122–134 (2019)

18. Swaileh, W., Lerouge, J., Paquet, T.: A unified French/English syllabic model for handwriting recognition. In: 2016 15th International Conference on Frontiers in Handwriting Recognition (ICFHR), pp. 536–541. IEEE (2016)

19. Toselli, A., Romero, V., Vidal, E., Sánchez, J.: Making two vast historical manuscript collections searchable and extracting meaningful textual features through large-scale probabilistic indexing. In: 2019 15th IAPR Int. Conf. on Document Analysis and Recognition (ICDAR) (2019)

20. Toselli, A.H., Vidal, E., Puigcerver, J., Noya-García, E.: Probabilistic multi-word spotting in handwritten text images. Pattern Anal. Appl. **22**(1), 23–32 (2019)

21. Toselli, A.H., Vidal, E., Romero, V., Frinken, V.: HMM word graph based keyword spotting in handwritten document images. Inf. Sci. **370–371**, 497–518 (2016)

22. Vidal, E., Toselli, A.H.: Probabilistic indexing and search for hyphenated words. In: Lladós, J., Lopresti, D., Uchida, S. (eds.) ICDAR 2021. LNCS, vol. 12822, pp. 426–442. Springer, Cham (2021). https://doi.org/10.1007/978-3-030-86331-9_28

23. Vidal, E., Toselli, A.H., Puigcerver, J.: A probabilistic framework for lexicon-based keyword spotting in handwritten text images. arXiv preprint arXiv:2104.04556 (2021). https://arxiv.org/abs/2104.04556

24. Villegas, M., Puigcerver, J., Toselli, A.H., Sánchez, J.A., Vidal, E.: Overview of the imageCLEF 2016 handwritten scanned document retrieval task. In: CLEF (Working Notes), pp. 233–253 (2016)

25. Ziran, Z., Pic, X., Innocenti, S.U., Mugnai, D., Marinai, S.: Text alignment in early printed books combining deep learning and dynamic programming. Patt. Recogn. Lett. **133**, 109–115 (2020)

A Unified Document-Level Chinese Discourse Parser on Different Granularity Levels

Weihao Liu[1], Feng Jiang[2], Yaxin Fan[1], Xiaomin Chu[1], Peifeng Li[1],
and Qiaoming Zhu[1(✉)]

[1] School of Computer Science and Technology, Soochow University, Suzhou, China
{whliu,yxfansuda}@stu.suda.edu.cn, {xmchu,pfli,qmzhu}@suda.edu.cn
[2] School of Data Science, The Chinese University of Hong Kong, Shenzhen, China
jeffreyjiang@cuhk.edu.cn

Abstract. Discourse parsing aims to comprehend the structure and semantics of a document. Some previous studies have taken multiple levels of granularity methods to parse documents while disregarding the connection between granularity levels. Additionally, almost all the Chinese discourse parsing approaches concentrated on a single granularity due to lacking annotated corpora. To address the above issues, we propose a unified document-level Chinese discourse parser based on multi-granularity levels, which leverages granularity connections between paragraphs and Elementary Discourse Units (EDUs) in a document. Specifically, we first identify EDU-level discourse trees and then introduce a structural encoding module to capture EDU-level structural and semantic information. It can significantly promote the construction of paragraph-level discourse trees. Moreover, we construct the Unified Chinese Discourse TreeBank (UCDTB), which includes 467 articles with annotations from clauses to the whole article, filling the gap in existing unified corpus resources on Chinese discourse parsing. The experiments on both Chinese UCDTB and English RST-DT show that our model outperforms the SOTA baselines.

Keywords: Discourse parsing · Information integration · Unified corpus

1 Introduction

Discourse parsing is to discover the internal structure of a document and identify the nuclearity and relationship between discourse units. It has been widely used in various natural language processing tasks, such as question-answering [1], machine translation [2], and sentiment analysis [3].

As one of the most influential theories in discourse parsing, Rhetorical Structure Theory (RST) [4] represents documents by labeled hierarchical structures, called Discourse Trees (DTs). Generally, discourse parsing is mainly divided into two levels: EDU-level (intra-paragraph) and paragraph-level (inter-paragraph).

G. A. Fink et al. (Eds.): ICDAR 2023, LNCS 14187, pp. 286–303, 2023.
https://doi.org/10.1007/978-3-031-41676-7_17

The former studies the intra- or inter-sentence relationship, while the latter studies the discourse relationship among paragraphs. An example of a discourse tree is shown in Fig. 1, its leaf nodes are Elementary Discourse Units (EDUs), and the neighbouring nodes are combined by relation and nuclearity labels to form higher-level discourse units. Since the paragraph-level unit contains more content, the structural information of the EDU level is often used to assist in determining the paragraph-level discourse tree in the annotation process.

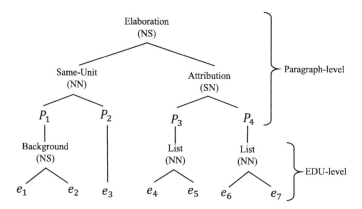

Fig. 1. Example of a discourse tree, where e and p denote EDU and paragraph, respectively.

However, due to the lack of annotated corpora, almost all the Chinese discourse parsing approaches have focused solely on obtaining better semantic representation at a single granularity level, ignoring the advantages of incorporating EDU-level structural information in enhancing paragraph-level discourse parsing. The absence of EDU-level structural information presents difficulties for the model in distinguishing components within a paragraph and extracting meaningful semantics of the paragraph. On the other hand, representing structural information is more complicated as it involves hierarchical information that cannot be easily captured through traditional semantic encoding. Therefore, representing and leveraging EDU-level structural and semantic information is a severe challenge. Moreover, the available corpora in Chinese have only ever been annotated at the EDU or paragraph level. Hence, another challenge is the lack of annotated corpus to connect the EDU-level and paragraph-level Chinese discourse parsing.

To solve the second challenge, we unify the existing two corpora with different levels of annotation to obtain a unified corpus with both the EDU-level and paragraph-level annotations. We first unify the EDU-level and paragraph-level annotation systems to form a document-level Chinese annotation system from clauses to the whole article. Then, we construct the Unified Chinese Discourse TreeBank (UCDTB) according to this annotation system which integrates CDTB [5] and MCDTB [6] and includes 467 articles.

To solve the first challenge, we propose a Document-level Chinese Discourse Parser (DCDParser) that leverages the connection between the EDU and paragraph granularity. Our model consists of two parts: the EDU-level parser and the paragraph-level parser. The EDU-level parser constructs the EDU-level DTs using EDU-level information, while the paragraph-level parser constructs the paragraph-level DTs using the EDU-level structural and semantic information and the paragraph-level information, which focuses on leveraging the connections between different granularity levels. To exploit the information of EDU-level DT, we introduce a structural encoding module, which uses label embedding method to embed the EDU-level structure into the text to obtain EDU-level structural and semantic information. Experiments on both Chinese UCDTB and English RST-DT show that our DCDParser outperforms the state-of-the-art baselines, especially the significant improvement at the paragraph level.

2 Related Work

In English, RST-DT [7] is one of the most popular corpora based on Rhetorical Structure Theory and contains 385 articles from the Wall Street Journal (WSJ). RST-DT annotates the discourse structure, nuclearity and relationship for the whole document. Unlike the study on English, research on Chinese lacks a unified corpus, resulting in a lack of research on identifying document-level discourse trees. Currently, the research of Chinese discourse parsing has two levels: EDU level and paragraph level. CDTB is a corpus annotated at the EDU level. It annotates each paragraph as a Connective-driven Discourse Tree. The corpus contains 500 articles annotated with elementary discourse units, connectives, EDU-level discourse structure, nuclearity and relationship. MCDTB is the only available corpus annotated at the paragraph level. It consists of 720 articles annotated with paragraph-level discourse structure, nuclearity, relations and additional discourse information.

On the RST-DT corpus, many works have successfully constructed document-level discourse trees [8–10]. Recently, discourse parsing utilizing multiple levels in a document has been proved to be an effective strategy. Feng et al. [11] used intra-sentence and inter-sentence parsers to parse documents. Joty et al. [12] divided the documents into two levels (intra-sentence and inter-sentence) and constructed a discourse structure parser using the dynamic conditional random field, respectively. Based on the study of Joty et al. [12], Feng et al. [13] proposed a bottom-up greedy parser with CRFs as a local classifier. Unlike previous studies at two levels, Kobayashi et al. [14] divided documents into three levels and parsed documents in a top-down method. However, none of the abovementioned works leverages the connection between granularity levels.

At the Chinese EDU level, Kong et al. [15] proposed an end-to-end Chinese discourse parser. Zhang et al. [9] used a pointer network to construct a discourse structure tree in a top-down manner. Based on this, Zhang et al. [10] used adversarial learning to consider global information and achieved the SOTA performance.

At the Chinese paragraph level, Zhou et al. [16] proposed a multi-view word-pair similarity model to capture the semantic interaction of two adjacent EDUs and constructed discourse structure trees through the shift-reduce algorithm. Jiang et al. [17] proposed the method of global backward reading and local reverse reading to construct discourse trees. Jiang et al. [18] proposed a hierarchical approach based on topic segmentation. Based on this, Fan et al. [19] proposed a method based on the dependency graph convolutional network to enhance semantic representation and interaction between discourse units and achieve the SOTA performance.

Table 1. Unified discourse relations. "EDU" means the unmergeable relations at the EDU level. "Para" means the unmergeable relations at the paragraph level.

Type	CDTB	MCDTB	UCDTB
Same	Background	Background	Background
	Coordination	Joint	Joint
	Continue	Sequence	Sequence
	Progression	Progression	Progression
	Inverse	Contrast	Contrast
	Elaboration	Elaboration	Elaboration
	Evaluation	Evaluation	Evaluation
Merged	Cause-Result	Cause-Result	Cause-Result
	–	Result-Cause	
	–	Behavior-Purpose	Purpose
	–	Purpose-Behavior	
	Purpose	–	
	–	Statement-Illustration	Example
	–	Illustration-Statement	
	Example	–	
EDU	Inference	–	Inference
	Hypothetical	–	Hypothetical
	Condition	–	Condition
	Selectional	–	Selectional
	Transition	–	Transition
	Concessive	–	Concessive
	Summary-Elaboration	–	Summary-Elaboration
Para	–	Supplement	Supplement
	–	Summary	Summary

3   Corpus Construction

As pointed out in the introduction, one crucial issue in constructing document-level Chinese discourse trees is the lack of a unified corpus. To alleviate this issue, we construct the Unified Chinese Discourse TreeBank (UCDTB) based on existing CDTB and MCDTB.

When we started to build the corpus, we found the differences in the annotation system between the above two corpora due to their different granularity. In CDTB, because EDUs are sentences or clauses with a small granularity, the connectives mainly determine the relations between discourse units. However, in MCDTB, since the EDUs are paragraphs with a larger granularity, some discourse relations used at the EDU level are redefined at the paragraph level.

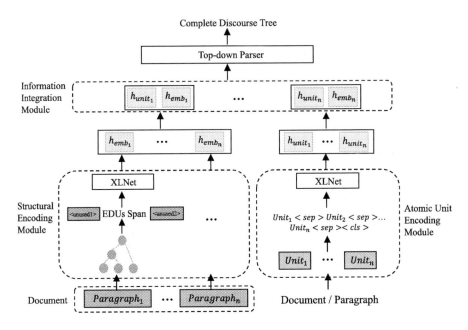

Fig. 2. The architecture of the paragraph-level parser.

As mentioned before, the different annotation systems lead to inconsistencies in the definition of discourse relations. To address this divergence, we redefine the discourse relations to form a unified Chinese annotation system from clauses to the whole article and finally obtain 19 kinds of discourse relations. As shown in Table 1, there are 7 relations with the same definition in both corpora, so we inherited them directly. For the relations with different definitions that can be merged, we merged them into 3 major categories. In addition, there are 7 and 2 unmergeable relations at the EDU level and paragraph level, respectively, which we have retained.

After unifying the annotation system, we merged the annotation information in the commonly annotated articles. Firstly, we selected articles that overlapped in MCDTB and CDTB. Secondly, we modified the text with differences. Finally, we take the paragraph-level text as the basis and insert EDU-level annotations under a paragraph if they exist. In this case, a complete EDU-level discourse tree is under each paragraph with EDU-level annotation.

At the paragraph level, UCDTB consisted of 467 articles with 2600 paragraphs. The average number of paragraphs is 5.57 per document, while the maximum number of paragraphs is 20 and the minimum number of paragraphs is 2. At the EDU level, the total number of EDUs is 10,324, with an average of 22.11 EDUs per document, while the maximum number of EDUs is 92 and the minimum number of EDUs is 2.

4   Chinese Discourse Parser on Different Granularity Levels

The connection between different granularity levels has been neglected in the existing studies on Chinese, restricting the information obtained at each level. Thus, we propose a document-level Chinese discourse parser that leverages the connection between different granularity levels. It consists of two parsers with different structures: the EDU-level parser and the paragraph-level parser. The former is to construct EDU-level discourse trees, while the latter is to construct paragraph-level DTs, which integrate the EDU-level structural and semantic information with paragraph-level information.

As shown in Fig. 2, the architecture of our paragraph-level parser consists of four components: 1) Atomic Unit Encoding Module, which is used to encode the discourse units sequences at each level and obtain their corresponding representations; 2) Structural Encoding Module, which uses label embedding to embed each EDU-level discourse tree individually. Then the embedding is encoded to obtain the structural and semantic information within the paragraph. It is worth noting that this module exists only in the paragraph-level parser. 3) Information Integration Module, which integrates the obtained EDU- and paragraph-level information and sends it to the top-down parser; 4) Top-down Parser, which identifies the discourse trees in a top-down method within each level by using a pointer network.

The EDU-level parser only contains three parts: the Atomic Unit Encoding Module, the Information Integration Module, and the Top-down Parser, i.e., without the Structural Encoding Module.

4.1   Atomic Unit Encoding Module

We encode the sequence of EDUs of different levels in the same way. For the convenience of representation, we follow Kobayashi et al. [14] and use the atomic unit sequence $Unit = \{unit_1, unit_2, ..., unit_n\}$ to represent the sequence of EDUs (paragraphs, sentences or clauses) at each level, where $unit_i$ is the i-th EDU, n is the number of EDUs. The atomic unit encoding module uses XLNet [20] to encode EDUs. To be consistent with the XLNet input, we represent the unit sequence $Unit$ into a single string S as follows.

$$S = unit_1 \langle sep \rangle \, unit_2 \langle sep \rangle \, ...unit_n \langle sep \rangle \langle cls \rangle \qquad (1)$$

Then, we feed the input S to XLNet for encoding and take out the vector h_{unit_i} at the $\langle sep \rangle$ position of each EDU as its semantic representation.

4.2   Structural Encoding Module

Since paragraph text is usually long, it is difficult for the model to grasp valuable information accurately. However, EDU-level structural information facilitates the model to better capture the core information by distinguishing the semantic relationships among the parts within a paragraph. In this paper, we aim to leverage the EDU-level structural information to facilitate our paragraph-level discourse tree construction. Since structural information is not convenient to encode directly, converting the structural information in the text to a serialized form is a sensible approach for a more straightforward representation. Inspired by Zhou et al. [21], we use label embedding to embed the structural information (i.e., structure, nuclearity) within the paragraph into the input text[1], allowing the paragraph model to obtain both EDU-level structural and semantic information when encoding.

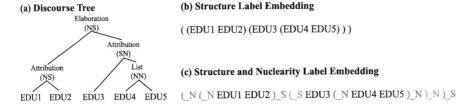

Fig. 3. Example of label embedding.

First, we refer to the form of the syntactic tree to embed the structure between discourse units into the input text, represented as a serialized form as shown in Fig. 3(b). In general, we embed the structural information in a bottom-up strategy. Specifically, we start at the bottom of the discourse tree and combine each pair of discourse units that can be merged with brackets to imply the boundary of the span. The EDU-level discourse tree shown in Fig. 3(a) is a three-level structure tree containing five EDUs. The bottom level includes two pairs of discourse units that can be merged, i.e., EDU1 and EDU2, EDU4 and EDU5. Then we combine them to obtain (EDU1 EDU2) and (EDU4 EDU5), two pairs of discourse units embedded with structural information. In the middle layer, EDU3 and (EDU4 EDU5) can be merged to obtain (EDU3 (EDU4 EDU5)). Keep searching upward and finally get the embedding representation, as shown in Fig. 3(b)[2].

---

[1] We tried to embed the relationship in, but it did not work well.
[2] In practice we use one pair of structure labels $\langle unused1 \rangle$ and $\langle unused2 \rangle$ to represent the left and right brackets.

Next, we add nuclearity information to make the model identify the core information. When merging discourse units, the beginning and end of the combined discourse units are labeled with $(_N$ and $)_S$ if their nuclearity relation is Nucleus-Satellite, $(_S$ and $)_N$ if their nuclearity relation is Satellite-Nucleus, and $(_N$ and $)_N$ if their nuclearity relation is Nucleus-Nucleus. The obtained label embedding representation is shown in Fig. 3(c)[3].

For paragraph i, the representation emb_i is obtained after label embedding. The structural encoding module uses another XLNet to encode structural information. Moreover, we do not fine-tune this XLNet during the training process. We take the vector at the $\langle sep \rangle$ position in XLNet as the output of this module as follows.

$$h_{emb_i} = XLNet(emb_i) \tag{2}$$

where $i = \{1, 2, ..., n\}$, h_{emb_i} is the output of the structural encoding module.

4.3   Information Integration Module

The information integration module can integrate the obtained paragraph- and EDU-level information. In the paragraph-level parser, for paragraph i, we sum up the paragraph- and EDU-level information to get the integrated information h_i as follows.

$$h_i = h_{unit_i} + h_{emb_i} \tag{3}$$

where $i = \{1, 2, ..., n\}$. The EDU-level parser uses the h_{unit_i} obtained from the atomic unit encoding module as h_i directly because there is no EDU-level structure information encoding module. Compared with LSTM, GRU has a smaller number of parameters and can achieve comparable performance to LSTM. Therefore, we use Bi-GRU to enhance the integrated paragraph representation. Sending $H = \{h_1, h_2, ..., h_n\}$ into the Bi-GRU, the output is $H' = \left\{h_1', h_2', ..., h_n'\right\}$ ($h_j' = \left[\overrightarrow{h_j'}; \overleftarrow{h_j'}\right]$), where $\overrightarrow{h_j'}$ and $\overleftarrow{h_j'}$ are the forward and backward outputs, respectively.

Although Bi-GRU can get the contextual information well, it cannot highlight the split point representation of the pointer network. To alleviate this issue, consistent with Zhang et al. [9], we use a convolutional neural network to fuse the EDU information at the left and right ends of the split point to enhance its semantic representation. The width of the convolution kernel is set to 2 and the activation function is Rectified Linear Unit (ReLU). For example, for the article $D = \{p_1, p_2, ..., p_n\}$, after feeding it into the Bi-GRU, we can obtain the paragraph representation $H' = \left\{h_1', h_2', ..., h_n'\right\}$. For the convenience calculation, we add two zero vectors for padding on the start and end of the EDU sequence, and the output is $I = \{i_0, i_1, ..., i_n\}$. In top-down parser, the first and last vectors are not involved in the calculation at each decoding step.

---

[3] In practice we use three pairs of labels $\langle unused1 \rangle$ and $\langle unused2 \rangle$, $\langle unused3 \rangle$ and $\langle unused4 \rangle$, $\langle unused5 \rangle$ and $\langle unused6 \rangle$ to correspond to the three pairs of tags $(_N$ and $)_S$, $(_S$ and $)_N$, $(_N$ and $)_N$.

4.4   Top-Down Parser

In constructing a discourse tree, the input sequence length constantly changes, and the pointer network [22] can solve this problem well. Thus, the top-down parser uses a pointer network to determine the split position of discourse units and build the discourse tree in a top-down method.

The pointer network is an encoder-decoder architecture and its encoder is a Bi-GRU. We input the split point vector $I = \{i_0, i_1, ..., i_m\}$ obtained from the information interaction module into the encoder of the pointer network to gain the output $E = \{i_0, i_1, ..., i_m\}$. Its decoder is a GRU that uses a stack to construct the discourse tree. In the initial state, the stack contains only one element $(0, N)$, i.e., the index pair of the start and end positions of the sequence to be split. We take the output E of the encoder as the input of the decoder. At the tth step of decoding, the top discourse unit (l, r) is popped out of the stack to the decoder, and then the decoder outputs d_t. d_t interacts with the encoder output $E' = \{e_{l+1}, e_l, ..., e_{r-1}\}$, and we use a biaffine attention function to determine the split point as follows.

$$s_j^i = e_i^T W d_j + U e_i + V d_j + b \tag{4}$$

where W, U, V denote the weight matrix, b is the bias, and s_j^i means the score of the i-th split point. In each step, the position with the highest score is selected as the split point, and the discourse unit is divided into two adjacent discourse units. Then the discourse units with lengths greater than 2 are pushed into the stack, and the process is repeated to build discourse tree until the stack is empty.

4.5   Model Training

It is worth mentioning that our parser uses the gold standard EDU-level discourse trees to achieve EDU-level information in the training phase. In contrast, it uses the predicted EDU-level discourse trees to obtain EDU-level information in the test phase.

We use Negative Log Likelihood Loss (NLL Loss) to calculate the split point prediction loss $L(\theta_s)$. Like Zhang et al. [10], we use one classifier to classify the nuclearity and relationship to obtain the N-R prediction loss $L(\theta_{N-R})$. The final loss L is shown as follows.

$$L(\theta_s) = - \sum_{i=1}^{batch} \sum_{t=1}^{T} \log P_{\theta_s}(y_t | y_{<t}, X) \tag{5}$$

$$L = \alpha_s L_{\theta_s} + \alpha_{N-R} L_{\theta_{N-R}} \tag{6}$$

where $y_{<t}$ is the discourse units generated before t-th step of the decoder, T is the number of discourse units in the stack, and X is the EDU sequence. Since the convergence speed of the split point prediction loss and the N-R prediction loss are different, we weighted and summed these two parts to obtain the final loss L.

5   Experimentation

5.1   Dataset and Experimental Settings

Our experiments are primarily evaluated on Unified Chinese Discourse TreeBank (UCDTB). Referring to previous studies [18], we transform the non-binary tree of the original data into the right-binary tree. There are 80% data (373 documents) for training, 10% data (47 documents) for validating, and 10% data (47 documents) for testing.

Following previous work [18], we report the micro-averaged F_1 score for predicting span attachments in discourse tree construction (Span), span attachments with nuclearity (Nuclearity), and span attachments with relation labels (Relation). Specifically, we evaluate the nuclearity with three classes (*Nucleus-Satellite*, *Satellite-Nucleus*, and *Nucleus-Nucleus*), and we use 19 finer-grained types for evaluation in relation classification as showned in Table 1. To get a more intuitive view of how our parser performs at different levels, we evaluate the performance of our parser at the paragraph level, the EDU level and the document level (i.e., EDU and paragraph level).

In the atomic unit encoding module and structural encoding module, we use XLNet-base for encoding. We set the number of GRU layers as 2. The dimension of the hidden layer is set as 512. We set the learning rate as 1e-4, the training epoch as 50, the batch size as 32, the α_S as 1 and the α_{N-R} as 0.1.

5.2   Baselines

To verify the effectiveness of our DCDParser, we compared it with the following strong baselines.

Paragraph-level Chinese discourse parser

MDParser-TS [18]: It proposed a hierarchical approach to constructing discourse structure trees based on topic segmentation.

DGCNParser [19]: It proposed a method based on GCN to enhance semantic representation and interaction between discourse units, which achieved the SOTA performance on MCDTB.

EDU-level Chinese discourse parser

Top-DownParser [9]: It proposed a top-down method and did not use additional manual features.

AdverParser [10]: It proposed an adversarial learning strategy based on Top-DownParser [9] and achieved the SOTA performance on CDTB.

5.3   Experimental Results

To prove the effectiveness of leveraging the connection between different granularity levels, we compare DCDParser with a model **Base-Model** without the structural encoding module, i.e., the granularity levels are independent of each other and are not connected. In addition, we used **DCDParser(S)** to denote the

method of embedding structure information and **DCDParser(S-N)** to denote
the method of embedding structure and nuclearity information, respectively.

Table 2 shows the performance comparison between our DCDParser and all
the baselines. Comparing our DCDParser with previous state-of-the-art parsers,
it performs better than all baselines due to its ability to leverage the connection
between different granularity levels. Furthermore, our DCDParser(S-N) shows a
significant improvement of 5.69%, 4.74% and 1.9% on span, nuclearity and relation at the paragraph level, respectively. Obviously, the utilization of EDU-level
structural and semantic information can greatly improve the parsing performance.

Table 2. The performance comparison on Chinese UCDTB. Superscript * indicates
we reproduce the model. DCDParser(S-N) was significantly superior to AdverParser
with a p-value <0.05 (t-test).

Level	Model	Span	Nuclearity	Relation
Paragraph-level	Top-DownParser	59.24	49.29	35.55
	DGCNParser*	60.19	46.92	29.38
	MDParser-TS	66.35	51.66	37.44
	AdverParser	66.82	59.24	48.34
	Base-Model	67.30	58.29	46.45
	DCDParser(S)	71.09	59.72	**50.71**
	DCDParser(S-N)	**72.51**	**63.98**	50.24
EDU-level	Top-DownParser	85.59	54.89	50.13
	DGCNParser*	83.58	56.52	54.26
	MDParser-TS	82.58	55.51	52.01
	AdverParser	86.59	63.91	60.65
	Base-Model	87.84	65.79	61.78
	DCDParser(S)	**88.72**	65.54	**63.91**
	DCDParser(S-N)	87.97	**67.29**	63.53
Document-level	Top-DownParser	80.08	53.72	47.08
	DGCNParser*	78.69	54.51	49.06
	MDParser-TS	79.19	54.71	48.96
	AdverParser	82.45	62.93	58.08
	Base-Model	83.55	64.22	58.57
	DCDParser(S)	**85.03**	64.32	**61.15**
	DCDParser(S-N)	84.74	**66.60**	60.75

In detail, compared to Base-Model, our DCDParser can improve the performance on all three indicators, especially at the paragraph level. DCDParser(S-N)
shows a significant improvement of 5.21% and 5.69% on span and nuclearity at
the paragraph level, respectively. This is due to the utilization of EDU-level
structural and semantic information, which allows the model to focus on the

core information of the paragraph while grasping the intra-paragraph structure. In contrast, DCDParser(S) has relatively little improvement on span and nuclearity at the paragraph level. This is because there is no embedded nuclearity information to capture the core semantics in a lengthy paragraph text accurately. Moreover, these two methods of embedding EDU-level information achieve similar results on relation. This is due to the global information that is considered more in paragraph-level relation recognition, and the addition of EDU-level information is of limited help to it.

These results indicate that there is a connection between the EDU-level structural and semantic information and the paragraph-level information, and that the lower-level structural information can assist the upper-level discourse parsing. There is also a performance improvement at the EDU level because the two models share the atomic encoding module. Ultimately, our parser achieves the best document-level performance by integrating EDU-level information.

Table 3. Performance on integrating different EDU-level information at the paragraph level.

Model	Span	Nuclearity	Relation
Base-Model	67.30	58.29	46.45
EDUs_TEXT	68.72	59.72	48.82
CORE_TEXT	70.62	59.24	49.76
DCDParser(S)	71.09	59.72	**50.71**
DCDParser(S-N)	**72.51**	**63.98**	50.24

6   Analysis

6.1   Analysis on Integrating Different EDU-Level Information

To investigate the effect of integrating different EDU-level information to the paragraph-level parsing, we replaced the structural encoding module using two different EDU-level information encoding modules.

The first one is to encode the whole EDU-level discourse tree's text to get the whole paragraph's semantic information, represented by **EDUs_TEXT**.

The second one is to obtain the core discourse unit of the paragraph according to the nuclearity between discourse units and encode it to get the semantic information of the paragraph, which is represented by **CORE_TEXT**.[4]

The experimental results are shown in Table 3. It can be seen that the performance is slightly improved after incorporating the whole EDU-level semantic information (EDUs_TEXT). In addition, the improvement brought by fusing core EDU-level semantic information (CORE_TEXT) is more prominent, indicating

---

[4] Same as the structural encoding module, we encode by using XLNet-base.

that enhancing text core information effectively understands the semantics of lengthy paragraphs. The above two approaches integrate EDU-level semantic information from different perspectives. Our model DCDParser(S-N) captures not only the core semantics but also the overall structure of the paragraph by explicitly embedding the structure information, which leads to a more accurate representation of the paragraph. The results show that it is more helpful to use EDU-level structural and semantic information than only EDU-level semantic information at the paragraph level.

6.2   Performance on Different Lengths of Documents at the Paragraph Level

According to Jiang et al. [18], existing methods perform worse on long documents due to the larger size and number of discourse units and fewer connectives between them. Hence, the ability of a parser to parse long documents is worth being focused on. We further analyze the performance of our DCDParser in terms of the number of EDUs. Figure 4 shows the Micro-F1 scores on span of our parser and several representative baselines. We can find that our DCDParser(S-N) achieves a significant improvement on long documents larger than eight paragraphs. The lower overall performance of the long text is due to the limited ability to exist methods to process long text, and it is difficult to capture the core semantics of the paragraph-level discourse unit that often contains much redundant information. These results indicate that our DCDParser (S-N) can effectively improve the ability to handle long texts by explicitly marking the nuclearity of EDU-level discourse units.

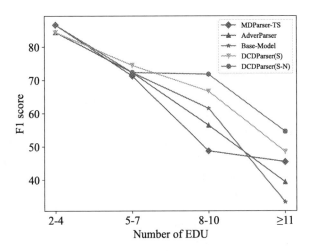

Fig. 4. Micro-F1 scores (span) on different lengths of documents at the paragraph level.

6.3 Performance on Different Layers of Discourse Tree at the Paragraph Level

To explore where the EDU-level information is helpful for the paragraph level, we divided the paragraph-level discourse tree node into three layers: top two layers, middle layers and bottom two layers. We analyzed the parsing ability of the nodes at different layers. Table 4 shows the performance of DCDParser and other baselines on different layers of paragraph-level discourse trees. We can find that DCDParser outperforms all the baselines at all three layers. It is worth noting that DCDParser(S-N) achieves a considerable improvement of 9.38% in the middle layers compared to the Base-Model. Since the structural tree with middle layers has more layers, its structure is more complex and the relationship between discourse units is more ambiguous. With the integration of EDU-level structure and nuclearity information, our model can obtain a more precise and effective paragraph representation to assist in parsing the complex structure of the middle layers.

6.4 Case Study

The examples of DCDParser(S-N) and the two baselines MDParser-TS and AdverParser parsing the chtb_0545 paragraph-level discourse tree are given in

Table 4. The performance of span on different layers of paragraph-level discourse trees.

Model	top two layers	middle layers	bottom two layers
MDParser-TS	81.55	28.13	66.96
AdverParser	78.64	46.88	62.61
Base-Model	81.55	43.75	65.22
DCDParser(S)	**84.47**	46.88	**67.83**
DCDParser(S-N)	**84.47**	**53.13**	**67.83**

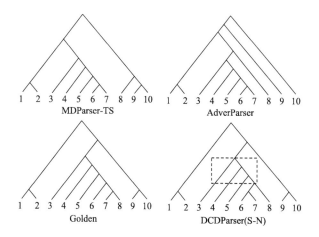

Fig. 5. The paragraph-level discourse tree of chtb_0545 parsed by various models.

Fig. 5. It shows that MDParser-TS and AdverParser do not accurately capture the structure and semantics of paragraphs due to the lack of using EDU-level information and make errors in constructing the upper-level complex discourse tree. Compared with them, the discourse tree built by our DCDParser(S-N) is more similar to the golden tree. With the help of EDU-level structure information, DCDParser(S-N) can get a more reasonable paragraph representation and obtain a more accurate bottom discourse unit, laying a good foundation for constructing the upper discourse tree.

6.5 Experimentation on English RST-DT

To verify the generalization of the proposed parser, we also evaluate our parser on the English RST-DT. Meanwhile, we processed some cross-paragraph EDUs in the corpus in the same way as Kobayashi et al. [14]. The upper bound of the performance at the EDU level is 95.15%. Following previous work [23], we binarize all non-binary trees with right-branching. Finally, we evaluated our model using the original Parseval and the results are shown in Table 5.

There are three baselines in Table 5 as follows: 1) **DynParser** [24]: It proposed a top-down parser with a dynamic oracle; 2) **AdverParser** [10]: It is based on the pointer network and proposed an adversarial learning strategy; 3) **Parser-EDUPLM** [25]: It proposed a second-stage EDU-level pre-training method and achieved the SOTA performance on RST-DT.

Table 5. The performance comparison on English RST-DT.

Level	Model	Span	Nuclearity	Relation
Paragraph-level	Base-Model	51.43	42.34	29.09
	DCDParser(S)	52.21	44.16	**31.17**
	DCDParser(S-N)	**55.06**	**45.71**	29.61
EDU-level	Base-Model	81.07	70.67	61.15
	DCDParser(S)	81.54	70.83	60.32
	DCDParser(S-N)	**82.42**	**71.40**	**62.09**
Document-level	DynParser	73.10	62.30	51.50
	AdverParser	76.30	65.50	55.60
	Parser-EDUPLM	76.40	66.10	54.50
	Base-Model	76.13	65.94	55.81
	DCDParser(S)	76.65	66.38	55.46
	DCDParser(S-N)	**77.86**	**67.11**	**56.67**

Compared with these baselines, our parser DCDParser(S-N) achieves the best performance at the document level, which proves the effectiveness of our method on English discourse parsing. Moreover, similar to the performance on UCDTB, there is a significant improvement at the paragraph level when integrating EDU-level structural and semantic information compared to Base-Model. It shows that integrating EDU-level structural and semantic information is also beneficial to English discourse parsing.

7   Conclusion

In this paper, we propose a unified document-level Chinese discourse parser that leverages the connection between granularity levels to construct better discourse trees. To the best of our knowledge, this is the first study on leveraging information at multiple levels of granularity in Chinese. In addition, we construct the Unified Chinese Discourse TreeBank (UCDTB), which contains 467 articles with annotations from clauses to the whole article, filling the gap in existing unified corpus resources on Chinese discourse parsing. Experimental results on both Chinese UCDTB and English RST-DT show that our DCDParser outperforms all baselines. In the future, we will explore the use of paragraph-level information to help EDU-level discourse parsing, thus further improving the overall performance of discourse parsing.

Acknowledgements. The authors would like to thank the three anonymous reviewers for their comments. This research was supported by the National Natural Science Foundation of China (Nos. 61836007 and 62276177), and Project Funded by the Priority Academic Program Development of Jiangsu Higher Education Institutions (PAPD).

References

1. Sadek, J., Meziane, F.: A discourse-based approach for Arabic question answering. In: ACM Transactions on Asian and Low-Resource Language Information Processing (TALLIP), vol. 16, no. 2, pp. 1–18 (2016)
2. Peldszus, A., Stede, M.: Joint prediction in MST-style discourse parsing for argumentation mining. In: Proceedings of the 2015 Conference on Empirical Methods in Natural Language Processing, pp. 938–948 (2015)
3. Bhatia, P., Ji, Y., Eisenstein, J.: Better document-level sentiment analysis from RST discourse parsing. arXiv preprint arXiv:1509.01599 (2015)
4. Mann, W.C., Thompson, S.A.: Rhetorical Structure Theory: A Theory of Text Organization. University of Southern California, Information Sciences Institute Los Angeles (1987)
5. Li, Y., Feng, W., Sun, J., Kong, F., Zhou, G.: Building Chinese discourse corpus with connective-driven dependency tree structure. In: Proceedings of the 2014 Conference on Empirical Methods in Natural Language Processing (EMNLP), pp. 2105–2114 (2014)
6. Jiang, F., Xu, S., Chu, X., Li, P., Zhu, Q., Zhou, G.: MCDTB: a macro-level Chinese discourse treebank. In: Proceedings of the 27th International Conference on Computational Linguistics, pp. 3493–3504 (2018)

7. Carlson, L., Marcu, D., Okurowski, M.E.: Building a discourse-tagged corpus in the framework of rhetorical structure theory. In: Current and New Directions in Discourse and Dialogue, pp. 85–112. Springer, Dordrecht (2003). https://doi.org/10.1007/978-94-010-0019-2_5

8. Wang, Y., Li, S., Wang, H.: A two-stage parsing method for text-level discourse analysis. In: Proceedings of the 55th Annual Meeting of the Association for Computational Linguistics (Volume 2: Short Papers), pp. 184–188 (2017)

9. Zhang, L., Xing, Y., Kong, F., Li, P., Zhou, G.: A top-down neural architecture towards text-level parsing of discourse rhetorical structure. arXiv preprint arXiv:2005.02680 (2020)

10. Zhang, L., Kong, F., Zhou, G.: Adversarial learning for discourse rhetorical structure parsing. In: Proceedings of the 59th Annual Meeting of the Association for Computational Linguistics and the 11th International Joint Conference on Natural Language Processing (Volume 1: Long Papers), pp. 3946–3957 (2021)

11. Feng, V.W., Hirst, G.: Text-level discourse parsing with rich linguistic features. In: Proceedings of the 50th Annual Meeting of the Association for Computational Linguistics (Volume 1: Long Papers), pp. 60–68 (2012)

12. Joty, S., Carenini, G., Ng, R., Mehdad, Y.: Combining intra-and multi-sentential rhetorical parsing for document-level discourse analysis. In: Proceedings of the 51st Annual Meeting of the Association for Computational Linguistics (Volume 1: Long Papers), pp. 486–496 (2013)

13. Feng, V.W., Hirst, G.: A linear-time bottom-up discourse parser with constraints and post-editing. In: Proceedings of the 52nd Annual Meeting of the Association for Computational Linguistics (Volume 1: Long Papers), pp. 511–521 (2014)

14. Kobayashi, N., Hirao, T., Kamigaito, H., Okumura, M., Nagata, M.: Top-down RST parsing utilizing granularity levels in documents. In: Proceedings of the AAAI Conference on Artificial Intelligence, vol. 34, pp. 8099–8106 (2020)

15. Kong, F., Zhou, G.: A CDT-styled end-to-end Chinese discourse parser. In: ACM Transactions on Asian and Low-Resource Language Information Processing (TALLIP), vol. 16, no. 4, pp. 1–17 (2017)

16. Zhou, Y., Chu, X., Li, P., Zhu, Q.: Constructing Chinese macro discourse tree via multiple views and word pair similarity. In: Tang, J., Kan, M.-Y., Zhao, D., Li, S., Zan, H. (eds.) NLPCC 2019. LNCS (LNAI), vol. 11838, pp. 773–786. Springer, Cham (2019). https://doi.org/10.1007/978-3-030-32233-5_60

17. Jiang, F., Chu, X., Li, P., Kong, F., Zhu, Q.: Chinese paragraph-level discourse parsing with global backward and local reverse reading. In: Proceedings of the 28th International Conference on Computational Linguistics, pp. 5749–5759 (2020)

18. Jiang, F., Fan, Y., Chu, X., Li, P., Zhu, Q., Kong, F.: Hierarchical macro discourse parsing based on topic segmentation. In: Proceedings of the AAAI Conference on Artificial Intelligence, vol. 35, pp. 13152–13160 (2021)

19. Fan, Y., Jiang, F., Chu, X., Li, P., Zhu, Q.: Chinese macro discourse parsing on dependency graph convolutional network. In: Wang, L., Feng, Y., Hong, Yu., He, R. (eds.) NLPCC 2021. LNCS (LNAI), vol. 13028, pp. 15–26. Springer, Cham (2021). https://doi.org/10.1007/978-3-030-88480-2_2

20. Yang, Z., Dai, Z., Yang, Y., Carbonell, J., Salakhutdinov, R.R., Le, Q.V.: XLNet: generalized autoregressive pretraining for language understanding. In: Advances in Neural Information Processing systems, vol. 32 (2019)

21. Zhou, J., Jiang, F., Chu, X., Li, P., Zhu, Q.: More Than One-Hot: Chinese macro discourse relation recognition on joint relation embedding. In: Mantoro, T., Lee, M., Ayu, M.A., Wong, K.W., Hidayanto, A.N. (eds.) ICONIP 2021. CCIS, vol. 1516, pp. 73–80. Springer, Cham (2021). https://doi.org/10.1007/978-3-030-92307-5_9

22. Vinyals, O., Fortunato, M., Jaitly, N.: Pointer networks. In: Advances in Neural Information Processing Systems, vol. 28 (2015)
23. Sagae, K., Lavie, A.: A classifier-based parser with linear run-time complexity. In: Proceedings of the Ninth International Workshop on Parsing Technology, pp. 125–132 (2005)
24. Koto, F., Lau, J.H., Baldwin, T.: Top-down discourse parsing via sequence labelling. arXiv preprint arXiv:2102.02080 (2021)
25. Yu, N., Zhang, M., Fu, G., Zhang, M.: RST discourse parsing with second-stage EDU-level pre-training. In: Proceedings of the 60th Annual Meeting of the Association for Computational Linguistics (Volume 1: Long Papers), pp. 4269–4280 (2022)

Document Analysis and Recognition 1:
Document Layout and Parsing

SwinDocSegmenter: An End-to-End Unified Domain Adaptive Transformer for Document Instance Segmentation

Ayan Banerjee[1], Sanket Biswas[1(✉)], Josep Lladós[1], and Umapada Pal[2]

[1] Computer Vision Center and Computer Science Department,
Universitat Autònoma de Barcelona, Barcelona, Spain
{abanerjee,sbiswas,josep}@cvc.uab.es
[2] CVPR Unit, Indian Statistical Institute, Kolkata, India
umapada@isical.ac.in

Abstract. Instance-level segmentation of documents consists in assigning a class-aware and instance-aware label to each pixel of the image. It is a key step in document parsing for their understanding. In this paper, we present a unified transformer encoder-decoder architecture for en-to-end instance segmentation of complex layouts in document images. The method adapts a contrastive training with a mixed query selection for anchor initialization in the decoder. Later on, it performs a dot product between the obtained query embeddings and the pixel embedding map (coming from the encoder) for semantic reasoning. Extensive experimentation on competitive benchmarks like PubLayNet, PRIMA, Historical Japanese (HJ), and TableBank demonstrate that our model with SwinL backbone achieves better segmentation performance than the existing state-of-the-art approaches with the average precision of **93.72**, **54.39**, **84.65** and **98.04** respectively under one billion parameters. The code is made publicly available at: github.com/ayanban011/SwinDocSegmenter.

Keywords: Document Layout Analysis · Instance-Level Segmentation · Swin Transformer · Contrastive Learning

1 Introduction

Document Intelligence (DI) systems help to provide solutions for automating large document processing workflows for information extraction and understanding its contents. Business intelligence processes like document retrieval, text recognition, content categorization, and others often require to extract the semantic information from documents when parsing the documents into a structured machine-readable format. This extracted data can be then integrated into document processing workflows in Robotic Process Automation tools. Thus, more efficient solutions have been developed in key industrial sectors (e.g. banking, finance, healthcare, and so on) [31,34]. Document layout analysis (DLA) has

---

A. Banerjee and S. Biswas—These authors contributed equally to this work.

© The Author(s), under exclusive license to Springer Nature Switzerland AG 2023
G. A. Fink et al. (Eds.): ICDAR 2023, LNCS 14187, pp. 307–325, 2023.
https://doi.org/10.1007/978-3-031-41676-7_18

become an important task in DI because any task related to document understanding entails the need of obtaining a structured representation that helps to localize the key information stored in them. Initially, remarkable progress has been observed with classical convolution-based algorithms (CNNs) such as Faster RCNN [43] for Document Object Detection (DOD), Mask RCNN [5] for instance segmentation, among other specialized architectures. These architectures are quite simple to implement and effective in some specific case studies (e.g. table detection [24], layout analysis of scientific articles [49] etc.) but they lack the generalization ability to address other similar tasks. Recently, Transformer-based architectures [2, 16] have achieved superior performance over CNNs with the help of a global attention mechanism. However, these models are not unified which prevents the mutual cooperation between the detection and segmentation tasks which affects their performance as the detection and segmentation modules cannot guide each other. Not only that, but those architectures were also biased toward their pre-trained datasets and failed to perform domain shifts for a similar task. As these transformer models are often pre-trained with massive amounts of data originating from a related source domain (i.e., large-scale industry documents [15] or scientific articles [49]), they fail to address relatively different tasks (e.g. layout extraction in magazines [10]). The introduction of this domain shift property to a DLA model has the potential to reduce computational expenses and help to create a more data-independent generic model.

To address the aforementioned issues, we propose *SwinDocSegmenter* framework to perform instance-level segmentation of complex document layouts, using content query embeddings on a high-resolution pixel embedding map obtained from the Swin Transformer feature extraction backbone [30] and Transformer encoder features. It helps define global semantic reasoning of the features at a higher level which overcomes the drawbacks of using the ResNet-FPN [28] backbone. Here, we initialize mask queries as anchors by utilizing the encoder dense prior to predicting the masks from the top-ranked tokens. It helps to perform pixel-wise segmentation at an early stage, which helps to enhance boxes. In the later stages, these boxes help to increase the segmentation performances by formulating dynamic anchor boxes. This phenomenon of mutual task cooperation helps to obtain a unified model for layout detection and segmentation. We introduce a contrastive denoising training inspired by [48] to accelerate segmentation training by focusing on low-level instances. It boosts the model performance a lot as one of the main drawbacks of Transformers to working with unlabeled data where it penalizes the classes that have a very low number of feature representations [4]. Last but not the least, we utilize a hybrid bipartite matching [21] for more consistent semantic matching which helps to perform *domain shift* and utilizes the pre-trained weights of the transformers from a completely different domain to perform similar tasks. In this case, we utilized the pre-trained weights of the MS-COCO Object Detection benchmark [29] for the instance segmentation of complex document layouts.

The overall contributions of this work can be summarized in three folds:

- A *unified Transformer-based framework* has been proposed to perform instance-level document layout segmentation, with a Swin Transformer backbone, anchor box-guided cross-attention, and enhanced query selection strategy.
- We introduce *contrastive denoising training* to enhance the low-level instances to boost the performance of the unlabeled dataset.
- We utilize *hybrid bipartite matching to invoke the domain shift property* to save the pretraining time and use the publicly available pre-trained weights from diverse domains for a similar task which improves model generalization.

The rest of the paper is organized in the following way: In Sect. 2 we review state-of-the-art approaches for document layout analysis. We describe the *Swin-DocSegmenter* in Sect. 3. We introduce our experimental evaluation as well as ablation studies in Sect. 4. Finally, Sect. 5 draws the conclusion and guides the future research directions.

2 Related Work

In order to extract the relevant information from digital documents, layout recognition methods obtain spatial understanding with relational reasoning between different layout components (e.g. table, text, figures, title, etc.). Mainstream layout analysis algorithms have been dominated by classical heuristic rule-based algorithms before the deep learning era. Later on, convolution frameworks play a leading role to solve this task until the transformers-based architectures achieve remarkable performance. This section is dedicated to obtaining an overview of the state-of-the-art for this task by analyzing different methodological schemes.

Heuristic Rule-based Document Layout Analysis. Document layout segmentation using heuristic methods can be further classified into three different categories: top-down, bottom-up, and hybrid strategies. Bottom-up approaches [3,38] perform basic operations like grouping and merging of pixels to create homogeneous regions for similar objects and separate them from the nonsimilar ones. Top-down strategies [17,19] split the document image into different regions iteratively, until a definite region has been obtained around similar objects. Although bottom-up approaches are able to tackle complex layouts, they are computationally expensive. Moreover, Top-down methods provide faster implementation but penalize the generalization, and perform effectively only on specific types of documents. To take advantage of both, hybrid methods [6,11] combine bottom-up and top-down cues to obtain fast and efficient results. Prior to the deep learning era, these methods were state-of-the-art for table detection.

Convolution-based Document Layout Analysis. Since 2012, deep learning algorithms replaced the rule-based algorithm and Convolutional Neural Networks (CNNs) became the prior strategy to solve instance document segmentation tasks. Faster-RCNN [37] provides a strong document object detection that can be utilized to solve page segmentation [25]. Later on, a similar network Mask-RCNN [1] provides the first layout segmentation benchmark for instance segmentation of newspaper elements. Another convolution benchmark has been provided by RetinaNet [27] for keyword detection in document images. This is a complex method and only helps to detect the text regions. In order to provide a new state-of-the-art benchmark for table detection and table structure recognition, DeepDeSRT [40] utilizes a novel image transformation strategy to identify the visual features of the table structures and feed them into a fully convolution network with skip pooling. Similarly, Oliviera et al. [33] used a similar FCNN-based framework for pixel-wise segmentation of historical document pages which outperforms the previous convolutional autoencoder-based benchmarks obtained by Chen et al. [7,8]. Saha et al. [39] provided ICDAR2017 POD (Page Object Detection) benchmark [12] to obtain state-of-the-art results by using transfer learning based Faster-RCNN backbone for detection of mathematical equations, tables, and figures. A new cross-domain DOD benchmark was established in [23] to apply domain adaptation strategies to solve the domain shift problem. Recently, A vision-based layout detection benchmark has been provided in [47] which utilized a recurrent convolutional neural network with VoVNet-v2 backbone [20] by generating synthetic PDF documents from ICDAR-2013 and GRO-TOAP dataset. It obtained a new benchmark to solve the scientific document segmentation task.

Transformer Based Document Layout Analysis. Nowadays Transformers which provide a more prominent performance with the utilization of positional embedding and self-attention mechanism [44]. Here, DiT [22] obtained a new baseline for document image classification, layout analysis, and table detection with self-supervised pretraining on large-scale unlabeled document images which cannot be applicable to small magazine datasets like PRIMA. Similarly, Li et al. [26] obtained a multimodal framework to understand the structured text in the documents. However, the model performs very poorly for similar semantics of textual content. In order to improve these performances a TILT [35] mechanism has been introduced which simultaneously learns textual semantics, visual features, and layout information with an encoder-decoder Transformer. A similar transformer encoder-decoder was utilized in [46] which provides a new baseline for the PubLayNet dataset (AP: 95.95) with the text information extracted through OCR. Recently, LayoutLmv3 [16] used joint learning of text, layout, and visual features to obtain state-of-the-art results in visual document understanding (VDU) tasks. It performs significantly well for large-scale datasets but fails for small-scale datasets. DocSegTr [4] utilized a ResNet-FPN backbone over the transformer layers with self attention mechanism, which helps it to converge faster for small scale datasets but unable to achieve state-of-the-art performances. Other recent approaches [2,13,14,18] also utilize this joint pretraining strategy to solve several VDU tasks including document visual question answering. These techniques are quite helpful to several downstream tasks by a unified pretraining. However, it comes with a pretraining bias which prevents them to perform a domain shift and they also unable to learn the class information with low number of instances as their is no weight prioritizing.

Motivated by the recent breakthrough of transformers and to improve its performance by solving the above-mentioned issues we are proposing an end-to-end unified domain adaptive document segmentation transformer benefitted with contrastive training that not only achieves superior performance on standard instance-level segmentation benchmarks but also provides the first transformer baseline for the newly proposed industrial document layout analysis dataset [34].

3 Method

The proposed *SwinDocSegmenter* is a unified end-to-end architecture that contains a Swin Transformer backbone [30], a Transformer encoder-decoder pair, and a segmentation branch obtained from multiple projection heads by class instance mapping. The proposed architecture is illustrated in Fig. 1 where the model first extracts multi-scale features with a Swin backbone. Then the features are flattened and downsampled before feeding them into the transformer encoder, otherwise it would generate a large number of trainable parameters which is impossible to train with limited resources. The Transformer encoder takes those features and their corresponding positional embeddings (obtained through several convolution layers with kernel size 3×3) as input to perform the feature enhancement. Here, a unified mixed query selection strategy has been obtained that passed through a low-level projection head to initialize the positional queries and anchors. The main advantage of this query selection strategy is that it does not initialize content queries but leaves them learnable which helps a lot in times of domain shift. Not only that, with the help of a low-level projection head it helps to focus on low-dimension image features which are often ignored in transformer training due to lack of data points. This also makes the decoder

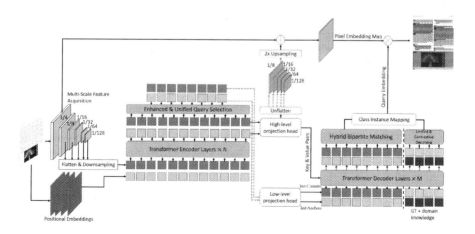

Fig. 1. Proposed SwinDocSegmenter Framework. Given an input document image from any domain, the model predicts the segmented document layout using a unified detection and segmentation branch

ready for contrastive denoising training (CDN) [48]. In the decoder, deformable attention [45] is utilized to combine the outputs of the encoder with layer-by-layer query updates. With CDN it is also considered the segmented/wrongly segmented region as hard negative samples and tries to rectify it with a look forward twice approach [48] where it passes the gradient between adjacent layers at early stages. We utilize a hybrid bipartite matching strategy to refine the segmented region based on the dynamic anchor boxes which help to generate an accurate segmented region. These two pieces of information are combined through class instance mapping to get the final query embedding. We perform a dot-product between the final query embedding and pixel embedding map to get the final instance segmentation output on document images.

3.1 Segmentation Branch

To perform mask classification, we utilize a key idea [9] to construct a pixel embedding map (PEM) by combining the multi-scale features (extracted by Swin backbone) and Transformer encoded features. As shown in Fig. 1, the PEM is constructed with a fusion between $1/4^{th}$ resolution feature map from the backbone (S_b) and upsampled $1/8^{th}$ resolution feature map from Transformer Encoder (T_e). The output mask M is computed by a dot-product between PEM and query embedding (Q_e) obtained from the decoder (see Eq. 1).

$$M = Q_e \otimes \delta(\Gamma(S_b) + \psi(T_e))$$ (1)

where Γ is the convolutional layer to map the channel dimension to the transformer dimension, ψ is the interpolation function for $2\times$ upsampling of T_e, and δ is the segmentation head. This mechanism is simple and easy to implement.

3.2 Feature Encoding Techniques

The feature encoding techniques consist of four important subparts: Query selection, low-level and high-level feature projection, and anchor initialization to boost the performance and simplify the decoding technique.

Query selection strategy. It has been observed that the output of the encoder contains dense features that can be used as better priors in the decoder. Here, we adopted one classification, one detection, and one segmentation head, in the encoder output followed by a low-level and high-level projection head. We obtained the classification score of each token as a confidence score and used them to select top-ranked features and feed them into the decoder as content queries. The selected features also regress boxes via detection and segmentation heads and passed through the high-level projection head to combine with the high-resolution feature map via dot product to predict the masks. These predicted masks and boxes are considered initial anchors for the decoder after passing it through the low-level projection head. It helps to make the decoder for contrastive training as both high-level and low-level class instances are present and we can improve the performance of low-level class instances without compromising the performance of high-level instances by adjusting contrastive loss.

Low-level projection head. It is a shallow multi-layer perceptron (MLP) that leverages the project features to low-level embeddings for contrastive learning in low-level views. It helps to learn more fine-grained invariances. Specifically, we apply a non-linear function $F = (f_1, f_2, ..., f_s)$ on low-level features to enhance them before initializing them as content queries and anchors in decoders. The objective function of this low-level projection head is defined in Eq. 2.

$$\mathcal{L}_{low} = \sum_{i=1}^{n} \sum_{j=1}^{n'} -log \frac{exp(f_i \cdot f_j / \tau)}{\sum_{c=1}^{k} exp(f_c \cdot f_j / \tau)} \qquad (2)$$

where, n and n' are the no. of features obtained from detection and segmentation heads, c is the no. of top-ranked features obtained from the classification heads. Here, we need a temperature hyperparameter τ to tune the layers to enhance the features based on the datasets we have used. $\tau = 0.02, 0.6, 0.1$, and 0.2 for PublayNet, Prima, HJ, and TableBank respectively. Note: all these hyperparameter values have been obtained experimentally.

High-level projection head. It is a deep MLP that preserves the high-level invariance of the high-level features. Basically, it set a different number of prototypes $P = (p_1, p_2, ..., p_m)$ to obtained different key-value pairs $k_1 v_1, k_2 v_2, ..., k_n v_n$ which also enriched the feature representation. The objective function of this prototyping has been defined in Eq. 3.

$$\mathcal{L}_{high} = \sum_{i=1}^{n} \sum_{j=1}^{n'} -log \frac{exp(f_i \cdot p_j / \phi_j)}{\sum_{c=1}^{k} exp(f_c \cdot p_j / \phi_j)} \qquad (3)$$

where, p_j is the prototype of the corresponding key-value pairs and ϕ_j is the concentration estimation indicator [32] for the distribution of representations around the prototype.

Anchor initialization. Document instance segmentation is a classification task at the pixel level whereas, object detection is a position regression task at the region level. Therefore, segmentation is more challenging due to its fine granularity than detection though it is simpler to learn in the beginning. Dot-producting queries using the high-resolution feature map, for instance, can predict masks by only comparing semantic similarity per pixel. However, the box coordinates must be directly regressed for detection in an image. As a result, mask prediction is significantly more accurate than box prediction in the initial stage. As a better anchor initialization for the decoder, therefore, we derive boxes from the predicted masks following unified query selection. The enhanced box initialization has the potential to significantly enhance the detection performance thanks to this efficient task cooperation.

3.3   Feature Decoding for Mask Prediction

At this stage, we introduced unified contrastive denoising training for effectively boosting the performance for the low-level instances and hybrid bipartite matching to perform domain shift. Below, we discuss both strategies in detail.

Unified Contrastive Denoising Training. Query denoising [48] is an effective technique to improve performance by accelerating convergence. However, it lacks the capability of separating two nearby class instances. CDN can tackle this issue by rejecting useless anchors. Here, noises are added to ground truth labels and boxes, and the Transformer decoder receives them as noised positional and content queries. Here, we have two hyperparameters λ_p and λ_e where, $\lambda_e > \lambda_p$ as depicted in Fig. 2. It helps to generate two types of queries (positive and negative). It is anticipated that positive queries within the inner square will reconstruct their corresponding ground truth boxes because their noise scale is less than λ_p. On the other hand, negative queries have a noise scale greater than λ_p and less than λ_e which are minimized through the focal loss. Generally, we keep λ_e very small as it helps to improve the performance by keeping the hard negative samples close to the ground truth anchors. Each CDN group has positive and negative queries (see Fig. 2). A CDN group will have $2 \times q$ queries for an image with q GT boxes, with each GT box producing a positive and negative query. To increase the efficiency we also employ multiple CDN groups.

In order to train the model, the noised versions of the object features have been utilized to reconstruct them. We also apply this method to tasks involving segmentation. Boxes and masks are naturally connected due to the fact that masks can be seen as a more finely detailed representation of boxes. As a result, we can train the model to predict masks given boxes as a denoising task and treat boxes as a noised version of masks. In order to train mask denoising more effectively, the boxes provided for mask prediction are also randomly noised. During

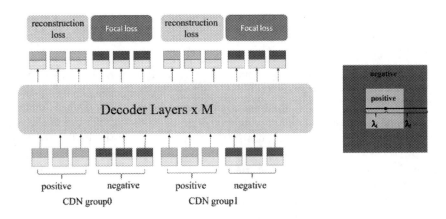

Fig. 2. Unified Contrastive Denoising Training Strategy. Similar to DINO [48] implementation, however, in place of no object detection we introduced a focal loss to optimize and enhance the low-level instances.

training, these noised objects will be added to the original decoder queries, but they will be removed during inference. We perform a lot of tunning to get the optimized value of λ_p and λ_e. However, it has been observed that, most generic performance has been achieved with $\lambda_p = 0.1$ and $\lambda_e = 0.02$ respectively.

Hybrid bipartite matching. This technique helps to remove the inconsistency between the pair of masks predicted from different heads by changing their corresponding weights. With this motivation, we utilize this concept in domain shift. Basically, we add an extra mask prediction loss in addition to the L1 and focal loss in bipartite matching. It encourages more accurate and consistent matching results for one query. So, when we utilized a pre-trained model from a different domain we penalize this loss more which forced us to make significant changes in their corresponding weights and slowly decrease the penalizing rate when it reached near the convergence. This loss is also optimized along with the L1 and focal loss to make this domain shift unified. Finally, a class instance mapping is performed between the classes and the predicted instances. It is a simple one-to-one mapping to perform query embedding which can be combined with pixel embedding map effectively through dot product in order to complete the instance segmentation process in and end-to-end manner.

4   Experimental Evaluation

Datasets. The Document Layout Analysis (DLA) community has always been concerned about the absence of standard public benchmarks. We use large-scale annotated datasets like PubLayNet [49], TableBank [36], and Historical Japanese (HJ) [41] as well as small-scale PRIMA [10] for evaluating our proposed segmentation approach in this work (Please refer to Table 1 for a detailed description). Besides that, we evaluate our model against a recently released standard industrial document layout segmentation benchmark **DocLayNet** [34]. It contains 91104 object instances of 11 distinct labels (Caption, Footnote, Formula, List-item, Page-footer, Page-header, Picture, Section-header, Table, Text, and Title) and covers a wide range of document object sizes (large to small).

Evaluation Metrics. The Intersection over Union (IoU) score is the most general way to assess the accuracy of the predicted instance (document category) for an instance-level segmentation task. Standard Microsoft COCO benchmark evaluation for instance segmentation uses the mean of APs at various IoU thresholds (0.5 to 0.95 with a step size of 0.05) to calculate the mean Average Precision (mAP) score for the entire model. Since all of them use a similar environment to compute the mAP, comparing the proposed approach to those that are already in use is helpful. In addition, the model performance for evaluating each categorical document instance has been calculated in accordance with [4,16,42].

Table 1. Experimental dataset description (instance level)

PublayNet			PRIMA			Historical Japanese			TableBank		
Object	Train	Eval	Object	Train	Eval	Object	Train	Eval	Object	Train	Eval
Text	2,343,356	88,625	Text	6401	1531	Body	1443	308	Table	2835	1418
Title	627,125	18,801	Image	761	163	Row	7742	1538	–	–	–
Lists	80,759	4239	Table	37	10	Title	33,637	7271	–	–	–
Figures	109,292	4327	Math	35	7	Bio	38,034	8207	–	–	–
Tables	102,514	4769	Separator	748	155	Name	66,515	7257	–	–	–
–	–	–	other	86	25	Position	33,576	7256	–	–	–
–	–	–	–	–	–	Other	103	29	–	–	–
Total	3,263,046	120,761	Total	8068	1891	Total	181,097	31,866	Total	2835	1418

The Choice of the Feature Extraction Backbone. In the context of instance-level document segmentation, extensive ablation studies were carried out to quantify the significance of each component of our model framework and to justify its use for segmenting various layout elements. All the ablations have been performed on the PRIMA dataset as it is the smallest dataset and it contains difficult layouts. In this study, different CNN and Vision Transformer backbones has been used (see Table 2). Among them, we take the SwinL Transformer backbone to multi-scale feature extraction. Though the no. of trainable parameters increases, it also improves the performance over ResNet, ResNeXt, and ViTs by 8%, and from Swin Tiny by 5%. The convolutional backbones provide attention to local features which are effective for small object detection however, there is no global attention that penalizes the cost for the large object. On the other hand, ViTs utilize self-attention but require a large amount of training data to learn the multi-scale features. Initially, Swin-Tiny will perform well but it is sensitive to noise so at a later stage, it penalizes the reconstruction which affects the overall performance. Due to its large size SwinL can eliminate noise very easily and achieves better performance than the rest.

Table 2. Ablation Study of different feature extraction backbones

Backbone	No. of Parameters	AP	AP@50	AP@75	APs	APm	APl
ResNet-50	52M	36.065	52.362	41.112	20.152	23.327	38.142
ResNet-101	102M	37.112	54.982	41.872	22.242	26.153	41.986
ResNext-101	104M	38.405	58.405	41.916	25.982	29.364	44.129
ViT-S	126M	40.342	59.763	42.158	29.176	33.129	48.526
ViT-B	164M	46.128	62.689	47.358	31.389	33.458	50.508
Swin-T	178M	49.349	65.956	50.317	34.128	36.909	52.049
Swin-L	223M	**54.393**	**69.313**	**52.965**	**39.327**	**42.061**	**60.142**

The Choice of the Input Image Resolution. Besides that, the image resolution also affects the model performance as the model is large and the number of trainable parameters is huge. So if we use the small image resolution then at the late stage, it only learns the noise which wastes the computational resources. Increasing the image resolution improves the performance (see Table 3), however, we are unable to increase beyond 1024 due to the limited resources. Moreover, from the trend, it can be concluded that increasing image resolution also improves the system performance until it meets the saturation point.

Table 3. How image resolution affects the instance segmentation performance

Image Resolution	AP	AP@50	AP@0.75	APs	APm	APl
256 × 256	45.022	60.189	46.258	28.372	32.458	53.568
512 × 512	50.132	66.235	52.317	32.242	36.909	54.148
1024 × 1024	54.393	69.313	52.965	39.327	42.061	60.142

The Choice of the number of Decoder Queries. Similarly, by taking a deep dive into the model we observe that, the no. of queries used for initializations in the decoder affects the overall performance. With a small number of queries it will be very difficult to generate the negative samples close to the performance, which not only penalizes the model performance but also increases the optimization time of the loss function. Also, dense queries stabilize the model and provide an opportunity to rectify the misclassified samples (see Table 4). With the SwinL backbone, we can extend it to 900–1200 but we have to restrict it to 300 due to the limited computational resources.

Table 4. Ablation Study on No. of queries generated from Transformer Encoder

No. of Queries	AP	AP@50	AP@0.75	APs	APm	APl
100	50.022	65.189	52.258	32.372	36.458	53.968
150	50.132	66.235	52.317	32.242	36.909	54.148
200	51.393	67.313	52.765	37.312	41.011	60.111
250	52.092	68.212	52.964	37.512	42.060	60.132
300	**54.393**	**69.313**	**52.965**	**39.327**	**42.061**	**60.142**

The Choice of the Learning Objectives. Last, but not the least an ablation study of the loss functions has been obtained to understand which combination of the reconstruction and classification loss is most optimized. From Fig. 3 it has been observed that the combination of L1 and focal loss is the most effective one for this task. The L1 loss tends to shrink coefficients to zero which is better for feature selection whereas L2 tends to shrink coefficients evenly. On the other hand, Focal Loss helps to scale the standard cross-entropy loss to down-weight loss corresponding to easily classifiable examples dynamically and focus more on hard examples to make the system perform better on hard examples as well.

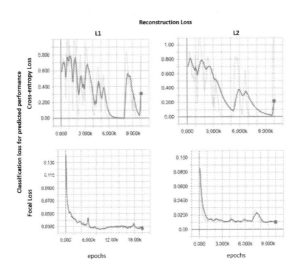

Fig. 3. The impact of learning objectives. The above graphs study the effectiveness of different combinations of loss functions

Moreover, the most interesting fact has been observed in Table 5 which shows how pre-training on similar dataset include biases and shrinks the overall performance of the model. Here, we have used one SwinL backbone pre-trained on the PubLayNet dataset and another one pre-trained on the MSCOCO dataset. We can observe that the model achieves very high performance on the class"Table" because it is a common class in both datasets. But as the pre-trained model is not familiar with the "Separator" region, it penalizes a lot decreasing the overall performance of the networks as it quickly converges the loss by looking at the similar classes and ignoring the others by taking them as negative samples in CDN. Whereas with the MSCOCO pre-training it achieves a generic performance due to the enhanced and unified query selection which let the content queries learnable, and hybrid bipartite matching helps to optimize the corresponding pre-training weights. This helps to eliminate the bias factor.

Table 5. How pre-training provides biases

pre-training	Overall Performance						Class-Wise Performance					
	AP	AP@50	AP@75	APs	APm	APl	Text	Image	Table	Math	Separator	Other
PubLayNet	49.36	64.43	51.45	32.94	34.07	54.21	85.55	72.51	**70.68**	56.05	8.55	2.83
Ms-COCO	**54.39**	69.31	52.96	39.32	42.06	60.14	87.72	75.92	49.89	**78.19**	**27.56**	**7.05**

Qualitative Insights. The layout segmentation results on the PRIMA dataset obtained by *SwinDocSegmenter* and state-of-the-art approaches are shown in Fig. 4. In this test case, *SwinDocSegmenter* is able to segment instances of different layout elements quite effectively.

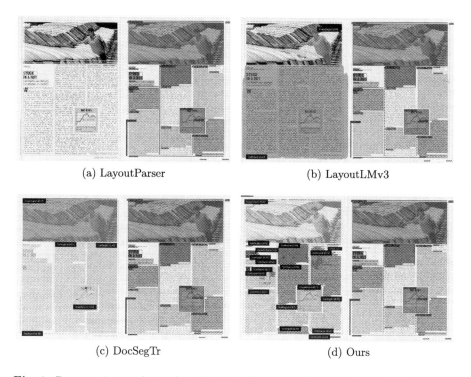

(a) LayoutParser

(b) LayoutLMv3

(c) DocSegTr

(d) Ours

Fig. 4. Comparative analysis of the SwinDocSegmenter framework with the state-of-the-art approaches (**Left:** Predicted layout **Right:** Ground-truth)

It can be observed from Fig. 4(a), though LayoutParser is quite effective for PRIMA dataset, it fails for this complex case as the bounding boxes are quite overlapping and not properly mapped with the ground truth. However, LayoutLMv3 (Fig. 4(b)) performs very poor in this case. It identifies text and figure instances but not the other class instances. DocSegTr (Fig. 4(c)) tries to improve the performance but it is still far away from the ground truth segmentation. On the other hand, our method segments complex document more satisfactorily and also maps the class instances with the ground truth (Fig. 4(d)).

Table 6. Performance on DocLayNet Benchmark

Classes	MaskRCNN	FasterRCNN	Yolov5	Ours
Caption	71.5	70.1	77.7	**83.56**
Footnote	71.8	73.7	**77.2**	64.82
Formula	63.4	63.5	**66.2**	62.31
List-item	80.8	81.0	**86.2**	82.33
Page-footer	59.3	58.9	61.1	**65.11**
Page-header	70.0	**72.0**	67.9	66.35
Picture	72.7	72.0	77.1	**84.71**
Section-header	69.3	68.4	**74.6**	66.5
Table	82.9	82.2	86.3	**87.42**
Text	85.8	85.4	88.1	**88.23**
Title	80.4	79.9	**82.7**	63.27
All	73.5	73.4	76.8	**76.85**

Quantitative Analysis. The final performance of the *SwinDocSegmenter* in terms of mAP is quite interesting and it has the ability to provide a new benchmark for Document Layout Segmentation. In Table 6 and 7, the method achieves a second position as both the LayoutLMv3 [16] and Layout Parser [42] use text information along with the visual information for instance segmentation task. In the case of PublayNet, we observe that the proposed is better identifying the text region than LayoutLMv3 and it provides comparable performance for other categories except for the "Title". Now "Title" also contains text so without textual information, it will be very difficult to solve these borderline cases. Also in terms of $AP@0.5$ and $AP@0.75$, it already surpasses LayoutLMv3 by only using visual information. Not only that, but it also outperforms the DiT (AP: 93.5) [22] and achieves comparable performance with UDoc (AP: 93.9) [13] which provide a standard benchmark on the PubLayNet dataset. The same observations have been noticed for the PRIMA dataset. It is already observed that LayoutLMv3 is not good enough to detect small objects. But Layout Parser can, as it has a convolution backbone instead of a Transformer backbone and it also uses Microsoft OCR to extract the textual information from the images to combine them with visual information. The proposed model surpasses this state-of-the-art for all categories except "Table" and "Others". The performance is mainly affected by the "others" category as there is no such particular definition and without proper text information it is very difficult to separate them from the other categories.

In Table 8 it has been observed that the proposed method outperforms all the previous state-of-the-art approaches. It outperforms the DocSegTr in the

Table 7. Performance Analysis on the PubLayNet and PRIMA Benchmark

PublayNet					PRIMA				
Object	Layout Parser	Doc SegTr	Layout LMv3	Ours	object	Layout Parser	Doc SegTr	Layout LMv3	Ours
Text	90.1	91.1	94.5	**94.55**	Text	83.1	75.2	70.8	**87.72**
Title	78.7	75.6	90.6	87.15	Image	73.6	64.3	50.1	**75.92**
Lists	75.7	91.5	95.5	93.03	Table	95.4	59.4	42.5	49.89
Figures	95.9	97.9	97.9	97.91	Math	75.6	48.4	46.5	**78.19**
Tables	92.8	97.1	97.9	97.25	Separator	20.6	1.8	9.6	**27.56**
					other	39.7	3.0	17.4	7.054
AP	86.7	90.4	**95.1**	93.72	AP	64.7	42.5	40.3	54.39
AP@0.5	97.2	97.9		**97.94**	AP@0.5	77.6	54.2		69.31
AP@0.75	93.8	95.8		**96.28**	AP@0.75	71.6	45.8		52.965

Historical Japanese dataset by a small margin (1%) but shows a significant improvement in the "Name" and "Position" categories. On the other hand, it shows a significant improvement (5%) in the Table Detection task on the Table-Bank dataset as it has only one category and comparatively less challenging layouts.

Table 8. Performance Analysis on HJ and TableBank Benchmark

Historical Japanese					TableBank				
object	Layout Parser	Doc SegTr	Layout LMv3	Ours	object	Layout Parser	Doc SegTr	Layout LMv3	Ours
Body	99.0	99.0	99.0	**99.72**	Table	91.2	93.3	92.9	**98.04**
Row	98.8	99.1	99.0	99.0					
Title	87.6	93.2	92.9	89.5					
Bio	94.5	94.7	94.7	86.26					
Name	65.9	70.3	67.9	**83.8**					
Position	84.1	87.4	87.8	**93.0**					
Other	44.0	43.7	38.7	40.57					
AP	81.6	83.1	82.7	**84.55**	AP	91.2	93.3	92.9	**98.04**
AP@0.5		90.1		**90.78**	AP@0.5		98.5		**98.95**
AP@0.75		88.1		**88.22**	AP@0.75		94.9		**98.90**

Last but not the least, we have obtained the first Transformer based baseline for a newly proposed dataset **DocLayNet** which contains industrial documents and the layouts are more challenging than PubLayNet benchmark. From Table 6 we conclude that our proposed SwinDocSegmenter outperforms the convolutional-based algorithms (MaskRCNN, FasterRCNN, etc.) by a significant margin.

5   Conclusion

In this paper we have presented *SwinDocSegmenter*, a powerful model to perform Document Layout Analysis by only utilizing the visual information. The improvement regarding the state-of-the-art is mainly constructed due to the enhanced and unified query selection, contrastive denoising training, and look forward twice approach. Also, the low-level projection head helps to enhance the low-level instances which makes a significant improvement in the overall performance as the Transformers are usually not good enough to detect small objects. However, there is still some scope for further improvement. The performance on PRIMA has still not reached the state-of-the-art with visual features as it contains a complex layout and very small training samples. A few-shot setting could help to improve the performance in the future.

Acknowledgment. This work has been partially supported by the Spanish project PID2021-126808OB-I00, the Catalan project 2021 SGR 01559 and the PhD Scholarship from AGAUR (2021FIB-10010). The Computer Vision Center is part of the CERCA Program / Generalitat de Catalunya.

References

1. Almutairi, A., Almashan, M.: Instance segmentation of newspaper elements using mask R-CNN. In: 2019 18th IEEE International Conference On Machine Learning And Applications (ICMLA), pp. 1371–1375. IEEE (2019)
2. Appalaraju, S., Jasani, B., Kota, B.U., Xie, Y., Manmatha, R.: Docformer: end-to-end transformer for document understanding. In: Proceedings of the IEEE/CVF International Conference on Computer Vision, pp. 993–1003 (2021)
3. Asi, A., Cohen, R., Kedem, K., El-Sana, J.: Simplifying the reading of historical manuscripts. In: Proceedings of the International Conference on Document Analysis and Recognition (2015)
4. Biswas, S., Banerjee, A., Lladós, J., Pal, U.: DocSegTr: an instance-level end-to-end document image segmentation transformer. arXiv preprint arXiv:2201.11438 (2022)
5. Biswas, S., Riba, P., Lladós, J., Pal, U.: Beyond document object detection: instance-level segmentation of complex layouts. Int. J. Doc. Anal. Recogn. (IJDAR) **24**(3), 269–281 (2021)
6. Chen, J., Lopresti, D.: Table detection in noisy off-line handwritten documents. In: ICDAR (2011)
7. Chen, K., Seuret, M., Hennebert, J., Ingold, R.: Convolutional neural networks for page segmentation of historical document images. In: 2017 14th IAPR International Conference on Document Analysis and Recognition (ICDAR), vol. 1, pp. 965–970. IEEE (2017)
8. Chen, K., Seuret, M., Liwicki, M., Hennebert, J., Ingold, R.: Page segmentation of historical document images with convolutional autoencoders. In: 2015 13th International Conference on Document Analysis and Recognition (ICDAR), pp. 1011–1015. IEEE (2015)
9. Cheng, B., Choudhuri, A., Misra, I., Kirillov, A., Girdhar, R., Schwing, A.G.: Mask2Former for video instance segmentation. arXiv preprint arXiv:2112.10764 (2021)
10. Clausner, C., Antonacopoulos, A., Pletschacher, S.: ICDAR 2019 competition on recognition of documents with complex layouts-RDCL2019. In: Proceedings of the International Conference on Document Analysis and Recognition, pp. 1521–1526 (2019)
11. Fang, J., Gao, L., Bai, K., Qiu, R., Tao, X., Tang, Z.: A table detection method for multipage pdf documents via visual seperators and tabular structures. In: ICDAR (2011)

12. Gao, L., Yi, X., Jiang, Z., Hao, L., Tang, Z.: ICDAR 2017 competition on page object detection. In: Proceedings of the International Conference on Document Analysis and Recognition, vol. 1, pp. 1417–1422 (2017)

13. Gu, J., et al.: Unidoc: unified pretraining framework for document understanding. Adv. Neural. Inf. Process. Syst. **34**, 39–50 (2021)

14. Gu, Z., et al.: XYLayoutLM: towards layout-aware multimodal networks for visually-rich document understanding. In: Proceedings of the IEEE/CVF Conference on Computer Vision and Pattern Recognition, pp. 4583–4592 (2022)

15. Harley, A.W., Ufkes, A., Derpanis, K.G.: Evaluation of deep convolutional nets for document image classification and retrieval. In: 2015 13th International Conference on Document Analysis and Recognition (ICDAR), pp. 991–995. IEEE (2015)

16. Huang, Y., Lv, T., Cui, L., Lu, Y., Wei, F.: LayoutLMv3: pre-training for document AI with unified text and image masking. arXiv preprint arXiv:2204.08387 (2022)

17. Journet, N., Eglin, V., Ramel, J.Y., Mullot, R.: Text/graphic labelling of ancient printed documents. In: Proceedings of the International Conference on Document Analysis and Recognition, pp. 1010–1014 (2005)

18. Kim, G., et al.: OCR-free document understanding transformer. In: Computer Vision-ECCV 2022: 17th European Conference, Tel Aviv, Israel, October 23–27, 2022, Proceedings, Part XXVIII, pp. 498–517. Springer (2022). https://doi.org/10.1007/978-3-031-19815-1_29

19. Kise, K., Sato, A., Iwata, M.: Segmentation of page images using the area voronoi diagram. Comput. Vis. Image Underst. **70**(3), 370–382 (1998)

20. Lee, Y., Hwang, J.W., Lee, S., Bae, Y., Park, J.: An energy and GPU-computation efficient backbone network for real-time object detection. In: Proceedings of the IEEE/CVF Conference on Computer Vision and Pattern Recognition Workshops (2019)

21. Li, F., et al.: Mask DINO: towards a unified transformer-based framework for object detection and segmentation. arXiv preprint arXiv:2206.02777 (2022)

22. Li, J., Xu, Y., Lv, T., Cui, L., Zhang, C., Wei, F.: DiT: self-supervised pre-training for document image transformer. In: Proceedings of the 30th ACM International Conference on Multimedia, pp. 3530–3539 (2022)

23. Li, K., et al.: Cross-domain document object detection: benchmark suite and method. In: Proceedings of the IEEE Conference on Computer Vision and Pattern Recognition (2020)

24. Li, M., Cui, L., Huang, S., Wei, F., Zhou, M., Li, Z.: TableBank: table benchmark for image-based table detection and recognition. In: Proceedings of the Twelfth Language Resources and Evaluation Conference, pp. 1918–1925 (2020)

25. Li, X.-H., Yin, F., Liu, C.-L.: Page segmentation using convolutional neural network and graphical model. In: Bai, X., Karatzas, D., Lopresti, D. (eds.) DAS 2020. LNCS, vol. 12116, pp. 231–245. Springer, Cham (2020). https://doi.org/10.1007/978-3-030-57058-3_17

26. Li, Y., et al.: StrucTexT: structured text understanding with multi-modal transformers. In: Proceedings of the 29th ACM International Conference on Multimedia, pp. 1912–1920 (2021)

27. Lin, G.S., Tu, J.C., Lin, J.Y.: Keyword detection based on retinanet and transfer learning for personal information protection in document images. Appl. Sci. **11**(20), 9528 (2021)

28. Lin, T.Y., Dollár, P., Girshick, R., He, K., Hariharan, B., Belongie, S.: Feature pyramid networks for object detection. In: Proceedings of the IEEE Conference on Computer Vision and Pattern Recognition, pp. 2117–2125 (2017)

29. Lin, T.-Y., et al.: Microsoft COCO: common objects in context. In: Fleet, D., Pajdla, T., Schiele, B., Tuytelaars, T. (eds.) ECCV 2014. LNCS, vol. 8693, pp. 740–755. Springer, Cham (2014). https://doi.org/10.1007/978-3-319-10602-1_48

30. Liu, Z., et al.: Swin transformer: hierarchical vision transformer using shifted windows. In: Proceedings of the IEEE/CVF International Conference on Computer Vision, pp. 10012–10022 (2021)

31. Mathur, P., et al.: LayerDoc: layer-wise extraction of spatial hierarchical structure in visually-rich documents. In: Proceedings of the IEEE/CVF Winter Conference on Applications of Computer Vision, pp. 3610–3620 (2023)

32. Mo, S., Sun, Z., Li, C.: Multi-level contrastive learning for self-supervised vision transformers. In: Proceedings of the IEEE/CVF Winter Conference on Applications of Computer Vision, pp. 2778–2787 (2023)

33. Oliveira, S.A., Seguin, B., Kaplan, F.: dhSegment: a generic deep-learning approach for document segmentation. In: ICFHR (2018)

34. Pfitzmann, B., Auer, C., Dolfi, M., Nassar, A.S., Staar, P.: DocLayNet: a large human-annotated dataset for document-layout segmentation. In: Proceedings of the 28th ACM SIGKDD Conference on Knowledge Discovery and Data Mining, pp. 3743–3751 (2022)

35. Powalski, R., Borchmann, L, Jurkiewicz, D., Dwojak, T., Pietruszka, M., Pałka, G.: Going Full-TILT boogie on document understanding with text-image-layout transformer. In: Lladós, J., Lopresti, D., Uchida, S. (eds.) ICDAR 2021. LNCS, vol. 12822, pp. 732–747. Springer, Cham (2021). https://doi.org/10.1007/978-3-030-86331-9_47

36. Prasad, D., Gadpal, A., Kapadni, K., Visave, M., Sultanpure, K.: CascadeTabNet: an approach for end to end table detection and structure recognition from image-based documents. In: CVPRW, pp. 572–573 (2020)

37. Ren, S., He, K., Girshick, R., Sun, J.: Faster R-CNN: towards real-time object detection with region proposal networks. In: NIPS (2015)

38. Saabni, R., El-Sana, J.: Language-independent text lines extraction using seam carving. In: Proceedings of the International Conference on Document Analysis and Recognition (2011)

39. Saha, R., Mondal, A., Jawahar, C.: Graphical object detection in document images. In: ICDAR (2019)

40. Schreiber, S., Agne, S., Wolf, I., Dengel, A., Ahmed, S.: DeepDeSRT: deep learning for detection and structure recognition of tables in document images. In: 2017 14th IAPR international conference on document analysis and recognition (ICDAR), vol. 1, pp. 1162–1167. IEEE (2017)

41. Shen, Z., Zhang, K., Dell, M.: A large dataset of historical Japanese documents with complex layouts. In: Proceedings of the IEEE Conference on CVPRW, pp. 548–549 (2020)

42. Shen, Z., Zhang, R., Dell, M., Lee, B.C.G., Carlson, J., Li, W.: LayoutParser: a unified toolkit for deep learning based document image analysis. In: Lladós, J., Lopresti, D., Uchida, S. (eds.) ICDAR 2021. LNCS, vol. 12821, pp. 131–146. Springer, Cham (2021). https://doi.org/10.1007/978-3-030-86549-8_9

43. Sun, N., Zhu, Y., Hu, X.: Faster R-CNN based table detection combining corner locating. In: 2019 International Conference on Document Analysis and Recognition (ICDAR), pp. 1314–1319. IEEE (2019)

44. Vaswani, A., et al.: Attention is all you need. In: Advances in Neural Information Processing Systems, vol. 30 (2017)

45. Xia, Z., Pan, X., Song, S., Li, L.E., Huang, G.: Vision transformer with deformable attention. In: Proceedings of the IEEE/CVF Conference on Computer Vision and Pattern Recognition, pp. 4794–4803 (2022)
46. Yang, H., Hsu, W.: Transformer-based approach for document layout understanding. In: 2022 IEEE International Conference on Image Processing (ICIP), pp. 4043–4047. IEEE (2022)
47. Yang, H., Hsu, W.H.: Vision-based layout detection from scientific literature using recurrent convolutional neural networks. In: 2020 25th International Conference on Pattern Recognition (ICPR), pp. 6455–6462. IEEE (2021)
48. Zhang, H., et al.: DINO: DETR with improved denoising anchor boxes for end-to-end object detection. arXiv preprint arXiv:2203.03605 (2022)
49. Zhong, X., Tang, J., Yepes, A.J.: PubLayNet: largest dataset ever for document layout analysis. In: Proceedings of the International Conference on Document Analysis and Recognition, pp. 1015–1022 (2019)

BaDLAD: A Large Multi-Domain Bengali Document Layout Analysis Dataset

Md. Istiak Hossain Shihab[1(✉)], Md. Rakibul Hasan[2],
Mahfuzur Rahman Emon[2], Syed Mobassir Hossen[1], Md. Nazmuddoha Ansary[1],
Intesur Ahmed[1,4], Fazle Rabbi Rakib[1,2], Shahriar Elahi Dhruvo[1,2],
Souhardya Saha Dip[1,2], Akib Hasan Pavel[1], Marsia Haque Meghla[1],
Md. Rezwanul Haque[1], Sayma Sultana Chowdhury[2], Farig Sadeque[1,3],
Tahsin Reasat[1,4], Ahmed Imtiaz Humayun[1,5], and Asif Sushmit[1,6]

[1] Bengali.AI, Dhaka, Bangladesh
istiak@proton.me
[2] Shahjalal University of Science and Technology, Sylhet, Bangladesh
[3] BRAC University, Dhaka, Bangladesh
[4] Vanderbilt University, Nashville, USA
[5] Rice University, Houston, USA
[6] RPI, Troy, USA

Abstract. While strides have been made in deep learning based Bengali Optical Character Recognition (OCR) in the past decade, absence of large Document Layout Analysis (DLA) datasets has hindered the application of OCR in document transcription, e.g., transcribing historical documents and newspapers. Moreover, rule-based DLA systems that are currently being employed in practice are not robust to domain variations and out-of-distribution layouts. To this end, we present the first multi-domain large **Bengali Document Layout Analysis Dataset: BaDLAD**. This dataset contains $33,695$ *human annotated document samples from six domains* - i) books and magazines ii) public domain govt. documents iii) liberation war documents iv) new newspapers v) historical newspapers and vi) property deeds; with $710K$ polygon annotations for four unit types: text-box, paragraph, image, and table. Through preliminary experiments benchmarking the performance of existing state-of-the-art deep learning architectures for English DLA, we demonstrate the efficacy of our dataset in training deep learning based Bengali document digitization models.

Keywords: Handwritten Document Images · Layout Analysis (Physical and Logical) · Mobile/Camera-Based · Other Domains · Typeset Document Images

1 Introduction

Understanding the layout of amorphous digital documents is a crucial step in parsing documents into organized machine-readable formats that are usable in

Md. I. H. Shihab, Md. R. Hasan, M. R. Emon, A. I. Humayun and A. Sushmit—Equal contribution.
Project website: https://bengaliai.github.io/badlad.

real-world applications. Despite tremendous developments in machine learning (ML) methods and deep neural networks (DNNs) in recent decades, transcription of documents, e.g., historical books, remains a difficult challenge [18]. Document layout analysis (DLA) is a preprocessing phase of a document transcription pipeline that detects and parses the structure of a document [4] by segmenting it into semantic units such as paragraphs, text-boxes, images and tables. Such segmented units are then transcribed via Optical Character Recognition (OCR) methods, for which robust algorithms have been proposed in literature [12,13]. The preprocessing step performed by DLA systems is often challenging due to different factors, e.g., free-writing style, deteriorating and faded text, ink spilling, and artistic lettering. Antiquated property documents, stained and torn papers and vague handwritten scripts make this task even more difficult [4]. Robust DLA methods are therefore a major requirement for the digitization of handwritten records.

A DLA pipeline comprises a number of steps that may differ among approaches based on the layout of the specific document category and analysis goals [4]. Although rule-based algorithms and heuristic approaches were the standard for DLA in its earlier days [1], recent decades have seen a major push towards solutions that use object detection models. Especially with the inception of DNNs, the accuracy and speed of such frameworks have greatly improved [5,9,11,19] paving the way for DNN based DLA methods [18]. While datasets like *DocBank* [16] and *PubLayNet* [20] are large enough to cater to the sample complexity of DNN based DLA frameworks, the datasets lack diversity in the orientation of annotations - which are mostly axes aligned. Moreover, such datasets contain data from a single domain, e.g., pdf articles from PubMed for *PubLayNet*. Therefore, DNNs trained on such homogeneous sources, risk being vulnerable towards domain or distribution shifts [15].

In this paper, we present a dataset of documents collected from the wild, from multiple domains containing text with diverse layouts and orientations. Our dataset is the first large scale multi-domain document layout analysis dataset for Bengali. Our main contributions are as follows:

- We present a human-annotated dataset of 33,693 documents collected in the wild "**BaDLAD**", for document layout analysis in Bengali. BaDLAD is the largest organic dataset for Bengali DLA to the best of our knowledge. Our dataset contains 710K polygon annotations for four unit/segment types: i) text-box, ii) paragraph, iii) images, and iv) table.
- BaDLAD comprises data collected from six different domains, i) books and magazines, ii) public domain govt. documents, iii) liberation war documents, iv) new newspapers, v) historical newspapers, and vi) property deeds. To the best of our knowledge, BaDLAD is also the first multi-domain DLA dataset for Bengali.
- We present preliminary results benchmarking the performance of popular DNN based DLA methods on BaDLAD. We show that existing English DLA state-of-the-art models, fine-tuned on BaDLAD, exhibit improved performance on Bengali document layout analysis tasks in the multi-domain setting.

Apart from this, we also present an additional *4 million* un-annotated images including captured, scanned and printed documents that can be used for unsupervised DLA. The following sections are organized as follows. In Sect. 2 we discuss related work on Document Layout Analysis that is present in literature. In Sect. 3 we discuss the challenges present in Bengali DLA, also motivating the need for documents collected from the wild. In Sect. 4 we present discussions on our collection protocols, annotation pipeline and statistics of our collected dataset. In Sect. 5 we present preliminary benchmarks on our dataset and following that in Sect. 6 we present conclusions and future directions. We make the codes for our benchmarking models and the corresponding data analysis publicly available under the CC BY-SA 4.0 license.

2   Related Work

Document Layout Analysis. According to [20], Zhong et al. generated and distributed the *PubLayNet* dataset for document layout analysis, which includes automatically annotated data through matching with XML representations. Using an implementation of the Detectron algorithm, they trained an F-RCNN model and an M-RCNN model using PubLayNet. This dataset is claimed to be the largest one out there, containing 1 million pdf pictures of PMCOA (PubMed Centre Open Access) articles. PubLayNet data represent only scientific papers, which is topic-specialized and reduces layout diversification.

Li et al. presented a dataset *DocBank* which contains 500K document-level images in English with fine-grained token-level annotations for structure analysis. They performed experiments on this dataset using four baseline models (BERT, RoBERT, LayoutLM, and Faster R-CNN) and claimed that the dataset can be utilized in any sequence labeling model [16]. However, this dataset is based on automatically annotated English documents, which hurts its generalizability.

Pfitzmann et al. presented a manually annotated document layout dataset *DocLayNet* in COCO format containing data from diverse sources [18]. They presented benchmark accuracies for a collection of standard object detection models (MASK R-CNN, Faster R-CNN and YOLOv5) and analyzed models trained on PubLayNet, DocBank, and DocLayNet [18]. Non-overlapping, vertically oriented, rectangular boxes were permitted during the annotation process. According to them, human-annotated datasets provide more credible layout ground truth on a diverse range of publication and typesetting styles compared to DocBank and PubLayNet. Oliveira et al. [3] proposed a block based classification method to detect the layout of structured image documents rapidly and automatically through one dimensional CNN approach with a bi-dimensional CNN to compare performance and demonstrate their work.

Bengali Document Layout Analysis. Clausner et al. [7] presented several methods including four open-source SOTA systems for the evaluation of page analysis and identification algorithms for ancient manuscripts written in Bengali through their comparative assessment on this topic. This dataset is available in

ICDAR challenges. For SOTA methods, they used Tesseract 3.04 and 4.0 with internal binarization and long short-term memory units (LSTMs). Bangla OCR-I used Google's Tesseract OCR engine for text classification and only works on printed scripts, whereas Bangla OCR-II's primary classification engine is a feature-based SVM and cannot handle intricate frames [7]. Some current datasets for Bengali document layout analysis are already being utilized in document processing tasks, although their size is rather limited [20].

3 Challenges of Bengali Document Layout Analysis

Bengali, one of the most widely spoken languages globally, is characterized by a large number of native speakers, estimated at almost 300 million, with 37 million international speakers. Despite its extensive usage, the field of Document Language Analysis (DLA) in Bengali remains in its nascent stages, with limited research and resources available on the subject. The synthetic data generation approach, commonly adopted by well-known datasets such as PubLayNet and DocBank, does not apply to Bengali given the majority of the publicly accessible Bengali documents are either scanned images or captured photographs of the original document and thus cannot be annotated using automatic algorithms. Moreover, such datasets are comprised of synthetic, born-digital documents and are carefully curated, resulting in annotations with exclusively horizontal and vertical boundaries. In contrast, our dataset incorporates irregularly-shaped polygon annotations and preserves their original boundaries. It is our belief that this approach will enhance the precision of layout detection and related challenges, such as optical character recognition and form detection.

The Bengali script, being a non-Latin-based script, possesses another challenge for DLA tasks. Bengali has an intricate writing system encompassing inflections, multiple script forms, and character composites. This is because individual characters can exhibit different forms based on their position within a word or the preceding and succeeding letters. Furthermore, certain characters in Bengali may be represented through a combination of multiple characters [2], which presents a challenge for models to identify them accurately. These complexities can result in inaccuracies in layout analysis, as the models may not be capable of discerning between the text elements and the interconnections among them.

The historical nature of printed Bengali documents, dating back to the early 1800s, coupled with the prevalence of typographical variations and the printing styles of ancient literature present significant difficulties for the document layout analysis (DLA) task in this language. Additionally, the complexity of the layout, frequently unintelligible handwriting, deteriorating paper quality, and non-standard formatting of modern Bengali legal documents further exacerbate the state of this area of research. Given the recent advances in massively data-driven deep learning techniques, development of a machine-trainable, hand-annotated dataset with sufficient diversity to address these challenges should be a priority—which is precisely what is proposed in the present paper.

4   Bengali Document Layout Analysis Dataset: *BaDLAD*

aDLAD comprises of data collected from six different domains. The dataset contains annotations for four semantic unit types via polygon annotations. In this section we first provide descriptions for the selected data domains and justifications for the semantic unit types. Following that we discuss our annotation pipeline and statistics of the collected data.

4.1   Semantic Units for Layout Segmentation

We started by scraping ∼ 20, 000 Bengali PDF files from publicly available online repositories for books. To explore the layout diversity of these books , we trained a self-supervised SwAV [6] model which generates prototypes that can be considered as the cluster centers of the model's embedding space. Upon inspecting the cluster centers, and manually inspecting a number of representatives from each cluster, we noticed four major semantic categories in which the layout can be partitioned:

- **Text-box** : A small isolated collection of letters, numbers, word or group of words, e.g., page number, book name, chapter name, headline/ title, or incomplete non-contiguous sentences.
- **Paragraph** : A collection of text that is made up of one or more sentences and deals with a single topic or idea and is separated from other paragraphs by a line break or indentation. A single word can be considered as a paragraph when it is in context and makes a meaningful point or statement on its own, e.g., in a dialogue.
- **Image**: Representation of any visual object that is not only comprised of text, e.g., logo, pictures, graphical handwritten signatures.
- **Table** : Structured set of data made up of rows and columns, which may or may not have table headers or borders.

We did not find a significant number of list elements in the clusters. Hence we did not include the list category as a semantic unit in our dataset. In our dataset, we have annotated lists as a collection of text-boxes or paragraphs, depending on which of the aforementioned definitions the list elements are closest to.

4.2   Domain Categories and Sources

To make the dataset diverse and complex, we collected documents from a wide range of domains, e.g., Novels, Magazines, Poems, Newspapers, Government Documents, Property Deeds, Liberation War Documents, which we have binned into the following categories based on sources. We have also presented representative samples in Fig. 1.

Magazine and Books. This domain comprises of samples from ∼ 20, 000 Bengali PDFs scraped from publicly available online repositories, as mentioned in

(a) Newspaper (Hist.)   (b) Book Cover   (c) Comic

(d) Magazine   (e) Liberation War Doc   (f) Poem

(g) Govt Doc   (h) 1-page-2-Column   (i) 1-page-1-Column

(j) Newspaper(New)   (k) 2-page-1-Scan   (l) 2-page-2-column-1-scan

Fig. 1. Different layout categories present in the BaDLAD dataset. Annotations are color coded as: ▓ Text-box, ▓ Paragraph, ▓ Image, ▓ Table. We do not present examples from the *Property Deeds* domain to ensure confidentiality. (Color figure online)

Sect. 4.1. The collection comprised of books, magazines (Fig. 1d), poems (Fig. 1f) and comics (Fig. 1c) with a very diverse set of layouts. We also take into account the book covers while sampling from the collected PDFs. All of the PDFs are scanned or photo captured versions of the original document, without any digital transcription. Literary works comprising mostly of text, e.g., novels, contain three major layout types - single page single column (Fig. 1i), single page double column (Fig. 1h) and double page single scan (Fig. 1k and 1l).

Historical Newspapers. Historical Newspapers that have been published before December 1971 that were manually scanned. The typesetting of such newspapers are significantly different from new newspapers, e.g., in terms of font size, font style, glyphs of consonant conjuncts (Fig. 1a).

New Newspapers. Recently published newspapers manually captured by scanners and cameras (Fig. 1j).

Liberation War Documents. Taken from a 15-part collection of liberation war documents, manually scanned (Fig. 1e).

Government Documents. We have collected publicly available government documents by scraping from online repositories and by manually collecting and scanning. These documents comprise of both handwritten and printed characters along with logos, seals, tables, headers and graphical elements (Fig. 1g).

Property Deeds. Confidential documents collected with consent via social media crowd-sourcing campaigns. We have anonymized the documents by removing sensitive and identifiable information and include them only in the hidden test dataset. These documents generally contain a lot of handwritten notes, signatures and free-form text, posing a challenging DLA task.

4.3   Annotation and Validation

To ensure diversity of samples, we chose 2 pages randomly from each scanned document, since pages from the same document have higher probability of being similar. A team of 13 annotators were trained to annotate document layout on the "Labelbox" platform. Polygon labeling was used because of the complex orientation of texts and images in our dataset (as can also be seen in Fig. 1). Each annotator was tasked to segment all the semantic units in a given sample. We also kept track of the time required to annotate each sample as metadata, which can be considered as a segmentation hardness measure for each sample. The annotators annotated 33,693 samples in total over a course of four months. During annotation, three curators were assigned the task of annotation verification and curation. Any document with wrong annotations were resent to the original annotator for correction. The annotation guidelines were also dynamically updated during this process. In Fig. 2 we provide a brief overview of our data collection, annotation and validation process.

Brief Annotation Guideline. In order to obtain more effective data, we developed objective guidelines for the annotators which applied in a domain-unit specific manner. All plots or graphs were considered as images. For samples with

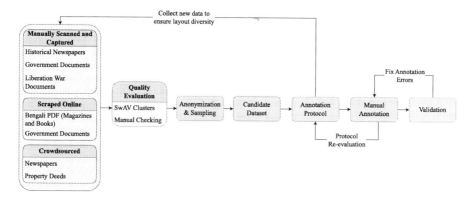

Fig. 2. Data collection and annotation pipeline for the BaDLAD dataset. Candidate samples from the un-annotated dataset are collected and curated dynamically with annotation and validation tasks to ensure layout diversity and quality of images.

double page scans, if any portion of the content of one page went to another, then the divided portions were annotated separately. In the case of poetry, if there were extra white spaces between lines then the lines were considered separate paragraphs. Bullets or numbered lists were separately annotated as text-boxes for one line sentences and paragraphs in the case of multi-line sentences. Hand-written texts were considered as texts except for signatures, signatures were marked as images. If there were any extra notes (e.g. URLs, post-scripts) along with a paragraph, then the extra portion was annotated as a text-box. Vertical lines, advertisements/links were marked as text-boxes. Tables were annotated as tables, but the contents were marked according to the definitions as images, text-boxes or paragraphs. If for a sample, the contents from the other side of a scanned page were visible due to transparency, the text from the opposite were ignored and only the main text from the correct side was annotated.

4.4   BaDLAD Statistics

After annotation and curation, the BaDLAD dataset comprises a total of 33,693 samples; of which 30054 samples are from *Magazines and Books*, 1285 samples from *Govt. Documents*, 1004 samples from *Liberation War Documents*, 861 samples from *Historical Newspapers*, 328 samples from *Property Deeds* and 161 samples from *New Newspapers*. While *Magazines and Books* is the most prevalent domain, as discussed in Sect.4.2 and presented in Fig.1, the domain contains a large diversity of layouts from multiple sources. In Table 1 and Table 2, we present the domain-wise number of annotations for every unit type, along with the time elapsed for annotation. We present it for the train and test splits separately as specified in Sect. 5.1. If we consider the average time required to annotate a sample as a hardness measure for each sample, we can see that samples from the *Historical Newspapers* and *New Newspapers* domains are the most challenging. On the other hand, samples from the *Liberation War Documents* domain is considerably easier, which can be attributed to the relatively larger volume of paragraph annotations. This distinction is also present in the number of polygon per page histogram, presented in Fig. 4.

Table 1. Domain-wise annotation statistics for BaDLAD (Train)

Domain	Samples	Text-box	Paragraph	Image	Table	Total Annotation Time (Hours)	Avg. Annotation Time (Minutes)
Historical Newspapers	516	25452	38990	1252	67	211.52	24.60
New Newspapers	96	3978	2507	494	36	27.35	17.10
Govt Documents	771	44017	2260	762	514	78.44	6.10
Magazines and Books	18380	123099	162570	7734	594	948.58	3.10
Property Deeds	0	0	0	0	0	0.0	0.0
Lib War Docs	602	7330	3262	55	142	22.43	2.24
Total	20365	203876	209589	10297	1353	1288.33	3.80

Table 2. Domain-wise annotation statistics for BaDLAD (Test)

Domain	Samples	Text-box	Paragraph	Image	Table	Total Annotation Time (Hours)	Avg. Annotation Time (Minutes)
Historical Newspapers	345	17611	26571	838	54	146.85	25.54
New Newspapers	65	3542	1902	237	24	22.57	20.83
Govt Documents	514	27903	1497	482	301	52.63	6.14
Magazines and Books	11674	80581	103390	4949	376	625.02	3.21
Property Deeds	328	6012	599	930	117	17.03	3.11
Lib War Docs	402	4733	2370	16	104	16.06	2.40
Total	13328	140382	136329	7452	976	880.17	3.96

In Fig. 5 we present the area covered by each unit type as a percentage of the total area of samples per domain, for the whole dataset. Here, area is calculated in pixel units. We can see that for Liberation War Documents, Magazines and Books and Newspapers, a large fraction of the images are covered in paragraphs. On the other hand, for Govt. Documents, a larger fraction of area is covered by tables. For all of these graphs, the percentage sum might be larger than 1 since there can be significant overlaps in the area covered by two different annotations. For example, in Fig. 3ii we present examples of different overlaps which are frequently present in the dataset.

In Fig. 6, we present the spatial distribution of different unit types for the whole dataset. The table polygons exhibit a highly concentrated localization within a distinct square shape, which is a result of the tendency for placing tables away from the borders of the document. The text-boxes are less overlapped, as is visible in the distribution, with the exception of the header section. Conversely, paragraphs are distributed evenly throughout the body of the page and exhibit a characteristic horizontal dark bar in the center, indicative of the presence of a significant number of double-page layouts within the original dataset. We generate the spatial distribution by resizing each image to a 128×128 square and counting for every pixel, the number of annotations for each unit type. For the Images and Tables unit types we use all the samples from the dataset. For the text-box and paragraphs unit types, we randomly sample 50K annotations for each, to generate the figures.

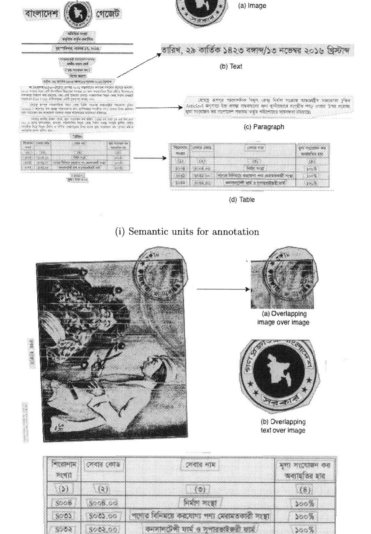

(i) Semantic units for annotation

(ii) Overlapping annotations

Fig. 3. Annotated samples from the BaDLAD dataset with semantic units (i). Annotation overlaps between different semantic units (ii).

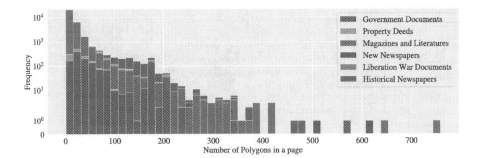

Fig. 4. Histogram of Polygons per page stacked and colored by the domain presented in logarithmic scale. Samples from the *Government Documents* domain contain a lower number of polygons in every page. Both *Historical Newspapers* and *New Newspapers* contains a higher number of polygons per page, which correlates with their higher avg. annotation time requirement according to Table 1. Samples from the *Magazines and Books* domain contain a large diversity in number of polygons per page.

5 Benchmark

In this section, we evaluate the performance of object detection and segmentation models that are prevalent in DLA literature. We detail our methodology for generating a standard training and testing split from our dataset, report performance of benchmark models and show prediction results with qualitative analysis.

5.1 Dataset Split

The dataset was split into a train and test partition to perform our benchmarks. The split was done in a stratified method where a 60:40 train-test ratio was maintained for each domain listed in Sect. 4.2 except for property deed which was kept entirely in the test set. Also we ensured that the pages coming from the same book was kept in the same split to prevent data leakage. Previously as the authors of PubLaynet [20] and HJDataset [16] claimed that segmentation masks are the quadrilateral regions for each block, Compared to the rectangular bounding boxes, they delineate the text region more accurately. The resulting train and test set had 20,365 and 13,328 samples respectively. Brief statistics of the train and test split can be found in Tables 1 and 2.

5.2 Model Description and Results

In this section we compare the performance of state-of-the-art DLA and object recognition methods on our dataset. We trained an F-RCNN and an M-RCNN model on BaDLAD utilizing the Detectron [10] implementation and a YOLOv8 model utilizing the Ultralytics [14] implementation. The R-CNN models were trained for 10,000 iterations with default hyperparameters; a learning rate of

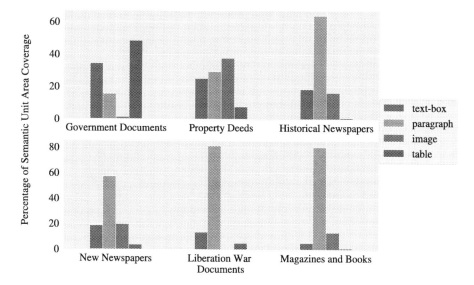

Fig. 5. Area covered by polygons for every unit type, normalized by the total area of samples per domain. Except for *Govt. Documents*, all the domains have a larger area covered by paragraphs. For the *Govt. Documents* domain, even though the number of paragraphs is higher than the number of tables, the area covered by tables is significantly higher than that of paragraphs.

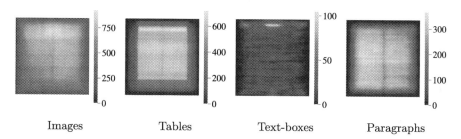

Fig. 6. Un-normalized spatial distribution of annotations for different unit types. Each sample from BaDLAD is resized to a square 128×128 image, and pixel-wise density for each annotation type is presented. While for all cases there is uniformity in spatial distribution, for Text-box annotations, we see a spike in the distribution around the top, indicating high density of headers annotated as text-boxes.

0.001 with a decay of 0.1, a minibatch size of 48, and a warm-up iteration of 5. The YOLO models were trained for 100 epochs. For this, we used a batch size of 8, an initial learning rate of 0.01, a weight decay of 0.0005, and a warm-up iteration of 3. As a feature extractor, the RCNN models employ a ResNet-50 model, except for the M-RCNN pretrained on PubLayNet, which employs ResNet101. The performance of the benchmark models utilized in this study is presented in Table 3. Note that while BaDLAD was labeled via polygon annotations, we also

Table 3. Comparison of mAP (50-95) for different DLA architectures on BaDLAD. Models are pre-trained on ImageNet (ImgNet), PubLayNet (PLNet), and COCO datasets. We present domainwise results for each unit type, categorized as P (Paragraph), Tx (Textbox), I (Image), and Tb (Table). M-RCNN has a ResNet50 backbone whereas, M-RCNN* has a ResNet101 backbone.

Arch.	Pretrain.	Annot.	Historical Newspapers				New Newspapers				Government Documents			
			P	Tx	I	Tb	P	Tx	I	Tb	P	Tx	I	Tb
F-RCNN	ImgNet	BBox	57.87	17.49	59.05	0.0	39.08	12.47	47.60	2.08	43.96	18.68	22.64	10.70
F-RCNN	PLNet	BBox	64.94	22.10	67.96	2.38	46.74	16.15	60.68	14.70	46.95	20.03	28.47	64.35
YOLOv8	COCO	BBox	**97.50**	**73.30**	**91.50**	**45.50**	**79.70**	**45.10**	**87.50**	**64.90**	**85.10**	**82.60**	**85.70**	**98.70**
F-RCNN	ImgNet	Mask	58.30	18.68	59.59	0.0	40.92	13.46	47.34	7.29	37.72	18.87	20.48	7.00
M-RCNN	ImgNet	Mask	60.33	18.29	57.30	0.0	41.39	13.15	45.22	1.91	39.06	18.73	19.43	3.73
M-RCNN*	PLNet	Mask	**68.63**	22.34	64.08	3.67	48.06	17.12	55.56	20.21	41.57	21.43	25.88	49.68
YOLOv8	COCO	Mask	64.40	**27.10**	**77.20**	**10.80**	**55.0**	**18.0**	**64.80**	14.20	**54.90**	**22.10**	**34.80**	**45.00**

Arch.	Pretrain.	Annot.	Magazine and Books				Liberation War Documents				Property Deeds			
			P	Tx	I	Tb	P	Tx	I	Tb	P	Tx	I	Tb
F-RCNN	ImgNet	BBox	65.16	24.71	48.11	1.61	78.60	26.37	1.83	36.20	0.54	0.55	1.61	0.58
F-RCNN	PLNet	BBox	68.91	26.29	58.36	15.61	79.63	27.64	1.00	69.60	0.34	0.82	1.24	0.95
YOLOv8	COCO	BBox	**93.90**	**68.20**	**79.40**	**35.00**	**98.70**	**78.50**	5.53	**91.30**	**58.00**	**51.90**	**36.40**	**57.00**
F-RCNN	ImgNet	Mask	61.59	25.52	46.81	2.86	70.40	26.63	1.34	40.94	0.70	0.58	1.05	0.86
M-RCNN	ImgNet	Mask	61.76	25.34	44.90	2.27	71.15	26.80	0.98	40.13	0.60	0.66	2.08	0.61
M-RCNN*	PLNet	Mask	65.77	**27.24**	52.03	11.96	72.36	**28.87**	2.11	**66.21**	0.51	0.69	2.73	5.68
YOLOv8	COCO	Mask	**65.20**	23.80	**58.30**	**12.20**	**72.90**	24.50	1.50	29.70	**38.70**	**16.20**	**19.10**	**6.27**

provide best fit bounding box, and segmentation masks as annotation. Therefore, models trained with an object detection target used bounding box as the ground truth, whereas, models trained with a segmentation target were evaluated on mask annotations. In accordance with the standard established by the COCO Competition [17], the Mean Average Precision (mAP) was computed utilizing the intersection over union (IoU) metric for bounding boxes. The YOLOv8 segmentation model employs a custom CNN feature extractor CSPDarknet53, in combination with a YOLO detection backbone to achieve superior accuracy in bounding box predictions across all domains and unit types. However, when it comes to mask prediction, the M-RCNN pre-trained on PubLayNet, exhibit better performance in predicting paragraphs in historical newspapers, as well as text boxes, images, and tables in Liberation war documents. YOLOv8 outperforms M-RCNN and F-RCNN in all other cases. YOLOv8 obtains an average mAP of 70.46% and 35.69% in the object recognition and segmentation settings respectively. The M-RCNN model pre-trained on PubLayNet acquired average mAP of 32.27% in the segmentation setting. The models are generally more accurate in detecting paragraphs and images than text-boxes and tables. As our dataset contains a low number of table annotations, our benchmark models seem to under-perform for that unit type. However, even after being the second most frequent unit type, the accuracy of detecting text-boxes is surprisingly low. The results show that there is major scope for improvement in the DLA tasks using our dataset.

In Fig. 7, we show performance of the M-RCNN model with ResNet101 backbone, on seven test samples. For the first three samples (top-left to bottom-right)

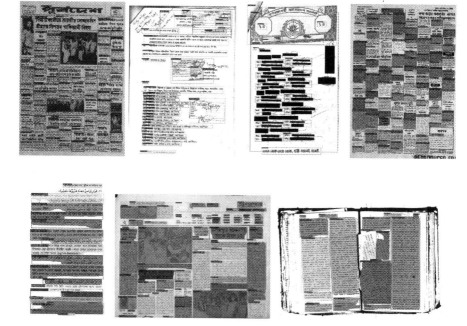

Fig. 7. Predictions of M-RCNN-101 model on BaDLAD Test samples. The contents of the third sample (from the *Property deeds* domain) has been redacted for confidentiality. The first 3 samples show only bounding box predictions and the rest show segmentation boundaries.

we show only the bounding box predictions while for the rest of the samples, we show both bounding box and segmentation masks predicted. We see that the model performs significantly bad for the third samples. This is due to the sample coming from the *Property Deeds* domain which was absent in training. The fourth sample also contains a number of paragraphs which are not correctly detected, possibly due to the boldface headers. In the fifth sample, we can see the network perform well even in the presence of code-switched text. In sample six, we see the model perform very well, especially since the sample is axes aligned. For sample seven, the network exhibits robustness to noisy images and partially torn pages in the scanned document.

6   Conclusion

In this paper, we introduced the BaDLAD Dataset on Bengali document layout analysis and presented preliminary benchmarking results using RCNN-based and Yolo-based approaches. Unlike many prominent datasets from this domain, this is a human-annotated large dataset for layout analysis and presents an unique set of challenges. As the creation of synthetic layout analysis datasets are challenging for Bengali, this work can serve as a foundation to the field of Bengali Optical

Character Recognition and also digitization of historical documents. As we are also releasing 4 million unannotated samples along with the dataset, future work can focus on utilizing unsupervised methods for training better models.

Although the dataset has diversity, it is imbalanced both in the source domains and the semantic units. This dataset will be a stepping stone in analyzing this imbalance. It can also be used as a fine-tuning dataset for a pretrained model. There are missing domains such as, shopping receipts, application form, id card etc. These domains can be added in future iterations of the dataset. We will utilize active learning methods for annotating more domain-diversified samples un-annotated portion of the dataset. One current trend in the development of DLA datasets is the use of having textual information along with layout information for Language Model based layout segmentation modeling [16]. It is possible to get word segmentation by using a word detection algorithm [8] and use a word recognition model to detect the text content. We leave this as a future work, which can convert the current dataset from a segmentation one to a LM based layout analysis dataset and hence, improve the quality of segmentation performance. [1]

Acknowledgement. We are thankful to Center for Bangladesh Genocide Research - CBGR for sharing some invaluable historical documents for this dataset. We also thank the Department of Software Engineering in Shahjalal University of Science and Technology, for their support.

References

1. Ahmad, R., Afzal, M.T., Qadir, M.A.: Information extraction from pdf sources based on rule-based system using integrated formats. In: Sack, H., Dietze, S., Tordai, A., Lange, C. (eds.) SemWebEval 2016. CCIS, vol. 641, pp. 293–308. Springer, Cham (2016). https://doi.org/10.1007/978-3-319-46565-4_23
2. Alam, S., et al.: A large multi-target dataset of common Bengali handwritten graphemes. In: Lladós, J., Lopresti, D., Uchida, S. (eds.) ICDAR 2021. LNCS, vol. 12824, pp. 383–398. Springer, Cham (2021). https://doi.org/10.1007/978-3-030-86337-1_26
3. Augusto Borges Oliveira, D., Palhares Viana, M.: Fast CNN-based document layout analysis. In: Proceedings of the IEEE International Conference on Computer Vision Workshops, pp. 1173–1180 (2017)
4. Binmakhashen, G.M., Mahmoud, S.A.: Document layout analysis: a comprehensive survey. ACM Comput. Surv. (CSUR) **52**(6), 1–36 (2019)
5. Carion, N., Massa, F., Synnaeve, G., Usunier, N., Kirillov, A., Zagoruyko, S.: End-to-end object detection with transformers. In: Vedaldi, A., Bischof, H., Brox, T., Frahm, J.-M. (eds.) ECCV 2020. LNCS, vol. 12346, pp. 213–229. Springer, Cham (2020). https://doi.org/10.1007/978-3-030-58452-8_13
6. Caron, M., Misra, I., Mairal, J., Goyal, P., Bojanowski, P., Joulin, A.: Unsupervised learning of visual features by contrasting cluster assignments. Adv. Neural. Inf. Process. Syst. **33**, 9912–9924 (2020)

---

[1] https://www.cbgr1971.org/.

7. Clausner, C., Antonacopoulos, A., Derrick, T., Pletschacher, S.: ICDAR2019 competition on recognition of early Indian printed documents-REID2019. In: 2019 International Conference on Document Analysis and Recognition (ICDAR), pp. 1527–1532. IEEE (2019)
8. Du, Y., et al.: PP-OCR: a practical ultra lightweight OCR system. arXiv preprint arXiv:2009.09941 (2020)
9. Girshick, R.: Fast R-CNN. In: Proceedings of the IEEE International Conference on Computer Vision, pp. 1440–1448 (2015)
10. Girshick, R., Radosavovic, I., Gkioxari, G., Dollár, P., He, K.: Detectron. https://github.com/facebookresearch/detectron (2018)
11. He, K., Gkioxari, G., Dollár, P., Girshick, R.: Mask R-CNN. In: Proceedings of the IEEE International Conference on Computer Vision, pp. 2961–2969 (2017)
12. Huang, J., et al.: A multiplexed network for end-to-end, multilingual OCR. In: Proceedings of the IEEE/CVF Conference on Computer Vision and Pattern Recognition, pp. 4547–4557 (2021)
13. Islam, N., Islam, Z., Noor, N.: A survey on optical character recognition system. arXiv preprint arXiv:1710.05703 (2017)
14. Jocher, G., et al.: ultralytics/yolov5: v3. 0. Zenodo (2020)
15. Khodabandeh, M., Vahdat, A., Ranjbar, M., Macready, W.G.: A robust learning approach to domain adaptive object detection. In: Proceedings of the IEEE/CVF International Conference on Computer Vision, pp. 480–490 (2019)
16. Li, M., et al.: DocBank: a benchmark dataset for document layout analysis. arXiv preprint arXiv:2006.01038 (2020)
17. Lin, T.-Y., et al.: Microsoft COCO: common objects in context. In: Fleet, D., Pajdla, T., Schiele, B., Tuytelaars, T. (eds.) ECCV 2014. LNCS, vol. 8693, pp. 740–755. Springer, Cham (2014). https://doi.org/10.1007/978-3-319-10602-1_48
18. Pfitzmann, B., Auer, C., Dolfi, M., Nassar, A.S., Staar, P.: DocLayNet: a large human-annotated dataset for document-layout segmentation. In: Proceedings of the 28th ACM SIGKDD Conference on Knowledge Discovery and Data Mining, pp. 3743–3751 (2022)
19. Ren, S., He, K., Girshick, R., Sun, J.: Faster R-CNN: towards real-time object detection with region proposal networks. In: Advances in Neural Information Processing Systems 28 (2015)
20. Zhong, X., Tang, J., Yepes, A.J.: PubLayNet: largest dataset ever for document layout analysis. In: 2019 International Conference on Document Analysis and Recognition (ICDAR), pp. 1015–1022. IEEE (2019)

SelfDocSeg: A Self-supervised Vision-Based Approach Towards Document Segmentation

Subhajit Maity[1] , Sanket Biswas[2]([✉]) , Siladittya Manna[3] ,
Ayan Banerjee[2] , Josep Lladós[2] , Saumik Bhattacharya[4] ,
and Umapada Pal[3]

[1] Technology Innovation Hub, Indian Statistical Institute, Kolkata, India
subhajit_t@isical.ac.in
[2] Computer Vision Center, Computer Science Department,
Universitat Autònoma de Barcelona, Barcelona, Spain
{sbiswas,abanerjee,josep}@cvc.uab.es
[3] CVPR Unit, Indian Statistical Institute, Kolkata, India
{siladittya_r,umapada}@isical.ac.in
[4] Department of Electronics and Electrical Communication Engineering,
Indian Institute of Technology Kharagpur, Kharagpur, India
saumik@ece.iitkgp.ac.in

Abstract. Document layout analysis is a known problem to the documents research community and has been vastly explored yielding a multitude of solutions ranging from text mining, and recognition to graph-based representation, visual feature extraction, *etc.* However, most of the existing works have ignored the crucial fact regarding the scarcity of labeled data. With growing internet connectivity to personal life, an enormous amount of documents had been available in the public domain and thus making data annotation a tedious task. We address this challenge using self-supervision and unlike, the few existing self-supervised document segmentation approaches which use text mining and textual labels, we use a complete vision-based approach in pre-training without any ground-truth label or its derivative. Instead, we generate pseudo-layouts from the document images to pre-train an image encoder to learn the document object representation and localization in a self-supervised framework before fine-tuning it with an object detection model. We show that our pipeline sets a new benchmark in this context and performs at par with the existing methods and the supervised counterparts, if not outperforms. The code is made publicly available at: github.com/MaitySubhajit/SelfDocSeg.

Keywords: Document Layout Analysis · Document Segmentation · Document Understanding · Self-supervised Learning

S. Maity and S. Biswas—These authors contributed equally to this work.

1 Introduction

From the early days of computer vision and document understanding research, document layout analysis (DLA) had been a primitive problem and had been conquered with a multitude of strategies [7] from classical methods [1,26,46] to state-of-the-art learning-based models [8,36,55]. While intelligent document processing (IDP) has emerged as an essential step toward automatic document understanding, bearing the rise of convolutional neural networks (CNN) and sequence models, researchers have achieved near-perfect accuracy in the context of DLA with models having vast deployability and reliability. With a rapidly growing population, and the span of digital personal lives, documents are becoming more unconventional and business applications for IDP are encountering complex and never-before-seen layouts taking the already solved challenge a level up with the requirement of deep features exploration and exploitation. To meet that end, the state-of-the-art DLA strategies had improved a lot over time and still continue to do so, while remaining one of the most trendy topics for the research community [47].

The challenge of document segmentation or DLA had been a research interest for nearly a decade and thus had been solved with classical heuristic rule-based methods [1,26] in the early days, while modern documents understanding community treats it as a document object detection (DOD) problem and is typically approached by a standard vision based object detector models [33,43,52]. With the breakthrough in sequence models and language models, researchers had been using the same along with object detection to solve the problem with better accuracy [36]. However, the research community has vastly ignored the fact that the growth in the number of unconventional documents raises the need for tedious annotation tasks to exploit the knowledge for better document understanding through the conventional supervised setting, and thus the self-supervised approaches toward the problem are highly relevant in this context.

Deploying self-supervision in DOD is quite non-trivial as the task inherently cannot use the power of contrastive learning as the images contain multiple document objects of different classes and thus naively using each picture as its own class is not going to help. Apart from handling the class information, the encoder needs to consider object localization which cannot be realized through self-supervision without preliminary knowledge of the locations of layout objects. Moreover, it had been observed that although deep visual features extracted from document images are rich in information, they are difficult to train and are usually guided with learned textual [29] or layout [8] cues or both [36] from existing text spotter and detector models. And thus the efficacy of self-supervision for document segmentation faces a big question mark, which we address in this paper.

The core of our design consists of a self-supervised framework designed following 'Bootstrap Your Own Latent' (BYOL) [28] which actively tries to fulfill both the objectives of representation and localization. As self-supervision is employed for pre-training of the encoder, we do not have access to the ground-truth document object locations and categories. Thus we use classical image processing

344   S. Maity et al.

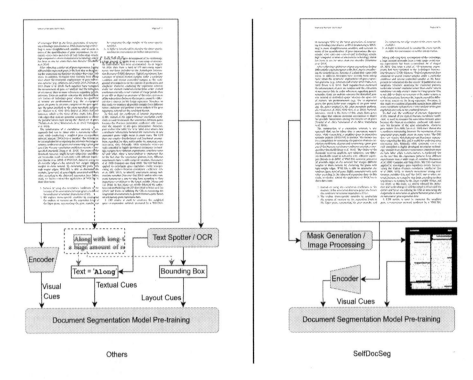

Fig. 1. Scope and Motivation. A basic methodological distinction between Self-DocSeg and existing approaches. While earlier works utilize information from visual, layout, and textual modalities for large-scale pre-training, we deal with visual cues only for boosting representation learning.

to approximate a rough physical layout mask for each document image and use the same as guidance for both document object localization and representation learning purpose. We use a backbone image encoder to obtain feature maps, followed by a mask pooling operation to extract encodings of all the possible relevant physical layout objects and train them in a self-supervised representation learning framework using negative cosine similarity. In parallel, we use a layout predictor module on the encoded feature maps, l and the module is tasked with a classification task of predicting if the salient features at every pixel of the feature map belong to a document object or not. This module is trained using focal loss [43] with the supervision of the generated physical layout mask.

The evidence had not been in favour of self-supervised vision-based approaches for DLA tasks as visual representations needed to be guided with learned textual and layout embeddings. However, we suggest that self-supervised visual representation learning can still be explored as visual features prove to be useful in the supervised settings for the task at hand. This had become the core motivating factor for our work as we try to explore and exploit the rich visual features extracted from the document images using the powerful CNN backbones.

The intuition behind the proposed framework is relatively simple as we try to guide the learning of the backbone encoder using approximated visual layouts for both document object localization and representation learning. Unlike, the layout cues in the existing self-supervised strategies, we do not use any layout information from any pre-trained text recognition model. Instead, we devise the inherent visual information as a layout to guide the visual representation learning. A clear distinction in working principles between the proposed framework, SelfDocSeg, and the existing self-supervised strategies can be realized in Fig. 1. In summary, our contributions in this paper can be divided into three folds:

(a) A novel *vision-based self-supervised framework*, specifically designed to pre-train an image encoder for DLA task;

(b) A *pseudo physical layout guided strategy for self-supervision* in the region of interest localization for document segmentation;

(c) A *data efficient pre-training strategy* to learn multiple document object representations simultaneously in the self-supervised setting.

The rest of the paper is organized as follows. A comprehensive literature review is provided in Sect. 2. The proposed methodology is described in Sect. 3. In Sect. 4 we discuss the experiments and results. Finally, the conclusions are drawn in Sect. 5.

2 Related Work

2.1 Self-supervision for Visual Representation

As the world of computer vision evolved it demanded undivided attention toward the exploration and exploitation of complex visual features to learn representations from images from a multitude of sources. Thus, data-centric machine-learning models with immense capability of feature extraction and correlation have emerged as modern-day solutions to the growing sophisticated requirements. However, with the huge amount of data required for modern network architectures, the need for data annotation has increased giving rise to the plethora of self-supervised strategies. MoCo [32] provided the research community with the idea of weight update using the exponential moving averages and large memory banks in contrastive learning settings. SimCLR [16] improved on it and introduced a large batch size as an alternative to the memory bank. DINO [15] introduced self-supervision in vision transformers [23]. Following suit, MoCov2 [17], SwAV [14] had achieved wonderful performance in the self-supervised paradigm. On the other hand, BYOL [28], SimSiam [18] treat two crops from the same image as similar pairs instead of contrastive learning, while masked autoencoders [31] introduced a masking strategy to the age-old autoencoders for learning representation via reconstruction.

Among this plethora of self-supervised methods, self-supervised object detection and document segmentation remained vastly unexplored. Having significantly superior performances from supervised object detection methods like Mask RCNN [33], Yolo [52], Retinanet [43], DETR [13] *etc.*, their self-supervised counterparts had been comparatively paler in diversity until recently. In the last few years, we have seen end-to-end self-supervised object detection models like UP-DETR [21], DETReg [6] and backbone pre-training strategies like Self-EMD [44], Odin [35]. However, being closely related to object detection, instance-level document layout analysis has barely adopted self-supervision. Although there had been a few self-supervised document segmentation approaches, none of them had been explicitly using self-supervised visual representation or guidance. We address this lack of vision-based self-supervision for document understanding in this paper.

2.2   Document Understanding

Document Understanding (DU) has been reformulated in the current document analysis literature [11] as a landscape term covering different problems and tasks related to Document Intelligence systems which majorly includes Key Information Extraction [37, 49, 56], Classification [30], Document Layout Analysis [54, 63], Question Answering [48, 59], and Machine Reading Comprehension [58] whenever they involve visually rich documents (VRDs) in contrast to plain texts or image-text pairs. Recent state-of-the-art DU systems majorly rely on large-scale pre-training to combine both visual and textual modalities as in [3, 27, 29, 36, 41] while methods like Donut [39] and Dessurt [22] mostly rely on combining more effective visual features by synthetic generation techniques [10, 20, 38, 61] for learning important layout representation during document pre-training. In this work, we aim to identify a new direction toward boosting visual representation through a self-supervision strategy for document layout understanding.

2.3   Document Layout Analysis

DLA has evolved as a significant DU application, dedicated towards the optimization of storage and large-scale information workflows [7]. Since the advent of deep learning and CNN-based approaches [53, 55], the segmentation of document layouts has been reformulated as DOD. With the launch of large-scaled annotated DLA benchmarks [54, 63] opened a new direction for deep-learning-based approaches. Later, Biswas et.al. [9] redefined the DLA task as an instance-level segmentation task to detect both bounding boxes along with segmentation masks, especially in layouts of pages containing overlapped objects. More recently, transformer-based approaches [4, 8, 36] claimed the recent state-of-the-art on DLA, especially on large-scaled document datasets. However, there is still a huge scope for improvement of transformer methods, especially in smaller annotated datasets like [19]. Recently, language-based approaches like LayoutLMv3 [36] and UDoc [29] have been tested on PubLayNet [63] benchmark to get the best performance. But they fail in document benchmarks with more complex layouts and much smaller annotated samples. In this work, we strive to propose SelfDocSeg a pure vision-based self-supervised strategy to mitigate the aforementioned issues.

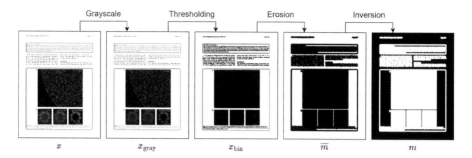

Fig. 2. Layout Mask Generation Pipeline. The figure illustrates a stepwise approach with different strategies adapted for generating the final layout mask for a given document image.

3 Methodology

3.1 Problem Formulation

In the context of document layout segmentation we have access to a relevant dataset $\mathcal{D} = \{x, y\}$ where $x \in \mathcal{I}^{3 \times H \times W}$ is a standard RGB document image with height H and width W, and $y = \{y_1, \ldots, y_p\}$ with y_l having a set of coordinates y_l^{mask} of object pixels and y_l^{label} for the object $l = \{1, \ldots, p\}$ assuming p objects in the image x. However, according to the core principle of self-supervised learning, we do not use this dataset as it contains the ground-truth annotations, and thus we derive a dataset $\mathcal{D}' = \{x, m\}$ from dataset $\mathcal{D}$ such that it contains only the images without annotated data and a rough binary mask m extracted from x as described in Sect. 3.2 depicting the physical layout of the document image.

The pre-training strategy for the image encoder briefly described in Sect. 3.3, has been adapted to our self-supervised framework, SelfDocSeg. Formally, we train an image encoder F_θ parameterized by θ in a self-supervised setting specifically designed to cater to document object recognition. Once the pre-training is done we apply these weights to initialize the backbone of an object detector model to be later fine-tuned for the segmentation of document objects as described in Sect. 3.4.

3.2 Layout Mask Generation

The layout mask generation is a crucial step for the proposed pipeline as our self-supervised framework depends on it for some visual guidance. For any document image x we generate a mask m using classical image processing techniques as depicted in Fig. 2. At first, we convert the RGB image x to grayscale x_{gray} as defined by CIE, followed by global thresholding to obtain a binarized output x_{bin}. We then perform an erosion operation over x_{bin} to make the logical layout elements of the document like characters, sub-figures, and plot lines a little thicker and preferably a blob, so that the blobs get connected to form a rough physical layout $\overline{m}$. The final layout mask m is generated by an inversion operation of this eroded rough layout $\overline{m}$.

3.3 Encoder Pre-training

Overview: The pre-training of the image encoder is quite difficult to achieve as we are specifically considering the object detection problem, a multi-task challenge that enforces the encoder to learn important salient features contributing towards inter-class variance of different logical layout components and simultaneously localizing the object regions. To overcome this problem, we use the layout masks as visual guidance. However, there is no option to use layout masks directly as we have two distinct yet related tasks at hand. Thus, the challenge and also the scope of innovation remain on how the guidance is provided for the two aforementioned tasks.

The proposed architecture of SelfDocSeg is developed inspired by the popular BYOL [28] self-supervised framework. It mainly consists of two branches, online and momentum, parameterized by two sets of weights θ and ξ respectively. The overall learning strategy of the architecture is to use a self-distillation technique that involves the two branches of the network which exploits the similarity in salient features extracted from two different views or variations of the same input. The feature similarity exploitation principle demands the generation of significantly different views of the input document image x. Thus we augment x randomly to generate two different versions v_1, v_2 and feed them to the two branches of the SelfDocSeg framework as illustrated in Fig. 3. Each of these branches employs multiple modules to generate meaningful semantic embeddings from the image through a mask pooling operation. This operation is designed in a way that simultaneously extracts embeddings for all possible layout objects as a batch from the input image x using the guidance of the layout mask m as we roughly extract separate masks of each layout object to obtain an average pooled vector for each.

In parallel, the layout prediction module ensures that the online network learns to locate regions of interest in the document image feature map. We use both the feature maps from v_1, v_2 to feed the layout predictor module and generate pixel-wise probability scores of a logical document object being present at that pixel. This whole module is trained using m with the help of focal loss [43]. The overall architecture is depicted in Fig. 3 and the overall loss function used to optimize the training is given by Eq. 1 where $\mathcal{L}_{\text{Det}}$ comes from the *Layout Prediction Module* and $\mathcal{L}_{\text{Sim}}$ comes from *Layout Object Representation Learning* as discussed later in the following subsections.

$$\mathcal{L}_{\text{total}} = \mathcal{L}_{\text{Det}} + \mathcal{L}_{\text{Sim}} \qquad (1)$$

Augmentation: The augmentations used to synthesize v_1, v_2 from the input document image x are well crafted to match the task and are very similar to the augmentations used in SimCLR [16]. From the set of augmentations used

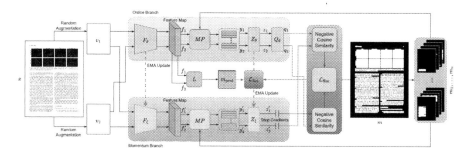

Fig. 3. Model Overview. A simple architectural design of SelfDocSeg pre-training modeled after the BYOL [28] framework along with a layout prediction module.

by SimCLR, we carefully exclude random cropping and random horizontal flip. As we are performing mask pooling on the feature map, it simultaneously tries to learn all the possible physical layout object representations. And thus, taking two patches from the image and using them as positive or contrastive pairs is not required and is inherently done for multiple pairs simultaneously. Moreover, the layout predictor module is designed to predict layout all over the document image and thus the task does not align well with taking randomly cropped patches from the document image. On the other hand, the inputs, being document images, contain a large variety of texts and glyphs which could break the clearly established pattern when flipped at random and thus can hinder the training of our module. So, we exclude these two from the set of augmentations and the final set includes the following: (a) Gaussian blurring, (b) color jittering, (c) color dropping, and (d) solarization. The augmentations are not manually controlled and are randomly applied on input image x with random parameters which provide a wide range of variations as required to learn the similarity between the two views v_1, v_2.

Image Encoder and Mask Pooling: We define the image encoder $F :$ $\mathcal{I}^{3 \times H \times W} \to f$ which takes an input image and encodes it to a salient feature map $f \in \mathbb{R}^{c \times h \times w}$ with c channels and height and width of h and w respectively. Considering both the online and the momentum branches, we have two image encoders F_θ and F_ξ providing feature maps f and f' respectively.

The mask pooling operation is performed on feature maps f and f' separately on the two branches to get all the layout object representation vectors simultaneously. The individual object layout masks $m_{k=1,...,n}$ are obtained by detecting individual contours from m assuming n separate contours in m. Once we have all the possible layout object masks, we get the average pooled vector $y^{(k)} \in \mathbb{R}^d$ with dimension d for each of the objects from the feature map f according to the established definition of Mask Pooling [35] operation $MP : f \to y$ as per Eq. 2 where i, j are pixel coordinates. Once we get $y^{(k)}$ for all $k = 1, \ldots, n$ estimated objects from m_k in layout mask m, we neatly stack the vectors to form $y \in \mathcal{R}^{k \times d}$.

$$y^{(k)} = \frac{1}{\sum_{i,j} m_k[i,j]} \sum_{i,j} m_k[i,j] f[i,j] \qquad (2)$$

Layout Object Representation Learning: The overall framework of Self-DocSeg is designed after BYOL [28] and thus follows a similar strategy to learn the representations from document images using cosine similarity. The online branch comprises an image encoder F_θ, projector module Z_θ, and a predictor module Q_θ while the momentum branch consists of the image encoder F_ξ and a projector module Z_ξ where θ and ξ are the sets of parameters for online and momentum branch respectively.

For inputs v_1, v_2 we get feature maps f_1, f_2 and f_1', f_2' from F_θ and F_ξ respectively and we use mask pooling MP as a strategy to get the encoded vector from the feature map which is designed to pool out all the approximate physical layout segments into individual encoded vectors simultaneously as discussed earlier and neatly stack the encodings in a batch resulting in y_1, y_2 and y_1', y_2', which contain average pooled vectors for all $k = 1, \ldots, n$ objects stacked in batches. The encoded vectors y_1, y_2 and y_1', y_2' are passed through projector modules Z_θ and Z_ξ to yield z_1, z_2 and z_1', z_2' respectively. As the online branch of the framework has a predictor layer Q_θ, it takes input z_1, z_2 and produces q_1, q_2.

$$\theta \leftarrow \text{optimizer}\left(\theta, \nabla_\theta \mathcal{L}_{\text{total}}, \eta\right)$$
$$\xi \leftarrow \tau\xi + (1 - \tau)\theta \qquad (3)$$

SelfDocSeg is designed to learn object representations without labels in a self-distillation approach and uses exponentially moving average (EMA) to update the weights of momentum branch ξ from the online branch weights θ. Thus, F_ξ and Z_ξ are updated using EMA from F_θ and Z_θ while F_θ and Z_θ are updated using back-propagation on the loss function as given in Eq. 3 where η is the learning rate, τ is the momentum of EMA and $\nabla_\theta \mathcal{L}_{\text{total}}$ denotes the gradient corresponding to the total loss $\mathcal{L}_{\text{total}}$ from Eq. 1. It is also to note that the back-propagation does not happen in the momentum branch as denoted by the 'stop gradient' sign in Fig. 3. The similarity loss $\mathcal{L}_{\text{Sim}}$ uses cosine similarity to compare representations from online and momentum branches as per Eq. 4.

$$\mathcal{L}_{\text{Sim}} = 4 - 2\left(\frac{\langle q_1, z_2' \rangle}{||q_1||_2 \cdot ||z_2'||_2} + \frac{\langle q_2, z_1' \rangle}{||q_2||_2 \cdot ||z_1'||_2}\right) \qquad (4)$$

Layout Prediction Module: The layout prediction module L is an auxiliary module facilitating the layout object localization. This module helps the encoder to learn better representation specifically for the detection task. The module takes input from both the feature maps f_1, f_2 and predicts a mask layout m_{pred}, which is then compared to the approximated mask m using focal loss [43], $\mathcal{L}_{\text{Det}}$ given in Eq. 5 with hyper-parameters α and γ. The intuition behind this module is to instill a notion of localization of the layout objects, of which the encoder is learning the representation from $\mathcal{L}_{\text{Sim}}$.

$$\mathcal{L}_{\text{Det}} = -\frac{\alpha}{\sum_{i,j} m[i,j]} \cdot \sum_{i,j} (m[i,j](1 - m_{\text{pred}}[i,j])^\gamma \log m_{\text{pred}}[i,j]$$
$$+ (1 - m[i,j])m_{\text{pred}}[i,j]^\gamma \log(1 - m_{\text{pred}}[i,j])) \tag{5}$$

3.4   Fine-Tuning

Once the image encoder is trained we proceed towards the main task of document segmentation and for this purpose, we take the pre-trained weights of the image encoder F_θ and use it to initialize the backbone weights of an object detector model. This model is trained with the supervision of ground-truth labels using whole annotated dataset $\mathcal{D}$ and we use Mask RCNN [33] as our object detector model equipped with a feature-pyramid network (FPN) [42] from intermediate layers for multi-scale detection.

4   Experiments

Datasets: For the training and evaluation of our framework for document segmentation, we used several datasets specifically designed for the task. We use the DocLayNet [51] dataset for our pre-training experiments. It contains 69,375 training, 6,489 validation, and 4,999 test images from six different domains with annotated ground-truth labels for 11 separate classes. However, we only use the training split without the ground-truth labels for the pre-training phase as mentioned in Sect. 3.

Once the pre-training is complete we move on to the document layout analysis task to evaluate the efficacy of our pre-training strategy and for the same, we use PRImA [2] dataset with 305 labeled images in total; Historic Japanese [54] dataset, having 181,097 training, 39,410 validation and 39,109 test layout objects of seven different categories spanning over 2,271 document images; and the extensive PubLayNet [63] dataset containing 335,703 training, 11,245 validation, and 11,405 test images with labeled layout masks of five different layout object classes, along with the DocLayNet [51] dataset used in pre-training.

Implementation Details: The complete self-supervised framework is developed using a self-supervised library named Lightly-AI [57], that is built on PyTorch Lightning [25] and PyTorch [50]. All the models are trained on a 48 GB Nvidia RTX A40 GPU.

The mask generation process using classical image processing, described in 3.2, is developed using OpenCV [12]. We empirically found the global threshold value of 239 working well enough for the pre-training dataset, DocLayNet [51] given that the grayscale images are having 8-bit unsigned integer datatype. For the erosion operation, we use a 5×5 rectangular kernel as the structuring element.

Firstly, for the image encoders F_θ, F_ξ we use a standard ResNet50 [34] and we use the last residual block for extracting encoded feature maps, leaving out the global average pooling and fully-connected layers at the end, and thus the average pooled object vectors have dimension $d = 2048$ same as c in the output of last residual block. The projectors Z_θ and Z_ξ, and the predictor Q_θ, are implemented as two-layer multi-layer perceptrons (MLPs) with 4096 hidden and 256 output dimensions. The auxiliary layout prediction module L is implemented by a 1×1 convolution block to predict the layout mask. The focal loss $\mathcal{L}_{\mathrm{Det}}$ mentioned in Eq. 5 uses two hyper-parameters, a weighing factor $\alpha = 0.25$ and a focusing parameter $\gamma = 2$.

In the pre-training phase, the optimal model parameters are learned by optimizing the loss $\mathcal{L}_{\mathrm{total}}$ from Eq. 1 using LARS [62] optimizer with an initial learning rate of $\eta = 0.2$, and a weight decay of 0.0005. The learning rate is decayed following a cosine annealing schedule [45] for 800 epochs. For the momentum branch, the value of the momentum hyper-parameter τ is set to 0.99.

The object detection model in fine-tuning with task-specific ground-truth annotation uses the same backbone, ResNet50 [34] for Mask RCNN [33] with FPN [42] and is trained with Detectron2 [60]. We use this framework because it is built on top of PyTorch, is extremely easy to use, and is reliable for deploying models. We have used Nvidia RTX A40 GPUs for all of our fine-tuning purposes. Our trained models have been based on ResNet-50 weights pre-trained as per our proposed framework, SelfDocSeg, on the DocLayNet [51] dataset. The initial learning rate of the model was 0.0025, and it is trained for a total of 300,000 iterations. A multitude of anchor scales and aspect ratios are considered to cover the input image, resulting in the number of anchor boxes $k = 64$. Nesterov Momentum was used to train with the Stochastic Gradient Descent (SGD) optimizer [24], and the batch size per image in the RoI heads was 128. Our task had a detection non-maximum suppression (NMS) threshold of 0.4 and a detection minimum confidence score of 0.6. The data loader's workforce was set at four. We set the model's testing threshold in the RoI heads to 0.6 following the completion of the fine-tuning process. The values mentioned are carefully chosen after reviewing empirical data.

4.1   Competitors

State-of-the-Art: In this work, our contribution is towards a novel self-supervised pre-training strategy, which cannot be directly compared to its counterparts without having to look at the performance of the downstream task. Thus, we compare the methodologies with existing models after fine-tuning the downstream task using a document object detector.

In the context of self-supervised DLA task, the research community has yet to see various strategies and thus we have only a few strategies to compare with. We compared our methods with state-of-the-art (SOTA) self-supervised methodologies: (a) LayoutLMv3 [36] uses a multi-modal transformer that uses a masked language modeling and masked image modeling-based strategy using linear image patch embeddings from the document and textual embeddings extracted from a pre-trined optical character recognition (OCR) and also tries to align word patches with image patches jointly with an image and word token

Table 1. Quantitative analysis of the performance of the document object detection task along with the guidances used during supervision. V, L, and T stand for visual, layout, and textual cues respectively.

	Methods	Cues V	L	T	# Data	DocLayNet mAP	PubLayNet mAP	PRImA mAP	HJ mAP
Supervised	DocSegTr [8]	✓	✗	✗	-	-	90.4	42.5	**83.1**
	LayoutParser [55]	✓	✓	✓	-	-	86.7	**64.7**	81.6
	Biswas *et al.* [9]	✓	✗	✗	-	-	89.3	56.2	82.0
	Mask RCNN [33]	✓	✗	✗	-	72.4	88.6	56.3	80.1
Self-Supervised	LayoutLMv3$_{Base}$ [36]	✓	✓	✓	11M	-	**95.1**	40.3	82.7
	UDoc [29]	✓	✓	✓	1M	-	93.9	-	-
	DiT$_{Base}$ [40]	✓	✗	✗	42M	-	93.5	-	-
Proposed	BYOL [28]	✓	✗	✗	81k	63.5	79.0	28.7	59.8
	SelfDocSeg	✓	✗	✗	81k	**74.3**	89.2	52.1	78.8

classification task. (b) UDoc [29] employs a region of interest alignment strategy that aligns the visual features from the document images with the textual embeddings and location information from a pre-trained OCR to generate joint embeddings at the sentence level to use them with contextual masking to learn with visual contrastive learning and visual-textual alignment through a cross attention module. The noticeable difference between these existing SOTA with the proposed pipeline can be identified as the use of pre-trained OCR to guide the pre-training with textual and layout cues along with visual features while we use only visual features for the purpose. (c) DiT [40] uses only visual features like the proposed strategy using masked image modeling with BEiT [5] pre-training over a massive dataset of size 42 million approximately.

In parallel, we show a comparative analysis with supervised SOTA methodologies: (a) DocSegTr [8] implements a transformer encoder-decoder architecture, being the first of its kind, along with a convolutional backbone for document segmentation. (b) LayoutParser [55] uses a universal framework built on top of CNN-based approaches for DOD evaluation (c) Biswas *et al.* [9] had used a Mask RCNN-styled architecture employing a fully convolutional network, for instance, level segmentation and class prediction and bounding box regression empowered by a region proposal network to suggest significant regions of interest from the features extracted at multiscale by FPN on a backbone.

Comparable Baselines: In terms of DOD, we also establish a few baselines, both supervised and unsupervised. We used BYOL [28] as a pre-training strategy to train an image encoder and used it to perform the fine-tune training on the document object detector, Mask RCNN [33] with ground-truth and consider it as our potential self-supervised baseline. On the other hand, we train Mask RCNN [33] in a fully supervised setting with ground-truth annotations.

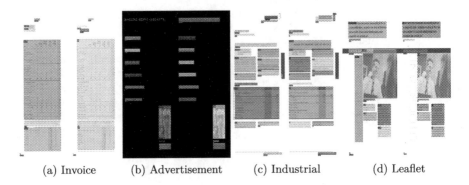

(a) Invoice   (b) Advertisement   (c) Industrial   (d) Leaflet

Fig. 4. Qualitative analysis of the SelfDocSeg framework on the DocLayNet datasets (**Left:** Predicted layout **Right:** Ground-truth)

4.2   Performance Analysis

The performance of our pre-training strategy is compared with the existing methodologies in Table 1 in terms of mean average precision (mAP) score at the pixel level segmentation of document objects along with the depiction of guidance/cues used for each of the methods and amount of data used for self-supervised pre-training. (a) LayoutLMv3 [36] and UDoc [29] have superior performance on PubLayNet [63] dataset being the current self-supervised SOTAs. LayoutLMv3 [36] also has superior performance over other works. However, both methods use pre-trained OCR for textual and layout cues which provides them an additional advantage in learning instance-level document layout component embeddings coupled with the superior backbone of a multi-modal transformer. In comparison, the proposed SelfDocSeg performs at par without any guidance from pre-trained OCR. (b) DiT$_{Base}$ [40] has a superior performance compared to the proposed strategy due to its powerful Vision Transformer [23] backbone coupled with the huge pre-training dataset of 42 million. (c) DocSegTr [8] uses a convolutional backbone as well as a transformer architecture with multi-headed attention that is inherently more powerful for extraction of useful features and thus supervision with ground truth provides superior performance in this scenario. It is noticed that transformer-based models could not perform well in small datasets like PRImA [2]. (d) LayoutParser [55] is a comprehensive layout analysis toolkit that offers state-of-the-art performance with common CNN-based approaches, which shows decent results owing to a powerful OCR module along with document image augmentations. (e) The work of Biswas *et al.* [9] employs a modified Mask RCNN architecture with end-to-end supervision and performs decently in the context of Mask RCNN [33], compared to vanilla Mask RCNN in the fully supervised setting. (f) The vanilla BYOL pre-training fails to learn meaningful document object representations and thus performs poorly in Mask RCNN. The possible justification for the performance could be attributed to unrestricted random cropping and flipping in augmentations, which leads to two different classes present in the same input document image being treated as

Table 2. Semi-supervised scenario. We evaluate the performance of Mask RCNN fine-tuning with varying % of labeled data used in DocLayNet.

% Annotations	mAP
10%	41.3
50%	65.1
100%	74.3

Table 3. Ablation Study. Contribution of individual learning objectives during Self-DocSeg pre-training on DocLayNet.

Loss	mAP
w/o $\mathcal{L}_{\text{Sim}}$	39.1
w/o $\mathcal{L}_{\text{Det}}$	69.7
Combined ($\mathcal{L}_{\text{total}}$)	74.3

positive pairs. (g) Vanilla Mask RCNN performs decently in the fully supervised setting. (h) It is evident that although Mask RCNN does not have all-over superiority in performance, it easily outperforms all the transformer-based models in small datasets like PRImA [2]. (i) It is also noticed that modern self-supervised strategies use a large amount of data for pre-training. In comparison, the proposed framework has achieved similar performance with significantly less data in pre-training.

To this end, the SelfDocSeg framework proves to be performing at par with the Mask RCNN competitors if not outperforming. And at the same time, it establishes that self-supervised pre-training with only visual features using a limited amount of data is efficient and effective for document segmentation tasks. Some visual results of our proposed pipeline are depicted in Fig. 4.

4.3 Ablation Study

How Effective and Generalizable is SelfDocSeg Pretraining? The importance of pre-training can be realized in the performance of the fine-tuned document segmentation model and the natural question lingers around the efficacy of the SelfDocSeg pre-training strategy. To answer the same, we perform several experiments on the fine-tuning stage in a semi-supervised setting. Meaning, we use multiple smaller sets of annotated training data in fine-tuning Mask RCNN and record the performance in Table 2. Although using all the available training data has the best result, it is evident that the model generalizes well enough due to the learning of extensive and useful visual feature extraction during pre-training as decreasing the amount of labeled data during fine-tuning does not seem to affect the model accuracy much. We see the mAP value drops 9% only while annotated training data is reduced to half. However, it starts getting affected as the quantity of labeled data drops drastically as we see more than 30% drop when the annotated dataset is reduced to 10%.

How Well do the Individual Losses Contribute in SelfDocSeg? We use two loss functions for the proposed pre-training strategy, focal loss in Eq. 5 and a representation loss with cosine similarity in Eq. 4. To dissect how much these losses individually contribute towards the pre-training, we train two separate encoders from scratch, one with focal loss only and another with representation loss only. The outcomes are depicted in Table 3. It is evident that the focal loss although helps in localization, it alone is not capable of learning meaningful object representations from the document images. On the other hand, the representation loss is capable enough to learn document object embeddings, but it falls short in the case of the layout analysis tasks. And both the losses together help to pre-train an image encoder for the document segmentation task.

5 Conclusion

While self-supervision is fairly new in the documents research community, in this paper, we studied the exploitation of rich visual features present in the document images using self-supervision. To this end, we designed a self-supervised strategy to pre-train the image encoder in the document layout analysis context. Our extensive experiments show that the proposed strategy performs decently and generalizes well enough in document images despite having no textual or layout guidance from pre-trained text recognition models. In a word, we present a complete vision-based self-supervised approach towards document segmentation which involves two specific strategies, *i.e.* a pseudo physical layout guided technique for document object localization and a document object representation learning in a self-supervised setting. We further intend to explore the scope of performance improvements in DLA using self-supervision.

Acknowledgment. This work has been partially supported by IDEAS - Technology Innovation Hub, Indian Statistical Institute, Kolkata; the Spanish project PID2021-126808OB-I00, the Catalan project 2021 SGR 01559 and the PhD Scholarship from AGAUR (2021FIB-10010). The Computer Vision Center is part of the CERCA Program/Generalitat de Catalunya.

References

1. Agrawal, M., Doermann, D.: Voronoi++: a dynamic page segmentation approach based on voronoi and docstrum features. In: Proceedings of the International Conference on Document Analysis and Recognition, pp. 1011–1015 (2009)
2. Antonacopoulos, A., Bridson, D., Papadopoulos, C., Pletschacher, S.: A realistic dataset for performance evaluation of document layout analysis. In: 2009 10th International Conference on Document Analysis and Recognition, pp. 296–300. IEEE (2009)

3. Appalaraju, S., Jasani, B., Kota, B.U., Xie, Y., Manmatha, R.: Docformer: end-to-end transformer for document understanding. In: Proceedings of the IEEE/CVF International Conference on Computer Vision, pp. 993–1003 (2021)

4. Banerjee, A., Biswas, S., Lladós, J., Pal, U.: SwinDocSegmenter: an end-to-end unified domain adaptive transformer for document instance segmentation. In: Document Analysis and Recognition-ICDAR 2023: 17th International Conference, San Jose, California, August 21–26, 2023, Proceedings. Springer (2023)

5. Bao, H., Dong, L., Piao, S., Wei, F.: BEit: BERT pre-training of image transformers. In: International Conference on Learning Representations (2022)

6. Bar, A., et al.: DETReg: unsupervised pretraining with region priors for object detection. In: Proceedings of the IEEE/CVF Conference on Computer Vision and Pattern Recognition (2022)

7. Binmakhashen, G.M., Mahmoud, S.A.: Document layout analysis: a comprehensive survey. ACM Comput. Surv. (CSUR) **52**(6), 1–36 (2019)

8. Biswas, S., Banerjee, A., Lladós, J., Pal, U.: DocSegTr: an instance-level end-to-end document image segmentation transformer. arXiv preprint arXiv:2201.11438 (2022)

9. Biswas, S., Riba, P., Lladós, J., Pal, U.: Beyond document object detection: instance-level segmentation of complex layouts. Int. J. Doc. Anal. Recogn. (IJDAR) **24**(3), 269–281 (2021)

10. Biswas, S., Riba, P., Lladós, J., Pal, U.: DocSynth: a layout guided approach for controllable document image synthesis. In: Lladós, J., Lopresti, D., Uchida, S. (eds.) ICDAR 2021. LNCS, vol. 12823, pp. 555–568. Springer, Cham (2021). https://doi.org/10.1007/978-3-030-86334-0_36

11. Borchmann, Ł., et al.: Due: End-to-end document understanding benchmark. In: Thirty-fifth Conference on Neural Information Processing Systems Datasets and Benchmarks Track (Round 2) (2021)

12. Bradski, G.: The OpenCV Library. Dr. Dobb's Journal of Software Tools (2000)

13. Carion, N., Massa, F., Synnaeve, G., Usunier, N., Kirillov, A., Zagoruyko, S.: End-to-end object detection with transformers. In: ECCV (2020)

14. Caron, M., Misra, I., Mairal, J., Goyal, P., Bojanowski, P., Joulin, A.: Unsupervised learning of visual features by contrasting cluster assignments. Adv. Neural Inf. Process. Syst. **33**, 9912–9924 (2020)

15. Caron, M., et al.: Emerging properties in self-supervised vision transformers. In: Proceedings of the IEEE/CVF International Conference on Computer Vision (2021)

16. Chen, T., Kornblith, S., Norouzi, M., Hinton, G.: A simple framework for contrastive learning of visual representations. In: Proceedings of the 37th International Conference on Machine Learning (2020)

17. Chen, X., Fan, H., Girshick, R., He, K.: Improved baselines with momentum contrastive learning. arXiv preprint arXiv:2003.04297 (2020)

18. Chen, X., He, K.: Exploring simple siamese representation learning. In: Proceedings of the IEEE/CVF Conference on Computer Vision and Pattern Recognition (2021)

19. Clausner, C., Antonacopoulos, A., Pletschacher, S.: ICDAR 2019 competition on recognition of documents with complex layouts-RDCL2019. In: Proceedings of the International Conference on Document Analysis and Recognition, pp. 1521–1526 (2019)

20. Coquenet, D., Chatelain, C., Paquet, T.: Dan: a segmentation-free document attention network for handwritten document recognition. IEEE Transactions on Pattern Analysis and Machine Intelligence (2023)

21. Dai, Z., Cai, B., Lin, Y., Chen, J.: Up-detr: Unsupervised pre-training for object detection with transformers. In: Proceedings of the IEEE/CVF Conference on Computer Vision and Pattern Recognition (2021)
22. Davis, B., Morse, B., Price, B., Tensmeyer, C., Wigington, C., Morariu, V.: End-to-end document recognition and understanding with Dessurt. In: Karlinsky, L., Michaeli, T., Nishino, K. (eds) Computer Vision – ECCV 2022 Workshops. ECCV 2022. Lecture Notes in Computer Science, vol 13804. Springer, Cham (2023). https://doi.org/10.1007/978-3-031-25069-9_19
23. Dosovitskiy, A., et al.: An image is worth 16x16 words: transformers for image recognition at scale. arXiv preprint arXiv:2010.11929 (2020)
24. Dozat, T.: Incorporating nesterov momentum into Adam (2016)
25. Falcon, F.N., et al.: W.: PyTorch lightning. GitHub. Note: https://github.com/PyTorchLightning/pytorch-lightning **3** (2019)
26. Fang, J., Gao, L., Bai, K., Qiu, R., Tao, X., Tang, Z.: A table detection method for multipage PDF documents via visual seperators and tabular structures. In: ICDAR (2011)
27. Gemelli, A., Biswas, S., Civitelli, E., Lladós, J., Marinai, S.: Doc2Graph: a task agnostic document understanding framework based on graph neural networks. In: Karlinsky, L., Michaeli, T., Nishino, K. (eds) Computer Vision – ECCV 2022 Workshops. ECCV 2022. Lecture Notes in Computer Science, vol 13804. Springer, Cham (2023). https://doi.org/10.1007/978-3-031-25069-9_22
28. Grill, J.B., et al.: Bootstrap your own latent-a new approach to self-supervised learning. Adv. Neural Inf. Process. Syst. **33**, 21271–21284 (2020)
29. Gu, J., et al.: Unidoc: unified pretraining framework for document understanding. Adv. Neural. Inf. Process. Syst. **34**, 39–50 (2021)
30. Harley, A.W., Ufkes, A., Derpanis, K.G.: Evaluation of deep convolutional nets for document image classification and retrieval. In: 2015 13th International Conference on Document Analysis and Recognition (ICDAR), pp. 991–995. IEEE (2015)
31. He, K., Chen, X., Xie, S., Li, Y., Dollár, P., Girshick, R.: Masked autoencoders are scalable vision learners. In: Proceedings of the IEEE/CVF Conference on Computer Vision and Pattern Recognition (2022)
32. He, K., Fan, H., Wu, Y., Xie, S., Girshick, R.: Momentum contrast for unsupervised visual representation learning. In: Proceedings of the IEEE/CVF Conference on Computer Vision and Pattern Recognition (2020)
33. He, K., Gkioxari, G., Dollár, P., Girshick, R.: Mask R-CNN. In: Proceedings of the IEEE International Conference on Computer Vision (2017)
34. He, K., Zhang, X., Ren, S., Sun, J.: Deep residual learning for image recognition. In: Proceedings of the IEEE Conference on Computer Vision and Pattern Recognition (2016)
35. Hénaff, O.J., et al.: Object discovery and representation networks. In: ECCV (2022)
36. Huang, Y., Lv, T., Cui, L., Lu, Y., Wei, F.: LayoutLMv3: pre-training for document ai with unified text and image masking. In: Proceedings of the 30th ACM International Conference on Multimedia, pp. 4083–4091 (2022)
37. Jaume, G., Ekenel, H.K., Thiran, J.P.: FUNSD: a dataset for form understanding in noisy scanned documents. In: 2019 International Conference on Document Analysis and Recognition Workshops (ICDARW), vol. 2, pp. 1–6. IEEE (2019)
38. Kang, L., Riba, P., Rusinol, M., Fornes, A., Villegas, M.: Content and style aware generation of text-line images for handwriting recognition. IEEE Trans. Pattern Anal. Mach. Intell. **44**(12), 8846–8860 (2021)
39. Kim, G., et al.: Donut: document understanding transformer without OCR. arXiv preprint arXiv:2111.15664 (2021)

40. Li, J., Xu, Y., Lv, T., Cui, L., Zhang, C., Wei, F.: DiT: self-supervised pre-training for document image transformer. In: Proceedings of the 30th ACM International Conference on Multimedia, pp. 3530–3539 (2022)

41. Li, P., et al.: SelfDoc: self-supervised document representation learning. In: Proceedings of the IEEE/CVF Conference on Computer Vision and Pattern Recognition, pp. 5652–5660 (2021)

42. Lin, T.Y., Dollár, P., Girshick, R., He, K., Hariharan, B., Belongie, S.: Feature pyramid networks for object detection. In: Proceedings of the IEEE Conference on Computer Vision and Pattern Recognition, pp. 2117–2125 (2017)

43. Lin, T.Y., Goyal, P., Girshick, R., He, K., Dollár, P.: Focal loss for dense object detection. In: Proceedings of the IEEE International Conference on Computer Vision (2017)

44. Liu, S., Li, Z., Sun, J.: Self-EMD: self-supervised object detection without ImageNet. arXiv preprint arXiv:2011.13677 (2020)

45. Loshchilov, I., Hutter, F.: SGDR: stochastic gradient descent with warm restarts. arXiv preprint arXiv:1608.03983 (2016)

46. Marinai, S., Gori, M., Soda, G.: Artificial neural networks for document analysis and recognition. IEEE Trans. Pattern Anal. Mach. Intell. **27**, 23–35 (2005)

47. Markewich, L., et al.: Segmentation for document layout analysis: not dead yet. Int. J. Doc. Anal. Recogn. (IJDAR) **25**, 1–11 (2021). https://doi.org/10.1007/s10032-021-00391-3

48. Mathew, M., Karatzas, D., Jawahar, C.: DocVQA: a dataset for VQA on document images. In: Proceedings of the IEEE/CVF Winter Conference on Applications of Computer Vision, pp. 2200–2209 (2021)

49. Park, S., et al.: Cord: a consolidated receipt dataset for post-OCR parsing. In: Workshop on Document Intelligence at NeurIPS 2019 (2019)

50. Paszke, A., et al.: PyTorch: an imperative style, high-performance deep learning library. In: Advances in Neural Information Processing Systems 32, pp. 8024–8035. Curran Associates, Inc. (2019), http://papers.neurips.cc/paper/9015-pytorch-an-imperative-style-high-performance-deep-learning-library.pdf

51. Pfitzmann, B., Auer, C., Dolfi, M., Nassar, A.S., Staar, P.: DocLayNet: a large human-annotated dataset for document-layout segmentation. In: Proceedings of the 28th ACM SIGKDD Conference on Knowledge Discovery and Data Mining, pp. 3743–3751 (2022)

52. Redmon, J., Divvala, S., Girshick, R., Farhadi, A.: You only look once: unified, real-time object detection. In: Proceedings of the IEEE Conference on Computer Vision and Pattern Recognition (2016)

53. Schreiber, S., Agne, S., Wolf, I., Dengel, A., Ahmed, S.: DeepdeSRT: deep learning for detection and structure recognition of tables in document images. In: 2017 14th IAPR International Conference on Document Analysis and Recognition (ICDAR), vol. 1, pp. 1162–1167. IEEE (2017)

54. Shen, Z., Zhang, K., Dell, M.: A large dataset of historical Japanese documents with complex layouts. In: Proceedings of the IEEE/CVF Conference on Computer Vision and Pattern Recognition Workshops, pp. 548–549 (2020)

55. Shen, Z., Zhang, R., Dell, M., Lee, B.C.G., Carlson, J., Li, W.: LayoutParser: a unified toolkit for deep learning based document image analysis. In: Lladós, J., Lopresti, D., Uchida, S. (eds.) ICDAR 2021. LNCS, vol. 12821, pp. 131–146. Springer, Cham (2021). https://doi.org/10.1007/978-3-030-86549-8_9

56. Stanisławek, T., et al.: Kleister: key information extraction datasets involving long documents with complex layouts. In: Lladós, Josep, Lopresti, Daniel, Uchida, Seiichi (eds.) ICDAR 2021. LNCS, vol. 12821, pp. 564–579. Springer, Cham (2021). https://doi.org/10.1007/978-3-030-86549-8_36

57. Susmelj, I., Heller, M., Wirth, P., Prescott, J., et al.: Lightly. GitHub. Note: https://github.com/lightly-ai/lightly (2020)

58. Tanaka, R., Nishida, K., Yoshida, S.: VisuaLMRC: machine reading comprehension on document images. In: Proceedings of the AAAI Conference on Artificial Intelligence, vol. 35, pp. 13878–13888 (2021)

59. Tito, R., Karatzas, D., Valveny, E.: Hierarchical multimodal transformers for multi-page docVQA. arXiv preprint arXiv:2212.05935 (2022)

60. Wu, Y., Kirillov, A., Massa, F., Lo, W.Y., Girshick, R.: Detectron2. https://github.com/facebookresearch/detectron2 (2019)

61. Yim, M., Kim, Y., Cho, H.-C., Park, S.: SynthTIGER: synthetic text image generator towards better text recognition models. In: Lladós, J., Lopresti, D., Uchida, S. (eds.) ICDAR 2021. LNCS, vol. 12824, pp. 109–124. Springer, Cham (2021). https://doi.org/10.1007/978-3-030-86337-1_8

62. You, Y., Gitman, I., Ginsburg, B.: Large batch training of convolutional networks. arXiv preprint arXiv:1708.03888 (2017)

63. Zhong, X., Tang, J., Yepes, A.J.: PubLayNet: largest dataset ever for document layout analysis. In: Proceedings of the International Conference on Document Analysis and Recognition, pp. 1015–1022 (2019)

Diffusion-Based Document Layout Generation

Liu He[1(✉)], Yijuan Lu[2(✉)], John Corring[2(✉)], Dinei Florencio[2],
and Cha Zhang[2]

[1] Purdue University, West Lafayette, IN 47906, USA
`he425@purdue.edu`
[2] Microsoft Cloud and AI, Bellevue, WA 98004, USA
`{yijlu,jocorrin,dinei,chazhang}@microsoft.com`

Abstract. We develop a diffusion-based approach for various document layout sequence generation. Layout sequences specify the contents of a document design in an explicit format. Our novel diffusion-based approach works in the sequence domain rather than the image domain in order to permit more complex and realistic layouts. We also introduce a new metric, Document Earth Mover's Distance (*Doc-EMD*). By considering similarity between heterogeneous categories document designs, we handle the shortcomings of prior document metrics that only evaluate the same category of layouts. Our empirical analysis shows that our diffusion-based approach is comparable to or outperforming other previous methods for layout generation across various document datasets. Moreover, our metric is capable of differentiating documents better than previous metrics for specific cases.

Keywords: Structured document generation · Document layout · Diffusion methods · Generative models

1 Introduction

Document creation involves many steps from generating textual content, organizing additional media, and producing a layout that makes the information comprehensible. Layout generation is a key step in document creation. Layouts differ to best convey the appropriate kind of information for different domains of documents. To model many different domains of document layouts, general yet powerful methods need to be developed. We undertake that goal in this paper.

Methodological and domain considerations for layout generation have arisen as a topic of interest recently in the Computer Vision and Machine Learning communities [19,33,45]. Fixed length generative methods leveraging adversarial training were shown effective at producing realistic but limited documents in LayoutGAN [33]. In READ [45] an approach to resolve shortcomings of the prior work was developed, namely permitting more complex structures in the layout, as well as introducing document metric considerations in the generative model literature. This work proposed a recursive autoencoder to iteratively

---

L. He—Work done while a research intern at Microsoft Cloud and AI.

© The Author(s), under exclusive license to Springer Nature Switzerland AG 2023
G. A. Fink et al. (Eds.): ICDAR 2023, LNCS 14187, pp. 361–378, 2023.
https://doi.org/10.1007/978-3-031-41676-7_21

extend the layout based on past generated layouts. However, READ [45] relies on a hierarchical document model that may not apply well in a wide range of document types, and is dependent on hyperparameters that dictate document layout length. This work can be seen as a segway to autoregressive methods such as Layout Transformer [19] where the modern generative techniques of language modeling have been applied to the field. Autoregressive generation is outlined in more detail below. This method allows for adaptive stopping that comes from the encoder states themselves, which change progressively during the generation. The result is simpler and improved modeling of layouts.

In this paper, we develop a new approach to layout generation using the recently emerging area of diffusion probabilistic models. The key idea is that when a diffusion process consists of small steps of Gaussian noise conditioned on the data, then the reversing process can be approximated by a conditional Gaussian as well. To use the conventional diffusion methods for discrete sequence generation, rounding and embedding steps have to be introduced [36].

Our main contributions are introducing a novel document comparison metric with several useful properties and being the first work to employ discrete-sequence diffusion for layout generation. We show that the proposed metric behaves well qualitatively and quantitatively by comparing the performance of different algorithms on well-known datasets. We also show how synthetic layout training data compares to real document data on an end-to-end task: layout detection. We compare across different synthetic data generation algorithms by the mean average precision (mAP) of a trained layout detector of a fixed architecture. Finally, we provide ablation of the proposed method to several variables.

2   Related Works

2.1   Generative Networks

Generative adversarial networks [15] launched the generative revolution in image generation [9,25,27,47], and text generation [8,18,61]. This self-supervised training scheme enables the networks to consume large unlabeled realistic dataset, and provides a powerful baseline in various downstream tasks like image colorization [42], image compositing [53–55], miscellaneous segmentation [21,44], 3D modeling [5,24,62], multi-modal recognition [12,22,39,60] and layout synthesis [4,32]. Variational autoencoder [30] is a counterpart framework in generative network domain. The network excludes the burden of discriminator and using only variational loss and regularization loss on latent space in the bottleneck. The concise of both network and training scheme support its generalized representation of the real dataset [20,59].

Modern generative methods have a root in autoencoding pretext tasks from the NLP literature [11]. Masked language modeling is a category of pretraining tasks in which pieces of a token sequence are hidden with a [MASK] token and the model fills in the missing piece. In the autoregressive setting, this approach is simplified to a forward-looking token regression and so models generally have a backward-looking attention emphasis, generating the newest token sequentially. This autoregressive scheme achieves success in computer vision tasks like ViT [13], large language models [48] like GPT-3 [6].

2.2   Layout Generation

Numerous works focusing layout generation have been proposed recently. LayoutGAN [34] and LayoutVAE [26] provide general 2D layout planning for natural images. More works focus on document layout generation [29,38,58]. In particular, [7,28] worked on document image generation. The work of [45] introduced recursive autoencoding for layout generation, as well as a layout similarity *Doc-Sim*, which looks at the geometric similarity of layouts weighted by size of the elements. [19] is an example, in which the transformer architecture is applied in this autoregressive fashion to generate layout tokens (class labels and bounding boxes). In [3] ideas from the transformer-based methods mentioned above are combined with variational autoencoders to produce a more controllable and predictable generator. This work also implements the Wasserstein sequence distance as a metric into the layout literature. Our work builds on this idea and gives the Wasserstein distance more geometric significance, giving us the *Doc-EMD* metric that compares well with the Wasserstein and the *DocSim*.

2.3   Diffusion Generative Methods

Diffusion models use a sequential denoising model as an objective to generate realistic objects from Gaussian noise [23,43,56]. A domain-specific mean-estimator is used to model a markov process that is similar in spirit to deconvolution, but can be trained in an end-to-end manner. This approach is primarily applied in the image domain due to the approximately continuous nature of images and the denoising approach relying on using a Gaussian estimator for the inversion step. The diffusion model has achieved great success in text-image synthesis [17,49,51], 3D neural rendering [46], 3D point cloud generation [40], image compositing [57], audio generation [1,31], and video generation [41]. Recent works have shown that to extend this generative mechanism to discrete spaces, a rounding network can be applied that maps the denoised token embedding to a dictionary [36]. Our work builds on this and applies this approach to document layout, producing realistic document layouts that can be conditioned on partial layouts.

3   Methods

3.1   Diffusion for Layout Generation

Our approach leverages recent advances in denoising diffusion applied to discrete sequences [36,56]. As Fig. 1 shows, we consider a document layout of N layout elements given by the $5-$tuples $S = \{s_i\}_{i=1}^{N} = \{(c_i, x_i, y_i, w_i, h_i)\}_{i=1}^{N}$, where x_i is the upper-left $x-$coordinate, y_i is the upper-left $y-$coordinate, w_i, h_i specify the

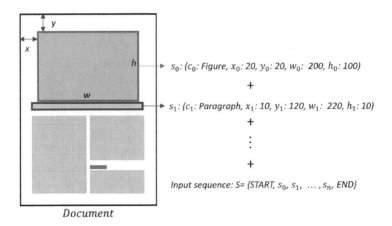

Fig. 1. Sequence representation of document layout.

width and height of the box (all in pixel coordinates), and c_i specifies the class (table, figure, formula, etc.) of the $i-$th box. Note that $x_i, y_i, w_i, h_i \in \mathbb{N}$ and $c \in [K]$ where K is the number of layout classes. To process this data structure as a sequence, we serialize S by flattening the sequence $\{c_0, x_0, y_0, w_0, h_0, c_1, \ldots\}$. Then the geometric entries are discretized and quantized to a fixed vocabulary G of size $|G|$. We can offset the class index entries by $|G|$ to keep the class coordinates distinct, or simply use a unique token outside of the vocabulary for G. This yields a sequential representation of length $5N$ with a fixed vocabulary $V = G \cup [K]$.

The framework of our method is illustrated in Fig. 2. An embedding step E is introduced to map the discrete sequence $s \in V^N$ to a feature vector $E(s) \in \mathbb{R}^{d \times N}$. We extend the conventional denoising transition model $x_t \to x_{t-1} \to \ldots x_0$ with another transition $x_0 \to E(s)$. Recall that the learned parameters refine the transition probability estimates:

$$p_\theta(x_{t-1}|x_t) = \mathcal{N}(x_{t-1}; \mu_\theta(x_t; t), \Sigma_\theta(x_t, t)) \tag{1}$$

by minimizing the variational upper bound of the negative log likelihood of the image over θ. The analogy to sequential learning is the log likelihood of the sequence embedding, so we take $x_0 \sim E(s)$ following [36]. To close the loop with the original sequence, a rounding module is introduced $p_\theta(s|x_0)$ which estimates the token sequence from the embedding $E(s) \approx x_0$. This is done by a rounding function with learned biases, call it R. So the (bidirectional) Markov chain is:

$$s \xrightarrow{E} w \xrightarrow{q} x_1 \xrightarrow{q} \ldots x_T,$$
$$s \xleftarrow{R} w \xleftarrow{p_\theta} x_1 \xleftarrow{p_\theta} \ldots x_T. \tag{2}$$

To solve for the reverse-process parameters several key simplifications were devised by [23]. The end result was to reduce the full variational bound objective

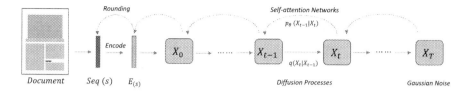

Fig. 2. Diffusion for Layout Generation.

to a simpler mean squared loss. The initial variational bound objective is written

$$L = \min_{\theta} \mathbb{E}\left(-\log(p_{\theta}(x_T)) - \sum_{t \geq 1} \log \frac{p_{\theta}(x_{t-1}|x_t)}{q(x_t|x_{t-1})} \right). \tag{3}$$

This can be rewritten as a series of KL-Divergences [23]

$$\min_{\theta} \mathbb{E}\left(D_{KL}(q(x_T|x_0)||p_{\theta}(x_T)) + \sum_{t>1} D_{KL}(q(x_{t-1}|x_t,x_0)||p_{\theta}(x_{t-1}|x_t)) \right) - \log(p_{\theta}(x_0|x_1)) \tag{4}$$

and each term is a divergence between two Gaussian distributions except for the last. By regrouping terms and throwing out the constant $\log(x_T)$ term, the closed form divergence for each term results in a mean squared error loss

$$L_2 = \sum_{t=1}^{T} \mathbb{E}\|\mu_{\theta}(x_t, t) - \hat{\mu}(x_t, x_0)\|^2. \tag{5}$$

$\hat{\mu}$ is the mean of the posterior and μ_{θ} is the predicted mean of the reverse process computed by a model parameterized by θ. In our implementation we use a transformer originated from BERT model ("BERT-Base" module) [11]. The maximum input and output length of the transformer is 256. In our diffusion framework, the transformer is μ_{θ}, and we set $T = 2000$. Note that for $\hat{\mu}$ a closed form gaussian, the derivation of which can be found in [23].

3.2   *Doc-EMD*: Earth Mover's Distance as a Document Metric

Document metrics are nontrivial because of the complex nature of documents [45]. Many metrics and similarities have been proposed, all with variation shortcomings that defy intuitive reasoning about the nature of document layouts. We propose to leverage the Earth-Mover's distance, which has been deployed successfully in contour matching [16] as well as image matching [52], to provide an underlying distance for document layouts. This allows us to leverage some useful properties of this well-established distance and we can leverage high quality existing open-source implementations to implement the Earth-Mover's distance [14]. Detailed analysis and description of our fast approximation of Earth-Mover's distance can be found in [10]. The runtime is not prohibitive and can be tuned by subsampling the points used in the distance calculation.

Definitions and Formal Distance. Consider a layout of N layout elements given by the 5−tuples $S = \{s_i\}_{i=1}^N$. We call this the source layout, and $T = \{t_i\}_{i=1}^M$ is the target layout. In this metric, we consider layouts as consisting of the 2−d points in the integer pixel coordinates falling inside of each layout box. Consider a single layout box $s = (c, x, y, w, h)$ we take $s \cong \rho(S) = \{p : x < p_1 < x + w, y < p_2 < y + h\}$ which we take as equivalent to the uniform pointwise density generated by those points. Let $|\cdot|$ denote the number of point elements in this set. Then we define the earth-mover's distance as

$$EMD(s,t) = EMD(\rho(s), \rho(T)) = \min_F \sum_{i,j} F_{ij} d(s_i, t_j) \tag{6}$$

$$s.t. \sum_j F_{i,j} \le 1/|s|, \sum_i F_{i,j} \le 1/|t|, \sum_{i,j} F_{i,j} = 1.$$

We conflate the EMD between class bounding box representation of s with $\rho(s)$ so that the cost of matching the element scales up with the size of the element. So now we have defined the data structure of the overarching layout S, each layout element s, and an element-to-element distance EMD which allows us to compare elements from different layouts.

Now we define the distance between two layouts S and T. First define the class function $C(s = (c, x, y, w, h)) = c$ and the set function $\hat{C}(S, c) = \{s \in S : C(s) = c\}$. Let $\kappa(cls; S, T)$ be the indicator function as to whether only one of S and T has elements of class cls (the exclusive or). The *Doc-EMD* is defined as

$$DOC_\lambda(S,T) = \sum_{cls} EMD(\cup_{\hat{C}(S,cls)} s, \cup_{\hat{C}(T,cls)} t) + \sum_{cls:\kappa(cls;S,T)=1} \lambda, \tag{7}$$

where λ is some positive factor used to penalize missing classes (we use $\lambda = 1$). In language, DOC is the sum of the earth-mover's distances between the sets of elements in S and T belonging to the same class plus a penalty term for each class that only appears in S or T. Note that this is very different from the *DocSim* [45] since we can avoid the computation of the size weightings in favor of the pointwise pmf contributions as well as skip the Hungarian matching step in favor of the earth-movers matching. Meanwhile, our method adds substance to the layout semantics which is missing from the Wasserstein sequence metrics [3] as they only capture exact matchings in the location and weight location and class in a disproportionate manner.

DOC is clearly reflexive, symmetric, and positive. Note that each term consisting of the EMD on subsets of S, T is a metric. Then for the second term, note that this is the discrete metric on the projection of the pointset to the class it belongs to scaled by the penalty term λ. So DOC_λ is a sum of metrics, which is also a metric, so it obeys the triangle inequality. This proof sketch shows that it enjoys all of the formal advantages of the Wasserstein distance.

Qualitative Comparisons with Other Document Metrics and Similarities. In this work we use several metrics to evaluate the performance of our proposed algorithm. We begin with a comment that document similarity (or distance) is a nontrivial problem. Documents are complex objects that can be represented in a number of ways and typically have no canonical underlying space from which they are drawn. Modeling that is the goal of generative layout algorithms, but how to measure the quality is directly related to the complexity of this problem. There is no perfect solution, so we develop an approach that covers some of the shortcomings of existing metrics, which we discuss now.

First, *DocSim* [45] proposes a Hungarian-matching based algorithm that uses a weighting term that scales linearly in the minimum area between boxes and exponentially in the size and shape difference. It has several major shortcomings

1. Not having an open-sourced common implementation;
2. Not well normalized so may not compare well between datasets;
3. The "similarity" is not an inner-product, so may not behave well in some cases.

We show qualitative examples highlighting each of these shortcomings (Fig. 3).

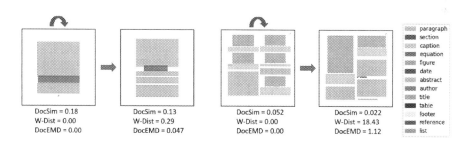

Fig. 3. DocSim Similarities. Visualization of how the *DocSim*, Wasserstein Distance, and *Doc-EMD* Distance vary across pairs of documents. The left image represents the source and the subsequent images are the targets. The metrics applied between them are shown below the target. The metrics shown below the source are the self-comparisons. *DocSim* is not normalized, has no common "self-similarity" value (this depends on the document), and does not always vary intuitively across structures. Wasserstein Distance is overly sensitive to class mismatch and does not capture structural similarities well. Note that this metric is primarily meant for distribution comparison. Also, note that as it is a distance higher values mean farther away (unlike the similarity of *DocSim*). For *Doc-EMD* Distance, there is a clear separation between single- and multi-column when classes are similar, the distance is well bounded, and the self-comparison is standardized.

Second, the Wasserstein sequence distance is applied to the layout sequences as opposed to the layout geometry. This acts as a distributional distance between the empirical discrete class label (categorical) distribution and the continuous bounding box distribution (Gaussian). First, this does not explicitly model similarity between two documents well (here it suffers from confusion across variations of box instances in the individual layouts), but rather between the distributions of documents given from a corpus. This is the key strength of this distance: it is a natural measure of distributional match. This is why we see this used commonly in generative modeling [2]. Note that it does not allow for weighting by area in any conventional sense, an advantage of the *DocSim* which is lost here.

Finally, the *Doc-EMD* distance is applicable both at the document pair level and at the distribution level. To compute the distributional similarity we take a similar approach to *DocSim*. We first compute the pairwise distance matrix with our metric applied to each pair of images. Then we obtain a negative matching score to which we apply the Hungarian matching algorithm. Key advantages over the former approaches are:

1. It is well normalized, scaling from 0 to K (the class number) if the appropriate pixel coordinates are used;
2. It keeps the distance properties of the Wasserstein without losing the geometric specificity of *DocSim*;
3. It behaves predictably when comparing structurally and semantically different layouts due to the geometric specificity of the per-class EMD.

Perhaps the greatest weakness of our method is that it requires greater runtime than the previous two in the pair-wise distance stage. However, as mentioned above, using open-sourced packages with hardware acceleration and optional speedups from approximations make it feasible even for very large datasets [14].

4   Experiments

4.1   Dataset

We evaluate our method on various public available document datasets from journal articles, tables, to magazines.

PubLayNet [64] consists of 330K document layouts by matching XML representations of public PDF articles from PubMed Centra^{TM}. It has 5 semantic categories including *Text, Title, List, Figure,* and *Table.* Semantic elements on each layout are annotated by categorized bounding box in COCO [37] format. Typically there is no overlapping between semantic units. We utilized its official splits: 335,703 for training, 11,245 for validation.

DocBank [35] consists of 500K document layouts by weak supervision of articles available on the arXiv.com. It contains 12 categories including *Abstract, Author, Caption, Equation, Figure, Footer, List, Paragraph, Reference, Section, Table,* and *Title.* Its annotated bounding boxes are created by merging extraction results of text lines. In general, this dataset has more (number) and fragmented (size) annotated bounding boxes compared to PubLayNet [64]. And overlapping exists between semantic units (*Paragraph* within *Figure* etc.). We use 399,811 layouts for training, and 49,980 layouts for validation.

Magazine [63] has 4K magazine layouts classified into 6 semantic categories including *Text, Image, Headline, Text-over-image, Headline-over-image,* and *Background.* The dataset itself holds natural overlaps between categories.

4.2   Comparisons

We compare our approach with 3 related methods on document layout generation task, including LayoutVAE [26], Gupta *et al.* [19], and VTN [3]. We utilize the code provided by the author's repository of [19] for the first two approaches. In particular, we privilege the ground truth bounding box count to LayoutVAE and only train its BBoxVAE portion. For Gupta *et al.* [19], the original inference code use the first bounding box as input prior and sample the top-k ($k = 5$) predicted layout bounding boxes to enable diversity. We modify the inference code to generate from initial token rather than the first bounding box. During inference, we find that results' diversity vanishes fast with $k < 5$. Thus the top-5 sampling is kept to ensure fairness. For VTN [3], we add the variational training scheme to the code of [19]. For fairness, all methods are trained for the same epochs on each dataset. Their models are based on default settings in the code or recommended settings in the original papers.

Quantitative Results. For all three datasets we list, we generate 1000 (391 for Magazine dataset since its validation dataset is small.) document layouts by ours and other competitors. Then we compare the generated results with the same amount of real document layouts by series of metrics. Specifically, the LayoutVAE [26] is conditioned on an input bounding box count. Gupta *et al.* [19] and VTN [3] generation is fully random from initial token.

Our set-by-set comparison is evaluated by 4 quantitative metrics. We already discuss the capability of *Doc-EMD* in Sect. 3.2, and we also include *DocSim* as reference. In theory, these two metrics should indicate reverse pattern (Smaller *DocSim* corresponds to larger *Doc-EMD*. And our results illustrate this pattern.). *Overlap* is the percentage of total overlapping area among the generated layout bounding boxes. Generally, less overlap indicates better performance a method achieves. However, we notice that there is a certain amount of reasonable overlapping existing in DocBank (Legend texts in a figure are recognized as paragraph. Thus its bounding box overlaps with figure's bounding box, etc.) and Magazine dataset (The categories of *Text-over-image*, and *Headline-over-image* naturally overlap with *Image* category.). Thus reasonable amount ($< 2\%$, etc.) of overlapping area will not affect the realism of generated layouts in terms of DocBank and Magazine dataset. Finally, *Coverage* is the percentage of total bounding box area over the document extent area. The closer value to the real data one, the better performance a method achieves.

As shown in Tables 1, 2, 3, our method outperforms competitors in most metrics on 3 datasets. Specifically, our method achieves the best performance in *DocSim* and *Doc-EMD*, and the second in other two metrics. For DocBank, our method achieves the best performance in *Coverage* and *Doc-EMD*, and the second in *DocSim*. For Magazine, our method dominates the *DocSim*, *Doc-EMD*, and *Overlap*. Moreover, our *Doc-EMD* metric keeps stable scalability and capability across three datasets, which verify the contributions in Sect. 3.2.

Table 1. Benchmark performance on PubLayNet Dataset

Approaches	DocSim ↑	Doc-EMD ↓	Overlap↓	Coverage
Layoutvae [26]	0.129	0.191	2.02%	**56.21%**
Gupta *et al.* [19]	0.137	0.063	**0.065%**	51.62%
VTN [3]	0.141	0.068	0.083%	53.49%
Ours	**0.163**	**0.053**	0.062%	55.30%
Real Data			0.026%	56.09%

Table 2. Benchmark performance on DocBank Dataset

Approaches	DocSim↑	Doc-EMD↓	Overlap↓	Coverage
Layoutvae [26]	0.087	0.592	2.02%	56.21%
Gupta *et al.* [19]	0.078	0.518	**0.56%**	44.01%
VTN [3]	**0.096**	0.353	0.61%	44.27%
Ours	0.093	**0.319**	2.04%	**45.49%**
Real Data			0.45%	46.20%

Table 3. Benchmark performance on Magazine Dataset

Approaches	DocSim↑	Doc-EMD↓	Overlap	Coverage
Layoutvae [26]	0.260	0.143	2.23%	80.93%
Gupta *et al.* [19]	0.176	0.227	12.6%	81.64%
VTN [3]	0.232	0.138	5.29%	**79.88%**
Ours	**0.302**	**0.117**	**1.23%**	70.55%
Real Data			1.36%	76.00%

Qualitative Results. Figures 4, 5, and 6 show qualitative comparison results in PubLayNet, DocBank, and Magazine dataset, respectively. The visual quality indicates that our method is able to produce diverse and realistic document layouts across three datasets.

For PubLayNet (Fig. 4), LayoutVAE [26] generates layouts that is poor in alignment, and containing noticeable overlaps. There are significant amount of "list" categorized bounding boxes appearing in most of the generated layouts (Fig. 4, the second to the rightmost column in LayoutVAE row), which is not realistic to open access academic papers that PubLayNet sampled from. The similar unrealistic pattern also happens in Gupta *et al.* [19] (Fig. 4, the second column in Gupta row) and VTN [3] (Fig. 4, the rightmost column in VTN row). This abnormality will not only reduce the performance of similarity evaluation in Sect. 4.2, but also negatively impact the downstream tasks using the generated layouts such as the detection task we discussed in Sect. 4.3. For Gupta *et al.* [19] and VTN [3], their performance are similar quantitatively and qualitatively.

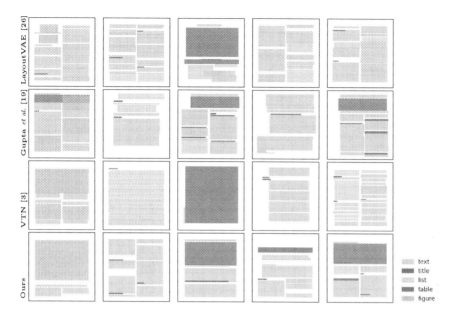

Fig. 4. Qualitative Results for PubLayNet. Competitors are worse in alignment, repeat similar patterns, contain abnormal bounding box categories, or have noticeable overlaps. Our method appears to represent PubLayNet better.

We find that if we generate layouts from the initial token rather than inputting the first bounding box (default setting by original codes), the diversity and realism of the generated results are rather repeating the same pattern (Fig. 4, the leftmost, the second, and the forth column in VTN [3] row), or poor in alignment and overlap (Fig. 4, Gupta *et al.* [19] row). However, our method outperforms other methods and generates both realistic and well-aligned document layouts without category abnormity. Our method illustrates plausible capability in both single and double column document layouts. The quality of our results will also support better performance in downstream detection task.

DocBank is the largest and the most complex dataset (more categories, diverse-sized bounding boxes) in our comparison. In this case, LayoutVAE [26] is unable to provide reasonable document layouts. For Gupta *et al.* [19] and VTN [3], most of generated results are unreal and repeating patterns of permuted "equation" and "paragraph" categories (Fig. 5, the leftmost, middle, and right most column in Gupta *et al.* [19] row; the middle column in VTN [3] row). There is also improper permutations of "reference" and 'paragraph' bounding boxes (Fig. 5, the second column in VTN [3] row). Our method shows reasonable patterns of permuted "equation" and "paragraph" categories (Fig. 5, the leftmost, and the rightmost column in Ours row). Moreover, our method is able to handle complex design of the cover page (Fig. 5, the second column in Ours row). And all generated layouts achieve plausible alignment.

372   L. He et al.

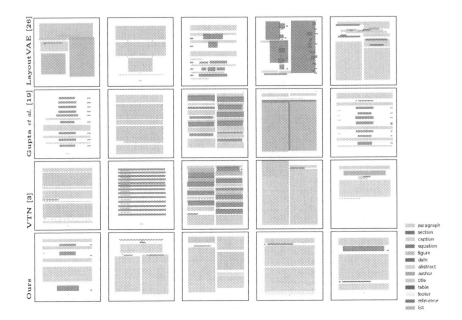

Fig. 5. Qualitative Results for DocBank. Competitors are poor in alignment, repeating similar patterns, containing abnormity of bounding box categories. Our method outperforms other competitors

Magazine dataset is the smallest dataset. But it has the most diverse alignment among bounding boxes (1 to 4 text columns within a single page), because the dataset is based on magazine design rather than academic papers. In this case, our competitors are unable to handle complex alignment and result in severe unreasonable overlaps between bounding boxes (Fig. 6, the first three rows). Note that there are natural overlaps in magazine dataset since two of the categories are defined to be on top of the image category. But most overlaps are between the same category (Fig. 6, the middle column in VTN [3] row, "text-over-image" category, etc.), or unreasonable overlaps (Fig. 6, the second column in VTN [3] row, several "text" rather than "text-over-image" bounding boxes are on top of the "image" bounding boxes, etc.). However, our method is able to handle complex text alignment (Fig. 6, the forth and the rightmost column in Ours row), and also prevent unreasonable overlaps. Our method also outperforms quantitatively in Table 3.

4.3 Layout Detection Task

The best way to illustrate benefits of a generative networks is to utilize its results for downstream tasks, especially for data augmentation. Document layout detection task is a subdomain of Optical Character Recognition (OCR). The detection network is trained to segment and label each layout bounding box

Fig. 6. Qualitative Results for Magazine. Competitors are poor in alignment, or with unreasonable overlaps. Our method outperforms other competitors.

in the given input document images. Practically, the annotation of document layouts is tedious and time-consuming, and the accuracy of the ground truth annotation contains inevitable ambiguity. To solve these problems, we can augment the dataset by generating document images with our generated document layout designs and labels. In this way, the quality of the annotation is ensured. And we will have unlimited amount of augmented dataset for better subtask model training and evaluation.

Similar to [3], as shown in Fig. 7, we develop a synthetic document image generation framework. First, we utilize our pretrained diffusion networks to randomly generate the same amount of document layouts as original training dataset (PubLayNet). Second, for each bounding box we generate, we find the nearest matching bounding box in the training dataset by matching categories, aspect ratio, size, etc. Then we slice the corresponding group of pixels from the original images in PubLayNet dataset, to synthetically mosaic a document image. Finally, we train a faster R-CNN model [50] as our document layout detector.

Table 4. Detection accuracy comparison between the detector trained by synthetic generated layouts, and by original PubLayNet.

	Ours	VTN [3]	PubLayNet
mAP (IoU=0.5)	0.795	0.769	0.9646

374   L. He et al.

We compare the detection performance with the one trained by our competitor results and the original dataset for evaluation.

In Table 4, we show the mean average precision (mAP) at IoU = 0.5. The values for VTN [3] and PubLayNet are reported by its original paper. Our diffusion-based method enable plausible detection accuracy and outperforms VTN.

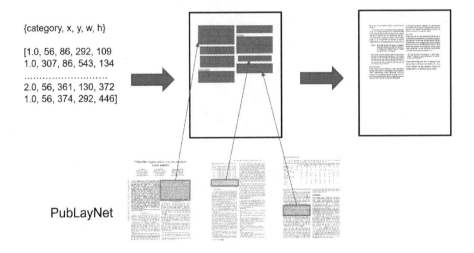

Fig. 7. Synthetic generation of document images. Utilize our method to create a training dataset based on PubLayNet images for a layout detector. We use our generated layouts, find the most similar bounding box in real dataset (by matching category, aspect ratio, size, etc.). Then crop pixels in the original images and render a new image dataset.

Table 5. Ablations on Magazine Dataset

Ablations	$DocSim\uparrow$	$Doc\text{-}EMD\downarrow$	Overlap	Coverage
lr = 0.0001, steps = 500	0.156	0.315	5.34%	79.80%
lr = 0.0001, steps = 1000	0.203	0.245	3.67%	**75.42%**
lr = 0.0002, steps = 2000	0.282	0.172	1.56%	72.34%
lr = 0.00001, steps = 2000	0.274	0.133	2.33%	71.21%
Ours (lr = 0.0001, steps = 2000)	**0.302**	**0.117**	**1.23%**	70.55%
Real Data			1.36%	76.00%

4.4   Parameter Ablations

In our model training, we conduct an ablation study on both learning rate and diffusion steps, as shown in Table 5. In general, more diffusion steps will improve network performance. But more diffusion steps also lead to extremely longer training time. We found that more diffusion steps above 2000 have no explicit benefits to the performance. Thus we choose diffusion steps as 2000 for all our trainings on Magazine and other datasets. Meanwhile, a proper learning rate is also the key to good performance. Though we found that diffusion model performance is quite robust to different learning rate. An arbitrary learning rate may still negatively influence the network performance.

5   Conclusion

In this work, we develop a new approach for document layout generation and a novel metric for document layouts evaluation. Our diffusion-based document layout generation approach shows outperforming and competitive results on several well known datasets in document layout generation. We have also shown how our approach can be applied to downstream applications, such as pretraining a document layout detector. In this paper, we provide extensive qualitative analysis and examples, so our readers understand both limitations and advantages of our method and our proposed metric.

In the future work, we plan to explore domain generalization and conditional generation. For domain generalization, higher specificity layouts as well as full OCR generation at the line level would provide a more complete document generation system. Meanwhile, conditioning generation on the right set of factors allows the proposed method to be used more effectively in document generation pipelines.

References

1. Agostinelli, A., et al.: Musiclm: generating music from text. arXiv preprint arXiv:2301.11325 (2023)
2. Arjovsky, M., Chintala, S., Bottou, L.: Wasserstein generative adversarial networks. In: International Conference on Machine Learning, pp. 214–223. PMLR (2017)
3. Arroyo, D.M., Postels, J., Tombari, F.: Variational transformer networks for layout generation. In: Proceedings of the IEEE/CVF Conference on Computer Vision and Pattern Recognition, pp. 13642–13652 (2021)
4. Benes, B., Zhou, X., Chang, P., Cani, M.P.R.: Urban brush: intuitive and controllable urban layout editing. In: The 34th Annual ACM Symposium on User Interface Software and Technology, pp. 796–814 (2021)
5. Bhatt, M., et al.: Design and deployment of photo2building: a cloud-based procedural modeling tool as a service. In: Practice and Experience in Advanced Research Computing, pp. 132–138 (2020)
6. Brown, T., et al.: Language models are few-shot learners. Adv. Neural. Inf. Process. Syst. **33**, 1877–1901 (2020)
7. Bui, Q.A., Mollard, D., Tabbone, S.: Automatic synthetic document image generation using generative adversarial networks: application in mobile-captured document analysis. In: 2019 International Conference on Document Analysis and Recognition (ICDAR), pp. 393–400. IEEE (2019)
8. Che, T., et al.: Maximum-likelihood augmented discrete generative adversarial networks. arXiv preprint arXiv:1702.07983 (2017)

9. Denton, E.L., Chintala, S., Fergus, R., et al.: Deep generative image models using a laplacian pyramid of adversarial networks. In: Advances in Neural Information Processing Systems, vol. 28 (2015)

10. Deshpande, I., et al.: Max-sliced Wasserstein distance and its use for GANs. In: Proceedings of the IEEE/CVF Conference on Computer Vision and Pattern Recognition, pp. 10648–10656 (2019)

11. Devlin, J., Chang, M.W., Lee, K., Toutanova, K.: BERT: pre-training of deep bidirectional transformers for language understanding. arXiv preprint arXiv:1810.04805 (2018)

12. Ding, Y., Huang, Y., He, L.: Pavement crack detection using directional curvature. Technical report (2017)

13. Dosovitskiy, A., et al.: An image is worth 16x16 words: transformers for image recognition at scale. arXiv preprint arXiv:2010.11929 (2020)

14. Flamary, R., et al.: Pot: Python optimal transport. J. Mach. Learn. Res. **22**(78), 1–8 (2021). http://jmlr.org/papers/v22/20-451.html

15. Goodfellow, I., et al.: Generative adversarial networks. Commun. ACM **63**(11), 139–144 (2020)

16. Grauman, K., Darrell, T.: Fast contour matching using approximate earth mover's distance. In: Proceedings of the 2004 IEEE Computer Society Conference on Computer Vision and Pattern Recognition, 2004. CVPR 2004, vol. 1, p. I. IEEE (2004)

17. Gu, S., et al.: Vector quantized diffusion model for text-to-image synthesis. In: Proceedings of the IEEE/CVF Conference on Computer Vision and Pattern Recognition, pp. 10696–10706 (2022)

18. Guo, J., Lu, S., Cai, H., Zhang, W., Yu, Y., Wang, J.: Long text generation via adversarial training with leaked information. In: Proceedings of the AAAI Conference on Artificial Intelligence, vol. 32 (2018)

19. Gupta, K., Lazarow, J., Achille, A., Davis, L.S., Mahadevan, V., Shrivastava, A.: Layouttransformer: layout generation and completion with self-attention. In: Proceedings of the IEEE/CVF International Conference on Computer Vision, pp. 1004–1014 (2021)

20. He, K., Chen, X., Xie, S., Li, Y., Dollár, P., Girshick, R.: Masked autoencoders are scalable vision learners. In: Proceedings of the IEEE/CVF Conference on Computer Vision and Pattern Recognition, pp. 16000–16009 (2022)

21. He, L., Shan, J., Aliaga, D.: Generative building feature estimation from satellite images. IEEE Trans. Geosci. Remote Sens. **61**, 1–13 (2023)

22. He, L., Yang, H., Huang, Y.: Automatic pole-like object modeling via 3D part-based analysis of point cloud. In: Remote Sensing Technologies and Applications in Urban Environments, vol. 10008, pp. 233–248. SPIE (2016)

23. Ho, J., Jain, A., Abbeel, P.: Denoising diffusion probabilistic models. arXiv preprint arxiv:2006.11239 (2020)

24. Huang, Y., Ma, P., Ji, Z., He, L.: Part-based modeling of pole-like objects using divergence-incorporated 3-d clustering of mobile laser scanning point clouds. IEEE Trans. Geosci. Remote Sens. **59**(3), 2611–2626 (2020)

25. Isola, P., Zhu, J.Y., Zhou, T., Efros, A.A.: Image-to-image translation with conditional adversarial networks. In: Proceedings of the IEEE Conference on Computer Vision and Pattern Recognition, pp. 1125–1134 (2017)

26. Jyothi, A.A., Durand, T., He, J., Sigal, L., Mori, G.: Layoutvae: stochastic scene layout generation from a label set. In: Proceedings of the IEEE/CVF International Conference on Computer Vision, pp. 9895–9904 (2019)

27. Karras, T., Laine, S., Aila, T.: A style-based generator architecture for generative adversarial networks. In: Proceedings of the IEEE/CVF Conference on Computer Vision and Pattern Recognition, pp. 4401–4410 (2019)

28. Kieu, V., Journet, N., Visani, M., Mullot, R., Domenger, J.P.: Semi-synthetic document image generation using texture mapping on scanned 3d document shapes. In: 2013 12th International Conference on Document Analysis and Recognition, pp. 489–493. IEEE (2013)

29. Kikuchi, K., Simo-Serra, E., Otani, M., Yamaguchi, K.: Constrained graphic layout generation via latent optimization. In: Proceedings of the 29th ACM International Conference on Multimedia, pp. 88–96 (2021)

30. Kingma, D.P., Welling, M.: Auto-encoding variational Bayes. arXiv preprint arXiv:1312.6114 (2013)

31. Kong, Z., Ping, W., Huang, J., Zhao, K., Catanzaro, B.: Diffwave: a versatile diffusion model for audio synthesis. arXiv preprint arXiv:2009.09761 (2020)

32. Li, C., Wand, M.: Precomputed real-time texture synthesis with Markovian generative adversarial networks. In: Leibe, B., Matas, J., Sebe, N., Welling, M. (eds.) ECCV 2016. LNCS, vol. 9907, pp. 702–716. Springer, Cham (2016). https://doi.org/10.1007/978-3-319-46487-9_43

33. Li, J., Yang, J., Hertzmann, A., Zhang, J., Xu, T.: LayoutGAN: generating graphic layouts with wireframe discriminators. In: 7th International Conference on Learning Representations, ICLR 2019, New Orleans, LA, USA, 6–9 May 2019. OpenReview.net (2019). https://openreview.net/forum?id=HJxB5sRcFQ

34. Li, J., Yang, J., Hertzmann, A., Zhang, J., Xu, T.: LayoutGAN: synthesizing graphic layouts with vector-wireframe adversarial networks. IEEE Trans. Pattern Anal. Mach. Intell. **43**(7), 2388–2399 (2020)

35. Li, M., et al.: Docbank: a benchmark dataset for document layout analysis. arXiv preprint arXiv:2006.01038 (2020)

36. Li, X.L., Thickstun, J., Gulrajani, I., Liang, P., Hashimoto, T.B.: Diffusion-LM improves controllable text generation (2022). https://doi.org/10.48550/ARXIV.2205.14217, https://arxiv.org/abs/2205.14217

37. Lin, T.-Y., et al.: Microsoft COCO: common objects in context. In: Fleet, D., Pajdla, T., Schiele, B., Tuytelaars, T. (eds.) ECCV 2014. LNCS, vol. 8693, pp. 740–755. Springer, Cham (2014). https://doi.org/10.1007/978-3-319-10602-1_48

38. Lin, Z., Winata, G.I., Xu, P., Liu, Z., Fung, P.: Variational transformers for diverse response generation. arXiv preprint arXiv:2003.12738 (2020)

39. Liu, Y., Huang, Y., Qiu, X., He, L.: Automatic guardrail inventory using mobile laser scanning (MLS). Technical report (2017)

40. Luo, S., Hu, W.: Diffusion probabilistic models for 3D point cloud generation. In: Proceedings of the IEEE/CVF Conference on Computer Vision and Pattern Recognition, pp. 2837–2845 (2021)

41. Molad, E., et al.: Dreamix: video diffusion models are general video editors. arXiv preprint arXiv:2302.01329 (2023)

42. Nazeri, K., Ng, E., Ebrahimi, M.: Image colorization using generative adversarial networks. In: Perales, F.J., Kittler, J. (eds.) AMDO 2018. LNCS, vol. 10945, pp. 85–94. Springer, Cham (2018). https://doi.org/10.1007/978-3-319-94544-6_9

43. Nichol, A.Q., Dhariwal, P.: Improved denoising diffusion probabilistic models. In: International Conference on Machine Learning, pp. 8162–8171. PMLR (2021)

44. Patel, P., Kalyanam, R., He, L., Aliaga, D., Niyogi, D.: Deep learning-based urban morphology for city-scale environmental modeling. PNAS Nexus **2**(3), pgad027 (2023)

45. Patil, A.G., Ben-Eliezer, O., Perel, O., Averbuch-Elor, H.: Read: recursive autoencoders for document layout generation. In: Proceedings of the IEEE/CVF Conference on Computer Vision and Pattern Recognition Workshops, pp. 544–545 (2020)
46. Poole, B., Jain, A., Barron, J.T., Mildenhall, B.: Dreamfusion: text-to-3D using 2D diffusion. arXiv preprint arXiv:2209.14988 (2022)
47. Radford, A., Metz, L., Chintala, S.: Unsupervised representation learning with deep convolutional generative adversarial networks. arXiv preprint arXiv:1511.06434 (2015)
48. Radford, A., Narasimhan, K., Salimans, T., Sutskever, I., et al.: Improving language understanding by generative pre-training (2018)
49. Ramesh, A., Dhariwal, P., Nichol, A., Chu, C., Chen, M.: Hierarchical text-conditional image generation with clip latents. arXiv preprint arXiv:2204.06125 (2022)
50. Ren, S., He, K., Girshick, R., Sun, J.: Faster R-CNN: Towards real-time object detection with region proposal networks. In: Advances in Neural Information Processing Systems, vol. 28 (2015)
51. Rombach, R., Blattmann, A., Lorenz, D., Esser, P., Ommer, B.: High-resolution image synthesis with latent diffusion models. In: Proceedings of the IEEE/CVF Conference on Computer Vision and Pattern Recognition, pp. 10684–10695 (2022)
52. Rubner, Y., Tomasi, C., Guibas, L.J.: The earth mover's distance as a metric for image retrieval. Int. J. Comput. Vision **40**(2), 99–121 (2000)
53. Sheng, Y., et al.: Controllable shadow generation using pixel height maps. In: Avidan, S., Brostow, G., Cissé, M., Farinella, G.M., Hassner, T. (eds.) ECCV 2022. LNCS, vol. 13683, pp. 240–256. Springer, Cham (2022). https://doi.org/10.1007/978-3-031-20050-2_15
54. Sheng, Y., Zhang, J., Benes, B.: SSN: soft shadow network for image compositing. In: Proceedings of the IEEE/CVF Conference on Computer Vision and Pattern Recognition, pp. 4380–4390 (2021)
55. Sheng, Y., et al.: Pixht-lab: pixel height based light effect generation for image compositing. arXiv preprint arXiv:2303.00137 (2023)
56. Song, J., Meng, C., Ermon, S.: Denoising diffusion implicit models (2020). https://doi.org/10.48550/ARXIV.2010.02502, https://arxiv.org/abs/2010.02502
57. Song, Y., et al.: Objectstitch: generative object compositing. arXiv preprint arXiv:2212.00932 (2022)
58. Tabata, S., Yoshihara, H., Maeda, H., Yokoyama, K.: Automatic layout generation for graphical design magazines. In: ACM SIGGRAPH 2019 Posters, pp. 1–2 (2019)
59. Van Den Oord, A., Vinyals, O., et al.: Neural discrete representation learning. In: Advances in Neural Information Processing Systems, vol. 30 (2017)
60. Wang, L., Huang, Y., Shan, J., He, L.: Msnet: multi-scale convolutional network for point cloud classification. Remote Sens. **10**(4), 612 (2018)
61. Yu, L., Zhang, W., Wang, J., Yu, Y.: SeqGAN: sequence generative adversarial nets with policy gradient. In: Proceedings of the AAAI Conference on Artificial Intelligence, vol. 31 (2017)
62. Zhang, X., Ma, W., Varinlioglu, G., Rauh, N., He, L., Aliaga, D.: Guided pluralistic building contour completion. Vis. Comput. **38**(9–10), 3205–3216 (2022)
63. Zheng, X., Qiao, X., Cao, Y., Lau, R.W.: Content-aware generative modeling of graphic design layouts. ACM Trans. Graph. (TOG) **38**(4), 1–15 (2019)
64. Zhong, X., Tang, J., Yepes, A.J.: Publaynet: largest dataset ever for document layout analysis. In: 2019 International Conference on Document Analysis and Recognition (ICDAR), pp. 1015–1022. IEEE (2019)

Frontiers in Handwriting Recognition 2
(Historical Documents)

DTDT: Highly Accurate Dense Text Line Detection in Historical Documents via Dynamic Transformer

Haiyang Li[1], Chongyu Liu[1], Jiapeng Wang[1], Mingxin Huang[1], Weiying Zhou[1], and Lianwen Jin[1,2(⊠)]

[1] South China University of Technology, Guangzhou, China
eelwjin@scut.edu.cn
[2] SCUT-Zhuhai Institute of Modern Industrial Innovation, Zhuhai, China

Abstract. Text detection in historical documents is challenging owing to the dense distribution of texts with diverse scales and complex layouts, resulting in low detection accuracy under high Intersection over Union (IoU) conditions. Historical document digitization requires highly accurate detection results to preserve the contents completely. In this paper, we present an end-to-end text detection framework, namely **D**ynamic **T**ext **D**etection **T**ransformer (**DTDT**), for dense text detection in historical documents under high accuracy requirements. We introduce a deformable convolution-based dynamic encoder to strengthen the text representation ability at different scales. In addition, the parallel dynamic attention heads are designed to facilitate better interaction between the box and mask branches to obtain accurate text detection results. Experiments on the MTHv2 and ICDAR 2019 HDRC-CHINESE (short for "IC19 HDRC") datasets show that the proposed DTDT method achieves state-of-the-art performance. Furthermore, our DTDT achieves competitive results in layout analysis on SCUT-CAB benchmark, demonstrating its excellent generalization capabilities.

Keywords: Text Detection · Detection Transformer · Historical Document Understanding

1 Introduction

Historical document digitization, which facilitates the preservation and understanding of the knowledge and insights that are contained in ancient books, has attracted increasing research attention [2,6,10,29,38]. The aim of text line detection, which is a critical step of historical document digitization, is to locate text instances. Accurate text detection is beneficial for subsequent tasks such as text recognition and ancient book restoration. Moreover, accurate text line detection results can effectively reduce the difficulty of layout analysis, which aims to locate and categorize document elements such as figures, tables and paragraphs.

With the rapid development of deep learning, scene text detection methods have made significant success on various benchmarks [19,47,51,57]. However, it is difficult for these methods to perform well on complex historical documents

© The Author(s), under exclusive license to Springer Nature Switzerland AG 2023
G. A. Fink et al. (Eds.): ICDAR 2023, LNCS 14187, pp. 381–396, 2023.
https://doi.org/10.1007/978-3-031-41676-7_22

with dense text alignment. Figure 1 (a) presents the results of the scene text detection methods DBNet++ [19], PSENet [47] and FCENet [57] for historical documents. It can be observed that many of the detection results of these methods overlap with neighboring texts and do not closely match the texts, and also suffer from missed and false detections. We summarize the reasons for the insufficient generalization ability of scene text detection methods for these historical documents as follows: (1) As illustrated in Fig. 1 (b), the text distribution in the historical documents is denser than scene text images. For example, MTHv2 [29] contains an average of 33 text instances, while there are only seven text instances per image on SCUT-CTW1500 [52]. (2) Significant degradation of historical documents, including stains, seal noise, ink seepage, and breakage, makes it difficult for scene text detection methods [19,24,26,47,48,54,57] to obtain accurate detection results, which are essential for the subsequent text recognition. Figure 1 (c)-(f) show examples of the degradation of ancient documents.

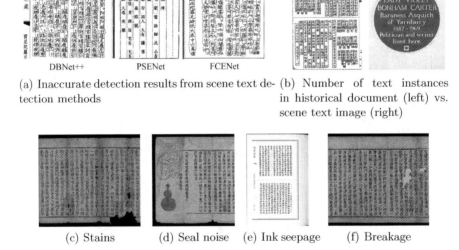

(a) Inaccurate detection results from scene text detection methods

(b) Number of text instances in historical document (left) vs. scene text image (right)

(c) Stains   (d) Seal noise   (e) Ink seepage   (f) Breakage

Fig. 1. (a) Inaccurate detection results of scene text detection methods on historical document images, (b) comparison of the number of texts of historical document and scene text image, and (c)–(f) degradation phenomena such as stains, seal noise, ink seepage, and breakage.

In this paper, to alleviate the problem of insufficient detection accuracy and difficulty in generalizing to complex layout scenarios with dense text distribution by previous methods, we propose the **Dynamic Text Detection Transformer** (**DTDT**) to adapt to the dense and multi-scale characteristics of historical document texts and to meet the requirements of high accuracy. Firstly, for the dense and multi-scale text arrangement, we present a deformable convolution-based dynamic encoder to fuse the adjacent scale features of the feature pyramid with dynamic attention, which leverages spatial attention, channel attention, and multi-scale feature aggregation to pay attention to text features at different scales. Second, to meet the high accuracy detection requirements, we introduce a parallel dynamic attention head using a dynamic attention module to fuse the Region of Interest (RoI) and image features, and make the box and mask branches interact effectively. The parallel dynamic attention head facilitates the mutual interaction of dual-path branch information and precisely detects text regions in a continuously refined manner. Furthermore, we employ the spatial attention transform (SAT) mask head [30] to suppress background noise in the feature maps. Discrete cosine transform (DCT) is also used to encode the text masks as compact vectors for the accurate representation of text in arbitrary shapes. We conduct experiments on the historical document datasets MTHv2, IC19 HDRC and SCUT-CAB, illustrating the strong robustness and generalization ability of our model.

The contributions of this paper are summarized as follows:

- We propose an end-to-end text detection model named DTDT, which is based on a dynamic Transformer for the accurate detection of dense texts in historical documents with complex layouts.
- We introduce a deformable convolution-based dynamic encoder using dynamic attention to improve the detection performance of text at different scales, and present parallel dynamic attention heads with shared image features for joint detection and segmentation.
- We adopt the SAT mask head to suppress the background noise and employ DCT to encode arbitrary-shaped text masks while maintaining a low training complexity.
- DTDT achieves state-of-the-art results with F-measure of 97.90% and 96.62% for MTHv2 and IC19 HDRC, respectively. Furthermore, it obtains competitive results for layout analysis on SCUT-CAB, illustrating its outstanding generalization capabilities.

2 Related Work

2.1 Regression-Based Methods

Regression-based methods directly regress the bounding boxes of the text. [17] modified the aspect ratios of anchors based on SSD [23] to accommodate the scale characteristics of text lines. TextBoxes++ [32] regressed the quadrilateral vertices to detect multi-oriented text. EAST [54] generated rotated rectangles and quadrilaterals directly at the pixel level. To avoid the learning confusion caused by the order of points, OBD [24] decomposed the order of the quadrilateral label points into key edges comprising four invariant points and included a key edge module for learning the bounding boxes. To prevent entangled vertices from interfering with the learning process, DCLNet [1] regressed each side that is disentangled from the quadrilateral contour. The above methods are mainly for horizontal and multi-oriented text, and their performance degrades when dealing with irregular text. To tackle the issue of irregular text detection, TextRay [46] represented arbitrary-shaped text in the polar system using a uniform geometric encoding. FCENet [57] mapped the text border to the Fourier domain to obtain Fourier contour embedding that fits curved text contours. Regression-based methods enjoy simple post-processing algorithms, but a complex representation design is required to fit arbitrary-shaped text. The one-stage methods [17,32,54] are slightly less accurate because they only regress once, and the two-stage methods [10,24,29] usually require the manual setting of the anchor to accommodate the multi-scale text distribution. In contrast, our method performs multiple iterations of the learnable query boxes to obtain more accurate results and proposes a dynamic encoder to fuse multi-scale features to better adapt to the textual characteristics of ancient documents.

2.2 Segmentation-Based Methods

In segmentation-based methods, text detection is considered as a segmentation problem. TextSnake [26] described the text as a series of ordered overlapping disks. PAN [48] adopted a lightweight segmentation head and a learnable post-processing method known as pixel aggregation. DBNet [18] provided differentiable binarization by adding the binarization step to the network for training. DBNet++ [19] extended DBNet by introducing an adaptive scale fusion module to enhance the scale robustness. To better distinguish adjacent text, PSENet [47] generated text segmentation maps in a progressive scale expansion manner. SAE [43] mapped pixels to an embedding space, drawing closer to pixels belonging to the same text and vice versa to divide the adjacent text more effectively. Although segmentation-based methods can be adapted to curved text, they require complex post-processing and are sensitive to background noise, and are more computationally intensive for ancient text detection owing to the dense text. Therefore, our method uses DCT to encode individual text instances to obtain a lightweight mask to reduce computational complexity. The SAT mask head is used to suppress noise in historical documents with complex layouts.

2.3 Transformer-Based Methods

Transformer [44] has attracted increasing attention in scene text detection. Raisi et al. [34] proposed a Transformer-based architecture for detecting multi-oriented text in scene images and a loss function for the rotated text detection problem. Tang et al. [41] adopted Transformer to model the relationship between a few sampled features to decode control points. DPText-DETR [51] used explicit box coordinates to generate and subsequently dynamically update position queries. The lack of interaction between the branches of the decoding the control points and those for detecting the bounding boxes prevents them from achieving better performance. Our DTDT explicitly establishes the interaction of the box and mask information for accurate text detection using the dynamic attention module.

3   Methodology

3.1   Overall Architecture of DTDT

As illustrated in Fig. 2, our proposed DTDT consists of three components: Backbone, Dynamic Encoder and Dynamic Decoder. The backbone network is composed of Swin Transformer (Swin-T) [25] and feature pyramid network (FPN) [20] to extract feature maps at different stages of the input image. The dynamic encoder applies dynamic attention to the features at different scales and fuses adjacent layer features to enhance multi-scale feature representation. The sum of the image features P extracted from x^{DE} and position embeddings E is fed into the Transformer encoder for self-attention learning to obtain enhanced features Z. Based on Sparse R-CNN [40], the RoI features U_t^{box} and U_t^{mask} together with the enhanced image features Z_{t-1} are fed into the dynamic attention module [9] of the box and mask branches, respectively, to obtain the object features O_t^{box} and O_t^{mask} for the prediction of the class, bounding box, and mask of each text instance. Finally, the output of the previous layer will be continuously refined in the dynamic decoder with parallel dynamic attention heads to obtain accurate results.

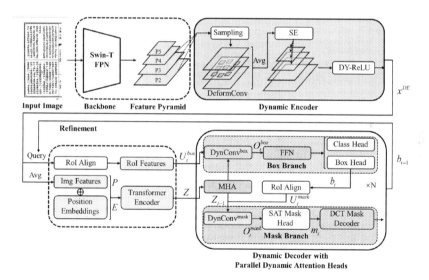

Fig. 2. Framework of proposed DTDT model. Our model consists of three components: the backbone, the dynamic encoder, and the dynamic decoder with parallel dynamic attention heads. MHA denotes the multi-head attention and FFN denotes the feedforward network.

3.2   Dynamic Encoder

In general, large and small objects are assigned to high-level and low-level feature maps to extract the RoI features, respectively. However, this may not be optimal

[22] as other unused feature maps may contain information that helps to improve the final prediction. Therefore, inspired by recent research on dynamic encoder [7,33], we introduce a dynamic encoder to perform multi-scale feature fusion on adjacent feature maps, which is depicted in the upper right part of Fig. 2. The process is divided into three steps. First, given a set of features $P = \{P_2, ..., P_k\}$ $(k = 5)$ from the feature pyramid, deformable aggregation, which consists of several deformable convolution layers [55] on each feature map and an averaging operator, is performed to simulate the spatial attention for specific regions on P_i.

This process can be formulated as follows:

$$s_i = Offset_i(P_i) \tag{1}$$

$$P_i^* = \{DeformConv_{i-1}(Downsample(P_{i-1}), s_i),$$
$$DeformConv_i(P_i, s_i), \tag{2}$$
$$DeformConv_{i+1}(Upsample(P_{i+1}), s_i)\}$$

$$P_i^{'} = Avg(P_i^*), \tag{3}$$

where the offset s_i that corresponds to the feature map P_i is learned using a 3×3 convolution $Offset_i$ for deformed sampling locations. The neighboring feature maps P_{i-1} and P_{i+1} are downsampled and upsampled, respectively, to the same size as P_i. Deformable convolution is performed on the sampled feature maps and P_i, and each feature map focuses on the specific position s_i that is learned from the middle layer to avoid conflicts during feature aggregation. $P_i^{'}$ is obtained by averaging each term of P_i^*.

Second, $P_i^{'}$ is used for channel attention learning with the squeeze and excitation (SE) module [13]:

$$P_i^{''} = SE(P_i^{'}). \tag{4}$$

Finally, we use the DY-ReLU [5] activation function, whose parameters are dynamically generated from the input elements to improve the feature representation capability:

$$P_i^o = DY\text{-}ReLU(P_i^{''}). \tag{5}$$

3.3   Parallel Dynamic Attention Heads

The feature maps from the dynamic encoder are cropped and aligned using RoIAlign [12] to obtain the RoI features $U \in \mathbb{R}^{k \times d \times l \times l}$ via k learnable query boxes b_t $(t = 0)$, where d is the channel dimension, and l denotes the output resolution after the pooling. The feature maps of each layer are averaged and summed to obtain the image features $P \in \mathbb{R}^{k \times d}$, which are summed with the learnable position embeddings $E \in \mathbb{R}^{k \times d}$ to be fed into the Transformer encoder and MHA module to obtain $Z_{t-1} \in \mathbb{R}^{k \times d}$. We design parallel dynamic attention heads with the RoI features U and enhanced image features Z_{t-1}, as indicated in the bottom right part of Fig. 2.

Existing methods [24,28,29] use the RoI features that are obtained from the box branch to predict the mask directly, which ignores the interaction between the box and mask branches. As illustrated in Fig. 3 (b), we use the dynamic attention module, namely $DynConv$, for more effective interaction of the box and mask branches, thereby enabling improved results. The box branch employs $DynConv_t^{box}$ to fuse the RoI features U_t^{box} and enhanced image features Z_{t-1} to extract object features O_t^{box} for classification and bounding box regression. The mask branch leverages the RoI features U_t^{mask} that are extracted from the predicted box b_t and the enhanced image features Z_{t-1} for further fusion in $DynConv_t^{mask}$ to obtain the final detection results m_t. The above process is expressed by Eqs. 6 and 7, where $\mathcal{P}^{box}$ and $\mathcal{P}^{mask}$ denote a pooling operator for the extraction of RoI features U_t^{box} and U_t^{mask}, respectively. $\mathcal{B}_t$ denotes the box head that is stacked by three linear layers. $\mathcal{M}_t$ indicates the SAT mask head. x^{DE} is the output feature map of the dynamic encoder.

$$U_t^{box} = \mathcal{P}^{box}(x^{DE}, b_{t-1}),$$
$$O_t^{box} = DynConv_t^{box}(U_t^{box}, Z_{t-1}), \qquad (6)$$
$$b_t = \mathcal{B}_t(FFN(O_t^{box})),$$

$$U_t^{mask} = \mathcal{P}^{mask}(x^{DE}, b_t),$$
$$O_t^{mask} = DynConv_t^{mask}(U_t^{mask}, Z_{t-1}), \qquad (7)$$
$$m_t = \mathcal{M}_t(O_t^{mask}).$$

The above process offers two advantages: (1) it provides the mask information obtained from the supervision of the mask branch to the box branch, and (2) the collaborative interaction between the box and the mask branches is improved. Moreover, we employ the SAT [30] mask head, which has been demonstrated as effective for dense instance segmentation and exploits spatial attention to suppress noise. The implementation details of the SAT mask head are illustrated in Fig. 3 (a). Average and max pooling operations are carried out along the channel axis of the object features $O_t^{mask} \in \mathbb{R}^{14 \times 14 \times C}$ that are obtained by $DynConv_t^{mask}$ to generate the pooling features $P_{avg}, P_{max} \in \mathbb{R}^{14 \times 14 \times 1}$, which are stacked along the channel, where C denotes the channel dimension. Subsequently, a 3×3 convolution layer is applied and the features are normalized with a sigmoid function. Finally, element-wise multiplication is performed on the object feature O_t^{mask}. A mask feature of length 40 is obtained using two convolution and linear layers.

3.4 DCT Mask Representation

The direct prediction of the two-dimensional binary grid incurs a high computational cost for large resolutions. However, fine-grained features cannot be captured on a small scale. Therefore, we apply DCT [39] to transform the text mask encoding into the frequency domain. As the energy is concentrated in the low-frequency part, we keep this part to produce a compact vector as a predictive object to accurately represent the text shape. The flow of the DCT encoding and inverse DCT (IDCT) decoding is depicted in Fig. 4.

We resize the ground truth mask $M_{gt} \in \mathbb{R}^{H \times W}$ to $M \in \mathbb{R}^{K \times K}$ during training, where H and W are the height and width of M_{gt}, and K denotes the mask size. We apply two-dimensional DCT transforms M to obtain $M_{DCT} \in \mathbb{R}^{K \times K}$.

$$M_{DCT}(u,v) = \frac{2}{K}C(u)C(v)\sum_{x=0}^{K-1}\sum_{y=0}^{K-1}M(x,y)cos\frac{(2x+1)u\pi}{2K}cos\frac{(2y+1)v\pi}{2K}, \quad (8)$$

where $C(w) = \frac{1}{\sqrt{2}}$ for $w = 0$ and $C(w) = 1$ otherwise.

The first N-dimensional vector V is sampled from the M_{DCT} in a "zig-zag" manner to obtain the one-dimensional mask representation. We extend V to $M_{dct} \in \mathbb{R}^{K \times K}$ by filling in zeros at the end during inference and apply two-dimensional IDCT processes V to obtain $M_{IDCT} \in \mathbb{R}^{K \times K}$.

$$M_{IDCT}(x,y) = \frac{2}{K}C(u)C(v)\sum_{u=0}^{K-1}\sum_{v=0}^{K-1}M_{dct}(u,v)cos\frac{(2x+1)u\pi}{2K}cos\frac{(2y+1)v\pi}{2K}$$

$$(9)$$

Finally, M_{IDCT} is resized to $M_{rec} \in \mathbb{R}^{H \times W}$ using bilinear interpolation. It is worth noting that the time complexity of DCT and IDCT is $O(nlogn)$ [11].

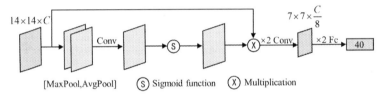

(a) Implementation details of SAT mask head

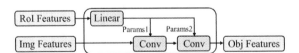

(b) Implementation details of dynamic attention module

Fig. 3. (a) Structure of SAT mask head. (b) Dynamic attention module applied to box and mask branches.

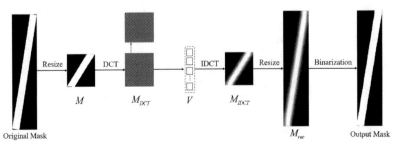

Fig. 4. DCT encoding and IDCT decoding.

3.5   Loss Function

We adopt the Hungarian algorithm [15] to match the predicted and ground truth boxes. DTDT applies a set prediction loss to the set of predictions of the categories, box coordinates, and mask representations. The total loss function can be formulated as follows:

$$\mathcal{L} = \lambda_{cls}\mathcal{L}_{cls} + \lambda_{box}\mathcal{L}_{box} + \lambda_{mask}\mathcal{L}_{mask}. \tag{10}$$

$\mathcal{L}_{box}$ is defined as:

$$\mathcal{L}_{box} = \lambda_{L1}\mathcal{L}_{L1} + \lambda_{giou}\mathcal{L}_{giou}. \tag{11}$$

$\mathcal{L}_{mask}$ is defined as:

$$\mathcal{L}_{mask} = \mathcal{L}_{L2} + \mathcal{L}_{dice}. \tag{12}$$

In the above equations, $\mathcal{L}_{cls}$ is the focal loss [21], and $\mathcal{L}_{L1}$ and $\mathcal{L}_{giou}$ are the L1 loss and the generalized IoU loss [37], respectively. $\mathcal{L}_{L2}$ is the L2 loss of the one-dimensional mask embedding before DCT decoding and $\mathcal{L}_{dice}$ is the dice loss [31] of the two-dimensional mask after IDCT decoding. λ_{cls}, λ_{box}, λ_{mask}, λ_{L_1} and $\lambda_{L_{giou}}$ are set to 2, 1, 5, 5 and 2, respectively.

4   Experiments

4.1   Datasets

MTHv2 [29] is a Chinese historical document dataset consisting of 2,399 training images and 800 testing images. The dataset includes character-level and line-level quadrilateral annotations.

ICDAR 2019 HDRC-CHINESE [38] is a large historical documents dataset of structured Chinese family records that are annotated using line-level quadrilaterals. We randomly used 10,715 images for training and 1,000 for testing among the 11,715 available images.

SCUT-CAB [6] is a complex layout analysis dataset of Chinese historical documents containing 3,200 training images and 800 testing images. SCUT-CAB contains two subsets: SCUT-CAB-Logical and SCUT-CAB-Physical, which have 27 and 4 categories, respectively. All text instances are annotated using quadrilaterals.

4.2   Implementation Details

We used Swin-T [25], pre-trained on ImageNet [8] as the backbone. The number of learnable proposal boxes was set to 500. The number of iterations was set to four to improve the accuracy. We selected a mask size of 80 × 80 and a 40-dimensional DCT mask vector. We trained DTDT for 90k iterations with a batch size of eight on two NVIDIA RTX A6000 GPUs. We used AdamW [27] as the optimizer and set an initial learning rate of $2.5e^{-5}$ and a weight decay of $1e^{-4}$. The learning rate was divided by 10 at 50% and 70% of the total number of iterations. We applied data augmentation methods including random cropping and multi-scale training. The maximum image scale was set to 1333 × 800.

4.3 Comparison with Previous Methods

We compared our method with previous state-of-the-art methods on MTHv2 and IC19 HDRC. Tables 1 and 2 display the quantitative experimental results. Figure 5 shows the qualitative results for MTHv2. Furthermore, by modifying the number of categories in the class head, we applied DTDT to the SCUT-CAB dataset to validate the potential of our method in the task of ancient book layout analysis.

Text Line Detection. The results in Tables 1 and 2 demonstrate the high accuracy and robustness of our method on these two datasets. Our method achieved an F-measure of 97.90% on MTHv2, which was 0.18% higher than the

Table 1. Detection results on MTHv2 dataset. "P", "R", and "F" indicate the precision, recall, and F-measure, respectively. **Bold** indicates the best performance. Underline indicates second best.

Method	IoU=0.5			IoU=0.6	IoU=0.7	IoU=0.8	Post-processing
	P	R	F	F	F	F	
Projection analysis [29]	–	–	69.22	66.87	60.97	–	–
EAST [54]	–	–	95.04	91.55	80.35	–	–
Ma et al. [29]	–	–	<u>97.72</u>	97.26	<u>96.03</u>	–	–
Mask R-CNN [12]	**98.17**	95.98	97.06	96.67	95.51	90.23	–
FCENet [57]	95.16	92.82	93.97	91.30	86.51	73.86	–
OBD [24]	97.83	<u>97.43</u>	97.63	<u>97.32</u>	**96.31**	<u>90.78</u>	–
Deformable DETR [56]	97.92	94.64	96.25	95.62	93.80	84.22	–
DBNet++ [19]	96.20	94.93	95.56	77.01	36.15	18.70	0.015s
PSENet [47]	93.97	87.84	90.80	88.65	83.68	70.96	0.022s
PAN [48]	97.18	93.14	95.12	92.55	84.63	62.74	<u>0.011s</u>
TextSnake [26]	95.07	89.00	91.94	90.92	89.36	84.58	0.497s
DTDT(Ours)	<u>97.94</u>	**97.86**	**97.90**	**97.41**	95.98	**91.18**	**0.008s**

Table 2. Detection results on IC19 HDRC dataset.

Method	IoU=0.5			IoU=0.6	IoU=0.7	IoU=0.8	Post-processing
	P	R	F	F	F	F	
Mask R-CNN [12]	<u>96.54</u>	96.21	<u>96.37</u>	<u>94.66</u>	<u>88.80</u>	70.01	–
FCENet [57]	93.63	91.50	92.55	87.74	77.25	52.12	–
OBD [24]	94.56	**97.02**	95.77	93.91	86.83	64.18	–
Deformable DETR [56]	94.43	95.72	94.57	92.55	86.27	**71.96**	–
DBNet++ [19]	96.37	95.73	96.05	90.64	75.57	48.51	0.021s
PSENet [47]	91.57	88.57	90.04	83.02	68.42	42.19	0.026s
PAN [48]	95.11	92.84	93.96	88.68	71.27	31.65	**0.012s**
TextSnake [26]	82.90	72.22	77.19	73.40	68.41	51.54	0.512s
DTDT(Ours)	**96.89**	<u>96.35</u>	**96.62**	**95.15**	**90.10**	<u>71.42</u>	<u>0.016s</u>

second best score when the IoU threshold was 0.5. Only three methods maintained performance above 90% when the IoU was 0.8, and our method is the best. Analogous results were obtained for IC19 HDRC. Our method obtained an F-measure of 96.62%, outperforming the second best method by 0.25%. Our method remained robust under high IoU requirements without much performance degradation compared to other methods. Our DTDT still yielded high accuracy when the IoU threshold was between 0.5 and 0.8. The post-processing times for the segmentation-based methods and our DTDT are given in Tables 1 and 2, and the results illustrate the rapidity of IDCT decoding.

Layout Analysis Experiments. Table 3 presents the experimental results for the ancient book layout analysis on SCUT-CAB dataset [6]. The results show that our method could achieve results that are comparable to those of other methods in the physical and logical layout analysis tasks. Our model achieved the best AP75 and AP results on the physical layout analysis task, demonstrating the effectiveness of DTDT. In the logical analysis task, DTDT yielded the second best performance, which was slightly lower than that of Deformable DETR.

Table 3. AP50, AP75, and AP of each model on SCUT-CAB testing sets. AP refers to average precision, AP50 and AP75 are the average precision at IoU = 0.5 and 0.75, respectively.

Method	Physical						Logical					
	Objection Detection			Instance Segmentation			Object Detection			Instance Segmentation		
	AP_{50}	AP_{75}	AP	AP_{50}	AP_{75}	AP	AP_{50}	AP_{75}	AP	AP_{50}	AP_{75}	AP
Anchor-based one-stage												
RetinaNet [21]	91.5	82.9	74.7	91.5	81.6	73.8	78.3	61.2	55.1	78.3	61.7	55.0
YOLOv3 [35]	87.6	82.5	75.9	87.1	79.1	73.1	71.4	59.3	52.7	71.4	59.3	52.7
GFL [16]	92.6	74.8	73.7	92.6	73.2	72.4	78.1	57.8	54.1	78.1	58.8	53.5
Anchor-free one-stage												
FCOS [42]	83.2	76.0	68.9	83.1	74.7	68.1	74.1	54.4	50.2	74.0	53.4	49.1
FoveaBox [14]	91.3	82.5	74.6	91.3	80.0	73.1	80.4	60.2	54.9	80.3	60.3	54.3
Anchor-based multi-stage												
Faster R-CNN [36]	91.3	86.1	77.5	91.0	83.4	75.3	77.4	61.3	54.9	77.3	60.6	54.2
Cascade R-CNN [3]	91.4	87.8	79.9	91.4	84.8	77.4	77.5	62.3	55.9	77.5	60.9	55.4
Mask R-CNN [12]	92.1	87.7	79.1	91.7	87.2	79.5	78.5	61.9	55.1	77.7	63.1	55.3
Cascade Mask R-CNN [3]	92.1	88.6	80.9	92.1	88.4	81.0	78.0	62.7	56.8	77.9	61.8	56.3
HTC [4]	92.8	_89.4_	_81.4_	92.8	88.8	81.0	80.1	65.2	58.3	80.0	63.1	58.0
SCNet [45]	**94.1**	89.0	81.3	**94.1**	_89.1_	82.0	_83.6_	67.3	60.2	_83.6_	_68.0_	60.3
Pure Instance Segmentation												
SOLO [49]	90.7	81.6	75.2	91.2	84.3	76.7	73.8	57.7	51.6	73.2	57.8	51.5
SOLOv2 [50]	91.5	81.6	75.1	92.2	85.1	78.7	76.4	53.2	50.5	77.0	59.7	53.9
Query-based												
Deformable DETR [56]	92.7	87.9	81.0	92.5	85.1	78.8	**84.6**	**69.8**	**61.6**	**84.6**	**69.9**	**61.1**
QueryInst [9]	91.7	87.1	79.3	91.2	86.7	79.2	80.4	65.7	58.5	80.4	65.3	58.1
Multi-modality based												
VSR [53]	90.4	85.5	78.5	90.4	84.5	78.2	78.3	61.6	55.7	78.2	61.1	55.1
DTDT(Ours)	_94.0_	**90.0**	**83.0**	_94.0_	**89.6**	**82.7**	81.1	_68.0_	_60.8_	81.1	67.8	_60.4_

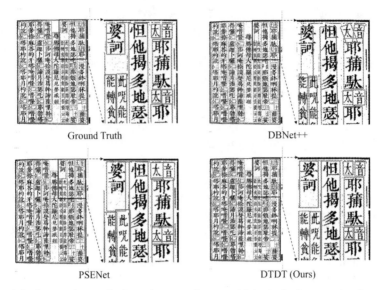

Ground Truth   DBNet++

PSENet   DTDT (Ours)

(a) Comparison of detection results on historical documents from MTHv2 dataset

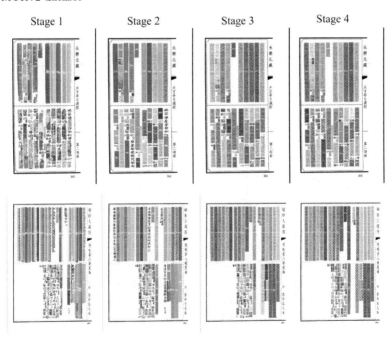

Stage 1   Stage 2   Stage 3   Stage 4

(b) Qualitative results at each stage: masks from DTDT on MTHv2 dataset

Fig. 5. (a) Visualization results of our method and other scene text detection methods. Our method achieved a higher detection accuracy. (b) Qualitative experimental results for the four stages of two example images. The different colors are used to distinguish the detection results of each text instance of the model.

4.4   Ablation Study

We performed an ablation study on MTHv2 to verify the effectiveness of our proposed method. The quantitative results for different settings are presented in Table 4. The DCT resulted in a 2.78% improvement, indicating that the text shape can be more represented accurately using DCT masks. The dynamic encoder achieved performance improvements of 0.12% and 0.23% in the precision and recall, respectively, on the MTHv2 dataset, indicating its ability to improve the network's adaptation to multi-scale text. The parallel dynamic attention heads resulted in a 0.12% improvement in the F-measure. The design of the parallel dynamic attention heads provides better interaction and collaboration between the box and mask branches, facilitating the benefits of the two branches. The SAT mask, which achieved an F-measure of 97.90%, has a certain ability to suppress noise.

Table 4. Detection results for different settings of DCT, dynamic encoder, parallel dynamic attention heads, and SAT mask head on MTHv2 dataset. "DE" indicates dynamic encoder and "PDAH" indicates parallel dynamic attention heads.

DCT	DE	PDAH	SAT	P	R	F	$\triangle$F
–	–	–	–	92.36	97.33	94.78	–
✓	–	–	–	97.83	97.28	97.56	↑2.78
✓	✓	–	–	97.95	97.51	97.73	↑0.17
✓	✓	✓	–	97.89	97.80	97.85	↑0.12
✓	✓	✓	✓	**97.94**	**97.86**	**97.90**	↑0.05

5   Conclusions

We proposed DTDT, which is a highly accurate text line detection method for dense text distribution of historical documents. We introduced a dynamic encoder to improve the representation ability of multi-scale text and parallel dynamic attention heads to facilitate the mutual benefits of the box and mask branches for generating more accurate text masks. The experiments demonstrated that our method achieved state-of-the-art results on historical document datasets such as MTHv2 and IC19 HDRC, and achieved comparable results on the layout analysis dataset SCUT-CAB. The potential of DTDT for text detection in modern documents and other scenarios will be explored further in future research.

Acknowledgements. This research is supported in part by NSFC (Grant No.: 61936003), Zhuhai Industry Core and Key Technology Research Project (no. 2220004002350), and Science and Technology Foundation of Guangzhou Huangpu Development District (No. 2020GH17) and GD-NSF (No.2021A1515011870).

References

1. Bi, Y., Hu, Z.: Disentangled contour learning for quadrilateral text detection. In: WACV, pp. 909–918 (2021)
2. Boillet, M., Kermorvant, C., Paquet, T.: Robust text line detection in historical documents: learning and evaluation methods. IJDAR **25**(2), 95–114 (2022)
3. Cai, Z., Vasconcelos, N.: Cascade R-CNN: delving into high quality object detection. In: CVPR, pp. 6154–6162 (2018)
4. Chen, K., et al.: Hybrid task cascade for instance segmentation. In: CVPR, pp. 4974–4983 (2019)
5. Chen, Y., Dai, X., Liu, M., Chen, D., Yuan, L., Liu, Z.: Dynamic ReLU. In: Vedaldi, A., Bischof, H., Brox, T., Frahm, J.-M. (eds.) ECCV 2020. LNCS, vol. 12364, pp. 351–367. Springer, Cham (2020). https://doi.org/10.1007/978-3-030-58529-7_21
6. Cheng, H., Jian, C., Wu, S., Jin, L.: SCUT-CAB: a new benchmark dataset of ancient Chinese books with complex layouts for document layout analysis. In: Porwal, U., Fornés, A., Shafait, F. (eds.) ICFHR 2022. LNCS, vol. 13639, pp. 436–451. Springer, Cham (2022). https://doi.org/10.1007/978-3-031-21648-0_30
7. Dai, X., Chen, Y., Yang, J., Zhang, P., Yuan, L., Zhang, L.: Dynamic DETR: end-to-end object detection with dynamic attention. In: ICCV, pp. 2988–2997 (2021)
8. Deng, J., Dong, W., Socher, R., Li, L.J., Li, K., Fei-Fei, L.: ImageNet: a large-scale hierarchical image database. In: CVPR, pp. 248–255. IEEE (2009)
9. Fang, Y., et al.: Instances as queries. In: ICCV, pp. 6910–6919 (2021)
10. Grüning, T., Leifert, G., Strauß, T., Michael, J., Labahn, R.: A two-stage method for text line detection in historical documents. Int. J. Doc. Anal. Recogn. (IJDAR) **22**(3), 285–302 (2019). https://doi.org/10.1007/s10032-019-00332-1
11. Haque, M.: A two-dimensional fast cosine transform. IEEE Trans. Acoust., Speech, Signal Process. **33**(6), 1532–1539 (1985)
12. He, K., Gkioxari, G., Dollar, P., Girshick, R.: Mask R-CNN. In: ICCV, pp. 2961–2969 (2017)
13. Hu, J., Shen, L., Sun, G.: Squeeze-and-excitation networks. In: CVPR, pp. 7132–7141 (2018)
14. Kong, T., Sun, F., Liu, H., Jiang, Y., Li, L., Shi, J.: FoveaBox: beyound anchor-based object detection. IEEE Trans. Image Process. **29**, 7389–7398 (2020)
15. Kuhn, H.W.: The Hungarian method for the assignment problem. Nav. Res. Logist. Q. **2**(1–2), 83–97 (1955)
16. Li, X., et al.: Generalized focal loss: learning qualified and distributed bounding boxes for dense object detection. In: Larochelle, H., Ranzato, M., Hadsell, R., Balcan, M., Lin, H. (eds.) NIPS 2020. LNCS, vol. 33, pp. 21002–21012. Curran Associates Inc, Red Hook, NY, USA (2020). https://doi.org/10.5555/3495724.3497487
17. Liao, M., Shi, B., Bai, X., Wang, X., Liu, W.: Textboxes: a fast text detector with a single deep neural network. In: AAAI (2017)
18. Liao, M., Wan, Z., Yao, C., Chen, K., Bai, X.: Real-time scene text detection with differentiable binarization. In: AAAI, pp. 11474–11481 (2020). https://doi.org/10.1609/aaai.v34i07.6812
19. Liao, M., Zou, Z., Wan, Z., Yao, C., Bai, X.: Real-time scene text detection with differentiable binarization and adaptive scale fusion. TPAMI (2022)
20. Lin, T.Y., Dollar, P., Girshick, R., He, K., Hariharan, B., Belongie, S.: Feature pyramid networks for object detection. In: CVPR, pp. 2117–2125 (2017)

21. Lin, T.Y., Goyal, P., Girshick, R., He, K., Dollar, P.: Focal loss for dense object detection. In: ICCV, pp. 2980–2988 (2017)

22. Liu, S., Qi, L., Qin, H., Shi, J., Jia, J.: Path aggregation network for instance segmentation. In: CVPR, pp. 8759–8768 (2018)

23. Liu, W., et al.: SSD: single shot multibox detector. In: Leibe, B., Matas, J., Sebe, N., Welling, M. (eds.) ECCV 2016. LNCS, vol. 9905, pp. 21–37. Springer, Cham (2016). https://doi.org/10.1007/978-3-319-46448-0_2

24. Liu, Y., Zhang, S., Jin, L., Xie, L., Wu, Y., Wang, Z.: Omnidirectional scene text detection with sequential-free box discretization. In: IJCAI, pp. 3052–3058 (2019)

25. Liu, Z., et al.: Swin transformer: hierarchical vision transformer using shifted windows. In: ICCV, pp. 10012–10022 (2021)

26. Long, S., Ruan, J., Zhang, W., He, X., Wu, W., Yao, C.: TextSnake: a flexible representation for detecting text of arbitrary shapes. In: ECCV, pp. 20–36 (2018)

27. Loshchilov, I., Hutter, F.: Decoupled weight decay regularization. In: ICLR (2019)

28. Lyu, P., Liao, M., Yao, C., Wu, W., Bai, X.: Mask TextSpotter: an end-to-end trainable neural network for spotting text with arbitrary shapes. In: Ferrari, V., Hebert, M., Sminchisescu, C., Weiss, Y. (eds.) Computer Vision – ECCV 2018. LNCS, vol. 11218, pp. 71–88. Springer, Cham (2018). https://doi.org/10.1007/978-3-030-01264-9_5

29. Ma, W., Zhang, H., Jin, L., Wu, S., Wang, J., Wang, Y.: Joint layout analysis, character detection and recognition for historical document digitization. In: ICFHR, pp. 31–36 (2020)

30. Mao, Q., Sun, L., Wu, J., Gao, Y., Wu, X., Qiu, L.: SATMask: spatial attention transform mask for dense instance segmentation. In: DSC, pp. 592–598 (2022)

31. Milletari, F., Navab, N., Ahmadi, S.A.: V-Net: fully convolutional neural networks for volumetric medical image segmentation. In: 3DV, pp. 565–571 (2016)

32. Minghui Liao, B.S., Bai, X.: Textboxes++: a single-shot oriented scene text detector. IEEE Trans. Image Process. **27**(8), 3676–3690 (2018)

33. Mishra, S.K., Sinha, S., Saha, S., Bhattacharyya, P.: Dynamic convolution-based-encoder decoder framework for image captioning in Hindi. ACM Trans. Asian Low-Resour. Lang. Inf. Process. **22**(4), 1–18 (2023)

34. Raisi, Z., Naiel, M.A., Younes, G., Wardell, S., Zelek, J.S.: Transformer-based text detection in the wild. In: CVPR Workshops, pp. 3162–3171 (2021)

35. Redmon, J., Farhadi, A.: YOLOv3: an incremental improvement. arXiv preprint arXiv:1804.02767 (2018)

36. Ren, S., He, K., Girshick, R., Sun, J.: Faster R-CNN: towards real-time object detection with region proposal networks. In: Cortes, C., Lawrence, N., Lee, D., Sugiyama, M., Garnett, R. (eds.) NIPS 2015. LNCS, vol. 28. Curran Associates, Inc. (2015)

37. Rezatofighi, H., Tsoi, N., Gwak, J., Sadeghian, A., Reid, I., Savarese, S.: Generalized intersection over union: a metric and a loss for bounding box regression. In: CVPR, pp. 658–666 (2019)

38. Saini, R., Dobson, D., Morrey, J., Liwicki, M., Simistira Liwicki, F.: ICDAR 2019 historical document reading challenge on large structured Chinese family records. In: ICDAR, pp. 1499–1504. IEEE (2019)

39. Shen, X., et al.: DCT-Mask: discrete cosine transform mask representation for instance segmentation. In: CVPR, pp. 8720–8729 (2021)

40. Sun, P., et al.: Sparse R-CNN: end-to-end object detection with learnable proposals. In: CVPR, pp. 14454–14463 (2021)

41. Tang, J., et al.: Few could be better than all: feature sampling and grouping for scene text detection. In: CVPR, pp. 4563–4572 (2022)

42. Tian, Z., Shen, C., Chen, H., He, T.: FCOS: fully convolutional one-stage object detection. In: ICCV, pp. 9627–9636 (2019)

43. Tian, Z., Shu, M., Lyu, P., Li, R., Zhou, C., Shen, X., Jia, J.: Learning shape-aware embedding for scene text detection. In: CVPR, pp. 4234–4243 (2019)

44. Vaswani, A., et al.: Attention is all you need. In: Guyon, I., et al. (eds.) NIPS 2017. LNCS, vol. 30, pp. 5998–6008. Curran Associates, Inc. (2017). https://doi.org/10.5555/3295222.3295349

45. Vu, T., Kang, H., Yoo, C.D.: SCNet: training inference sample consistency for instance segmentation. In: AAAI, pp. 2701–2709 (2021)

46. Wang, F., Chen, Y., Wu, F., Li, X.: TextRay: contour-based geometric modeling for arbitrary-shaped scene text detection. In: ACM MM, pp. 111–119 (2020)

47. Wang, W., Xie, E., Li, X., Hou, W., Lu, T., Yu, G., Shao, S.: Shape robust text detection with progressive scale expansion network. In: CVPR, pp. 9336–9345 (2019)

48. Wang, W., et al.: Efficient and accurate arbitrary-shaped text detection with pixel aggregation network. In: ICCV, pp. 8440–8449 (2019)

49. Wang, X., Kong, T., Shen, C., Jiang, Y., Li, L.: SOLO: segmenting objects by locations. In: Vedaldi, A., Bischof, H., Brox, T., Frahm, J.-M. (eds.) ECCV 2020. LNCS, vol. 12363, pp. 649–665. Springer, Cham (2020). https://doi.org/10.1007/978-3-030-58523-5_38

50. Wang, X., Zhang, R., Kong, T., Li, L., Shen, C.: SOLOv2: dynamic and fast instance segmentation. In: Larochelle, H., Ranzato, M., Hadsell, R., Balcan, M., Lin, H. (eds.) NIPS 2020. LNCS, vol. 33, pp. 17721–17732. Curran Associates Inc, Red Hook, NY, USA (2020)

51. Ye, M., Zhang, J., Zhao, S., Liu, J., Du, B., Tao, D.: DPText-DETR: towards better scene text detection with dynamic points in transformer. In: AAAI (2023)

52. Yuliang, L., Lianwen, J., Shuaitao, Z., Sheng, Z.: Detecting curve text in the wild: new dataset and new solution. arXiv preprint arXiv:1712.02170 (2017)

53. Zhang, P., et al.: VSR: a unified framework for document layout analysis combining vision, semantics and relations. In: Lladós, J., Lopresti, D., Uchida, S. (eds.) ICDAR 2021. LNCS, vol. 12821, pp. 115–130. Springer, Cham (2021). https://doi.org/10.1007/978-3-030-86549-8_8

54. Zhou, X., et al.: East: an efficient and accurate scene text detector. In: CVPR (2017)

55. Zhu, X., Hu, H., Lin, S., Dai, J.: Deformable convnets v2: more deformable, better results. In: CVPR, pp. 9308–9316 (2019)

56. Zhu, X., Su, W., Lu, L., Li, B., Wang, X., Dai, J.: Deformable DETR: deformable transformers for end-to-end object detection. In: ICLR (2021)

57. Zhu, Y., Chen, J., Liang, L., Kuang, Z., Jin, L., Zhang, W.: Fourier contour embedding for arbitrary-shaped text detection. In: CVPR, pp. 3123–3131 (2021)

The Bullinger Dataset: A Writer Adaptation Challenge

Anna Scius-Bertrand[1,2]([✉]), Phillip Ströbel[3], Martin Volk[3], Tobias Hodel[4],
and Andreas Fischer[1,2]

[1] iCoSys Institute, University of Applied Sciences and Arts Western Switzerland,
Fribourg, Switzerland
[2] DIVA Group, University of Fribourg, Fribourg, Switzerland
{anna.scius-bertrand,andreas.fischer}@hefr.ch
[3] Department of Computational Linguistics, University of Zurich, Zurich, Switzerland
{pstroebel,volk}@cl.uzh.ch
[4] Walter Benjamin Kolleg, University of Bern, Bern, Switzerland
tobias.hodel@unibe.ch

Abstract. One of the main challenges of automatically transcribing large collections of handwritten letters is to cope with the high variability of writing styles present in the collection. In particular, the writing styles of non-frequent writers, who have contributed only few letters, are often missing in the annotated learning samples used for training handwriting recognition systems. In this paper, we introduce the Bullinger dataset for writer adaptation, which is based on the Heinrich Bullinger letter collection from the 16th century, using a subset of 3,622 annotated letters (about 1.2 million words) from 306 writers. We provide baseline results for handwriting recognition with modern recognizers, before and after the application of standard techniques for supervised adaptation of frequent writers and self-supervised adaptation of non-frequent writers.

Keywords: Handwriting Recognition · Writer Adaptation · Historical Documents · Handwritten Letters

1 Introduction

Handwriting recognition remains a mostly unsolved problem and a very active field of research, because it challenges pattern recognition and machine learning techniques in various ways: Even when considering samples written in the same language and time period, there is a high variability in character shapes and character connections to model, especially in the case of cursive handwriting. When changing the language, there is a distribution shift regarding language models, even when the same set of characters are used. For historical documents [4], additional difficulties include the absence of timing information, which is only available for modern on-line handwriting with an electronic pen, degraded paper or parchment due to old age, which leads to artifacts on the scanned page images,

© The Author(s), under exclusive license to Springer Nature Switzerland AG 2023
G. A. Fink et al. (Eds.): ICDAR 2023, LNCS 14187, pp. 397–410, 2023.
https://doi.org/10.1007/978-3-031-41676-7_23

and a large number of different languages, scripts, and time periods to consider. Therefore, handwritten text recognition (HTR) usually targets a very specific type of handwriting, e.g. a historical manuscript written by only a few different hands, with similar imaging conditions across the scans, the same language, etc. and is trained with a large amount of annotated learning samples from the same type of handwriting.

In this paper, we introduce a novel challenge for HTR in the context of a comprehensive digitization project [2] in Switzerland that aims to create a digital edition of a large collection of historical letters, namely the Bullinger letters, which include about 12,000 letters written or received by Heinrich Bullinger (1504–1575), an important Swiss Reformer. He was in contact with over 1,000 persons, which introduces a high variability of writing styles, as well as differences in writing support and writing instruments. Furthermore, the letters are not only written in Latin but also in a premodern form of German, and sometimes the two languages are mixed, while the language might change either from paragraph to paragraph or even mid-sentence (i.e., code-switching). Over the past years, transcription and transcription alignment efforts have been focused on the most frequent writers, i.e. Bullinger himself and persons who have written a considerable number of letters to him. However, there are thousands of letters from non-frequent writers, whose writing styles are not present in the annotated training material. Therefore, one of the most intriguing problems is that of writer adaptation: *"Is it possible to adapt a generic HTR system to the specific writing style of a non-frequent writer, who is not represented in the training data, such that the HTR performance is improved?"* The same question, although less challenging, can also be asked for frequent writers, who are represented in the training set and may also profit from an adaptation to their particular style of writing.

1.1   Related Work

There is a rich body of literature on the topic of writer adaptation for HTR. To name just a few, early examples include [17], where the adaptation is performed based on writer-specific allographs that are used to re-evaluate the output of an HTR system, and [3], where unsupervised clustering is used to estimate Gaussian mixture models that are specific to a writing style. In [6], a self-training approach is pursued to improve the performance of an HTR system by adapting it to the recognition output of unlabeled samples. The work presented in [7] employs a keyword spotting strategy to adapt an HTR system trained for modern handwriting to historical handwriting. More recent attempts to perform transfer learning are reported in [8,10]. A competition organized on the READ dataset specifically included the problem of writer adaptation with respect to 22 different hands, 5 of which are used both in the training and the test set to investigate supervised adaptation [21]. The best results are obtained when adapting both the optical and the language models, and when including data augmentation [19]. Targeting the more difficult case of unsupervised adaptation, a style adaptation at multiple abstraction layers of a deep convolutional model is proposed in [22]. Another recent unsupervised adaptation scheme is based on fully synthetic training data [11].

1.2 Contribution

In the present work, we do not introduce a novel method for writer adaptation. Instead, we introduce the Bullinger dataset for writer adaptation [1] and establish baseline results using state-of-the-art HTR systems with standard adaptation strategies, i.e. fine-tuning a generic HTR system on the training data of the frequent writers, and fine-tuning the models on confidently transcribed text lines of non-frequent writers, following a self-training methodology [6]. The dataset is publicly and freely available for developing and comparing novel approaches to writer adaptation.

When comparing the Bullinger dataset with other datasets used for writer adaptation research, we can highlight the difficulty of the handwriting itself (cf. Fig. 1), which is also difficult to read for human experts, and the large number of over one million words, which is suitable for experiments with deep learning models from the current state of the art. Table 1 provides a comparison with other related research datasets. Note that only about a quarter of all letters are currently included in the Bullinger dataset, representing the progress of the digitalization project. The total number of writers is over 1,000 for the entire letter collection.

Table 1. Related research datasets for writer adaptation research.

Dataset	Number of words	Number of writers
Georges Washington [13]	4,860	2
Parzival [5]	23,478	3
Rimes [9]	66,978	1,300
READ [21]	98,239	22
CVL [12]	99,902	310
IAM Handwriting [15]	115,320	657
Bullinger	1,241,714	306

In the remainder, we describe the handwriting present in the Bullinger letters and how it has been transcribed so far, introduce the HTR systems and writer adaptation techniques considered for the experiments, present the baseline results, and draw some conclusions.

2   Dataset

The Bullinger Digital project [2] aims to bring together all available resources about the comprehensive letter correspondence of Heinrich Bullinger (1504–1575), a Swiss Reformer, in a single database. For this purpose, all available meta-information, e.g. about the writers of the letters, but also scanned page images and existing transcriptions are brought together. Furthermore, the goal is to automatically align existing transcriptions with the page images and to use HTR to perform an automatic transcription for the remaining letters. Heinrich Bullinger was a key actor during the Reformation in Switzerland and Europe. His letter correspondence includes about 2,000 letters written by himself and about 10,000 letters that he has received from over 1,000 persons.

Certain writers, including Bullinger, exhibit writing styles that are very difficult to read, even for human experts. Figure 1 provides an example of Bullinger's handwriting. We can observe a mix of Latin and a premodern form of German phrases, abbreviations, and words that are very difficult to decipher without intimate knowledge of the handwriting, or access to a transcription. At the beginning of the third line, we can also observe a missing word in the transcription. It is due to an error of the automatic transcription alignment, which was performed using the Text2Image module of the Transkribus platform [16]. In general, the quality of the alignment is high, and thus the quality of the ground truth for handwriting recognition, but especially at the beginning and at the end of the text lines errors may arise due to word breaks. Furthermore, the transcription is not necessarily character-accurate, e.g. abbreviations are often written out in full. This noise in the automatically generated ground truth is an additional difficulty for training HTR systems.

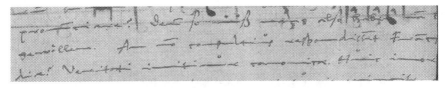

pronunciaret denn sparnuß michs also habe man
gewǒllen. An non consultius respondissent Franc
veritati invitimur canonicae. Hinc immor

Fig. 1. Text lines written by Bullinger with automatically aligned transcriptions.

Figure 2 illustrates the high variability of writing styles present in the Bullinger dataset. The first line shows three examples of how Bullinger writes his own name. They exhibit a considerable intra-writer variability. The remaining words are written by other persons, demonstrating changes in the writing style, writing support, and writing instrument.

Fig. 2. Different writing styles and different forms of the word "Bullinger". The writer IDs are indicated in the bottom-left corner of each automatically segmented word. All words of the first line are written by the same writer, Bullinger himself.

Note that the sample word images were automatically cut out from text lines that have been processed by an HTR system and may contain segmentation errors. The text lines themselves were cut out from the scanned page images according to polygonal boundaries provided by Transkribus' layout analysis system. The special background pattern around the text lines is added artificially instead of white background, to make the background more homogeneous for HTR.

3 Methods

3.1 Handwriting Recognition

We consider two state-of-the-art models for handwriting recognition, namely PyLaia [18] and HTR-Flor [20]. They both consider deep convolutional layers to extract features from text line images, followed by bidirectional recurrent layers with connectionist temporal classification (CTC) loss to analyze the features from left-to-right as well as right-to-left to recognize character sequences. They differ in the composition of the layers as illustrated in Fig. 3. To reduce the number of trainable parameters, HTR-Flor uses gated convolution and bidirectional gated recurrent units (BGRU) instead of standard convolution and bidirectional long short-term memory cells (BLSTM). In effect, HTR-Flor only has around

820 thousand parameters, which is significantly less than the 9.6 million parameters of PyLaia. Nevertheless, both models achieve similar results on several benchmark datasets [20].

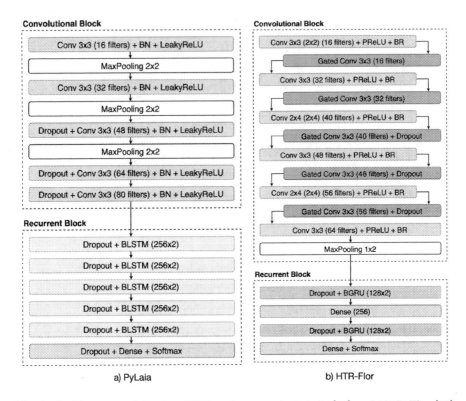

a) PyLaia   b) HTR-Flor

Fig. 3. Architectures of the two HTR systems used: PyLaia [18] and HTR-Flor [20]. Both figures are taken from [20].

3.2 Writer Adaptation

We consider two standard writer adaptation methods to study their impact on the HTR performance. The first method is used for frequent writers, for which some of the letters were transcribed and are part of the training set. In this case, we train a generic HTR system on all letters and then fine-tune it on the training letters of the frequent writer in order to adapt the model to the specific writing style.

The second method is used for non-frequent writers, for which none of the letters have been transcribed and therefore no training material is available. In this case, we follow the self-training approach proposed in [6]. We start again with a generic HTR system trained on all letters and apply it to the letters of the non-frequent writer. Afterwards, we compute a confidence measure $C(s)$ for the predicted characters sequences $s = c_0, \ldots, c_N$,

$$C(s) = \prod_{i=0}^{N} p(c_i) \, , \tag{1}$$

where $p(c_i)$ is the softmax probability of the character c_i according to the CTC decoding. Afterwards, we sort the character sequences of all text lines according to their confidence and use the most confident P percent of the text lines as a new training set for fine-tuning the generic HTR system.

4   Experimental Evaluation

4.1   Database Setup

To study the impact of writer adaptation, we consider text line images from a subset of 3,622 letters by 306 writers with automatically aligned transcriptions, which are used as ground truth for the HTR experiments.

As illustrated in Fig. 4, the database is split as follows for the Bullinger writer adaptation challenge: First, we sort the writers according to their number of letters, observing a Zipf distribution with only few frequent writers and a large number of non-frequent writers. Then, using a threshold of 5 letters, we distinguish two groups of writers:

– **Frequent writers:** Writers with at least 5 letters.
– **Non-frequent writers:** Writers with less than 5 letters.

There are 106 frequent writers in total. We use the first 80% of their letters for training, the next 10% for validation (optimization of hyper-parameters), and the final 10% for testing. For the non-frequent writers, we select the next 200 writers in the sorted list of writers to compose a second test set of similar size. In this experimental setup, the test set for frequent writers estimates how well HTR performs for known writers, where several of their letters have been transcribed for training, and the test set for non-frequent writers estimates how well HTR performs for unknown writers, whose writing styles are not present during training. This scenario reflects the real situation in the Bullinger Digital project [2], where the transcription efforts are directed towards the most important (most frequent) writers. Table 2 shows the exact repartition of writers, letters, pages, text lines, and words across the different sets. The training set has a considerable size of 109,627 text lines with 876,003 words. After removing some very rare characters, we retain a total of 78 distinct characters in the database including the space character.

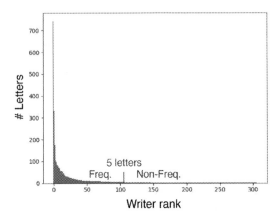

Fig. 4. Database setup. The graph shows the number of letters per writer. Frequent writers (Freq.) have five or more letters and non-frequent writers (Non-Freq.) have less than five letters.

Table 2. Database setup: Distribution of writers, letters, pages, text lines and words for frequent writers (Freq.) and non-frequent writers (Non-Freq.).

	Training	Validation	Test Freq.	Test Non-Freq.	Total
# of writers	106	106	106	200	306
# of letters	2,581	337	337	367	3,622
# of pages	5,927	806	787	873	8,393
# of lines	109,627	14,516	15,368	15,735	155,246
# of words	876,003	122,211	115,289	128,211	1,241,714

4.2   HTR Setup

The hyper-parameters of the HTR systems were optimized on the validation set during preliminary experiments. They have been fixed to the same values for HTR-Flor and PyLaia. The text line images are resized to a height of 128 pixels, keeping the aspect ratio, and in addition to the three RGB channels we add a fourth channel with a binary version of the image obtained by means of a global Otsu threshold. We consider 256 hidden units for the LSTMs/GRUs, a dropout of 0.3 in the recurrent layers, and a mini-batch size of 64. The learning rate is optimized with AdamW [14] using a weight decay of 0.0001, $\beta_1 = 0.9$, and $\beta_2 = 0.98$. The peak learning rate is 0.00055. The HTR systems are trained for 100 epochs until convergence and the best model epoch is chosen with respect to the character error rate on the validation set. We do not observe a significant overfitting effect. Training one epoch with two NVIDIA TITAN RTX cards took around 23 min for HTR-Flor and around 30 min for PyLaia.

For the CTC, we use 78 character tokens, decode greedily by considering the character with maximum probability at each time step, and remove repeated consecutive characters. Note that we do not use a lexicon for text recognition, because of the presence of both Latin and German texts, spelling variants, and abbreviations. Instead, the text is transcribed character by character. In the present baseline experiments, we only focus on the optical model and the implicit language model learned by the recurrent layers on the training set. No explicit character language model is used.

4.3 Writer Adaptation Setup

For writer adaptation, we first train a generic system on the entire training set. Afterwards, the system is adapted as follows:

– **Frequent writers:** The training letters of the writer are used to further fine-tune the generic system. Either the training is **continued** with the same, small learning rate or the training is **restarted** with the initial, high learning rate. Either 10 or 20 epochs of training are pursued and the best number of epochs is determined on the validation letters of the writer.
– **Non-frequent writers:** Self-training is performed by recognizing the test letters of the writer with the generic system. Afterwards, the automatic transcriptions are sorted by recognition confidence (see Sect. 3.2) and the top 50%, 75%, or 100% of the text lines are used as learning samples to further fine-tune the generic system. Training is continued for either 10 or 20 epochs (since the non-frequent writers have no annotated validation letters, it is not possible to determine the best epoch prior to 10 or 20).

4.4 Evaluation Measures

We use the standard measures of character error rate (CER) and word error rate (WER) to evaluate the HTR performance. They are calculated by computing the string edit distance between the recognition output and the ground truth, to obtain the number of substitution, deletion, and insertion errors. By dividing the number of character errors with the number of characters in the ground truth, we obtain the CER, and similarly the WER.

For measuring the impact of writer adaptation, we report absolute improvements, e.g. $CER_g - CER_a$, as well as relative improvements in percentage, e.g.

$$100 \cdot \frac{CER_g - CER_a}{CER_g} , \qquad (2)$$

with CER_g the error rate of the generic system and CER_a the (typically lower) error rate of the adapted system.

Table 3. HTR performance. The best results are highlighted in bold.

	Frequent writers		Non-Frequent writers	
	HTR-Flor	PyLaia	HTR-Flor	PyLaia
CER	9.56	**8.36**	10.67	**9.85**
WER	33.72	**29.64**	37.56	**34.39**

Table 4. Frequent writer adaptation. Relative improvement of the CER in percentage for continuing training during 10 or 20 epochs (C-10 and C-20) and for restarting training during 10 or 20 epochs (R-10 and R-20). The best results of each HTR system are highlighted in bold.

Frequent Writers		
Configuration	HTR-Flor	PyLaia
C-10	9.08	**3.70**
C-20	**9.76**	3.39
R-10	4.30	−0.05
R-20	6.80	1.56

4.5   Results

HTR Performance. Table 3 shows the CER of the generic (non-adapted) HTR systems for frequent and non-frequent writers, respectively. The best results are obtained with PyLaia, which achieves 8.36% CER for frequent writers and 9.85% CER for non-frequent writers. HTR-Flor performs about one percent CER worse, which may be due to the reduced number of model parameters when compared with PyLaia, taking into account the large size of the training set. For both systems, the error rate for non-frequent writers is significantly higher, which is expected because the writing styles of the non-frequent writers are not present in the training set. When comparing the overall HTR performance with the results for HTR-Flor on the IAM database reported in [20], namely 3.98% CER, the increased difficulty of the Bullinger database becomes evident.

Frequent Writer Adaptation. Table 4 shows the results of frequent writer adaptation for different fine-tuning strategies, in terms of relative improvements of the CER. For both HTR systems, restarting with a high learning rate is significantly worse than continuing the fine-tuning with a low learning rate. In the case of PyLaia, restarting 10 epochs even leads to an increase in the CER. The largest gain is observed for HTR-Flor, where the relative reduction of the CER is 9.76%.

Non-frequent Writer Adaptation. Table 5 shows the results of non-frequent writer adaptation for different self-training and fine-tuning strategies. The gain

Table 5. Non-frequent writer adaptation. Relative improvement of the CER in percentage for continuing training during 10 or 20 epochs (C-10 and C-20) and selecting the 50%, 75%, and 100% most confidently recognized text lines for self-training (S-50%, S-75%, and S-100%). The best results of each HTR system are highlighted in bold.

Non-Frequent Writers		
Configuration	HTR-Flor	PyLaia
C-10, S-50%	2.65	0.44
C-10, S-75%	2.68	0.51
C-10, S-100%	2.01	**0.71**
C-20, S-50%	**2.88**	−1.84
C-20, S-75%	2.54	−2.99
C-20, S-100%	0.79	−4.42

in performance is very limited for PyLaia, which clearly overfits to the self-labeled transcriptions when fine-tuning 20 epochs. The best results are achieved when selecting all self-labeled transcriptions and fine-tuning 10 epochs. HTR-Flor achieves the best result when selecting the 50% most confident text lines and fine-tuning 20 epochs. In this scenario, the relative improvement of the CER is 2.88%.

Table 6. Detailed adaptation results for the optimal system configurations. CER, WER, and improvements in percentage. The improvements for frequent writers and the improvements of HTR-Flor for non-frequent writers are significant ($p < 0.05$). The improvements of PyLaia for non-frequent writers are not significant.

	Frequent writers		Non-Frequent writers	
	HTR-Flor	PyLaia	HTR-Flor	PyLaia
Configuration	C-20	C-10	C-20, S-50%	C-10, S-100%
CER before	9.56	**8.36**	10.67	9.85
CER after	8.62	**8.05**	10.36	9.78
Improvement	**0.93**	0.31	0.31	0.07
Relative improvement	**9.76**	3.70	2.88	0.71
# writers improved	**97/106**	84/106	133/200	93/200
% writers improved	**91.51**	79.25	66.50	46.50
WER before	33.72	**29.64**	37.56	34.39
WER after	30.96	**28.72**	36.75	33.99
Improvement	**2.75**	0.93	0.81	0.39
Relative improvement	**8.17**	3.12	2.16	1.15
# writers improved	**96/106**	82/106	129/200	95/200
% writers improved	**90.57**	77.36	64.50	47.50

Detailed Adaptation Results. Table 6 provides a more detailed account for both frequent and non-frequent writer adaptation using the best fine-tuning and self-training strategies. Besides the improvements and relative improvements in CER and WER, we also indicate for how many writers the performance was improved. The improvements for frequent writers and the improvements of HTR-Flor for non-frequent writers are significant (p < 0.05). The improvements of PyLaia for non-frequent writers are not significant.

Overall, the adaptation results highlight a clear adaptation success for the frequent writers but only a limited success for the non-frequent writers. Even with the best fine-tuning and self-training configurations, the relative improvements in CER and WER remain very modest for the non-frequent writers.

5 Conclusion

The Bullinger dataset for writer adaptation introduced in this paper is a novel benchmark for developing and comparing writer adaptation methods. Its difficult handwriting, high variability in writing styles, and large size make it ideally suited for investigating writer adaptation with deep learning models from the current state of the art. The baseline results provided for the HTR-Flor and PyLaia architectures achieve up to 9.76% relative improvement of the CER for supervised adaptation of frequent writers, but only up to 2.88% relative improvement of the CER for self-supervised adaptation of non-frequent writers.

Promising lines of research to improve over these baseline results include a conjoint adaptation of optical models and explicit language models, writing style clustering, data augmentation, and synthetic data generation, to name just a few.

The Bullinger project is still ongoing and we expect to be able to release more versions of the challenge in the future, increasing the size of the database and improving the quality of the ground truth.

Acknowledgements. This work has been supported by the Hasler Foundation, Switzerland. We would also like to thank Anna Janka, Raphael Müller, Peter Rechsteiner, Dr. Patricia Scheurer, David Selim Schoch, Dr. Raphael Schwitter, Christian Sieber, Martin Spoto, Jonas Widmer, and Dr. Beat Wolf for their contributions to the dataset and ground truth creation.

References

1. Bullinger Dataset for Writer Adaptation. https://tc11.cvc.uab.es/datasets/BullingerDB_1. Accessed 30 Apr 2023
2. Bullinger Digital project. https://www.bullinger-digital.ch. Accessed 30 Apr 2023
3. Fink, G.A., Plötz, T.: Unsupervised estimation of writing style models for improved unconstrained off-line handwriting recognition. In: 10th International Workshop on Frontiers in Handwriting Recognition (IWFHR), pp. 1–6 (2006)

4. Fischer, A., Liwicki, M., Ingold, R. (eds.): Handwritten Historical Document Analysis, Recognition, and Retrieval - State of the Art and Future Trends. World Scientific (2020)
5. Fischer, A., Keller, A., Frinken, V., Bunke, H.: Lexicon-free handwritten word spotting using character HMMs. Pattern Recogn. Lett. **33**(7), 934–942 (2012)
6. Frinken, V., Bunke, H.: Evaluating retraining rules for semi-supervised learning in neural network based cursive word recognition. In: Proceedings of 10th International Conference on Document Analysis and Recognition (ICDAR), pp. 31–35 (2009)
7. Frinken, V., Fischer, A., Bunke, H., Manmatha, R.: Adapting BLSTM neural network based keyword spotting trained on modern data to historical documents. In: Proceedings of 12th International Conference on Frontiers in Handwriting Recognition (ICFHR), pp. 352–357 (2010)
8. Granet, A., Morin, E., Mouchère, H., Quiniou, S., Viard-Gaudin, C.: Transfer learning for handwriting recognition on historical documents. In: Proceedings of 7th International Conference on Pattern Recognition Applications and Methods (ICPRAM), pp. 1–8 (2018)
9. Grosicki, E., Carre, M., Brodin, J.M., Geoffrois, E.: Rimes evaluation campaign for handwritten mail processing. In: 11th International Conference on Frontiers in Handwriting Recognition (ICFHR), pp. 1–6 (2008)
10. Jaramillo, J.C.A., Murillo-Fuentes, J.J., Olmos, P.M.: Boosting handwriting text recognition in small databases with transfer learning. In: Proceedings 16th International Conference on Frontiers in Handwriting Recognition (ICFHR), pp. 429–434 (2018)
11. Kang, L., Rusinol, M., Fornes, A., Riba, P., Villegas, M.: Unsupervised writer adaptation for synthetic-to-real handwritten word recognition. In: Proceedings of the IEEE/CVF Winter Conference on Applications of Computer Vision (WACV), pp. 3502–3511 (2020)
12. Kleber, F., Fiel, S., Diem, M., Sablatnig, R.: CVL-database: an off-line database for writer retrieval, writer identification and word spotting. In: Proceedings 12th International Conference on Document Analysis and Recognition (ICDAR), pp. 560–564 (2013)
13. Lavrenko, V., Rath, T.M., Manmatha, R.: Holistic word recognition for handwritten historical documents. In: Proceedings of 1st International Workshop on Document Image Analysis for Libraries, pp. 278–287 (2004)
14. Loshchilov, I., Hutter, F.: Decoupled weight decay regularization. In: 7th International Conference on Learning Representations (ICLR), pp. 1–18 (2019)
15. Marti, U.V., Bunke, H.: The IAM-database: an English sentence database for offline handwriting recognition. Int. J. Doc. Anal. Recogn. **5**, 39–46 (2002)
16. Muehlberger, G., Seaward, L., Terras, M., et al.: Transforming scholarship in the archives through handwritten text recognition: Transkribus as a case study. J. Documentation **75**(5), 954–976 (2019)
17. Nosary, A., Heutte, L., Paquet, T.: Unsupervised writer adaptation applied to handwritten text recognition. Pattern Recogn. **37**(2), 385–388 (2004)
18. Puigcerver, J.: Are multidimensional recurrent layers really necessary for handwritten text recognition? In: Proceedings of 14th International Conference on Document Analysis and Recognition (ICDAR). vol. 1, pp. 67–72 (2017)
19. Soullard, Y., Swaileh, W., Tranouez, P., Paquet, T., Chatelain, C.: Improving text recognition using optical and language model writer adaptation. In: Proceedings of 15th International Conference on Document Analysis and Recognition (ICDAR), pp. 1175–1180 (2019)

20. de Sousa Neto, A.F., Bezerra, B.L.D., Toselli, A.H., Lima, E.B.: HTR-Flor: a deep learning system for offline handwritten text recognition. In: Proceedings of 33rd SIBGRAPI Conference on Graphics, Patterns and Images (SIBGRAPI), pp. 54–61 (2020)
21. Strauß, T., Leifert, G., Labahn, R., Hodel, T., Mühlberger, G.: ICFHR 2018 competition on automated text recognition on a read dataset. In: Proceedings of 16th International Conference on Frontiers in Handwriting Recognition (ICFHR), pp. 477–482 (2018)
22. Yang, H.M., Zhang, X.Y., Yin, F., Sun, J., Liu, C.L.: Deep transfer mapping for unsupervised writer adaptation. In: Proceedings of 16th International Conference on Frontiers in Handwriting Recognition (ICFHR), pp. 151–156 (2018)

Towards Writer Retrieval for Historical Datasets

Marco Peer[(✉)] [iD], Florian Kleber[iD], and Robert Sablatnig[iD]

Computer Vision Lab TU Wien, Vienna, Austria
{mpeer,kleber,sab}@cvl.tuwien.ac.at
https://github.com/marco-peer/icdar23

Abstract. This paper presents an unsupervised approach for writer retrieval based on clustering SIFT descriptors detected at keypoint locations resulting in pseudo-cluster labels. With those cluster labels, a residual network followed by our proposed NetRVLAD, an encoding layer with reduced complexity compared to NetVLAD, is trained on 32×32 patches at keypoint locations. Additionally, we suggest a graph-based reranking algorithm called SGR to exploit similarities of the page embeddings to boost the retrieval performance. Our approach is evaluated on two historical datasets (Historical-WI and HisIR19). We include an evaluation of different backbones and NetRVLAD. It competes with related work on historical datasets without using explicit encodings. We set a new State-of-the-art on both datasets by applying our reranking scheme and show that our approach achieves comparable performance on a modern dataset as well.

Keywords: Writer Retrieval · NetVLAD · Reranking · Document Analysis

1 Introduction

Writer retrieval is the task of retrieving documents written by the same author within a dataset by finding similarities in the handwriting [14]. In particular, writer retrieval enables experts in history or paleography to trace individuals or social groups across different time epochs [7]. Furthermore, it helps to identify documents of unknown writers and to detect similarities within those documents [5]. Due to the time-consuming process of analyzing large corpora of documents required by experts, image retrieval algorithms are applied to find all relevant documents of a specific writer.

State-of-the-art methods for writer retrieval consist of four parts: First, characteristics of the handwriting within the document are sampled, e.g., by using interest point detectors such as SIFT [14,20,22]. Then, traditional algorithms or deep-learning-based approaches are applied to extract features. In the end, those embeddings are encoded and aggregated to obtain powerful global page descriptors, which are then compared to retrieve a ranked list for each query document.

© The Author(s), under exclusive license to Springer Nature Switzerland AG 2023
G. A. Fink et al. (Eds.): ICDAR 2023, LNCS 14187, pp. 411–427, 2023.
https://doi.org/10.1007/978-3-031-41676-7_24

412   M. Peer et al.

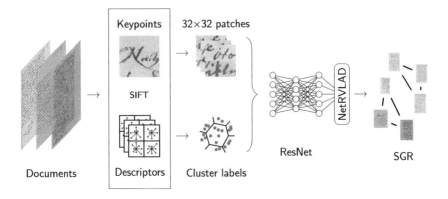

Fig. 1. Overview of our proposed pipeline.

Since the datasets contain a training and a test set with disjunct writers, the performance of writer retrieval approaches is evaluated by using each document of the test set as a query once.

While on modern datasets, neural networks trained in a supervised manner dominate [10,14,20,22], for historical datasets, training on writer label information [19,25] trails either unsupervised methods [5] or approaches based on handcrafted features [16]. Historical data introduces additional challenges, e.g., degradation, different languages, the amount of text, or even potential writer-label noise by external influences on handwriting, such as the pen used. However, a different strategy we investigate to improve the performance of writer retrieval is *reranking*: After the global descriptors are calculated and compared, reranking exploits the geometric relationships in the embedding space, as well as the information included in the ranked list to refine the final ranking [13].

Our paper presents an unsupervised approach illustrated in Fig. 1. It is based on a Convolutional Neural Network (CNN) trained on 32×32 patches extracted at SIFT keypoint locations. As a target label, 5000 classes are generated by clustering the corresponding descriptors via k-means [5]. We encode the embeddings of our neural network by Random NetVLAD (NetRVLAD), particularly designed for writer retrieval by removing normalization layers and the initialization, which, we show in our evaluation, harms the performance. In contrast to [5], we do not rely on external codebooks such as VLAD. Instead, we directly learn a codebook during the network training within the NetRVLAD layer. The global page descriptors are obtained by sum pooling. Secondly, we rerank our global page descriptors with our proposed Similarity Graph Reranking (SGR) and boost the performance of NetRVLAD. Our reranking is based on the work of Zhang et al. [27], who build a graph and aggregate its vertices to refine the features. Their method relies on two hyperparameters (k_1, k_2) dependent on the test set whose properties are usually unknown. We propose SGR, where an initial graph of the global page descriptors is built using cosine similarity and a weighting function. Afterward, a graph network refines and aggregates the node features, which are then considered as the reranked descriptors. SGR improves the work in [27] by eliminating k_1 and is also robust to the choice of k_2 across datasets. Additionally, our results show that when using NetRVLAD, the performance is significantly improved by removing complexity compared to the original NetVLAD. In our experiments, NetRVLAD performs stable, even when choosing a smaller codebook size, reducing computational resources. Combined with SGR, we outperform related work. Ultimately, we show that our approach is feasible for smaller modern datasets.

Summarizing, our contributions are:

– NetRVLAD, an encoding layer for writer retrieval based on NetVLAD [1],
– SGR, a reranking algorithm using a similarity graph,
– a thorough evaluation of our approach on two historical datasets, namely the Historical-WI [9], and HisIR19 [7], where we outperform State-of-the-art on both datasets.

The remaining part of our paper is structured as follows: In Sect. 2, we describe the related work regarding writer retrieval and reranking strategies used. We cover our approach, including NetRVLAD and SGR in Sect. 3 followed by our evaluation protocol and implementation details in Sect. 4. Our experiments and results are given in Sect. 5. We conclude our paper in Sect. 6.

2   Related Work

In the following, we give an overview of related work for writer retrieval as well as reranking strategies.

2.1   Writer Retrieval

Writer retrieval approaches are divided into codebook-based and codebook-free methods. Those codebooks are used as a model to calculate statistics of the handwriting, with Vector of Locally Aggregated Descriptors (VLAD) the most prominent one for writer retrieval [3,5,6,14]. Additionally, the characteristics of the handwriting are either extracted by traditional algorithms (handcrafted features) or deep learning.

For codebook-based methods on modern datasets such as ICDAR2013 [17] or CVL [15], the authors of [4] compute SURF features encoded by Gaussian mixture models. Christlein et al. [3] extract Zernike moments of the contours and build a codebook based on multiple VLAD encodings. In contrast to those handcrafted features, Fiel and Sablatnig [10] introduced CNNs to the domain of writer retrieval. Their codebook-free method relies on aggregating CNN activations via sum-pooling. Similarly, CNNs are applied in [6,14] as a feature extractor followed by VLAD. The authors of [20,22] investigate NetVLAD [1], a learnable version of VLAD, plugged in at the end of the network to directly learn the codebook during training. All of those networks are trained in a supervised manner with the writer label as target.

For historical datasets, Christlein et al. [5] show that training on pseudolabels generated by clustering the SIFT descriptors outperforms supervised methods such as [25]. Furthermore, for each descriptor, an Exemplar-SVM (ESVM) is trained to refine the encoding. Peer et al. [19] apply a self-supervised algorithm using morphological operations to generate augmented views without any labels. The winners of the HisIR19 competition [7] and the current state-of-the-art method on the Historical-WI dataset rely on handcrafted features (SIFT and pathlet) for retrieval. They encode both features via *bagged VLAD* (bVLAD) [16]. Our approach is mainly inspired by the work in [5], but we train our network with triplets and additionally encode our embeddings based on NetVLAD.

2.2   Reranking

While reranking is a method to improve the performance of image retrieval in general, two approaches [13,22] investigate reranking in the domain of writer retrieval.

In [22], Rasoulzadeh and Babaali propose an adaption to the standard reranking method Query Expansion (QE) [8]. They average each descriptor with their top k Reciprocal Nearest Neighbor (kRNN) and show that they can boost the retrieval performance by reducing the effect of false matches. Jordan et al. [13] extend the ESVMs of Christlein et al. [5] as a baseline for their reranking evaluation. They consider additional positive samples for the training ESVMs called *Pair* or *Triple SVM* and increase the performance of [5].

Recent methods in image retrieval apply neural networks to refine the ranking, e.g., Tan et al. [23] suggest reranking transformers, and Gordo et al. propose attention-based query expansion learning with a contrastive loss [11].

Our approach is based on the work of Zhang et al. [27]. They build a graph with the k_1 nearest neighbors and aggregate the nodes of the k_2 nearest neighbors by using a graph network, arguing the generality of their approach, e.g., including approaches like α-QE [21]. We suggest using the similarity of the embeddings to create the initial graph, which removes the requirement for selecting an appropriate value for k_1.

3   Methodology

In this section, we describe each aspect of our approach and explain the two main parts we propose for writer retrieval: NetRVLAD and SGR.

3.1   Patch Extraction

Our preprocessing is based on the approach of Christlein et al. [5]. Firstly, we detect keypoints for each document as well as the corresponding descriptors, both via SIFT. These descriptors are normalized with the Hellinger kernel (elementwise square root followed by l_1-normalization) and dimensionality reduction via PCA from 128 to 32. We cluster the descriptors via k-means in 5000 clusters

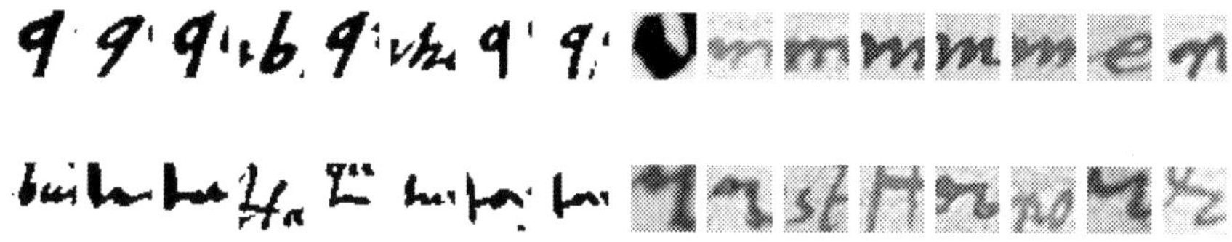

(a) Historical-WI   (b) HisIR19

Fig. 2. Examples of the clustered 32×32 patches.

[5]. As an additional preprocessing step, we filter keypoints whose descriptors **d** violate

$$\frac{||\mathbf{d} - \boldsymbol{\mu}_1||}{||\mathbf{d} - \boldsymbol{\mu}_2||} > \rho, \tag{1}$$

where $\boldsymbol{\mu}_i$ denotes the i-th nearest cluster of **d** and $\rho = 0.9$. By applying (1) we filter keypoints that lay near the border of two different clusters - those are therefore considered to be ambiguous. The 32×32 patch is extracted at the keypoint location, and the cluster membership is used as a label to train the neural network. In Fig. 2, we show eight samples of two clusters each for both datasets used. We observe clusters where characters written in a specific style dominate, e.g., 'q' or 'm' on top, and clusters containing general patterns included in the handwriting (bottom).

3.2 Network Architecture

Our network consists of two parts: a residual backbone and an encoding layer, for which we propose *NetRVLAD*. The output of NetRVLAD is used as a global descriptor of the 32×32 input patch.

Residual Backbone. Similar to [5,22], the first stage of our network is a ResNet to extract an embedding for each patch. The last fully connected layer of the network is dropped, and the output of the global averaging pooling layer of dimension $(64, 1, 1)$ is used. We evaluate the choice of the depth of the network in our results.

NetRVLAD. The traditional VLAD algorithm clusters a vocabulary to obtain N_c clusters $\{\mathbf{c}_0, \mathbf{c}_1, \ldots, \mathbf{c}_{N_c-1}\}$ and encodes a set of local descriptors $\mathbf{x}_i$, $i \in \{0, \ldots, N-1\}$ via

$$\mathbf{v}_k = \sum_{i=0}^{N-1} \mathbf{v}_{k,i} = \sum_{i=0}^{N-1} \alpha_k(\mathbf{x}_i)(\mathbf{x}_i - \mathbf{c}_k), \quad k \in \{0, \ldots, N_c - 1\}, \tag{2}$$

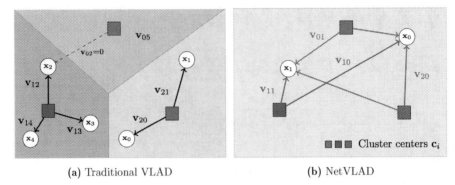

(a) Traditional VLAD   (b) NetVLAD

Fig. 3. Traditional VLAD and NetVLAD. While VLAD hard-assigns each descriptor x_i to its nearest cluster to compute the residual, NetVLAD directly learns 1) the cluster centers and 2) their assignments allowing to aggregate multiple residuals for one descriptor.

with $\alpha_k = 1$ if $\mathbf{c}_k$ is the nearest cluster center to $\mathbf{x}_i$, otherwise 0, hence making the VLAD encoding not differentiable. The final global descriptor is then obtained by concatenating the vectors $\mathbf{v}_k$. Arandjelović et al. [1] suggest the NetVLAD layer which tackles the non-differentiability of α_k in (2) by introducing a convolutional layer with parameters $\{\mathbf{w}_k, b_k\}$ for each cluster center $\mathbf{c}_k$ to learn a soft-assignment

$$\overline{\alpha}_k(\mathbf{x}_i) = \frac{e^{\mathbf{w}_k^T \mathbf{x}_i + b_k}}{\sum_{k'} e^{\mathbf{w}_{k'}^T \mathbf{x}_i + b_{k'}}}. \tag{3}$$

The cluster centers $\mathbf{c}_k$ are also learned during training. A schematic comparison is shown in Fig. 3. The input of NetVLAD is a feature map of dimension (D, H, W) handled as a $D \times N$ spatial descriptor with $N = HW$. Normalization and concatenation of the vectors $\mathbf{v}_k$

$$\mathbf{v}_k = \sum_{i=0}^{N} \mathbf{v}_{k,i} = \sum_{i=0}^{N} \overline{\alpha}_k(\mathbf{x}_i)(\mathbf{x}_i - \mathbf{c}_k), \quad k \in \{0, \dots, N_c - 1\} \tag{4}$$

yields the final NetVLAD encoding $\mathbf{V} \in \mathbb{R}^{N_c \times D}$.

For writer retrieval, the main idea of applying NetVLAD is learning a powerful codebook via its cluster centers, representing, e.g., features like characters or combinations of them or more high-level ones like slant directions of the handwriting. We generate a meaningful descriptor by concatenating the residuals between a patch embedding to the cluster centers. A page is then characterized by measuring differences between those features. In contrast to VLAD, the codebook is directly integrated into the network. For our approach, we reduce the complexity of NetVLAD and adapt two aspects which we call *NetRVLAD*:

1) Similar to RandomVLAD proposed by Weng et al. [26], we loosen the restriction of the embeddings $\mathbf{x}$ of the backbone as well as the cluster residuals $\mathbf{v}_k$ lying on a hypersphere. Since we only forward one descriptor per patch ($H = W = 1$), we argue that NetVLAD learns this during training on its own - therefore, we remove the pre- and intranormalization of NetVLAD.

2) Arandjelovic et al. [1] propose an initialization of the convolutional layer where the ratio of the two closest (maximum resp. second highest value of $\bar{a}_k$) cluster assignments is equal to $\alpha_{\text{init}} \approx 100$. To improve performance, we initialize the weights of the convolutional layer and the cluster centers randomly rather than using a specific initialization method, as this can increase the impact of the initialization of the cluster centers. Additionally, the hyperparameter α_{init} is removed. We compare NetRVLAD to the original implementation in Sect. 4.

3.3   Training

Our network is trained with the labels assigned while clustering the SIFT descriptors. Each patch is embedded in a flattened $N_c \times 64$ descriptor. We directly train the encoding space using the distance-based triplet loss

$$\mathcal{L}_{\text{Triplet}} = \max(0, d_{ap} - d_{an} + m), \tag{5}$$

with the margin m where a denotes the anchor, p the positive and n the negative sample. We only mine *hard* triplets [24] in each minibatch. Therefore, each triplet meets the criterion.

$$d_{an} < d_{ap} - m. \tag{6}$$

3.4   Global Page Descriptor

During inference, we aggregate all embeddings $\{\mathbf{V}_0, \mathbf{V}_1, \ldots, \mathbf{V}_{n_p-1}\}$ of a page using l_2 normalization followed by sum pooling

$$\mathbf{V} = \sum_{i=0}^{n_p-1} \mathbf{V}_i \tag{7}$$

to obtain the global page descriptor $\mathbf{V}$. Furthermore, to reduce visual burstiness [12], we apply power-normalization $f(x) = \text{sign}(x)|x|^\alpha$ with $\alpha = 0.4$, followed by l_2-normalization. Finally, a dimensionality reduction with whitening via PCA is performed.

3.5   Reranking with SGR

Writer retrieval is evaluated by a leave-one-out strategy: Each image of the set is once used as a query q, the remaining documents are called the gallery. For each q, the retrieval returns a ranked list of documents $L(q)$. Reranking strategies exploit the knowledge contained in $L(p_i)$ with $p_i \in L(q)$ and refine the descriptors

[13]. We can intuitively model those relationships by a graph $\mathcal{G} = (\mathcal{V}, \mathcal{E})$ with its vertices $\mathcal{V}$ and edges $\mathcal{E}$.

Our approach called SGR is conceptually simple and consists of two stages inspired by the work in [27]. The first stage is building the initial graph using the page descriptors to compute the vertices. Instead of only considering k nearest neighbours as described by Zhang et al. [27], we propose to use the cosine similarity $s_{i,j} = \mathbf{x}_i^{\mathrm{T}} \cdot \mathbf{x}_j$ and obtain the symmetric adjacency matrix by

$$\mathbf{A}_{i,j} = \exp\left(-\frac{(1 - s_{i,j})^2}{\gamma}\right) \tag{8}$$

with a hyperparameter γ which mainly determines the decay of edge weights when similarity decreases. Therefore, our approach additionally benefits from a continuous adjacency matrix by using the learned embedding space. We consider similarities while replacing the task-dependent hyperparameter k_1 in [27].

Furthermore, we compute the vertices by encoding the similarity of each descriptor instead of adopting the original page descriptors: The rows of the adjacency matrix $\mathbf{A}$ - we denote the i-th row as $\mathbf{h}_i$ in the following - are used as page descriptors which we refer to as a *similarity graph*. While Zhang et al. [27] propose a discrete reranked embedding space ($\mathbf{A}_{i,j} \in \{0, \frac{1}{2}, 1\}$), we argue that a continuous embedding space further improves the reranking process by using our weighting function to refine the embeddings. Thus, we are able to exploit not only the neighborhood of a page descriptor, but also its distances.

Secondly, each vertex is propagated through a graph network consisting of L layers via

$$\mathbf{h}_i^{(l+1)} = \mathbf{h}_i^{(l)} + \sum_j s_{i,j}\, \mathbf{h}_j^{(l)}, \quad j \in \mathcal{N}(i,k), \quad l \in \{1, \dots, L\}, \tag{9}$$

where $\mathcal{N}(i,k)$ denotes the k nearest neighbors of vertex i. Those neighbors are aggregated with their initial similarity $s_{i,j}$. During message propagation, we only consider the k (equal to k_2 in [27]) nearest neighbors to reduce the noise of aggregating wrong matches (k is usually small, e.g., $k = 2$) and also eliminate the influence of small weight values introduced by (8). The vertices are l_2-normalized after each layer. $\mathbf{h}_i^{(L)}$ is used as the final reranked page descriptor. In our evaluation, we report the performances of SGR, as well as the initial approach by Zhang et al. [27], and provide a study on the hyperparameters of our reranking.

4   Evaluation Protocol

In this section, we cover the datasets and metrics used and give details about our implementation.

4.1   Datasets

We use two historical datasets with their details stated in the following. In Fig. 4, examples of the two datasets used are shown.

Historical-WI. This dataset proposed by Fiel et al. [9] at the *ICDAR 2017 Competition on Historical Document Writer Identification* consists of 720 authors where each one contributed five pages, resulting in a total of 3600 pages. Originating from the 13th to 20th century, the dataset contains multiple languages such as German, Latin, and French. The training set includes 1182 document images written by 394 writers with an equal distribution of three pages per writer. Both sets are available as binarized and color images. To ensure a fair comparison, we follow related work and report our results on the binarized version of the dataset.

HisIR19. Introduced at the *ICDAR 2019 Competition on Image Retrieval for Historical Handwritten Documents* by Christlein et al. [7], the test set consists of 20 000 documents of different sources (books, letters, charters, and legal documents). 7500 pages are isolated (one page per author), and the remaining authors contributed either three or five pages. The training set recommended by the authors of [7] and used in this paper is the validation set of the competition, including 1200 images of 520 authors. The images are available in color.

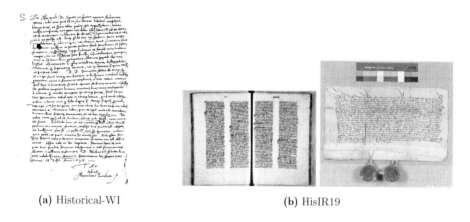

(a) Historical-WI   (b) HisIR19

Fig. 4. Example images of the datasets used.

4.2 Metrics

To evaluate performance, we use a leave-one-out retrieval method where each document is used as a query and a ranked list of the remaining documents is returned. The similarity is measured by using cosine distance between global page descriptors.

Our results are reported on two metrics. Mean Average Precision (mAP) and Top-1 accuracy. While mAP considers the complete ranked list by calculating the mean of the average precisions, Top-1 accuracy measures if the same author writes the nearest document within the set.

4.3   Implementation Details

Patch Extraction and Label Generation. For preprocessing, we rely on the experiments of Christlein et al. [5] and use 32×32 patches clustered into 5000 classes. We only use patches with more than 5% black pixels for binary images. To filter the patches of color images, a canny edge detector is applied, and only patches with more than 10% edge pixels are taken [25]. This value is chosen since the HisIR19 dataset contains multiple sources of noise (book covers, degradation of the page, or color palettes) we consider irrelevant for writer retrieval. To decrease the total number of patches of the test sets, we limit the number of patches on a single page to 2000.

Training. We train each network for a maximum of 30 epochs with a batch size of 1024, a learning rate of $l_r = 10^{-4}$ and a margin $m = 0.1$ for the triplet loss. Each batch contains 16 patches per class. 10% of the training set are used as the validation set. We stop training if the mAP on the validation set does not increase for five epochs. Optimization is done with Adam and five warmup epochs during which the learning rate is linearly increased from $l_r/10$ to l_r. Afterward, a cosine annealing is applied. As data augmentation, we apply erosion and dilation. All of our results on the trained networks are averages of three runs with the same hyperparameters but different seeds to reduce the effect of outliers due to initialization or validation split. If not stated otherwise, our default network is ResNet56 with $N_c = 100$.

Retrieval and Reranking. For aggregation, the global page descriptor is projected into a lower dimensional space (performance peaks at 512 for Historical-WI, 1024 for HisIR19) via a PCA with whitening followed power-normalization ($\alpha = 0.4$) and a l_2-normalization. For experiments in which the embedding dimension is smaller than 512, only whitening is applied.

5   Experiments

We evaluate each part of our approach in this section separately, starting with NetRVLAD and its settings, followed by a thorough study of SGR. In the end, we compare our results to state-of-the-art methods on both datasets.

5.1   NetRVLAD

Firstly, we evaluate the backbone of our approach. We choose four residual networks of different depth, starting with ResNet20 as in related work [5,22] up to ResNet110, and compare the performance of NetVLAD to our proposed NetRVLAD. As shown in Table 1, NetRVLAD consistently outperforms the original NetVLAD implementation in all experiments. Secondly, we observe deeper networks to achieve higher performances, although, on the Historical-WI dataset, the gain saturates for ResNet110. ResNet56 with our NetRVLAD layer is used for further experiments as a tradeoff architecture between performance and computational resources.

Table 1. Comparison of NetVLAD and NetRVLAD on different ResNet architectures with $N_c = 100$. Each result is an average of three runs with different seeds.

	Historical-WI				HisIR19			
	NetRVLAD		NetVLAD		NetRVLAD		NetVLAD	
	mAP	Top-1	mAP	Top-1	mAP	Top-1	mAP	Top-1
ResNet20	71.5	87.6	67.4	85.3	90.1	95.4	89.4	94.5
ResNet32	72.1	88.2	67.9	85.5	90.6	95.7	89.6	94.9
ResNet56	**73.1**	**88.3**	68.3	85.8	91.2	96.0	90.2	95.3
ResNet110	**73.1**	**88.3**	68.9	86.2	**91.6**	**96.1**	89.9	95.5

Cluster Centers of NetRVLAD. We study the influence of the size N_c of the codebook learned during training. In related work [20,22], the vocabulary size is estimated considering the total amount of writers included in the training set. However, this does not apply to our unsupervised approach. In Fig. 5, we report the performance of NetRVLAD while varying N_c. We report a maximum in terms of mAP when using a codebook size of 128 resp. 256 on Historical-WI and HisIR19. In general, a smaller codebook works better on Historical-WI; we think this is caused by a) HisIR19 is a larger dataset and b) it introduces additional content, e.g. book covers or color palettes as shown in Fig. 4 enabling a better encoding by learning more visual words. For HisIR19, performance is relatively stable over the range we evaluate. Since it also contains noise like degradation or parts of book covers, NetRVLAD seems to benefit when training with more cluster centers. It is also robust - with a small codebook ($N_c = 8$), the drop is only –3.6% resp. –1.9% compared to the peak performance.

5.2 Reranking

Once the global descriptors are extracted, we apply SGR to improve the performance by exploiting relations in the embedding space by building our similarity graph and aggregating its vertices. SGR relies on three hyperparameters: the k nearest neighbors which are aggregated, the number of layers L of the graph network, and γ, the similarity decay of the edge weights. While L and γ are parameters of the general approach and are validated on the corresponding training set, k is dependent on two aspects:

1. The performance of the retrieval on the baseline descriptors - if the top-ranked samples are false, the relevant information within the ranked list is either noise or not considered during reranking.
2. The gallery size n_G - the number of samples written by an author, either a constant or varies within the dataset.

We evaluate SGR by first validating L and γ and then studying the influence of k on the test set.

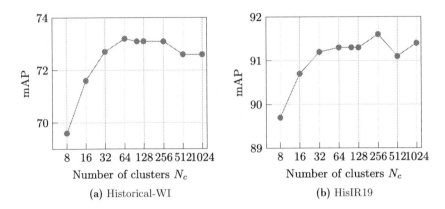

Fig. 5. Influence of N_c on the performance in terms of mAP on the Historical-WI and HisIR19 test dataset.

Hyperparameter Evaluation. For choosing L and γ, we perform a grid search on the global descriptors of the training set on both datasets where $\gamma \in [0.1, 1]$, $L \in \{1, 2, 3\}$. We fix $k = 1$ to concentrate on the influence of γ and L by prioritizing aggregating correct matches ($n_G = 3$ for the training set of Historical-WI and $n_G \in \{1, 3, 5\}$ for the training set of HisIR19). The results on both sets are shown in Fig. 6. Regarding γ, values up to 0.5 improve the baseline performance. Afterward, the mAP rapidly drops on both datasets - large values of γ also flatten the peaks in the similarity matrix. The influence of the number of layers is smaller when only considering $\gamma \leq 0.5$. However, the best mAP is achieved with $L = 1$. Therefore, for the evaluation of the test sets, we choose $\gamma = 0.4$ and $L = 1$.

Reranking Results. Finally, we report our results for different values of k on the test set as illustrated in Fig. 7. The gallery sizes are $n_G = 5$ for Historical-WI and

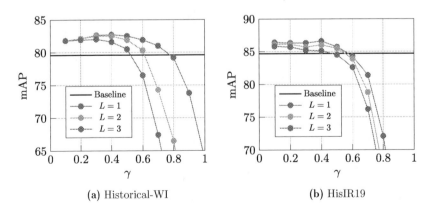

Fig. 6. Hyperparameter evaluation of SGR on the training sets with $k = 1$.

$n_G \in \{1, 3, 5\}$ for HisIR19. SGR boosts mAP and Top-1 accuracy, in particular the mAP when choosing small values for k. For the two datasets with different gallery sizes, the best mAP is obtained for $k = 2$, for which we achieve 80.6% and 93.2% on Historical-WI resp. HisIR19. Afterward, the mAP drops on the HisIR19 dataset - we think this is mainly due to the large number of authors contributing only a single document which may be reranked when considering too many neighbors. Interestingly, the Top-1 accuracy even increases for larger values peaking for both datasets at $k = 4$ with 92.8% and 97.3%.

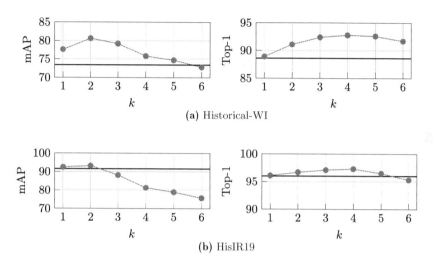

(a) Historical-WI

(b) HisIR19

Fig. 7. Reranking results of SGR for both datasets. Horizontal lines mark the baseline performance of NetRVLAD. ($\gamma = 0.4$, $L = 1$)

5.3   Comparison to State-of-the-art

We compare our approach concerning two aspects: the performance of the baseline (NetRVLAD) and the reranked descriptors (SGR). Our baseline is combined with the graph reranking method in [27] as well as the kRNN-QE proposed in [22], which is mainly designed for writer retrieval. For both methods [22,27], we perform a grid search and report the results of the best hyperparameters to ensure a fair comparison.

Our feature extraction is similar to Christlein et al. [5] and Chammas et al. [2] in terms of preprocessing and training. In contrast, the method proposed in [16] relies on handcrafted features encoded by multiple VLAD codebooks.

For the Historical-WI dataset, NetRVLAD achieves a mAP of 73.4% and, according to Table 2, our global descriptors are less effective compared to the work of [5]. Regarding reranking, SGR outperforms the reranking methods proposed by Jordan et al. [13], who use a stronger baseline with the mVLAD app-

Table 2. Comparison of state-of-the-art methods on Historical-WI. (*) denotes our implementation of the reranking algorithm, (+) reranking applied on the baseline method.

	mAP	Top-1
CNN+mVLAD [5]	74.8	88.6
Pathlet+SIFT+bVLAD [16]	77.1	90.1
CNN+mVLAD+ESVM [5]	76.2	88.9
+ Pair/Triple SVM [13]	78.2	89.4
NetRVLAD (ours)	73.4	88.5
+ kRNN-QE* $_{k=3}$ [22]	77.1	86.8
+ Graph reranking* $_{k_1=4,\ k_2=2,\ L=3}$ [27]	77.6	87.4
+ SGR $_{k=2}$ (ours)	**80.6**	**91.1**

roach of [5]. Additionally, SGR performs better than the graph reranking approach [27] our method is based on. When using SGR, our approach sets a new State-of-the-art performance with a mAP of 80.6% and a Top-1 accuracy of 91.1%. Compared to the other reranking methods, SGR is the only method that improves the Top-1 accuracy.

Table 3. Comparison of state-of-the-art methods on HisIR19. (*) denotes our implementation of the reranking algorithm, (+) reranking applied on the baseline method.

	mAP	Top-1
CNN+mVLAD [2]	91.2	97.0
Pathlet+SIFT+bVLAD [16]	92.5	**97.4**
NetRVLAD (ours)	91.6	96.1
+ kRNN-QE* $_{k=4}$ [22]	92.6	95.2
+ Graph reranking* $_{k_1=4,\ k_2=2,\ L=2}$ [27]	93.0	95.7
+ SGR $_{k=2}$ (ours)	**93.2**	96.7

Regarding the performance on the HisIR19 dataset shown in Table 3, NetRVLAD achieves a mAP of 91.6% and therefore slightly beats the traditional mVLAD method in [2]. SGR is better than the reranking methods proposed in [22,27] with a mAP of 93.2%, a new State-of-the-art performance. However, even with reranking, the Top-1 accuracy of NetRVLAD+SGR trails the VLAD methods in [2,16]. The improvements of SGR are smaller than on the Historical-WI dataset given that the baseline performance is already quite strong with over 90%, increasing the difficulty of the reranking process.

ICDAR2013. Finally, to show the versatility of our unsupervised method, we report the performance on the ICDAR2013 dataset [18], a modern dataset with

250/1000 pages including two English and two Greek texts per writer with only four lines of text each. Although we are limited to less data compared to historical datasets with a large amount of text included in a page, our approach achieves a notable performance (86.1% mAP), in particular Top-1 accuracy (98.5%), where it outperforms the supervised approach [22] as shown in Table 4.

Table 4. Comparison of state-of-the-art methods on ICDAR2013.

	mAP	Top-1
Zernike+mVLAD [3]	88.0	**99.4**
NetVLAD+kRNN-QE (supervised) [22]	**97.4**	97.4
NetRVLAD+SGR $_{k=1}$ (ours)	86.1	98.5

6 Conclusion

This paper introduced an unsupervised approach for writer retrieval. We proposed NetRVLAD to directly train the encoding space with 32×32 patches on labels obtained by clustering their SIFT descriptors. In our experiments, we showed that NetRVLAD outperforms the traditional implementation while also being relatively robust to the codebook's size and backbone architecture. Furthermore, our graph reranking method SGR was used to boost the retrieval performance. SGR outperformed the original graph reranking and reranking methods recently applied in the domain of writer retrieval. Additionally, we beat the State-of-the-art with our reranking scheme and showed the performance on a modern dataset.

Regarding future work, we think our approach is mainly limited due to the cluster labels used for training. We could overcome this by unlocking the potential of self-supervised methods and train the encoding space without any labels. Other approaches could include learnable poolings, e.g., instead of sum pooling to calculate the global page descriptors, a neural network invariant to permutation could be trained on the patch embeddings to learn a powerful aggregation. Finally, investigating learning-based reranking methods [11,23] are a considerable choice for further improving retrieval performance.

Acknowledgments. The project has been funded by the Austrian security research programme KIRAS of the Federal Ministry of Finance (BMF) under the Grant Agreement 879687.

References

1. Arandjelovic, R., Gronát, P., Torii, A., Pajdla, T., Sivic, J.: NetVLAD: CNN architecture for weakly supervised place recognition. In: 2016 IEEE Conference on Computer Vision and Pattern Recognition, CVPR 2016, Las Vegas, NV, USA, June 27–30, 2016, pp. 5297–5307 (2016)

2. Chammas, M., Makhoul, A., Demerjian, J.: Writer identification for historical handwritten documents using a single feature extraction method. In: 19th IEEE International Conference on Machine Learning and Applications, ICMLA 2020, Miami, FL, USA, December 14–17, 2020, pp. 1–6 (2020)
3. Christlein, V., Bernecker, D., Angelopoulou, E.: Writer identification using VLAD encoded contour-zernike moments. In: 13th International Conference on Document Analysis and Recognition, ICDAR 2015, Nancy, France, August 23–26, 2015, pp. 906–910 (2015)
4. Christlein, V., Bernecker, D., Hönig, F., Maier, A.K., Angelopoulou, E.: Writer identification using GMM supervectors and exemplar-SVMs. Pattern Recognit. **63**, 258–267 (2017)
5. Christlein, V., Gropp, M., Fiel, S., Maier, A.K.: Unsupervised feature learning for writer identification and writer retrieval. In: 14th IAPR International Conference on Document Analysis and Recognition, ICDAR 2017, Kyoto, Japan, November 9–15, 2017, pp. 991–997 (2017)
6. Christlein, V., Maier, A.K.: Encoding CNN activations for writer recognition. In: 13th IAPR International Workshop on Document Analysis Systems, DAS 2018, Vienna, Austria, April 24–27, 2018, pp. 169–174 (2018)
7. Christlein, V., Nicolaou, A., Seuret, M., Stutzmann, D., Maier, A.: ICDAR 2019 competition on image retrieval for historical handwritten documents. In: 2019 International Conference on Document Analysis and Recognition, ICDAR 2019, Sydney, Australia, September 20–25, 2019, pp. 1505–1509 (2019)
8. Chum, O., Philbin, J., Sivic, J., Isard, M., Zisserman, A.: Total recall: automatic query expansion with a generative feature model for object retrieval. In: IEEE 11th International Conference on Computer Vision, ICCV 2007, Rio de Janeiro, Brazil, October 14–20, 2007, pp. 1–8 (2007)
9. Fiel, S., et al.: ICDAR2017 competition on historical document writer identification (historical-WI). In: 14th IAPR International Conference on Document Analysis and Recognition, ICDAR 2017, Kyoto, Japan, November 9–15, 2017, pp. 1377–1382 (2017)
10. Fiel, S., Sablatnig, R.: Writer identification and retrieval using a convolutional neural network. In: Azzopardi, G., Petkov, N. (eds.) CAIP 2015. LNCS, vol. 9257, pp. 26–37. Springer, Cham (2015). https://doi.org/10.1007/978-3-319-23117-4_3
11. Gordo, A., Radenovic, F., Berg, T.: Attention-based query expansion learning. In: Vedaldi, A., Bischof, H., Brox, T., Frahm, J.-M. (eds.) ECCV 2020. LNCS, vol. 12373, pp. 172–188. Springer, Cham (2020). https://doi.org/10.1007/978-3-030-58604-1_11
12. Jégou, H., Douze, M., Schmid, C.: On the burstiness of visual elements. In: 2009 IEEE Computer Society Conference on Computer Vision and Pattern Recognition (CVPR 2009), 20–25 June 2009, Miami, Florida, USA, pp. 1169–1176 (2009)
13. Jordan, S., et al.: Re-ranking for writer identification and writer retrieval. In: Bai, X., Karatzas, D., Lopresti, D. (eds.) DAS 2020. LNCS, vol. 12116, pp. 572–586. Springer, Cham (2020). https://doi.org/10.1007/978-3-030-57058-3_40
14. Keglevic, M., Fiel, S., Sablatnig, R.: Learning features for writer retrieval and identification using triplet CNNs. In: 16th International Conference on Frontiers in Handwriting Recognition, ICFHR 2018, Niagara Falls, NY, USA, August 5–8, 2018, pp. 211–216 (2018)
15. Kleber, F., Fiel, S., Diem, M., Sablatnig, R.: CVL-DataBase: an off-line database for writer retrieval, writer identification and word spotting. In: 12th International Conference on Document Analysis and Recognition, ICDAR 2013, Washington, DC, USA, August 25–28, 2013, pp. 560–564 (2013)

16. Lai, S., Zhu, Y., Jin, L.: Encoding pathlet and SIFT features with bagged VLAD for historical writer identification. IEEE Trans. Inf. Forensics Secur. **15**, 3553–3566 (2020)

17. Louloudis, G., Gatos, B., Stamatopoulos, N., Papandreou, A.: ICDAR 2013 competition on writer identification. In: 12th International Conference on Document Analysis and Recognition, ICDAR 2013, Washington, DC, USA, August 25–28, 2013, pp. 1397–1401 (2013)

18. Louloudis, G., Gatos, B., Stamatopoulos, N., Papandreou, A.: ICDAR 2013 competition on writer identification. In: 12th International Conference on Document Analysis and Recognition, ICDAR 2013, Washington, DC, USA, August 25–28, 2013, pp. 1397–1401 (2013)

19. Peer, M., Kleber, F., Sablatnig, R.: Self-supervised vision transformers with data augmentation strategies using morphological operations for writer retrieval. In: Frontiers in Handwriting Recognition - 18th International Conference, ICFHR 2022, Hyderabad, India, December 4–7, 2022, Proceedings, pp. 122–136 (2022)

20. Peer, M., Kleber, F., Sablatnig, R.: Writer retrieval using compact convolutional transformers and NetMVLAD. In: 26th International Conference on Pattern Recognition, ICPR 2022, Montreal, QC, Canada, August 21–25, 2022, pp. 1571–1578 (2022)

21. Radenovic, F., Tolias, G., Chum, O.: Fine-tuning CNN image retrieval with no human annotation. IEEE Trans. Pattern Anal. Mach. Intell. **41**(7), 1655–1668 (2019)

22. Rasoulzadeh, S., BabaAli, B.: Writer identification and writer retrieval based on netVLAD with re-ranking. IET Biom. **11**(1), 10–22 (2022)

23. Tan, F., Yuan, J., Ordonez, V.: Instance-level image retrieval using reranking transformers. In: 2021 IEEE/CVF International Conference on Computer Vision, ICCV 2021, Montreal, QC, Canada, October 10–17, 2021, pp. 12085–12095 (2021)

24. Wang, X., Zhang, H., Huang, W., Scott, M.R.: Cross-batch memory for embedding learning. In: 2020 IEEE/CVF Conference on Computer Vision and Pattern Recognition, CVPR 2020, Seattle, WA, USA, June 13–19, 2020. pp. 6387–6396 (2020)

25. Wang, Z., Maier, A., Christlein, V.: Towards end-to-end deep learning-based writer identification. In: 50. Jahrestagung der Gesellschaft für Informatik, INFORMATIK 2020 - Back to the Future, Karlsruhe, Germany, 28. September - 2. Oktober 2020, vol. P-307, pp. 1345–1354 (2020)

26. Weng, L., Ye, L., Tian, J., Cao, J., Wang, J.: Random VLAD based deep hashing for efficient image retrieval. CoRR abs/2002.02333 (2020)

27. Zhang, X., Jiang, M., Zheng, Z., Tan, X., Ding, E., Yang, Y.: Understanding image retrieval re-ranking: A graph neural network perspective. arXiv preprint arXiv:2012.07620 (2020)

HisDoc R-CNN: Robust Chinese Historical Document Text Line Detection with Dynamic Rotational Proposal Network and Iterative Attention Head

Cheng Jian[1], Lianwen Jin[1,2]([✉]), Lingyu Liang[1,2], and Chongyu Liu[1,2]

[1] South China University of Technology, Guangzhou, China
eechengjian@mail.scut.edu.cn, lianwen.jin@gmail.com
[2] SCUT-Zhuhai Institute of Modern Industrial Innovation, Zhuhai, China

Abstract. Text line detection is an essential task in a historical document analysis system. Although many existing text detection methods have achieved remarkable performance on various scene text datasets, they cannot perform well because of the high density, multiple scales, and multiple orientations of text lines in complex historical documents. Thus, it is crucial and challenging to investigate effective text line detection methods for historical documents. In this paper, we propose a Dynamic Rotational Proposal Network (DRPN) and an Iterative Attention Head (IAH), which are incorporated into Mask R-CNN to detect text lines in historical documents. The DRPN can dynamically generate horizontal or rotational proposals to enhance the robustness of the model for multi-oriented text lines and alleviate the multi-scale problem in historical documents. The proposed IAH integrates a multi-dimensional attention mechanism that can better learn the features of dense historical document text lines while improving detection accuracy and reducing the model parameters via an iterative mechanism. Our HisDoc R-CNN achieves state-of-the-art performance on various historical document benchmarks including CHDAC (the IACC competition (http://iacc.pazhoulab-huangpu.com/shows/108/1.html) dataset), MTHv2, and ICDAR 2019 HDRC CHINESE, thereby demonstrating the robustness of our method. Furthermore, we present special tricks for historical document scenarios, which may provide useful insights for practical applications.

Keywords: Deep learning · Text line detection · Historical document analysis

1 Introduction

Historical documents form an important library of human history and are the valuable heritage of human civilization. Thus, the effective preservation of historical documents is considered to be an urgent research target. The digitization of historical document images is a vital component of cultural preservation

G. A. Fink et al. (Eds.): ICDAR 2023, LNCS 14187, pp. 428–445, 2023.
https://doi.org/10.1007/978-3-031-41676-7_25

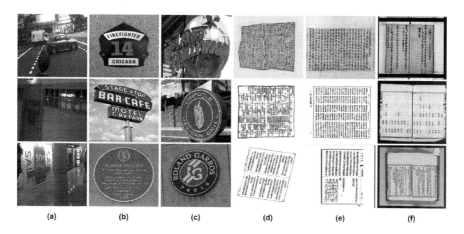

Fig. 1. Examples of scene texts and historical document text lines from different datasets. (a) ICDAR 2015, (b) CTW1500, (c) Total-Text, (d) CHDAC, (e) MTHv2, (f) ICDAR 2019 HDRC CHINESE.

because it facilitates the extraction and storage of valuable information from historical documents. Text line detection is one of the most important techniques for digitizing historical document images. It is a critical stage in the historical document analysis pipeline for locating the text lines and extracting them from the historical documents. Localization accuracy seriously affects the performance of downstream tasks, such as text recognition and document understanding in historical documents.

Additionally, the difficulties of historical document text line detection and scene text detection are of different natures, as shown in Fig. 1. The text instances in historical documents are arranged in a much denser way than those in natural scenes, with varied scales and closer intervals. Besides, various extents of degradation often occur in historical documents. Moreover, various layout structures, writing styles, textures, decorations, rotation, and image warping are other challenges for the text line detection task.

With the advancement of deep learning, text detection algorithms have achieved significant improvement in the field of scene text [12,29,30]. However, the existing methods may not be generic for historical document images. As shown in Fig. 2, some text instances are not detected by the scene text detector (second column), and the predicted text boundaries are easily sticky due to the high density of text lines. Most scene text detectors are prone to miss-detection and inaccurate localization because they are not optimized for dense multi-scale text lines in historical documents. Hence it is necessary to design a robust text line detection algorithm for historical documents.

In this paper, we design a text line detection method named HisDoc R-CNN, which consists of a Dynamic Rotational Proposal Network (DRPN) and an Iterative Attention Head (IAH), to handle historical document text lines under complex scenarios. To effectively detect dense text lines in historical documents,

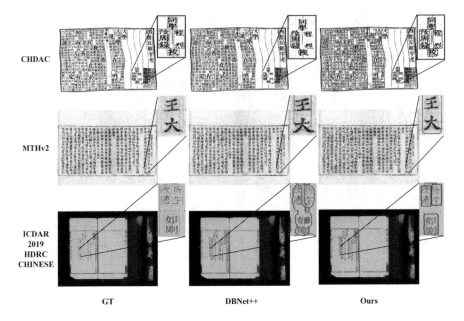

Fig. 2. Visualization of historical document text line detection results. The first column is the ground truth. The results of DBNet++ and our method are shown in the second and third columns, respectively.

we extend the anchor-based method, Mask R-CNN [4], as our overall framework. The anchor-based detection method generally has a higher recall value and is more appropriate for historical documents because it can detect text areas precisely by flattening the predefined anchors. In addition, rotated text lines exist in historical document images due to the camera angle, document scanning errors, and writing style, which are not well addressed by the general anchor-based methods because they only produce horizontal boxes [4,24]. Our proposed DRPN leverages the rotational proposal mechanism (RP) in [31] to detect multi-oriented text lines in historical documents. Besides, the DRPN replaces shared convolution in the region proposal network (RPN) with dynamic convolution (DC). Because the model should use different convolution kernels to generate proposals based on the feature pyramid network (FPN) features of different scales, it is beneficial to create proper candidate boxes for multi-scale text lines in historical documents. Finally, the DRPN can dynamically generate rotational proposals depending on multi-level features, thus alleviating the multi-orientation and multi-scale problem of historical document text lines. To improve the efficiency of the historical document text line detector, we also propose an Iterative Attention Head consisting of an iterative mechanism and a multi-dimensional attention mechanism. The iterative mechanism iteratively trains a box head and a mask head, which is more lightweight than the cascade structure and significantly improves the accuracy of text line detection in historical documents. Due to the small interval between historical text lines, the region of interest may contain the redundant features of neighboring text lines. The multi-dimensional attention mechanism makes IAH focus on the text instances in the foreground, thereby effectively reducing the interference of adjacent text lines in historical documents.

The contributions of this paper are summarized as follows:

- We propose a Dynamic Rotational Proposal Network (DRPN), which can efficiently handle horizontal and rotational Chinese historical documents with various layouts while alleviating the multi-scale problem of dense text lines in historical documents.
- We present an Iterative Attention Head (IAH) to significantly improve the accuracy of text line detection in historical documents while reducing the parameters of our text line detector.
- Our HisDoc R-CNN consistently achieves state-of-the-art performance on three historical document benchmarks and their rotated versions, demonstrating the robustness of our method on complex historical documents.
- The tricks used in the historical document scenario are presented, which may provide useful insights for practical applications.

2   Related Work

2.1   Scene Text Detection Methods

In recent years, the development of deep learning has led to remarkable results for text detectors. Scene text detection aims to locate text instances in scene images. Most approaches for scene text detection can be roughly divided into regression-based and segmentation-based methods.

Regression-Based Methods. Regression-based methods are often motivated by object detection methods such as Faster R-CNN [24] and SSD [14], which regarded text detection as a regression problem for bounding boxes. TextBoxes [9] handled text with extreme scales by refining the scales of anchors and convolution kernels in SSD. TextBoxes++ [8] and DMPNet [16] regressed quadrangles for multi-oriented text detection. RRD [11] used rotation-invariant features in the text classification branch and rotation-sensitive features in the text regression branch to improve the detection of long text. EAST [32] directly regressed text instances at pixel level and obtained the final result using non-maximum suppression (NMS). RRPN [19] introduced rotated region proposals to Faster R-CNN to detect titled text. OBD [15] combined an orderless box discretization block and Mask R-CNN to address the inconsistent labeling issue of regression-based methods. However, the above methods cannot deal with curved text effectively.

Segmentation-Based Methods. Segmentation-based methods usually start with pixel-level prediction and then obtain separate text instances using post-processing. TextSnake [18] represented text instances using center lines and ordered disks. PSENet [29] separated close text instances using a progressive scale expansion algorithm on multi-scale segmentation maps. PAN [30] proposed a learnable pixel aggregation algorithm to implement an arbitrary-shaped text detection network with low computational cost. DBNet [10] designed a differentiable binarization module to realize real-time scene text detection. FCENet [33] predicted the Fourier signal vector of text instances and then reconstructed the text contours via an Inverse Fourier Transformation. DBNet++ [12] proposed an efficient adaptive scale fusion module based on DBNet to improve the accuracy of text detection. However, these methods are unable to resolve overlapping text due to pixel-level segmentation.

2.2   Historical Document Text Line Detection Methods

The methods for text line detection of historical documents are also built based on object detection and semantic segmentation methods. Barakat et al. [1] modified FCN [17] to segment historical handwritten documents and extracted text lines using connected component analysis. Mechi et al. [21] proposed an adaptive U-Net [26] architecture to implement text line segmentation of historical document images with low computational cost. Renton et al. [25] defined text lines through their X-Heights, which can effectively separate overlapping lines in historical handwritten documents. Ma et al. [20] extended Faster R-CNN by adding a character prediction and a layout analysis branch and then grouped individual characters into text lines, which achieved high performance. Mechi et al. [22] investigated adequate deep architecture for text line segmentation in historical Arabic and Latin document images and used topological structural analysis to extract whole text lines. Prusty et al. [23] and Sharan et al. [28] adapted Mask R-CNN to segment complex text lines and layout elements for Indic historical manuscripts. Most of these methods rely on complex post-processing modules that lack robustness.

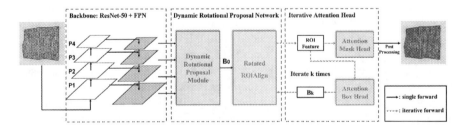

Fig. 3. The architecture of our proposed HisDoc R-CNN, where the Dynamic Rotational Proposal Module (DRPM) and the Attention Head are shown in Fig. 4. B_0 is the initial rotated box generated by DRPM, and B_k is the output rotated box produced by the Attention Head in the kth iteration.

3 Methodology

3.1 Overall Architecture

The proposed framework, as shown in Fig. 3, is an anchor-based text line detector that can be divided into three parts, including backbone, Dynamic Rotational Proposal Network (DRPN), and Iterative Attention Head (IAH). The backbone consists of a ResNet-50 [5] and a Feature Pyramid Network (FPN) [13] for extracting pyramid feature maps. To handle multi-oriented and multi-scale text lines, DRPN first dynamically predicts candidate rotated boxes (r-boxes) based on multi-level features. Then Rotated ROIAlign is used to extract the region proposal features of the r-boxes for r-box prediction. Finally, the IAH iteratively refines the r-boxes and corresponding masks of text instances to output the final results, which makes the r-box regression more accurate and reduces the model parameters.

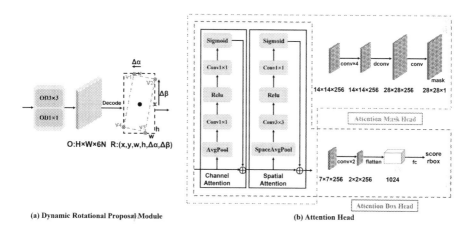

(a) Dynamic Rotational Proposal Module   (b) Attention Head

Fig. 4. (a) The architecture of the proposed Dynamic Rotational Proposal Module (DRPM). "OD" refers to ODConv. (b) The architecture of the proposed Attention Head. The multi-dimensional attention mechanism of the head contains channel attention and spatial attention. The attention module gives the output as input to the box head and the mask head.

3.2 Dynamic Rotational Proposal Network (DRPN)

Our DRPN consists of a Dynamic Rotational Proposal Module (DRPM) and a Rotated ROIAlign, as shown in Fig. 3. The standard RPN [24] has two drawbacks: (1) it can only generate horizontal rectangular proposals for multi-oriented dense texts that are easily suppressed by NMS, and (2) it is unfair to share convolutional layers for multi-level features when generating proposals because each layer is responsible for predicting text instances of different scales. For (1),

DRPM enables the generation of rotated text proposals for the subsequent IAH to predict r-boxes and masks. The r-boxes not only fit the multi-oriented text instances tighter than horizontal boxes but also remove the partial impacts of adjacent text lines to assist mask prediction while passing Rotated ROIAlign. For (2), DRPM introduces ODConv [7], which dynamically adjusts the convolutional kernel depending on different inputs. Therefore, the production of multi-scale r-boxes is closely associated with the features of different scales.

As shown in Fig. 4 (a), DRPM adopts a novel and efficient approach to output a set of rotated proposals. We assume that each anchor is denoted by $A = (A_x, A_y, A_w, A_h)$, where (A_x, A_y) is the center coordinate, A_w and A_h denote the width and height of the anchor, respectively. Given a feature map of shape $H \times W$ from FPN as input, we use a shared 3×3 ODConv[1] and two 1×1 ODConvs to output the classification score map S and offset map $O = (\delta_x, \delta_y, \delta_w, \delta_h, \delta_\alpha, \delta_\beta)$ of the anchors. The shapes of S and O are $H \times W \times N$ and $H \times W \times 6N$, respectively. N is the number of preset anchors for each spatial location. The DRPN combines A and O to decode the representation parameter $R = (x, y, w, h, \Delta\alpha, \Delta\beta)$ of the rotated proposal by the following equation:

$$\begin{cases} x = \delta_x \cdot A_w + A_x, \quad y = \delta_y \cdot A_h + A_y \\ w = A_w \cdot e^{\delta_w}, \qquad h = A_h \cdot e^{\delta_h} \\ \Delta\alpha = \delta_\alpha \cdot w, \qquad \Delta\beta = \delta_\beta \cdot h \end{cases} \tag{1}$$

where (x, y) is the center coordinate of the rotated proposal. w and h denote the width and height of the external rectangular box of the rotated proposal. $\Delta\alpha$ and $\Delta\beta$ denote offsets relative to the midpoints of the upper and right bounds of the external rectangular box. Finally, we define the coordinates of the four vertices of the rotated proposal as $v = (v1, v2, v3, v4)$, and express v with representation parameters R as follows:

$$\begin{cases} v_1 = (x, y - h/2) + (\Delta\alpha, 0) \\ v_2 = (x + w/2, y) + (0, \Delta\beta) \\ v_3 = (x, y + h/2) + (-\Delta\alpha, 0) \\ v_4 = (x - w/2, y) + (0, -\Delta\beta) \end{cases} \tag{2}$$

Compared with the original RPN [24], we only need to predict two additional parameters $(\Delta\alpha, \Delta\beta)$ which allows us to efficiently regress the rotational proposal with no significant increase in model parameters.

3.3   Iterative Attention Head (IAH)

The cascade paradigm proposed by Cai et al. [2] can stably improve the performance of the two-stage detector. However, its box head uses two fully connected layers (FC) and the used triple heads involve a large number of parameters with high storage costs. To alleviate this issue, we propose a new head structure

---

[1] ODConv is a dynamic convolution. For details, please refer to the original paper [7].

with an iterative mechanism and replace the original FC layers with convolution layers.

The proposed IAH has only one box head and mask head, as shown in Fig. 3. For training, the predicted r-boxes pass through the Rotated ROIAlign to extract features which are fed again into the same head to output new r-boxes, and we repeat this process three times. During each iteration, we select positive and negative samples according to a progressively increasing threshold of intersection over union (IoU). The IoU of the predicted r-box B and the ground truth G is denoted by $IoU(B, G)$, and the threshold of IoU is denoted by T. We select the positive and negative samples in the following manner:

$$I = \begin{cases} 1, & IoU(B, G) > T \\ 0, & IoU(B, G) \leq T \end{cases} \tag{3}$$

where I is the indicator function and 1 or 0 indicates whether the r-box is a positive or negative sample. We set threshold T to (0.5, 0.6, 0.7) for each iteration following [2]. Compared with Cascade R-CNN [2], the iterative strategy significantly reduces the model parameters by decreasing the number of heads.

To improve the precision of text line detection for historical documents, we introduce a multi-dimensional attention mechanism to the head, which includes a channel attention mechanism and spatial attention mechanism (See Fig. 4(b)). The multi-dimensional attention mechanism facilitates the flexible extraction of foreground text line features. We assume that the input feature maps of the head are $X \in \mathbb{R}^{N \times C \times H \times W}$, where N is the number of rotated proposals. First, we apply global average pooling to X and obtain the channel attention weights $W_c \in \mathbb{R}^{N \times C \times 1 \times 1}$ using two following 1×1 convolutional layers. Then we add W_c to X to obtain the intermediate features $X_c \in \mathbb{R}^{N \times C \times H \times W}$. Next, we use a spatial average pool on X_c, and obtain the spatial attention weights $W_s \in \mathbb{R}^{N \times 1 \times H \times W}$ using a following 3×3 convolutional layer and a 1×1 convolutional layer. Finally, we sum X_c with the spatial attention weights W_s to get the output features $X_o \in \mathbb{R}^{N \times C \times H \times W}$. The multi-dimensional attention is defined as follows:

$$\begin{aligned} W_c &= Channel\_Attention(X) \\ X_c &= X + W_c \\ W_s &= Spatial\_Attention(X_c) \\ X_o &= X_c + W_s \end{aligned} \tag{4}$$

where X_o is subsequently adopted for the r-box regression or mask prediction of text lines in historical documents. $Channel\_Attention$ and $Spatial\_Attention$ are the attention modules in the heads. The detailed structures of the box and mask heads are illustrated in Fig. 4(b). Since the head is mainly composed of convolutional layers, the model parameters are further decreased.

3.4   Loss Function

Our loss function is defined as:

$$L = L_{drpn} + \lambda_0 L_{iter_0} + \lambda_1 L_{iter_1} + \lambda_2 L_{iter_2} \tag{5}$$

where L_{drpn} is the loss of the DRPN containing binary cross entropy loss (BCE) for text classification and smooth L1 loss for r-box regression. L_{iter_n} is the loss of IAH for each iteration, which contains one more BCE loss for text mask prediction than L_{drpn}. λ_n is used to balance the importance of each iteration, and we set $(\lambda_0, \lambda_1, \lambda_2)$ to (1, 0.5, 0.25) in all experiments following the Cascade R-CNN setting.

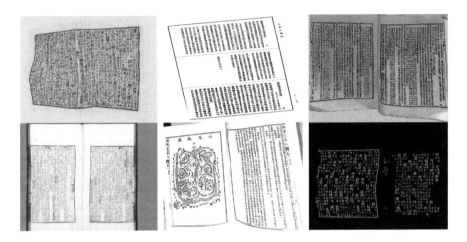

Fig. 5. Examples from the CHDAC dataset. The challenge of this dataset can be illustrated in above figures.

4   Experiment

4.1   Datasets

CHDAC. The IACC competition dataset, Chinese Historical Document Analysis Challenge (CHDAC), is a complex rotated and warped Chinese historical document analysis dataset including 2,000 images for training and 1,000 for testing[2], respectively. This dataset provides text line detection, recognition, and reading order annotation to inspire a practical and novel historical document analysis framework. The text lines in the dataset are numerous, dense, partially rotated, and warped, which complicates the challenge. Examples of this dataset are shown in Fig. 5.

---

[2] This paper uses the datasets provided by the organizers in the preliminaries and finals. The datasets voluntarily submitted by the teams in the final stage are not included.

MTHv2. MTHv2 is a Chinese historical document dataset consisting of the Tripitaka Koreana in Han (TKH) Dataset and Multiple Tripitaka in Han (MTH) Dataset [20]. It contains 3,199 pages of historical documents, including 2,399 pages for training and 800 for testing. The annotation of each text line consists of the four vertices of the quadrilateral bounding box and its transcription.

ICDAR 2019 HDRC CHINESE. ICDAR 2019 HDRC CHINESE is a large structured Chinese family record dataset used to overcome the historical document reading challenge [27]. It contains 11,715 training images and 1,135 test images, which are annotated with the four vertices of the bounding-box quadrilateral for the text line detection subtask. However, only 1,172 images are now publicly available online[3] and we randomly divide them into 587 training, 117 validation, and 468 testing samples.

4.2 Implementation Details

We use ResNet-50 pre-trained on ImageNet [3] with FPN as our backbone. Our model is optimized using AdamW with an initial learning rate of 0.0001. We train our model with batch size 8 for 160 epochs, and the learning rate is decayed by a factor of 0.1 at epochs 80 and 128.

We conduct comparison experiments over existing text detection methods and our HisDoc R-CNN on the above datasets. For the fairness of comparison, We train and test methods based on the MMOCR toolbox [6] without particular instructions[4]. All models are trained using scaling, color jittering, flipping, cropping, and rotation augmentations. For training, the longer sides of training images are resized to a fixed size of 1,333, and the shorter sides are randomly resized to different scales (704, 736, 768, 800, 832, 864, 896). For testing, the longer and shorter sides are resized to 1,333 and 800, respectively. Color jittering and horizontal flipping are also applied to the images. We randomly crop the images to the size of 640×640 and rotate them in the range of -10 to $10°$.

To evaluate the robustness of each model, we randomly rotate the testing images of MTHv2 and ICDAR 2019 HDRC CHINESE datasets from -15 to $15°$ so that all models would use the rotation augmentation with the same range. All experiments are conducted on two RTX 3090 GPUs.

As metrics to compare the performance of different methods, we usually use their precision, recall, and F-measure under the IoU threshold of 0.5, which is the convention for text detection.

4.3 Ablation Studies of the Proposed Method

Effectiveness of the DRPN. In Table 1, the rotational proposal mechanism (RP) brings a large performance improvement to the baseline (10.13% higher).

---

[3] https://tc11.cvc.uab.es/datasets/ICDAR2019HDRC_1,
[4] EAST and OBD are not implemented by MMOCR.

RP+DC improves the F-measure by 0.11% over RP, and RP+DC+IM+AH improves by 0.26% over RP+IM+AH, proving the effectiveness of dynamic convolution (DC). Finally, our proposed DRPN (RP+DC) significantly improves the performance on the CHDAC and promotes the Mask R-CNN baseline by 10.24% in terms of the F-measure. The effectiveness of our DRPN is verified, and the recall is substantially increased compared to Mask R-CNN (12.56% higher), avoiding the NMS suppression of rotated text lines with a high IoU. Moreover, the model parameters are increased by only 4.6%, showing that our method is accurate and efficient.

Table 1. Ablation studies for the effectiveness of our method on the CHDAC. The baseline is Mask R-CNN. "RP" indicates the rotational proposal. "DC" indicates dynamic convolution. "CM" and "IM" indicate the cascade and iteration mechanism of the head, respectively. "AH" indicates the proposed attention head. P: precision, R: recall, F:F-measure (IoU threshold: 0.5). Params: number of model parameters.

Baseline	RP	DC	CM	IM	AH	P	R	F	Params(M)
Mask R-CNN [4]						89.03	80.90	84.77	43.85
	√					97.08	92.82	94.90	43.99
	√	√				96.78	93.46	95.01	45.85
	√		√			98.15	92.73	95.36	77.04
	√			√		98.08	92.70	95.31	43.99
	√			√	√	**98.29**	93.16	95.66	32.39
	√	√		√	√	98.19	**93.74**	**95.92**	34.25

Effectiveness of the Iterative Mechanism. To demonstrate the effectiveness of the proposed iterative mechanism, we compare the cascade head (CM) and our iterative head (IM), which both have a consistent structure for fairness. As shown in Table 1, RP+IM drops the F-measure by 0.05% compared to RP+CM. The result shows that the iterative mechanism can almost replace the cascade mechanism with only a slight performance loss. However, the iterative mechanism eliminates the redundant box and mask heads so that the model parameters are significantly reduced (42.9% lower).

Effectiveness of the Attention Head. Although the iterative mechanism certainly streamlines the model, the original head contains fully connected layers with enormous parameters. Our proposed attention head is mainly composed of convolutional layers, so the model with attention head (RP+IM+AH) can reduce the model parameters by 26.4% compared with RP+IM (See Table 1), meanwhile compensating the performance loss from the iterative mechanism. RP+IM+AH has a better F-measure than RP+IM (0.35% higher) and RP+CM (0.3% higher), indicating that the multi-dimensional attention mechanism performs better in capturing the features of dense text lines in historical documents.

4.4 Comparison with Existing Methods

To compare the performance of different methods, we evaluate all text detection methods on three benchmarks. CHDAC is a Chinese historical document analysis competition dataset, then ICDAR 2019 HDRC CHINESE and MTHv2 are widely used benchmarks for text line detection on historical documents. The experiment results (given in Table 2, 3 and 4) show that our method achieves state-of-the-art performance on the three benchmarks and is also the strongest in rotated text line detection.

Table 2. Detection results for various text detection methods on CHDAC. * indicates the use of the proposed specific tricks described in Sect. 4.5 (With large scale and crop ratio(H:0.7, W:0.5)). Bold indicates SOTA. Underline indicates second best.

Method	P	R	F
EAST [32]	61.41	73.13	66.76
Mask R-CNN [4]	89.03	80.90	84.77
Cascade R-CNN [2]	92.82	83.63	87.98
OBD [15]	94.73	81.52	87.63
TextSnake [18]	96.33	89.62	92.85
PSENet [29]	76.99	89.62	82.83
PAN [30]	92.74	85.71	89.09
FCENet [33]	88.42	85.04	86.70
DBNet++ [12]	91.39	89.15	90.26
HisDoc R-CNN(ours)	**98.19**	93.74	95.92
HisDoc R-CNN(ours)*	98.16	**96.59**	**97.37**

Results on CHDAC. Most of the text lines in CHDAC are dense or small, and some are rotated or warped, which trials the capability of text line detectors. With the help of DRPN and IAH, HisDoc R-CNN achieves the state-of-the-art result of 98.19%, 93.74%, and 95.92% for precision, recall, and F-measure, respectively (See Table 2). Besides, our method outperforms Cascade R-CNN, which is a high-quality regression-based detector, by 7.94% for the F-measure. Moreover, the F-measure of HisDoc R-CNN is also 5.66% higher than DBNet++, which is a segmentation-based text detection method. As shown in Table 2, the performance of our approach significantly surpasses these previous methods, and most of them have relatively low recall indicating that the existing scene text detectors are insufficient to handle the dense and multi-scale text lines in historical documents. Notably, HisDoc R-CNN obtains both the highest precision and recall, showing that our method accurately locates text instances and effectively alleviates the problems of missed detection and small targets in historical documents. With our proposed tricks (Details in Sect. 4.5), the F-measure performance can be further improved by 1.45%.

Table 3. Detection results for various text detection methods on MTHv2 and its rotated version. * indicates the use of the proposed specific tricks described in Sect. 4.5 (With large scale and crop ratio(H:0.7, W:0.5)). Bold indicates SOTA. Underline indicates second best.

Method	MTHv2			Rotated MTHv2		
	P	R	F	P	R	F
EAST [32]	82.79	89.73	86.12	87.01	89.61	88.29
Mask R-CNN [4]	95.83	96.35	96.09	44.27	37.65	40.69
Cascade R-CNN [2]	**98.57**	96.52	97.53	60.63	44.32	51.21
OBD [15]	<u>98.17</u>	<u>97.19</u>	97.68	<u>97.49</u>	84.72	90.66
TextSnake [18]	94.31	91.77	93.02	94.46	88.45	91.36
PSENet [29]	96.87	95.82	96.34	90.16	89.70	89.93
PAN [30]	97.65	95.28	96.45	97.39	<u>91.58</u>	<u>94.40</u>
FCENet [33]	92.47	88.19	90.28	89.96	89.83	89.89
DBNet++ [12]	93.48	93.22	93.35	89.92	90.16	90.04
Ma et al. [20]	–	–	<u>97.72</u>	–	–	–
HisDoc R-CNN(ours)	**98.57**	97.05	97.80	**98.21**	96.01	97.10
HisDoc R-CNN(ours)*	98.14	**98.26**	**98.20**	97.94	**98.19**	**98.06**

Table 4. Detection results for various text detection methods on ICDAR 2019 HDRC CHINESE and its rotated version. * indicates the use of the proposed specific tricks described in Sect. 4.5 (With large scale and crop ratio(H:0.7, W:0.5)). Bold indicates SOTA. Underline indicates second best.

Method	ICDAR 2019 HDRC CHINESE			Rotated ICDAR 2019 HDRC CHINESE		
	P	R	F	P	R	F
EAST [32]	83.36	87.70	85.47	87.46	89.02	88.23
Mask R-CNN [4]	94.50	95.11	94.81	37.19	31.25	33.96
Cascade R-CNN [2]	<u>94.73</u>	<u>95.28</u>	<u>95.00</u>	42.73	33.17	37.35
OBD [15]	94.45	94.78	94.61	<u>93.94</u>	88.59	91.18
TextSnake [18]	81.70	72.95	77.07	83.96	69.96	76.32
PSENet [29]	92.83	93.68	93.25	86.67	<u>91.49</u>	89.01
PAN [30]	93.34	89.34	91.30	92.67	84.75	88.53
FCENet [33]	92.38	91.11	91.74	89.35	87.70	88.52
DBNet++ [12]	93.10	91.05	92.06	92.95	90.75	<u>91.84</u>
HisDoc R-CNN(ours)	94.61	**95.65**	95.13	**94.36**	94.35	94.36
HisDoc R-CNN(ours)*	**95.43**	95.27	**95.35**	93.83	**95.82**	**94.81**

Results on MTHv2. As shown in Table 3, our method also obtains the best results on MTHv2, with a slightly better F-measure than Ma et al. (0.08% higher) and a significant performance improvement over EAST (11.78% higher). The approach proposed by Ma et al. uses more fine-grained supervision to implement character detection and groups characters to generate text line results by post-processing. The drawback is that it may not work for historical documents of other types, while our method has better generalization capability. To evaluate the robustness of various text detection methods, we randomly rotate the test images of MTHv2 and train the models with random rotation augmentation. The results in Table 3 show that our method is quite robust on the rotated images, while most methods suffer from significant performance degradation. The F-measure of the anchor-based approach, Mask R-CNN, drops rapidly because the horizontal boxes for rotational text lines usually have a high IoU and are easily filtered out by the NMS operation. Thanks to the DRPN, we can still preserve a high F-measure (97.10%). As shown in Table 3, our proposed tricks also bring performance gains of 0.4% and 0.96% to HisDoc R-CNN on MTHv2 and its rotated version.

Results on ICDAR 2019 HDRC CHINESE. We evaluated our method on ICDAR 2019 HDRC CHINESE to test its performance on historical documents with other layouts. As shown in Table 4, the F-measure of our method is 96.60%, which is 0.73% higher than the result of Cascade R-CNN. The results verify that our method can also slightly outperform the rectangular box-based methods on horizontal historical document texts. Similar conclusions to MTHv2 can be obtained on the rotated ICDAR 2019 HDRC CHINESE. The F-measure of HisDoc R-CNN is 2.52% higher than DBNet++ on the rotated historical document images. Our method guarantees that the performance remains the best

Table 5. Ablation studies of practical tricks based on our method. Each variation is evaluated on the test set of CHDAC.

Method	P	R	F	Δ F
Baseline model(ResNet-50)	**98.19**	**93.74**	**95.92**	–
Tricks: flip + random crop size (640×640) + multi-scale training (Longer side: 1,333)				
Data augmentation				
With large scale (Longer side: 2,000)	97.90	94.42	96.13	↑0.21
With crop ratio(H:0.7, W:0.5)	97.96	93.91	95.89	↓0.03
With large scale and crop ratio(H:0.5, W:0.5)	97.74	95.68	96.70	↑0.78
With large scale and crop ratio(H:0.5, W:0.6)	97.71	91.73	94.63	↓1.29
With large scale and crop ratio(H:0.5, W:0.7)	98.00	94.93	96.44	↑0.52
With large scale and crop ratio(H:0.6, W:0.5)	97.89	96.32	97.10	↑1.18
With large scale and crop ratio(H:0.7, W:0.5)	98.16	96.59	97.37	↑1.45

(94.36%) on the rotated historical documents with different layouts. As shown in Table 4, our proposed tricks also result in performance gains of 0.22% and 0.45% to HisDoc R-CNN on ICDAR 2019 HDRC CHINESE and its rotated version. Visualization results on each dataset are shown in Fig. 2.

4.5   Ablation Studies of Practical Tricks on Historical Documents

In this section, we analyze practical tricks we used for model refinement in the historical document scenario based on HisDoc R-CNN. We conduct experiments on the CHDAC dataset. Our baseline model only uses the random flip, random crop ($crop\_size$: 640×640), and multi-scale resizing for data augmentation.

Effectiveness of the Image Size. We further investigate the effect of the image size of historical documents on our method. Due to the high density of text lines in historical documents, a larger image size may result in more details of dense text lines. We scale up the long sides of the input image to 2,000 and use a multi-scale strategy for the shorter sides for training. The image is rescaled to a size of $1,600 \times 1,600$ and kept in aspect ratio for testing. The results in Table 5 show that a larger image size improves the F-measure by 0.21% but the performance improvement is not significant. Therefore, we further explore the effect of crop ratio, which is another important factor determining the size of the input historical document image.

Effectiveness of the Crop Ratio. Since there are many long vertical text lines in historical documents (See Fig. 2 and Fig. 5), the common crop ratio may not be appropriate for historical documents. Therefore, we explore whether cropping images to long vertical bars is a better data augmentation approach for historical documents. As shown in Table 5, we set the crop ratios for the height and width to 0.7 and 0.5 respectively, and find that this strategy decreases the F-measure by 0.03% for a small image size while increasing the F-measure significantly by 1.45% for a large image size. In addition, using the crop ratio (H:0.6, W:0.5) only increases 1.18%, indicating that the higher the ratio the greater the improvement. By contrast, using the crop ratios (H:0.5, W:0.5) and (H:0.5, W:0.7) improve the F-measure by 0.78% and 0.52% respectively, while using the crop ratio (H:0.5, W:0.6) decreases it by 1.29%. The results show that adjusting the crop ratio according to the long vertical text lines in historical documents can avoid destroying the characteristics of text lines and help the model to detect such text lines better.

As shown in Tables 2, 3, 4, the tricks achieve consistent performance gains on CHDAC, MTHv2, and ICDAR 2019 HDRC CHINESE, as well as on the rotated datasets. In general, the tricks we propose can improve the performance of history document text line detection in most cases.

5   Conclusion

In this paper, we present a robust approach, HisDoc R-CNN, for detecting historical document text lines in complex scenarios, which improves the traditional anchor-based model from two aspects: (1) we proposed a Dynamic Rotational Proposal Network, which enhances the robustness of the model by dynamically generating rotational proposals according to multi-level features; (2) we present an Iterative Attention Head to efficiently improve the accuracy of text line detection in conjunction with the model parameters. Both modules significantly improve the performance of the proposed HisDoc R-CNN. The experiments demonstrate that our method consistently outperforms state-of-the-art methods on three historical document benchmarks and maintains strong performance on rotated text line detection. In particular, we investigate several useful tricks which may provide useful insights for practical applications in the historical document scenario.

Acknowledgement. This research is supported in part by NSFC (Grant No.: 61936003), Zhuhai Industry Core and Key Technology Research Project (No. 2220004002350), Science and Technology Foundation of Guangzhou Huangpu Development District (No. 2020GH17) and GD-NSF (No.2021A1515011870), and Fundamental Research Funds for the Central Universities.

References

1. Barakat, B., Droby, A., Kassis, M., El-Sana, J.: Text line segmentation for challenging handwritten document images using fully convolutional network. In: 2018 16th International Conference on Frontiers in Handwriting Recognition (ICFHR), pp. 374–379. IEEE (2018)
2. Cai, Z., Vasconcelos, N.: Cascade R-CNN: delving into high quality object detection. In: Proceedings of the IEEE Conference on Computer Vision and Pattern Recognition, pp. 6154–6162 (2018)
3. Deng, J., Dong, W., Socher, R., Li, L.J., Li, K., Fei-Fei, L.: ImageNet: a large-scale hierarchical image database. In: 2009 IEEE Conference on Computer Vision and Pattern Recognition, pp. 248–255. IEEE (2009)
4. He, K., Gkioxari, G., Dollár, P., Girshick, R.: Mask R-CNN. IEEE Trans. Pattern Anal. Mach. Intell. **42**(2), 386–397 (2018)
5. He, K., Zhang, X., Ren, S., Sun, J.: Deep residual learning for image recognition. In: Proceedings of the IEEE Conference on Computer Vision and Pattern Recognition. pp. 770–778 (2016)
6. Kuang, Z., et al.: MMOCR: a comprehensive toolbox for text detection, recognition and understanding. In: Proceedings of the 29th ACM International Conference on Multimedia, pp. 3791–3794 (2021)
7. Li, C., Zhou, A., Yao, A.: Omni-dimensional dynamic convolution. In: International Conference on Learning Representations (2022)
8. Liao, M., Shi, B., Bai, X.: Textboxes++: a single-shot oriented scene text detector. IEEE Trans. Image Process. Publ. IEEE Signal Process. Soc. **27**(8), 3676–3690 (2018)

9. Liao, M., Shi, B., Bai, X., Wang, X., Liu, W.: Textboxes: a fast text detector with a single deep neural network. In: Thirty-First AAAI Conference on Artificial Intelligence (2017)
10. Liao, M., Wan, Z., Yao, C., Chen, K., Bai, X.: Real-time scene text detection with differentiable binarization. In: Proceedings of the AAAI Conference on Artificial Intelligence, vol. 34, pp. 11474–11481 (2020)
11. Liao, M., Zhu, Z., Shi, B., Xia, G.s., Bai, X.: Rotation-sensitive regression for oriented scene text detection. In: Proceedings of the IEEE Conference on Computer Vision and Pattern Recognition, pp. 5909–5918 (2018)
12. Liao, M., Zou, Z., Wan, Z., Yao, C., Bai, X.: Real-time scene text detection with differentiable binarization and adaptive scale fusion. IEEE Trans. Pattern Anal. Mach. Intell. **45**, 919–931 (2022)
13. Lin, T.Y., Dollár, P., Girshick, R., He, K., Hariharan, B., Belongie, S.: Feature pyramid networks for object detection. In: Proceedings of the IEEE Conference on Computer Vision and Pattern Recognition, pp. 2117–2125 (2017)
14. Liu, W., et al.: SSD: single shot MultiBox detector. In: Leibe, B., Matas, J., Sebe, N., Welling, M. (eds.) ECCV 2016. LNCS, vol. 9905, pp. 21–37. Springer, Cham (2016). https://doi.org/10.1007/978-3-319-46448-0_2
15. Liu, Y., et al.: Exploring the capacity of an orderless box discretization network for multi-orientation scene text detection. Int. J. Comput. Vision **129**(6), 1972–1992 (2021)
16. Liu, Y., Jin, L.: Deep matching prior network: toward tighter multi-oriented text detection. In: 2017 IEEE Conference on Computer Vision and Pattern Recognition (CVPR), pp. 3454–3461. IEEE Computer Society (2017)
17. Long, J., Shelhamer, E., Darrell, T.: Fully convolutional networks for semantic segmentation. In: Proceedings of the IEEE Conference on Computer Vision and Pattern Recognition, pp. 3431–3440 (2015)
18. Long, S., Ruan, J., Zhang, W., He, X., Wu, W., Yao, C.: Textsnake: a flexible representation for detecting text of arbitrary shapes. In: Proceedings of the European Conference on Computer Vision (ECCV), pp. 20–36 (2018)
19. Ma, J., et al.: Arbitrary-oriented scene text detection via rotation proposals. IEEE Trans. Multimedia **20**(11), 3111–3122 (2018)
20. Ma, W., Zhang, H., Jin, L., Wu, S., Wang, J., Wang, Y.: Joint layout analysis, character detection and recognition for historical document digitization. In: 2020 17th International Conference on Frontiers in Handwriting Recognition (ICFHR), pp. 31–36. IEEE (2020)
21. Mechi, O., Mehri, M., Ingold, R., Amara, N.E.B.: Text line segmentation in historical document images using an adaptive U-Net architecture. In: 2019 International Conference on Document Analysis and Recognition (ICDAR), pp. 369–374. IEEE (2019)
22. Mechi, O., Mehri, M., Ingold, R., Essoukri Ben Amara, N.: A two-step framework for text line segmentation in historical Arabic and Latin document images. Int. J. Doc. Anal. Recogn. (IJDAR) **24**(3), 197–218 (2021)
23. Prusty, A., Aitha, S., Trivedi, A., Sarvadevabhatla, R.K.: Indiscapes: instance segmentation networks for layout parsing of historical Indic manuscripts. In: 2019 International Conference on Document Analysis and Recognition (ICDAR), pp. 999–1006. IEEE (2019)
24. Ren, S., He, K., Girshick, R., Sun, J.: Faster R-CNN: towards real-time object detection with region proposal networks. In: Proceedings of the 28th International Conference on Neural Information Processing Systems-Volume 1, pp. 91–99 (2015)

25. Renton, G., Soullard, Y., Chatelain, C., Adam, S., Kermorvant, C., Paquet, T.: Fully convolutional network with dilated convolutions for handwritten text line segmentation. Int. J. Doc. Anal. Recogn. (IJDAR) **21**(3), 177–186 (2018). https://doi.org/10.1007/s10032-018-0304-3

26. Ronneberger, O., Fischer, P., Brox, T.: U-Net: convolutional networks for biomedical image segmentation. In: Navab, N., Hornegger, J., Wells, W.M., Frangi, A.F. (eds.) MICCAI 2015. LNCS, vol. 9351, pp. 234–241. Springer, Cham (2015). https://doi.org/10.1007/978-3-319-24574-4_28

27. Saini, R., Dobson, D., Morrey, J., Liwicki, M., Liwicki, F.S.: ICDAR 2019 historical document reading challenge on large structured Chinese family records. In: 2019 International Conference on Document Analysis and Recognition (ICDAR), pp. 1499–1504. IEEE (2019)

28. Sharan, S.P., Aitha, S., Kumar, A., Trivedi, A., Augustine, A., Sarvadevabhatla, R.K.: Palmira: a deep deformable network for instance segmentation of dense and uneven layouts in handwritten manuscripts. In: Lladós, J., Lopresti, D., Uchida, S. (eds.) ICDAR 2021. LNCS, vol. 12822, pp. 477–491. Springer, Cham (2021). https://doi.org/10.1007/978-3-030-86331-9_31

29. Wang, W., et al.: Shape robust text detection with progressive scale expansion network. In: Proceedings of the IEEE/CVF Conference on Computer Vision and Pattern Recognition, pp. 9336–9345 (2019)

30. Wang, W., et al.: Efficient and accurate arbitrary-shaped text detection with pixel aggregation network. In: Proceedings of the IEEE/CVF International Conference on Computer Vision, pp. 8440–8449 (2019)

31. Xie, X., Cheng, G., Wang, J., Yao, X., Han, J.: Oriented R-CNN for object detection. In: Proceedings of the IEEE/CVF International Conference on Computer Vision, pp. 3520–3529 (2021)

32. Zhou, X., et al.: East: an efficient and accurate scene text detector. In: Proceedings of the IEEE Conference on Computer Vision and Pattern Recognition, pp. 5551–5560 (2017)

33. Zhu, Y., Chen, J., Liang, L., Kuang, Z., Jin, L., Zhang, W.: Fourier contour embedding for arbitrary-shaped text detection. In: Proceedings of the IEEE/CVF Conference on Computer Vision and Pattern Recognition, pp. 3123–3131 (2021)

Keyword Spotting Simplified: A Segmentation-Free Approach Using Character Counting and CTC Re-scoring

George Retsinas[1]([✉]), Giorgos Sfikas[2], and Christophoros Nikou[3]

[1] School of Electrical and Computer Engineering, National Technical University of Athens, Athens, Greece
`gretsinas@central.ntua.gr`
[2] Department of Surveying and Geoinformatics Engineering, University of West Attica, Aigaleo, Greece
`gsfikas@uniwa.gr`
[3] Department of Computer Science and Engineering, University of Ioannina, Ioannina, Greece
`cnikou@cse.uoi.gr`

Abstract. Recent advances in segmentation-free keyword spotting borrow from state-of-the-art object detection systems to simultaneously propose a word bounding box proposal mechanism and compute a corresponding representation. Contrary to the norm of such methods that rely on complex and large DNN models, we propose a novel segmentation-free system that efficiently scans a document image to find rectangular areas that include the query information. The underlying model is simple and compact, predicting character occurrences over rectangular areas through an implicitly learned scale map, trained on word-level annotated images. The proposed document scanning is then performed using this character counting in a cost-effective manner via integral images and binary search. Finally, the retrieval similarity by character counting is refined by a pyramidal representation and a CTC-based re-scoring algorithm, fully utilizing the trained CNN model. Experimental validation on two widely-used datasets shows that our method achieves state-of-the-art results outperforming the more complex alternatives, despite the simplicity of the underlying model.

Keywords: Keyword Spotting · Segmentation-Free · Character Counting

1 Introduction

Keyword spotting (or simply *word* spotting) has emerged as an alternative to handwritten text recognition, providing a practical tool for efficient indexing and searching in document analysis systems. Contrary to full handwriting text recognition (HTR), where a character decoding output is the goal, spotting approaches typically involve a soft-selection step, enabling them to "recover" words that could have been potentially been assigned a more or less erroneous

G. A. Fink et al. (Eds.): ICDAR 2023, LNCS 14187, pp. 446–464, 2023.
https://doi.org/10.1007/978-3-031-41676-7_26

decoding. Keyword spotting is closely related to text localization and recognition. The latter problems are typically cast in an "in-the-wild" setting, where we are not dealing with a document image, but rather a natural image/photo that may contain patches of text. The processing pipeline involves producing a set of candidate bounding boxes that contain text and afterwards performing text recognition [10] or localization and recognition are both part of a multi-task loss function [15,32]. The main difference between methods geared for in-the-wild detection and localization and document-oriented methods is that we must expect numerous instances of the same character on the latter; differences in character appearance in the wild relate to perspective distortions and font diversity, while in documents perspective distortion is expected to be minimal; also, information is much more structured in documents, which opens the possibility to encode this prior knowledge in some form of model inductive bias.

In more detail, keyword spotting (KWS) systems can be categorized w.r.t. a number of different taxonomies. Depending on the type of query used, we have either Query-by-Example (QbE) or Query-by-String (QbS) spotting. The two settings correspond to using a word image query or a string query respectively. Another taxonomy of KWS systems involves categorization into segmentation-based and segmentation-free systems. These differ w.r.t. whether we can assume that the document collection that we are searching has been pre-segmented into search target tokens or not; the search targets will usually be word images, containing a single word, or line images, containing a set of words residing in a single text line. Segmentation-based methods involve a simpler task than segmentation-free methods, but in practice correctly segmenting a document into words and lines can be a very non-trivial task, especially in the context of documents that involve a highly complex structure such as tabular data, or manuscripts that include an abundance of marginalia, etc.

The former category motivated a representation-driven line of research, with Pyramidal Histogram of Characters (PHOC) embeddings being the most notable example of attribute-based representation [2,13,19,20,22,26–28]. Sharing the same main concept, a different embedding was introduced by Wilkinson et al. [30], while recognition-based systems were also used to tackle the problem in a representation level [12,21]. Contrary to segmentation-based literature, modern systems capable of segmentation-free retrieval on handwritten documents are limited. A commonly-used methodology is the straightforward sliding window approach, with [5,6,24] being notable examples of efficient variations of this concept. Line-level segmentation-free detection uses essentially a similar concept in its core [17], whilst simplified due to the sequential nature of text-line processing. Another major direction is the generation of word region candidates before applying a segmentation-based ranking approach (e.g. using PHOC representations) [7,25,31,33] Candidate region proposal can be part of an end-to-end architecture, following the state-of-the-art object detection literature, as in [31,33], at the cost of a document-level training procedure that may require generation of synthetic data.

In this work, we focus on the segmentation-free setting, where no prior information over the word location is known over a document page. Specifically, we aim to bypass any segmentation step and provide a spotting method that works in document images without sacrificing efficiency. The main idea of this work is to effectively utilize a per-pixel character existence estimate. To this end, we first build a CNN-based system that transforms the document input image into a down-scaled map of potential characters. Then, by casting KWS as a **character counting** problem, we aim to find bounding boxes that contain the requested characters. In a nutshell, our contributions can be summarized as follows.

- We build a computationally efficient segmentation-free KWS system;
- We propose a training scheme for computing a character counting map and consequently a counting model; training is performed on segmented words, with no need for document-level annotation which may require synthetic generation of images to capture large variations over the localization of word images [31].
- We introduce a document scanning approach for the counting problem with several computational-improving modifications, including integral images of one-step sum computations for counting characters and binary search for efficient detection of bounding boxes.
- We further improve the spotting performance of the counting system; a finer KWS approach is used on the subset of detected bounding boxes to enhance performance and reduce counting ambiguities (i.e., "end" vs "den").
- In this work, we use the trained CNN model in two variations: a pyramidal representation of counting, akin to PHOC embeddings [1] and a CTC-based scoring approach, akin to forced alignment [29]. The latter method is also capable of enhancing the detection of the bounding box, notably improving performance.
- We propose a new metric to quantify bounding box overlap, replacing the standard Intersection over Union (IoU) metric. We argue that the new metric is more suitable than IoU in the context of KWS, as it does not penalize enlarged detected boxes that do no contain any misdetected word.

The effectiveness of the proposed method serves as a counter-argument to using object detection methods with complex prediction heads and a predefined maximum number of detections (e.g. [31]) and showcase that state-of-the-art results can be attained with an intuitively simple pipeline. Moreover, contrary to the majority of existing approaches, the proposed method enables sub-word or multi-word search since there is no restriction over the bounding box prediction.

2   Proposed Methodology

Our core idea can be summarized as: *"efficiently estimate the bounding box by counting character occurrences"*. In this section, we will describe how 1) to build and train such a character counting network, 2) use character counts to efficiently estimate a bounding box, and 3) provide an enhanced similarity score to boost performance. We will only explore the QbS paradigm, which is in line with a character-level prediction system, however extending this into a simple QbE system is straightforward (e.g., estimate character occurrences into an example image and use this as target).

2.1 Training of Character Counting Network

Method Formulation: The simplest way to treat the counting problem is through regression: given a per-character histogram of character occurrences, one can train a regression DNN with mean squared loss that takes as input word images. Nonetheless, such an approach lacks the ability to be easily applicable to page-level images. A key concept in this work is the per-pixel analysis of the character probabilities and their scale. This way, the desired counting operation can be decoupled into two sub-problems: 1) compute the character probability at each point of the feature map and 2) compute the scale corresponding to each point (i.e., the size of the point w.r.t. the whole character it belongs to), such that summation over the word gives us the requested counting histogram.

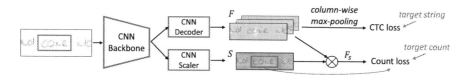

Fig. 1. Overview of the model's components and how they contribute to the losses.

Formally, we denote as F the feature map that contains character-level predictions of their probability, and is generated by a deep neural network. This 3D tensor F has size equal to $H_r \times W_r \times C$, where $H_r \times W_r$ is the downscaled size of the initial $H \times W$ size of the input image and C is the number of possible characters. We also generate a character-independent scale matrix S, also from a DNN, of size $H_r \times W_r$. Then the scaled feature map is denoted as F_s, where each spatial point of F is multiplied by the corresponding scale value, that is $F_s[i,j,k] = F[i,j,k] \cdot S[i,j]$. Given a bounding box that contains a word, denoted by the starting point (s_i, s_j) and the ending point (e_i, e_j), one can compute the character occurrences y_c as:

$$y_c = \sum_{i=s_i}^{e_i} \sum_{j=s_j}^{e_j} F_s(i,j), \; y_c \in \mathbb{R}^C \tag{1}$$

The latter formulation can be used to straightforwardly regress the models with respect to the target count histogram.

To assist this decoupling approach, we also constrain the feature map F to be in line with a handwritten text recognition system, using a CTC loss [8]. In other words, character probabilities are explicitly trained with this extra loss. Note that, even though we do not explicitly train the scale map S, it is actually implicitly learned to correspond to scale-like predictions through the elegant combination of the CTC and counting regression loss.

The previous formulation is depicted in Fig. 1. We first feed an image to a *CNN backbone* and obtain a feature tensor, which is then used by the *CNN Decoder* to predict the feature map F, and also by the *CNN Scaler*, to predict the scale map S. Then, the feature map is multiplied with the scale map S, and used to calculate the counting regression loss. The feature map is also independently used to calculate the CTC loss.

Module Architectures: Having described now the main idea, we outline in more detail the architecture of each CNN component. Note that we did not thoroughly explore these architectures, since it is not the core goal of our paper. In fact, since efficiency is the main point of this approach, we used a lightweight ResNet-like network [18], consisted of the following components:

- *CNN Backbone:* The CNN backbone is built by stacking multiple residual blocks [9]. The first layer is a 7×7 convolution of stride 2 with 32 output channels, followed by cascades of 3×3 ResNet blocks (2 blocks of 64, 4 blocks of 128 and 4 blocks of 256 output channels). All convolutions are followed by ReLU and batch-norm layers. Between cascades of blocks max-pooling downscaling of $\frac{1}{2}$ is applied, resulting to an overall downscale of $\frac{1}{8}$.
- *CNN Decoder:* The CNN Decoder consists of two layers; one is a simple 3×3 conv. operation of 128 output channels, and it is followed by an 1×5 convolution, since we assume horizontal writing. The latter conv. operation has C output channels, as many as the characters to be predicted. Between the two layers, we added ReLU, batch-norm, and Dropout. Note that we use only conv. operations so that the decoder can be applied to whole pages. If we used recurrent alternatives (e.g. LSTM), a sequential order should be defined, which is not feasible efficiently considering the 2D structure of a raw document page.
- *CNN Scaler:* The CNN Scaler also consists of two conv. layers, both with kernel of size 3×3. The first has 128 output channels and the second, as expected, only one, i.e. the scale value. Again, between the two conv. layers, we added ReLU, batch-norm and Dropout. Finally, we apply a sigmoid function over the output to constrain the range of the scale between 0 and 1 (when the scale $s = 1$ then the pixel corresponds to a whole character).

These components and their functionality are visualized in Fig. 1. As we highlighted earlier, the aforementioned architecture contains a novel and crucial modification; it includes a *separate* scale map that enables character counting and is trained *implicitly* as a auxiliary path.

Training Details/Extra Modifications: The feature map F of character probabilities is essentially the output of the CNN decoder, after applying a per-pixel softmax operation over the characters' dimension. This way, we can straightforwardly generate page-level detections of characters as seen in Fig. 2. Nonetheless, training using the CTC loss requires a sequence of predictions and not a 2D feature map. To this end, during word-level training, we apply a flattening operation before the aforementioned softmax operation. Specifically, we use the column-wise max-pooling used in [18].

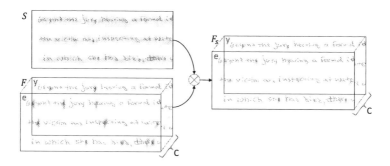

Fig. 2. Document-level extraction of feature maps. The scale map S is multiplied with character probability map F to generate F_s. For the 3D tensors F, F_s of C channels, we indicatively show the per-character activation of characters 'e' and 'y'.

We also used the following modifications to assist training: 1) we extract a larger region around a word to learn that different characters may exist outside the bounding box as neighboring "noise" and 2) we penalize pixels outside the bounding box before the column-wise maxpooling operation and thus force them to be ignored during the recognition step (we use constant negative values since we have a max operation)

Overall, the loss used for training was $L_{CTC} + 10 \cdot L_{count}$, where: **1)** $L_{CTC} = CTC(F_{max}, s_{target})$ is the recognition loss, with F_{max} being the column-wise max-pooled version of F with penalized values outside of the bounding box and s_{target} the target word text, *and* **2)** $L_{count} = ||y_c - t_c||_2$ is the counting regression loss, where y_c is the predicted counting histogram calculated as in Eq. 1 and t_c the target counting histogram. The weights of each individual loss were set empirically with no further exploration. Since the regression loss is more important for our case, it was assigned a larger weight.

2.2   Efficient Spotting Using Character Counting

Having defined the counting architecture and its training procedure, we proceed with our major goal: utilize the generated counting map to efficiently detect bounding boxes of queries. To this end, in what follows, we will present different sub-modules designed to provide fast and accurate predictions. Figure 3 contains a visualization of the proposed pipeline.

Problem Statement and Initial Complexity: Here, the input is the per-character scaled map F_s, as generated by the proposed architecture. The task is to *estimate a bounding box that contains a character count that is similar to the query*. Note that this way, the detection is size-invariant - the bounding box can be arbitrarily large. The simplest way to perform such actions is extremely ineffective; for each pixel of the (downscaled) feature map, one should check all the possible bounding boxes and compute their sum, resulting to an impractical complexity of $\mathcal{O}(N_r^3)$, where $N_r = H_r \times W_r$ is the number of pixels of the feature map. This can be considered as a naive sliding box approach.

Cost-Free Summation with Integral Images: Since the operation of interest is summation, the use of integral images can decrease this summation step of arbitrary-sized boxes to a constant $\mathcal{O}(1)$ step. The use of integral images is widespread in computer vision applications with SURF features as a notable example [3], while Ghosh et al. [6] used this concept to also speed-up segmentation-free word spotting. Moreover, as we will describe in what follows, the proposed detection algorithm heavily relies on integral images for introducing several efficient modifications.

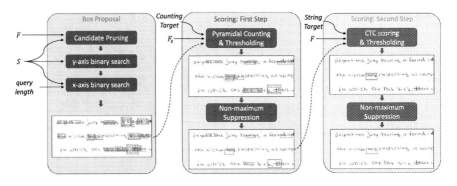

Fig. 3. Overview of the proposed spotting pipeline. The box proposal stage is described in Sect. 2.2, while the scoring steps are described in Sect. 2.3.

Bounding Box Estimation with Binary Search: Next, we focus on how to search through any window size without the need to actually parse the whole image. Specifically consider a starting point (x_s, y_s) on the feature map. We seek the ending point (x_e, y_e), where the counting result in the bounding box, defined by these two points, is as close as possible to the character count of the query. Here, we make use of a simple property of the generated feature map. The counting result should be **increasing** as we go further and further from the starting point. This property enables us to break this task into two simple increasing sub-tasks that can be efficiently addressed by *binary search* operations. Note that we compute the required counting value on-the-fly by using integral images (only the starting point and ending point suffice to compute the requested summation).

To further simplify the process we act only on the integral image of scale S and we perform the following steps: **1)** Find a rectangular area of count equal to 1. This is the equivalent of detecting one character. We assume that the side of the detected rectangle is a good estimation of a word height on this specific location of the document. This operation is query independent and can be pre-computed for each image. This step is depicted in Fig. 4(a). **2)** Given the height of the search area, find the width that includes the requested character count. This step is depicted in Fig. 4(b).

The aforementioned two-step procedure reasoning is two-fold; first it produces candidate bounding boxes with minor computational requirements and it also resolves possible axis ambiguity. The latter problem can be seen in Fig. 4(c), where we can find a "correct" bounding box, containing the requested counting histogram, across neighboring text-lines.

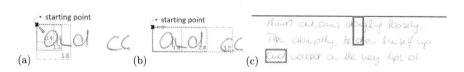

(a)   (b)   (c)

Fig. 4. (a,b): Binary search operations over the scale map S to find the height (a) and the width (b) of a possible word. (c): Visualization of the issue of correct counting across lines. Both boxes contain the requested character counting for the query 'and'.

Candidate Point Pruning: Up to this point, the proposed modifications notably decreased the algorithm complexity to $\mathcal{O}(N_r \log N_r)$. Nonetheless, traversing through all points is unnecessary and we can further decrease computational requirements via a *pruning* stage over possible starting points. We distinguish two useful (heuristic) actions that can considerably reduce the search space:

1) The starting point should have (or be close to a point that has) a non-trivial probability over the first character of the query. Implementation-wise, a max-pooling operation with kernel size 3 (morphological dilation) is used to simulate the proximity property, while a probability threshold is used to discard points. The threshold is relatively low (e.g., 0.1), in order to allow the method to be relatively robust to partial character misclassification.

2) Find only "well-centered" bounding boxes. This step is implemented through the use of integral images, where we compute the counting sum over a reduced window over y-axis (height) in the center of the initially detected bounding box. If the counting sum of the subregion divided by the sum of the whole word region is lower than a threshold ratio, we discard the point. In other words, we try to validate if the included information in the bounding box corresponds to "centered" characters, or it could be a cross-line summation of character parts.

2.3 Similarity Scoring

The proposed detection algorithm acts in a character-agnostic way with the exception of the first character of the query. Now, we have to find the most similar regions w.r.t. to the actual query. The most straightforward solution is to compute the per-character counting inside the bounding box and compare it to the query counting. The predicted counting can be computed with minor overhead by using the integral image rationale over the extracted scaled character probability map. Comparison is performed via cosine similarity.

A counting-based retrieval cannot distinguish between different permutations of the query characters. For example, "and" and "dan" of Fig. 5(left) have the same counting description, confusing the system. Therefore, the necessity to alleviate ambiguities comes in the limelight. Towards addressing this problem, we propose two different approaches that can be combined into a single method, as distinct steps, and have a common characteristic: the *already trained network* is appropriately utilized to effectively predict more accurate scores. Both steps are followed by a typical non-maximum suppression step. Figure 3 depicts these steps, as well as their outputs, in detail.

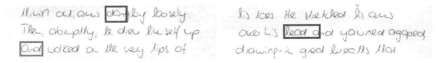

Fig. 5. Examples of two possible detection problems: character order ambiguity (left) and box extension to neighboring words (right).

First Step: Pyramidal Counting. A straightforward extension of the basic counting procedure is building a descriptor in a pyramidal structure, akin to PHOC [1,27]. The idea is simple: break the query string into uniform parts and compute the character count into each part. This is repeated for l levels, with l counting histograms at each vector, resulting to a descriptor of size $= l(l+1)C/2$. A Gaussian-based weighting is used for characters belonging to more than one segments, in order to distribute the value accordingly along the segments.

Given the pyramidal query descriptor, one must find its similarity to the respective predicted pyramidal counting over the detected boxes. This is also efficiently performed by using integral images over F_s to compute sub-area sums of horizontal segments in the interior of each detected box. Matching is then performed by computing cosine similarity. In its simplest form, this first step computes the cosine similarity of the count histograms with only 1 level.

Second Step: CTC Re-scoring. For this re-scoring step, we consider a more complex approach based on CTC loss with a higher computational overhead. Thus, we use the descriptors of the first step to limit the possible region candidates. This is performed by a non-maximum suppression step with a low IoU threshold ($= 0.2$) that prunes overlapping regions. Subsequently, the top $K = 30$ results per document page are considered for this re-scoring step.

Given a detected box, one can compute the CTC score from the already extracted map, resembling forced-alignment rationale. Implementation-wise, to avoid a max-pooling operation of different kernels at each region, we precomputed a vertical max-pooling of kernel $= 3$ over the character probabilities map. Subsequently, the sequence of character probabilities was extracted from the centered y-value of the bounding box under consideration.

Despite the intuitive concept this approach, as is, under-performed, often resulting to worse results compared to a pyramidal counting with many levels. A common occurring problem, responsible for this unexpected performance inferiority, was the inaccurate estimation of bounding boxes that extended to neighboring words, as seen in Fig. 5(right). Such erroneous predictions are typically found if a character existing in the query also appears in neighboring words. Note that the used counting approach cannot provide extremely accurate detections and we do not expect it to do so.

The solution to this issue is simple when considering the CTC algorithm: a score matrix $D \in \mathbb{R}^{T \times C}$ emerges, where the score $D[t, c]$ corresponds to step $t = 0, \ldots, T - 1$ assuming that the character $c = 0, \ldots, C - 1$ exists at this step. Therefore, instead of selecting the score of the last query character (alongside the blank character - but we omit this part for simplicity) at the last step, we search for the best score of the last character over all steps, meaning that the whole query sequence should be recognized and only redundant predictions are omitted. To assist this approach, an overestimation of the end point is considered over the x-axis. This approach is straightforwardly extended to correct the starting point of the box, by inverting the sequence of probabilities along with the query characters.

Summing this procedure up, we can use the recognition-based CTC algorithm over the detected box to not only provide *improved scores*, but also *enhance the bounding box prediction*. As one can deduct, the re-scoring step can be performed by an independent (large) model to further increase performance, if needed. In this work, however, it is of great interest to fully utilize the already existing model and simultaneously avoid adding an extra deep learning component which may add considerable computational overhead.

3   Revisiting Overlap Metrics

Objection detection is typically evaluated with mean Average Precision (mAP) if the detected bounding box considerably overlaps with the corresponding gt box. This overlap is quantified through the Intersection over Union (IoU) metric. Although the usefulness of such an overlap metric is indisputable, it may not be the most suitable for the word spotting application. Specifically, in documents, we have disjoint entities of words that do not overlap. A good detection could be defined as one that includes the word of interest and no other (neighboring) word. However, the IoU metric sets a relatively strict constraint of how the boxes interact. For example, in Fig. 6(left) we have an arguably good detection of the word, denoted by a green box, that has a very low IoU score of 0.33 with respect to the groundtruth blue box. This phenomenon is frequent when using our method, since we have not imposed constraints of how this box should be, only that it should contain the word that we are interested in. Therefore, for lower IoU thresholds we report very good performance that deteriorates quickly when increasing the threshold, even though the detections are notably spot-on.

To address this issue, we propose a different metric that does not penalize enlarged boxes as long as no other neighboring word is intersected by the detected box. Specifically, we want to penalize such erroneous intersections and we include their total area into the denominator of our new metric, along with the area of the groundtruth box. The numerator, as usual, is the intersection of the detected box with the groundtruth box. This concept is clearly visualized in Fig. 6(right).

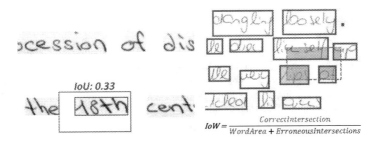

Fig. 6. Visualization of the IoU issue (left), where a good detection is overly penalized, and of the proposed IoW metric (right), where blue shaded is the original gt area, green shaded the overlap with the gt word and red shaded the erroneous overlaps.

4   Experimental Evaluation

Datasets: We evaluated the proposed system in two widely-used English datasets. The first one is the challenging **IAM** dataset [16], consisted of handwritten text from 657 different writers and partitioned into writer-independent train/val./test sets. IAM grountruth has indication of erroneous word-level annotation that was used to mask out these regions from the page-level images. The second dataset is a collection of manuscripts from George Washington (GW) [4], consisted of 20 pages with 4860 words in total. The split into training/test sets follows the protocol of [31] with a 4-fold validation scheme.

Following the standard paradigm for these two datasets, we ignore punctuation marks and we merge lowercase and uppercase characters. All possible words that do not contain any non-alphanumeric character are considered as queries. Only for the case of IAM, queries that belong to the official stopword list are removed from the query list [1, 27].

Metrics: We evaluate our approach using the standard metric for retrieval problems, Mean Average Precision (MAP). Since the task at hand is segmentation-free spotting, we also utilize a overlap metric between detections and ground-truth boxes. Following the discussion of Sect. 3 we consider three different overlap metrics: *1)* standard Intersection over Union (IoU) *2)* a modified IoU, dubbed as x-IoU, that focuses on the overlap over the x-axis by assuming the same y-coordinates between the detection and the groundtruth, while requires an initial overlap of over 0.1 IoU. This resembles a more line-focused metric. *3)* the proposed IoW metric of Sect. 3.

Implementation Details: Training of our model was performed in a single NVidia 1080Ti GPU using the Pytorch framework. We trained our model for 80 epochs with Adam optimizer [11] along with a cosine annealing scheduler, where learning rate started at $1e-3$. The proposed spotting method, applied over the feature map extracted by the CNN, is implemented with cpu-based Numba [14] functions in order to achieve efficient running time. GPU-based implementation of such actions has not been explored, since specific operations cannot be straightforwardly ported with efficient Pytorch functions (e.g., binary search). Nonetheless, implementation on GPU could considerably improve time requirements as a potential future extension. Code is publicly available at https://github.com/georgeretsi/SegFreeKWS.

4.1   Ablation Studies

Every ablation is performed over the validation set of IAM using a network of $\sim 6M$ parameters. The typical IoU threshold is used, unless stated otherwise.

Impact of Spotting Modifications: Here, we will explore the impact of hyper-parameters (thresholds) selected in the proposed spotting algorithm. Specifically, we focus on discovering the sensitivity of the candidate pruning thresholds, since this is the only step that introduces critical hyper-parameters. To this end, we perform a grid search over the probability threshold p_{thres} and the centering ratio threshold r_{thres}, as reported in Tables 1. Only the simple cosine similarity of the character count histogram has used in this experiment and thus the MAP scores are relatively low. As we can see, the 0.5 ratio threshold to be the obvious choice, while, concerning the probability threshold, both 0.01 and 0.05 values provide superior performance with minor differences and thus 0.05 value is selected as the default option for the rest of the paper, since it provides non-trivially faster retrieval times.

Impact of Scoring Methods: As we can see from the previous exploration, relying only on character counting leads to underperformance. Here, we will explore the impact of the different scoring methods proposed in Sect. 2.3. We distinguish different strategies according to the use of the CTC re-scoring step.

Table 1. Exploration of pruning thresholds with the character counting retrieval method - only cosine similarity over character histograms was used. We report MAP 25% IoU Overlap & time (in parenthesis) to retrieve bboxes for a (image, query) pair.

p_{thres}	r_{thres}				
	0.25	0.33	0.5	0.66	0.75
0.001	65.85 (2.422)	65.96 (2.209)	66.15 (1.243)	54.66 (0.499)	35.11 (0.363)
0.01	67.57 (1.101)	67.59 (1.006)	67.97 (0.628)	53.30 (0.317)	32.54 (0.234)
0.05	67.82 (0.643)	67.85 (0.624)	**67.99** (0.433)	52.02 (0.263)	31.17 (0.224)
0.1	67.53 (0.569)	67.55 (0.535)	67.67 (0.389)	51.49 (0.253)	30.82 (0.222)
0.2	67.09 (0.509)	67.10 (0.475)	67.16 (0.355)	51.00 (0.249)	30.47 (0.215)
0.5	66.03 (0.445)	66.04 (0.428)	66.04 (0.322)	49.38 (0.238)	29.66 (0.203)

Specifically we report results without the re-scoring step and with the one-way (adjusting the bound on the right of the box) or two-way (adjusting both horizontal bounds using a reverse pass over the sequence - see Sect. 2.3) CTC step. We also report results for the multilevel pyramidal representations (denoted as PCount) of character counting and the PHOC alternative, implemented as a thresholded version of the counting histogram that does not exceed 1 at each bin. These results are summarized in Table 2, where we also report the time needed for a query/image pair. The following observation can be made: • As expected, adding the CTC scoring step, the results are considerably improved. The two-way CTC score approach achieves the best results overall, regardless the initial step (e.g., # levels, PHOC/PCount). • PCount provides more accurate detections compared to PHOC when a single level is used, but this is not the case in many configurations for extra levels. • For the CTC score variant, we do not see any improvement when using more level on the first step. This can be attributed to the fact that extended box proposals was a common error, as shown in Fig. 5(right), and thus sub-partitioning of the box does not correspond to actual word partition. • Time requirements are increasing as we use more levels. Furthermore, as expected, the increase when using the CTC score approach and especially the two-way variant. Nonetheless, due to its performance superiority, we select the 1-level PCount first-step along with the two-way CTC score as the default option for the rest of the paper.

Comparison of Overlap Metrics: Even though we presented notable MAP results for the 25% IoU overlap, we noticed that the performance considerably decreases as the overlap threshold increases. Specifically, for our best-performing system, the MAP drops from 88.98% to only 54.68%. Preliminary error analysis showed that the main reason was the strict definition of IoU for word recognition, as described in Sect. 3, where many correct detections presented a low IoU metric. To support this claim, we devised two different overlap metrics (x-IoU and IoW) and we validated the attained MAP for a large range of threshold values, as shown in Fig. 7. As we can see, both the alternatives provide almost the

Table 2. Impact of different scoring approaches: PCount vs PHOC for different levels/use of CTC scoring step (one-way or two-way). We reported MAP 25% IoU Overlap & time (in miliseconds) to retrieve bboxes for a (image, query) pair.

levels	w/o ctc-score			w/ one-way ctc-score			w/ two-way ctc-score		
	PHOC	PCount	time	PHOC	PCount	time	PHOC	PCount	time
1	64.60	67.99	0.44	85.29	85.58	0.68	88.90	**88.98**	0.81
2	72.57	72.24	0.50	86.30	85.83	0.69	88.90	88.88	0.82
3	73.45	72.65	0.62	86.58	86.06	0.77	88.74	88.55	0.92
4	72.76	71.52	0.78	86.31	86.11	0.87	88.24	88.18	0.98
5	71.14	69.82	0.93	85.86	85.60	1.01	87.85	87.61	1.09

same performance up to 60% overlap threshold whereas rapidly decreases from early on. Overall the proposed IoW metric has very robust performance, proving that the main source of performance decrease for larger overlap thresholds was the "strict" definition of what a good detection is for word spotting applications (see Sect. 3 for details).

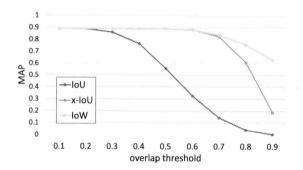

Fig. 7. MAP performance for different overlap thesholds over 3 different overlap metrics: IoU, x-IoU and IoW.

4.2 State-of-the-Art Comparisons

Having explored the different "flavors" of our proposed systems, we proceed to evaluate our best-performing approach (single-level counting similarity along with two-way CTC score) against state-of-the-art segmentation-free approaches. Apart form evaluating over the test set of IAM, for the GW dataset, we follow the evaluation procedure used in [23,31] where two different 4-fold cross validation settings are considered. The first assumes a train set of 15 pages, setting aside 5 for testing, while the second assumes a more challenging 5/15 split for training/testing, respectively. In each setting, the reported value is the average across the 4-fold setup. We trained our models only on the pages available and not pre-training was performed. The results are summarized in Table 3, where MAP values for both 25% and 50% overlap thresholds are reported. The presented results lead to several interesting observations: • the proposed method outperforms substantially the compared methods (even the complex end-to-end architecture of [31]), for the challenging cases of GW5-15 and IAM, when the 25% IoU overlap is used. In fact, the proposed method is robust even when using limited training data (e.g., GW5-15), since it does not learn both detection and recognition tasks, but only relies on recognition. • As we described before, performance is decreased for the 50% IoU overlap, but the different overlap metrics (x-IoU and IoW) show the effectiveness of the method for larger overlap thresholds, as shown in Table 3(b). Nonetheless, this is not directly comparable with the other compared methods that relied on IoU. • For the case of GW15-5, we do not report results on par with the SOTA, even though the same method has a considerable boost for GW5-15, where fewer data were used. Error analysis showed that for GW, proposals were extended in the y-axis also, resulting to low overlap scores. In other words, the "rough" localization of our method leads to this reduced performance (also discussed in limitations).

Table 3. (a) MAP comparison of state-of-the-art approaches for IAM and two variations of GW dataset. Both 25% and 50% overlap thresholds are reported. (b) We also report the performance of our method when the proposed overlap metric alternatives are considered (x-IoU, IoW).

		GW 15-5		GW 5-15		IAM	
	method	25%	50%	25%	50%	25%	50%
	BoF HMMs [23]	80.1	76.5	58.1	54.6	-	-
	BG index [5]	-	-	-	-	-	48.6
(a)	Word-Hypothesis [25]	90.6	84.6	-	-	-	-
	Ctrl-F-Net DCToW [31]	95.2	91.0	76.8	73.8	82.5	80.3
	Ctrl-F-Net PHOC [31]	93.9	90.1	68.2	65.6	80.8	78.8
	Resnet50 + FPN [33]	96.5	94.1	-	-	-	-
	Proposed (IoU)	91.6	66.4	85.9	66.3	85.8	59.2

		GW 15-5		GW 5-15		IAM	
	Proposed (IoU)	91.6	66.4	85.9	66.3	85.8	59.2
(b)	Proposed (x-IoU)	93.2	92.9	86.8	86.7	86.9	86.8
	Proposed (IoW)	92.7	87.6	86.8	83.0	86.9	86.3

Visual Examples: Figure 8 contains examples of retrieval for the GW dataset, where we can see that retrieved boxes are not tight and can be extended. Specifically, for the case of the query 'them', we retrieved erroneous boxes where the word 'the' appears followed by a word that starts with 'm'. In Fig. 9, we present some examples of successful multi-word and sub-word retrieval in the IAM dataset. Notably, in the sub-word case of the "fast" query, the fast suffix was detected for the word breakfast in both the handwritten text and also the typewritten reference text of different scale at the top of the image.

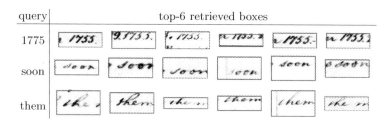

Fig. 8. Examples of QbS with top-6 retrieved words reported for the GW dataset. Green box corresponds to an overlap greater than 25% IoU, while red to lower.

(a) Query: "he came"   (b) Query: "fast"

Fig. 9. Examples of multi-word and sub-word detections.

Limitations: • The bounding box proposal stage of Sect. 2.2 tends to provide over-estimations of the actual box, as shown in Fig. 5(b). This is adequately addressed by the CTC re-scoring step, but such phenomena may cause correct regions to be dismissed before re-scoring. Therefore, a tightened box prediction could improve the overall performance, especially if we assist this with an appropriately designed model component. • The proposed approach does not learn to distinguish the space character, as a separator between words, and thus we the ability to detect sub-words can be also be seen as an issue that affects performance. In fact, if we let sub-words to be counted as correct predictions the MAP (at 25% IoU overlap) increases from 88.98% to 89.62%. Even though, detecting sub-words is a desirable property, it would useful if the user could select when this should happen. To add such property, we could include and train the space character as a possible future direction.

5   Conclusions

In this work we presented a novel approach to segmentation-free keyword spotting that strives for simplicity and efficiency without sacrificing performance. We designed an architecture that enables counting characters at rectangular subregions of a document image, whereas it is only trained on single word images. The box proposal and scoring steps are designed to speed-up the retrieval of relevant regions, utilizing integral images, binary search and re-ranking of retrieved images using CTC score. The reported results for both GW and IAM dataset prove the effectiveness of our method, while using a simple network of 6M parameters.

Acknowledgments. This research has been partially co - financed by the EU and Greek national funds through the Operational Program Competitiveness, Entrepreneurship and Innovation, under the call "OPEN INNOVATION IN CULTURE", project *Bessarion* (T6YBII - 00214).

References

1. Almazán, J., Gordo, A., Fornés, A., Valveny, E.: Word spotting and recognition with embedded attributes. IEEE Trans. Pattern Anal. Mach. Intell. **36**(12), 2552–2566 (2014)
2. Almazan, J., Gordo, A., Fornés, A., Valveny, E.: Handwritten word spotting with corrected attributes. In: Proceedings of the IEEE International Conference on Computer Vision, pp. 1017–1024 (2013)
3. Bay, H., Ess, A., Tuytelaars, T., Van Gool, L.: Speeded-up robust features (SURF). Comput. Vis. Image Underst. **110**(3), 346–359 (2008)
4. Fischer, A., Keller, A., Frinken, V., Bunke, H.: Lexicon-free handwritten word spotting using character HMMs. Pattern Recogn. Lett. **33**(7), 934–942 (2012)
5. Ghosh, S.K., Valveny, E.: Query by string word spotting based on character bigram indexing. In: 2015 13th International Conference on Document Analysis and Recognition (ICDAR), pp. 881–885. IEEE (2015)
6. Ghosh, S.K., Valveny, E.: A sliding window framework for word spotting based on word attributes. In: Paredes, R., Cardoso, J.S., Pardo, X.M. (eds.) IbPRIA 2015. LNCS, vol. 9117, pp. 652–661. Springer, Cham (2015). https://doi.org/10.1007/978-3-319-19390-8_73
7. Ghosh, S.K., Valveny, E.: R-PHOC: segmentation-free word spotting using CNN. In: 2017 14th IAPR International Conference on Document Analysis and Recognition (ICDAR), vol. 1, pp. 801–806. IEEE (2017)
8. Graves, A., Fernández, S., Gomez, F., Schmidhuber, J.: Connectionist temporal classification: labelling unsegmented sequence data with recurrent neural networks. In: Proceedings of the 23rd International Conference on Machine Learning, pp. 369–376. ACM (2006)
9. He, K., Zhang, X., Ren, S., Sun, J.: Deep residual learning for image recognition. In: Proceedings of the IEEE Conference on Computer Vision and Pattern Recognition, pp. 770–778 (2016)

10. Jaderberg, M., Vedaldi, A., Zisserman, A.: Deep features for text spotting. In: Fleet, D., Pajdla, T., Schiele, B., Tuytelaars, T. (eds.) ECCV 2014. LNCS, vol. 8692, pp. 512–528. Springer, Cham (2014). https://doi.org/10.1007/978-3-319-10593-2_34

11. Kingma, D.P., Ba, J.: Adam: A method for stochastic optimization. arXiv preprint arXiv:1412.6980 (2014)

12. Krishnan, P., Dutta, K., Jawahar, C.: Word spotting and recognition using deep embedding. In: 2018 13th IAPR International Workshop on Document Analysis Systems (DAS), pp. 1–6. IEEE (2018)

13. Krishnan, P., Jawahar, C.: HWNet v2: an efficient word image representation for handwritten documents. arXiv preprint arXiv:1802.06194 (2018)

14. Lam, S.K., Pitrou, A., Seibert, S.: Numba: a LLVM-based python JIT compiler. In: Proceedings of the Second Workshop on the LLVM Compiler Infrastructure in HPC, pp. 1–6 (2015)

15. Lyu, P., Liao, M., Yao, C., Wu, W., Bai, X.: Mask textspotter: an end-to-end trainable neural network for spotting text with arbitrary shapes. In: Proceedings of the European Conference on Computer Vision (ECCV), pp. 67–83 (2018)

16. Marti, U.V., Bunke, H.: The IAM-database: an English sentence database for offline handwriting recognition. Int. J. Doc. Anal. Recogn. **5**(1), 39–46 (2002)

17. Retsinas, G., Louloudis, G., Stamatopoulos, N., Sfikas, G., Gatos, B.: An alternative deep feature approach to line level keyword spotting. In: Proceedings of the IEEE/CVF Conference on Computer Vision and Pattern Recognition, pp. 12658–12666 (2019)

18. Retsinas, G., Sfikas, G., Gatos, B., Nikou, C.: Best practices for a handwritten text recognition system. In: Uchida, S., Barney, E., Eglin, V. (eds.) DAS 2022. LNCS, vol. 13237, pp. 247–259. Springer, Cham (2022). https://doi.org/10.1007/978-3-031-06555-2_17

19. Retsinas, G., Sfikas, G., Gatos, B., Nikou, C.: On-the-fly deformations for keyword spotting. In: Uchida, S., Barney, E., Eglin, V. (eds.) DAS 2022. LNCS, vol. 13237, pp. 338–351. Springer, Cham (2022). https://doi.org/10.1007/978-3-031-06555-2_23

20. Retsinas, G., Sfikas, G., Louloudis, G., Stamatopoulos, N., Gatos, B.: Compact deep descriptors for keyword spotting. In: 2018 16th International Conference on Frontiers in Handwriting Recognition (ICFHR), pp. 315–320. IEEE (2018)

21. Retsinas, G., Sfikas, G., Nikou, C., Maragos, P.: From Seq2Seq to handwritten word embeddings. In: British Machine Vision Conference (BMVC) (2021)

22. Retsinas, G., Sfikas, G., Stamatopoulos, N., Louloudis, G., Gatos, B.: Exploring critical aspects of CNN-based keyword spotting. a phocnet study. In: 2018 13th IAPR International Workshop on Document Analysis Systems (DAS), pp. 13–18. IEEE (2018)

23. Rothacker, L., Fink, G.A.: Segmentation-free query-by-string word spotting with bag-of-features HMMs. In: 2015 13th International Conference on Document Analysis and Recognition (ICDAR), pp. 661–665. IEEE (2015)

24. Rothacker, L., Rusinol, M., Fink, G.A.: Bag-of-features HMMs for segmentation-free word spotting in handwritten documents. In: 2013 12th International Conference on Document Analysis and Recognition, pp. 1305–1309. IEEE (2013)

25. Rothacker, L., Sudholt, S., Rusakov, E., Kasperidus, M., Fink, G.A.: Word hypotheses for segmentation-free word spotting in historic document images. In: 2017 14th IAPR International Conference on Document Analysis and Recognition (ICDAR), vol. 1, pp. 1174–1179. IEEE (2017)

26. Sfikas, G., Retsinas, G., Gatos, B.: A PHOC decoder for lexicon-free handwritten word recognition. In: 2017 14th IAPR International Conference on Document Analysis and Recognition (ICDAR), vol. 1, pp. 513–518. IEEE (2017)
27. Sudholt, S., Fink, G.A.: PHOCNet: a deep convolutional neural network for word spotting in handwritten documents. In: Proceedings of the 15th International Conference on Frontiers in Handwriting Recognition (ICFHR), pp. 277–282 (2016)
28. Sudholt, S., Fink, G.A.: Evaluating word string embeddings and loss functions for CNN-based word spotting. In: 2017 14th IAPR International Conference on Document Analysis and Recognition (ICDAR), vol. 1, pp. 493–498. IEEE (2017)
29. Toselli, A.H., Vidal, E., Romero, V., Frinken, V.: HMM word graph based keyword spotting in handwritten document images. Inf. Sci. **370**, 497–518 (2016)
30. Wilkinson, T., Brun, A.: Semantic and verbatim word spotting using deep neural networks. In: 2016 15th International Conference on Frontiers in Handwriting Recognition (ICFHR), pp. 307–312. IEEE (2016)
31. Wilkinson, T., Lindstrom, J., Brun, A.: Neural Ctrl-F: segmentation-free query-by-string word spotting in handwritten manuscript collections. In: Proceedings of the IEEE International Conference on Computer Vision, pp. 4433–4442 (2017)
32. Zhang, X., Su, Y., Tripathi, S., Tu, Z.: Text spotting transformers. In: Proceedings of the IEEE/CVF Conference on Computer Vision and Pattern Recognition (CVPR), June 2022, pp. 9519–9528 (2022)
33. Zhao, P., Xue, W., Li, Q., Cai, S.: Query by strings and return ranking word regions with only one look. In: Proceedings of the Asian Conference on Computer Vision (2020)

Document Analysis and Recognition 3: Text and Document Recognition

A Hybrid Model for Multilingual OCR

David Etter[1(✉)], Cameron Carpenter[2], and Nolan King[2]

[1] Human Language Technology Center of Excellence, Johns Hopkins University, Baltimore, USA
`detter2@jhu.edu`
[2] SCALE Participants, Laurel, USA

Abstract. Large-scale document processing pipelines are required to recognize text in many different languages. The writing systems for these languages cover a diverse set of scripts, such as the standard Latin characters, the logograms of Chinese, and the cursive right-to-left of Arabic. Multilingual OCR continues to be a challenging task for document processing due to the large vocabulary sizes and diversity of scripts. This work introduces a multilingual model that recognizes over nine thousand unique characters and seamlessly switches between ten different scripts. Our transformer-based encoder-decoder approach combines a CTC objective on the encoder with a cross-entropy objective on the full autoregressive decoder. The hybrid approach allows the fast non-autoregressive encoder to be used in standalone mode or with the full autoregressive decoder. We evaluate our approach on a large multilingual dataset, where we achieve state-of-the-art character error rate results in all thirteen languages. We also extend the encoder with auxiliary heads to identify language, predict font, and detect vertical lines.

Keywords: Multilingual OCR · Transformer · Hybrid · Encoder-Decoder · CTC · Auxiliary heads · Synthetic data

1 Introduction

The ISO 693-3 [3] international standard identifies over 7,800 languages. The standard includes languages classified as living, extinct, ancient, and artificially constructed. Approximately 4,000 [8] of these languages have some form of a writing system, with the most common scripts being Latin, Chinese, and Arabic. The complexities of these scripts and the size of the vocabulary have made multilingual OCR a challenging task.

Latin script is widely used for writing European languages, such as English, French, and Spanish. The relatively small character set and abundance of annotated data have made the script a popular choice for optical character recognition (OCR) research [16,30]. In contrast, Chinese (Han) script, which is used within writing the Chinese, Japanese, and Korean languages, includes over 70,000 characters or logograms. Chinese script also contains over 200 radicals that are added to characters and can change form based on their position. Finally, Arabic is the

G. A. Fink et al. (Eds.): ICDAR 2023, LNCS 14187, pp. 467–483, 2023.
https://doi.org/10.1007/978-3-031-41676-7_27

primary script used by the Arabic, Persian, Pashto, and Urdu languages. The Arabic script contains approximately 28 characters and is written from right to left using a cursive style. Arabic script also uses ligatures that join characters into a single glyph and includes several contextual forms that allow characters to change shape based on their position in a word.

Recent work in the OCR community has begun to focus on the challenges of multilingual text within a corpus or document. Datasets such as CAMIO [6], ICDAR MLT [24], and Text OCR [26], provide annotated training and evaluation data in a large number of scripts, languages, and domains. There has also been recent work on multilingual recognition, such as the script-specific head approach of [15] and the transformer-based encoder-decoder approach of [21].

Just as transformer [29] architectures have dominated the NLP [11], and Computer Vision communities [13], text recognition has seen a shift from CNN-LSTM based [25] architectures to those using transformers. Many of these approaches [9,12] continue to integrate a CNN backbone with the self-attention encoder to extract visual features. An end-to-end transformer-based encoder-decoder is used in [21], where the model is initialized using a pretrained Vision encoder with a pretrained NLP decoder.

The combination of connectionist temporal classification (CTC) [14] with attention has proven to be an effective approach for sequence-to-sequence problems. It has been applied to text recognition on recipts [9] and Chinese text [7]. In addition, CTC with attention has also been used in automatic speech recognition (ASR) [17]. A hybrid CTC/attention model that uses multi-objective learning was applied to ASR in [31].

Large-scale document processing pipelines are required to recognize text in many languages and genres. The documents can include images of web pages, newspapers, figures, maps, and presentation slides. In addition, they have challenging layouts, such as multi-column text, diverse fonts, and complex backgrounds. A solution often requires a separate recognizer for each language and genre of interest. However, training and maintaining multiple models adds significant complexity to document pipelines.

Given the challenges of the multilingual recognition problem, we introduce a hybrid transformer-based encoder-decoder model for OCR.Our hybrid model provides true multilingual decoding with a vocabulary of over nine thousand characters and representing ten diverse scripts. The approach combines Connectionist Temporal Classification (CTC) [14] objective function on the encoder with a cross-entropy objective function on the full autoregressive (Seq2Seq) encoder-decoder. This approach allows a single recognizer model to be used for line images from any of the supported languages and provides the ability to handle script-switching within a line image. It also eliminates the need for an explicit script identification step in an OCR pipeline prior to character recognition. Furthermore, the hybrid model can use the faster non-autoregressive encoder in standalone mode or the full autoregressive model with a beam decoder for improved accuracy in some challenging conditions.

Different from [21], we do not rely on a pretrained vision encoder, which requires a predefined fixed input image size and patch size. Instead, our input image uses a height that is more naturally associated with the height of text line images and allows a variable width input that matches the sequence-to-sequence problem. This approach also allows us to investigate patch sizes that best match image characters and generalize to multilingual input. Our extensive experiments with the multilingual CAMIO [6] dataset show that patch size is an essential parameter of multilingual recognition.

Recognition is often a single step in a more extensive document processing pipeline. The recognizer forwards its output to other analytics such as document reconstruction, machine translation (MT), visual question and answer (VQA), or named entity recognition (NER). These downstream systems can benefit from additional visual attributes such as language identification, font family identification, or line orientation. For example, in the case of document reconstruction, identifying a similar font and text orientation is essential for converting the document image into an editable document that preserves the original layout.

To capture visual attributes of the line image, we extend our robust visual encoder with auxiliary heads. The approach uses the outputs of the encoder to train additional heads for language identification, font prediction, and vertical text detection. The outputs from these auxiliary heads are critical to downstream tools and analytics in the document processing pipeline.

Our work makes the following contributions to the document analysis and recognition community:

1. We introduce a hybrid transformer-based encoder-decoder that allows the faster non-autoregressive encoder to be used in standalone mode or jointly with the full autoregressive decoder.
2. Our multilingual model includes a vocabulary of over nine thousand characters in ten unique scripts.
3. Our approach achieves state-of-the-art character error rate results on the multilingual CAMIO [6] dataset.
4. We include auxiliary heads on the encoder that can identify language, predict font family, and detect vertical text.

2   Approach

Our multilingual OCR architecture, shown in Fig. 1, combines a CTC-trained encoder with an autoregressive decoder. This approach allows the self-attention encoder to be used as a fast standalone non-autoregressive model or in combination with the decoder for an autoregressive model. During training, we combine a CTC loss from the self-attention encoder with a cross-entropy loss from the decoder. The encoder loss forces the model to generate aligned visual embeddings in left-to-right order while still allowing the combined encoder-decoder to be jointly optimized. We also take advantage of this robust visual encoder embedding to train additional auxiliary heads to identify language, font, rotation, and orientation.

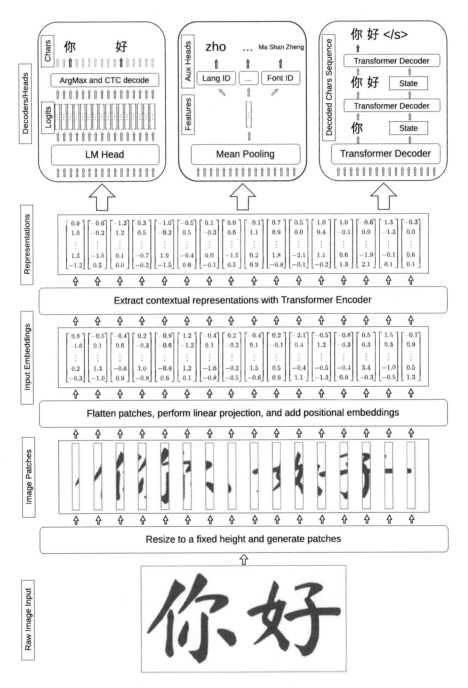

Fig. 1. A diagram of the hybrid CTC/autoregressive decoder design with auxiliary heads.

The encoder takes as input a line image of size (h, w, c) corresponding to height, width, and channel. The image is resized to a fixed height (h_{fix}) with variable width (w_{var}), which maintains the original image aspect ratio. The image is then divided into non-overlapping patches of size (p_h, p_w), where patch height and patch width are hyperparameters of the model. The result is a variable width sequence of image patches with length $s_l = \frac{h}{p_h} \times \frac{w}{p_w}$. Each patch of size $p_h \times p_w \times c$ is then projected to a patch embedding of dimension d. Finally, a positional embedding is added to each patch embedding to encode the original patch order of the line image.

The encoder follows a standard transformer architecture that maps the sequence of s_l patches with dimension d into its hidden states using a stack of residual encoder blocks. Each encoder block includes a multi-headed self-attention layer and two fully connected layers. The output from the final encoder hidden state provides a contextualized representation of the input image patches. This final layer is used both as input to the decoder and is passed to an encoder language model (LM) head. The encoder LM head projects the final encoder layer into a logit vector corresponding to the vocabulary size. A CTC loss function is applied to the encoder LM head outputs during training. The LM head can be used for inference as a standalone OCR decoder, where a character is predicted at each patch step.

In addition to the LM head, we train auxiliary encoder heads for several important visual attributes in an OCR pipeline. The auxiliary heads include language, font, and vertical detection. Each encoder head takes the final encoder layer as input and projects the mean of that embedding into a logit vector corresponding to the attribute vocabulary. The auxiliary heads are trained with a cross-entropy loss over the logit vector.

The decoder also follows a standard encoder-decoder transformer architecture and consists of a stack of residual decoder blocks. Each decoder block includes a self-attention layer, a cross-attention layer, and two fully connected layers. A decoder LM head maps the outputs from the last decoder block into a logits vector corresponding to the vocabulary size. A cross-entropy loss is applied to the decoder LM head during training. The decoder is used in an autoregressive fashion at inference time to predict the next character sequence using outputs from the previous timesteps.

3   Experiments

3.1   Train and Evaluation Corpus

We train the model using both real and synthetically generated line images. The real training images are drawn from the Linguistic Data Consortium (LDC) Corpus of Annotated Multilingual Images for OCR (CAMIO) [6]. The CAMIO dataset is a large corpus of text images covering a wide variety of languages that include text localization and transcript annotations. Figure 2 shows example documents from the training set split for Chinese and Arabic. The dataset is partitioned into 13 languages based on the primary or majority language written in those images. The languages are Arabic, Mandarin, English, Farsi, Hindi, Japanese, Kannada, Korean, Russian, Tamil, Thai, Urdu, and Vietnamese. These cover the 10 scripts of Arabic, Chinese (Simplified), Latin, Devanagari, Japanese, Kannada, Hangul, Cyrillic, Tamil, and Thai.

For each language partition, there are 1,250 images, split into 625 images for training, 125 images for validation, and 500 images for testing (50/10/40 split). The images exhibit a wide range of domains and artifacts, including dense text, unconstrained text, diagrams, tables, multiple text columns, fielded forms, other complex layouts, perspective differences, lighting, skew, and noise. These differing artifacts are generally consistent in proportion across the 13 language partitions. Furthermore, line images were cropped from those images based on the polygon regions labeled for those full images to be used to train and evaluate the decoder.

Fig. 2. Sample document images from the CAMIO [6] train split.

3.2   Synthetic Line Images

In addition to the real training data, we generated a set of synthetic line images for each of the thirteen CAMIO [6] languages. Our synthetic line generation approach inputs seed text, fonts, and background images and produce a collection of synthetic images and annotations. The seed text for our images is drawn from the September 2021 version of the Open Super-large Crawled Aggregated coRpus (OSCAR) [5]. This large multilingual set is filtered from the Common Crawl corpus and is often used for the pretraining of large language models. The fonts to render the line images are selected from Google Fonts, Noto Fonts, and GNU Unifont. Table 1 shows examples of synthetic line images using the seed text, Universal Declaration of Human Rights [28], translated [4] for each of the thirteen languages.

To identify fonts with the ability to render the given language, we sample a set of lines from the corresponding OSCAR language and then perform a mapping of unique character codepoints to each font glyph table. Fonts that match a selected percentage of the unique characters are added to the candidate fonts list for that language. Then, at the time of synthetic line image generation, we randomly select a font from the language list and again perform the codepoint-to-glyph match to verify the current line of text can be rendered. The GNU Unifont is used as the default fallback font in instances where a valid font can not be identified.

In real-world extraction scenarios, the decoder is often presented with noisy cropped lines. This noise includes complex backgrounds, cutoff words, and bounding boxes with inconsistent padding. To make our models more robust to the type of text found in these extraction scenarios, we generate noisy synthetic lines. This includes adding random padding to each line of rendered text's top, bottom, left, and right. We also randomly subsample the lines of OSCAR seed text to create lines with single characters, partial words, and cutoff lines. Finally, to provide complex backgrounds for the synthetic images, we overlay a random subset of the images onto crops of the COCO dataset [22] that has been determined not to include text.

3.3 Evaluation Metric

We evaluate our models using character error rate (CER) over each language in the CAMIO [6] test set after filtering out images annotated as illegible or low-quality. CER measures the percentage of incorrect predictions by the OCR model. Formally, the equation for CER is:

$$CER = (Ins + Sub + Del)/Total\_Char$$

The Levenshtein edit distance calculates character misses in the form of insertions, substitutions, and deletions. The algorithm provides the minimum number of edit operations needed to transform the hypothesis string into the reference string. The lower the score, the better a system performs, with a CER of 0.0 being the perfect output. Before calculating the CER, we perform a standard normalization on the hypothesis and reference strings. First, we normalize all white space characters to a Unicode space (U+0020) and remove repeated white space. Next, we remove punctuation and convert all characters to lowercase. Finally, we apply the standard Unicode Normalization Form: Compatibility (K) Composition (NFKC), which uses compatibility decomposition and canonical composition.

Table 1. Synthetic line image examples.

Language	Script	Image
Arabic	Arabic	الإعلان العالمي لحقوق الإنسان
English	Latin	Universal Declaration of Human Rights
Farsi	Arabic	اعلامیه‌جهانی حقوق بشر
Hindi	Devanagari	मानव अधिकारों की सार्वभौम घोषणा
Japanese	Japanese	『世界人権宣言』
Kannada	Kannada	
Korean	Hangul	세계 인권선언
Russian	Cyrillic	Всеобщая декларация прав человека
Tamil	Tamil	
Thai	Thai	ปฏิญญาสากลว่าด้วยสิทธิมนุษยชน
Urdu	Arabic	انسانی حقوق کا عالمی
Vietnamese	Latin	
Chinese	Simplified	世界人权宣言

3.4 Dictionary

The character dictionary is generated by selecting each unique character from the training set, converting it into a Unicode string representation, and storing it in a list. For example, the ARABIC LETTER WAW with Unicode representation U+0648 is stored as U0648. The index of the Unicode representation in the list is used for character mapping during model training and inference. Positions zero through three of the dictionary are reserved for the special characters start of sentence <s>, pad </pad>, end of sentence </s>, and mask </mask>. The CAMIO [6] thirteen language set data plus OSCAR synthetic data generates a dictionary containing 9,689 characters. Table 2 provides a summary of the scripts in our model dictionary.

3.5 Logical and Visual Ordering

Our multilingual model can train and perform inference on images that present text in both left-to-right and right-to-left horizontal reading order. Languages such as Arabic, Urdu, and Persian use a script that primarily displays right-to-left text, but also frequently includes interposed segments of left-to-right text. Examples include numerals, code switching, or Latin-script strings like e-mail addresses. For such lines of text, the in-memory storage order (called the *logical order* in the Unicode Standard) of the encoded bytes of the text content differs from the visual reading order of the corresponding glyphs. During training,

Table 2. Model scripts.

Script	Character Count
Arabic	153
Cyrillic	81
Devanagari	94
Han	6315
Hangul	1603
Hiragana	84
Kannada	76
Katakana	76
Latin	337
Tamil	60
Thai	82
Other	723

we convert bi-directional text from its logical ordering into a visual ordering using the International Components for Unicode (ICU) BiDi Algorithm [10]. This allows us to train and perform inference using the visual ordering for all languages and then convert back to logical ordering for display and storage.

3.6 Preprocessing

Line images are resized to a fixed height of $h_{\text{fix}} = 28$ and a variable width w_{var} that maintains the original aspect ratio. Images are sorted from largest to smallest width and batched to a fixed maximum width per batch. This provides a variable width batching size that results in smaller batches of wide images and larger batches of narrow images. Finally, the images within a batch are padded to the maximum line width of a given batch. Our batching strategy aims to provide maximum training efficiency on the GPU with a minimum amount of variance within batch widths. A fixed height of 28 was selected during parameter optimization over our validation set. Training images are converted to Arrow tables [1] to provide efficient in-memory access in our distributed training environment. During training, we filter images with a resized width greater than 1200.

In addition to resizing, we perform a number of augmentations to the line images during training. These augmentations include Gaussian noise, blur, JPEG image compression, patch dropout, sharpen, brightness and contrast, and changes to hue, saturation, and value. One random augmentation is selected from the set for half of all training images.

3.7 Hybrid Setup

After the input image has been resized to a fixed 28 height, we divide the input into s_l non-overlapping patches of 28 pixels high and 2 or 4 pixels wide. The result is an average of about three patches for each character displayed in the image. Each patch is then projected into a $d = 512$-dimensional embedding, resulting in a vector of size (batch size × patch count × 512). A sinusoidal positional embedding is then added to each patch embedding according to its position in the patch sequence.

The combined embedding is passed to the encoder, which consists of 16 encoder blocks with 8 attention heads. The encoder also takes as input a padding mask for images that need to be padded to the maximum batch width. Outputs from the encoder are a hidden state which maintains the input shape of (batch size × patch count × 512) and the CTC-head, which projects the hidden state dimension to the vocabulary size of (batch size × patch count × 9,689). The decoder takes as input the encoder hidden state of (batch size × patch count × 512) and adds a sinusoidal positional embedding. This embedding is passed to the decoder, which consists of 6 decoder blocks with 8 attention heads. The final output from the decoder has the shape batch size × patch count × 9,689.

The multilingual OCR model is trained in four stages consisting of an encoder-only (CTC), decoder with a frozen encoder (Seq2Seq), both encoder and decoder (Joint), and finally, the auxiliary encoder heads. Each pass is trained in distributed mode on an NVIDIA TITAN RTX with 4 GPU and 24 GB GPU RAM. The models are trained using a polynomial learning rate scheduler with a warmup period and an AdamW [23] optimizer.

The first training stage includes only the encoder using CTC loss with mean reduction. The encoder is trained for 50 epochs, with ten warm-up epochs and a maximum learning rate of 0.002. The batch size is set to a maximum of 800 line images and a maximum batch width of 80,000 pixels. Training progress is periodically monitored using CER on the CAMIO [6] validation split.

The second stage trains the autoregressive decoder using the frozen encoder from the initial training step and a cross-entropy loss with mean reduction. The decoder is again trained for 50 epochs, with ten warm-up epochs and a maximum learning rate of 0.002. The batch size is set to a maximum of 300 line images and a maximum batch width of 30,000 pixels.

The third stage performs a finetune train of both the encoder and decoder, beginning with the model checkpoints from steps one and two. The learning rate is reduced to a maximum of 0.0001 with a total of 50 epochs and one warm-up epoch. The batch size is set to a maximum of 200 line images and a maximum batch width of 20,000 pixels.

The final stage trains the auxiliary heads using a cross-entropy loss over the mean of the encoder hidden state output. We provide additional details in the section on Auxiliary Encoder Heads.

3.8   Results

We evaluate our multilingual hybrid model on the CAMIO [6] thirteen language test split. The Table 3 shows the results for the initial encoder only (CTC), the frozen encoder with decoder (Seq2Seq), and the jointly trained encoder (CTC-Joint) and encoder-decoder (Seq2Seq-Joint). We perform inference on the standalone encoder using a greedy argmax decoder and a beam search decoder with size five on the full encoder-decoder. The results are evaluated using CER and are shown for input patch sizes of 28×4 and 28×2.

The results show that jointly training with the encoder CTC loss and cross-Entropy loss improves both the standalone encoder and the autoregressive encoder-decoder. We also see that the smaller input patch width of 2 provides noticeable improvements over the larger patch width of 4. This improvement is due to the ability of the smaller patch size to capture small changes in character shapes for scripts like Chinese and Arabic. The largest difference we see is in Urdu, where the patch size of two helps to reduce the CER from 23 down to 10. This is because the Urdu collection contains many lines with Nastaliq fonts, and the smaller patch size helps capture more context. Overall, we see that the full encoder-decoder slightly outperforms the standalone encoder for most languages, at the cost of additional runtime.

We also provide a comparison to pretrained models for Tesseract [27], Easy-OCR [2], TrOCR [21], and MMOCR [18]. The Tesseract results are based on the best LSTM model for each language. The TrOCR model is only available for English and uses the TrOCR-Large-Printed model, available from Huggingface [32]. For MMOCR, we use the SAR [20] model for Chinese, and the SegOCR [18] for English.

Table 3. Results on the CAMIO [6] test split.

Approach	ara	zho	eng	fas	hin	jpn	kan	kor	rus	tam	tha	urd	vie
Patch size (28 × 4)													
CTC	12.1	6.9	2.6	9.1	7.5	7.5	6.1	3.6	2.8	2.1	6.6	28.9	5.5
CTC-Joint	11.0	6.5	2.3	8.1	6.6	6.9	5.6	3.2	2.5	1.9	5.8	23.9	5.0
Seq2Seq	9.9	8.3	2.5	7.1	6.1	8.1	5.1	3.9	3.3	2.0	4.9	11.7	4.9
Seq2Seq-Joint	8.4	6.7	2.1	6.0	5.0	6.8	4.5	3.2	2.6	1.5	4.1	8.7	4.0
Patch size (28 × 2)													
CTC	9.7	6.7	2.1	7.3	6.4	7.4	5.4	3.3	2.7	1.9	5.7	13.0	4.8
CTC-Joint	8.7	**5.9**	1.9	6.4	5.5	6.3	4.8	**2.8**	**2.6**	1.6	5.0	10.3	4.2
Seq2Seq	9.5	7.9	2.2	6.8	5.8	8.7	5.1	3.7	3.2	2.0	4.7	9.6	4.5
Seq2Seq-Joint	**7.8**	6.5	**1.8**	**5.4**	**4.7**	**6.2**	**4.2**	3.0	2.9	**1.4**	**3.6**	**7.3**	**3.5**
Pretrained models													
Paddle OCRv3 [19]	49.6	7.7	5.7	59.7	59.4	34.2	–	45.2	58.6	43.1	–	85.6	31.2
Tesseract [27]	23.4	31.0	13.2	21.4	22.4	35.9	17.0	22.3	15.9	30.1	24.4	49.8	19.9
EasyOCR [2]	18.3	29.5	13.6	19.9	16.7	38.5	16.1	22.3	17.4	10.6	18.1	34.0	17.9
MMOCR [18]	–	32.3	8.4	–	–	–	–	–	–	–	–	–	–
TrOCR [21]	–	–	10.9	–	–	–	–	–	–	–	–	–	–

To understand how the standalone encoder behaves at each patch step, we visualize the output for a Chinese and Arabic line image from the seed text, Universal Declaration of Human Rights [4,28]. Figure 3 displays three rows for each image, which correspond to the original image, an overlay of the patch steps with width four, and the encoder top one output. The red dot aligns the top one output to the corresponding patch step. The visualization shows that the CTC loss helps the encoder align to the patch near the center of a character in both scripts.

Fig. 3. Hybrid Encoder outputs for CAMIO [6] Chinese and Farsi.

For the multilingual setting, we investigate how the large mixed script vocabulary impacts character confusion by the model. For example, Table 4 shows five of the top most confused characters for the standalone encoder on the CAMIO [6] Japanese test split. We observe similar results across the thirteen evaluation languages, where visually similar characters make up the majority of top model confusion.

Table 4. Top confusion pairs for CAMIO [6] Japanese.

Ref Char	Hyp Char	Ref Unicode	Hyp Unicode
ブ	ブ	u+30d7	u+30d6
で	て	u+3067	u+3066
パ	パ	u+30d0	u+30d1
ボ	ポ	u+30dc	u+30dd
ピ	ビ	u+30d4	u+30d3

3.9 Auxiliary Encoder Heads

The language head is trained using a cross-entropy loss over the mean of the encoder hidden state output. We freeze the encoder and train only the language head using the CAMIO [6] and OSCAR synthetic training sets. Results are reported in Table 5 using top one, and top two classification accuracy over the cropped line images of the CAMIO [6] test set. The results show that the auxiliary head from the encoder provides an overall highly accurate language identification. We further investigate results for individual languages using the normalized confusion matrix shown in Fig. 4. We see that most confusions occur for languages with a common script, such as Arabic (ara), Farsi (fas), and Urdu (urd). Similar within-script confusions are found between Chinese (zho) and Japanese (jpn). Overall, the results along the diagonal of the matrix show that our model provides good accuracy in all of the CAMIO languages.

The font attribute for a line image is not available in the CAMIO [6] test dataset, so we create a synthetic data to demonstrate the auxiliary font head of the encoder. First, we hand-selected approximately 95 fonts from the Google Font set that provides full coverage of the thirteen languages and minimizes visual similarity within each language. Next, we generated approximately 25,000 synthetic line images for each language using OSCAR as the seed text. The synthetic lines were split into 80/10/10 for train, validation, and test. The font head was trained with the frozen encoder using a cross-entropy loss over the mean of the encoder's hidden state output. Finally, we evaluate the top 1 and top 2 accuracy results on the test split, shown in Table 6, to demonstrate the ability of the auxiliary head to identify line image font.

We also train an auxiliary head to identify vertical text in line images. The model is trained similarly to the other auxiliary heads using the mean of the encoder hidden state output. We train and evaluate using the CAMIO Chinese, Japanese, Korean, and Vietnamese languages. The results in Table 7, show the precision, recall, and F1-score for the evaluation split in these languages.

Table 5. Language ID accuracy for CAMIO [6].

Auxiliary Head	Top 1	Top 2
Language	0.96	0.99

Table 6. Font ID accuracy for synthetic test set.

Auxiliary Head	Top 1	Top 2
Font	0.95	0.98

Table 7. Vertical evaluation results for CAMIO [6] Chinese, Japanese, Korean, and Vietnamese.

Auxiliary Head	Precision	Recall	F1
Vertical	0.98	0.87	0.92

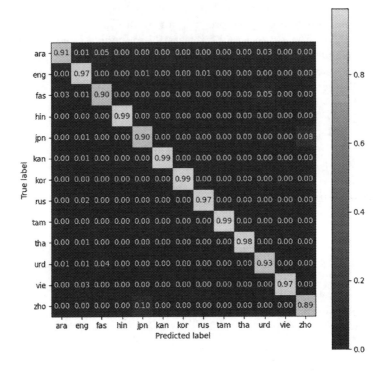

Fig. 4. Language ID confusion matrix on CAMIO [6].

Table 8. Hybrid OCR inference times on the CAMIO [6] Chinese test split (patch size 28×2).

Model	Total Time in seconds	Images per second
CTC-Encoder	1.1	7,280
Seq2Seq	135.9	58

3.10 Inference Speed

We evaluate the CTC, and Seq2Seq model inference speeds on the CAMIO [6] Chinese test set. The evaluation is performed on a TITAN RTX with 24GB of memory using CUDA 11.5. We follow our training preprocess approach by resizing all images to a fixed height of 28 and variable size width that maintains the original aspect ratio. The images are sorted by width from longest to shortest and then batched into groups with a maximum width of 10,000 and a maximum batch size of 100. For the CTC-Encoder, the inference time is calculated by summing the total time of the model forward calls. The Seq2Seq inference uses the Huggingface beam decoder with a beam size of 5. In this case, we calculate inference time by summing the calls of the beam decoder. Table 8 shows that the CTC-Encoder provides a fast and accurate alternative to the full autoregressive model and can average over 7,000 line images per second.

4   Conclusions

This work presents a multilingual approach to text recognition that provides a single model that can recognize over nine thousand characters from thirteen diverse scripts. The hybrid transformer-based encoder-decoder model allows the fast non-autoregressive encoder to be used in standalone mode or with the full autoregressive decoder. Our approach allows variable-width inputs that naturally match the sequence-to-sequence recognition problem. We also experimented with different encoder patch widths and showed how this parameter impacts recognition performance across scripts. In addition, we extend the encoder with auxiliary heads for several important visual attributes of the document processing pipeline. Finally, we evaluate our approach on a challenging multilingual evaluation dataset and show state-of-the-art results for character error rate.

References

1. Arrow. a cross-language development platform for in-memory data (2023). https://arrow.apache.org
2. Easyocr. https://github.com/JaidedAI/EasyOCR (2023)
3. Iso 639–3 code set. sil.org (2023)
4. Udhr in unicode. https://unicode.org/udhr/ (2023)
5. Abadji, J., Ortiz Suarez, P., Romary, L., Sagot, B.: Towards a Cleaner Document-Oriented Multilingual Crawled Corpus. arXiv e-prints arXiv:2201.06642 (Jan 2022)
6. Arrigo, M., Strassel, S., King, N., Tran, T., Mason, L.: CAMIO: a corpus for OCR in multiple languages. In: Proceedings of the Thirteenth Language Resources and Evaluation Conference, pp. 1209–1216 (2022)
7. Cai, S., Xue, W., Li, Q., Zhao, P.: HCADecoder: a hybrid CTC-attention decoder for chinese text recognition. In: Lladós, J., Lopresti, D., Uchida, S. (eds.) ICDAR 2021. LNCS, vol. 12823, pp. 172–187. Springer, Cham (2021). https://doi.org/10.1007/978-3-030-86334-0_12
8. Campbell, L.: Ethnologue: Languages of the world (2008)
9. Campiotti, I., Lotufo, R.: Optical character recognition with transformers and CTC. In: Proceedings of the 22nd ACM Symposium on Document Engineering, pp. 1–4 (2022)
10. Davis, M., et al.: The bidirectional algorithm. Unicode Stand. Annex **9**, 95170–0519 (2004)
11. Devlin, J., Chang, M.W., Lee, K., Toutanova, K.: BERT: Pre-training of deep bidirectional transformers for language understanding. arXiv preprint arXiv:1810.04805 (2018)

12. Diaz, D.H., Qin, S., Ingle, R., Fujii, Y., Bissacco, A.: Rethinking text line recognition models. arXiv preprint arXiv:2104.07787 (2021)
13. Dosovitskiy, A., et al.: An image is worth 16x16 words: Transformers for image recognition at scale. arXiv preprint arXiv:2010.11929 (2020)
14. Graves, A., Fernández, S., Gomez, F., Schmidhuber, J.: Connectionist temporal classification: labelling unsegmented sequence data with recurrent neural networks. In: Proceedings of the 23rd International Conference on Machine Learning, pp. 369–376. ACM (2006)
15. Huang, J., et al.: A multiplexed network for end-to-end, multilingual OCR. In: Proceedings of the IEEE/CVF Conference on Computer Vision and Pattern Recognition, pp. 4547–4557 (2021)
16. Karatzas, D., et al.: ICDAR 2015 competition on robust reading. In: 2015 13th International Conference on Document Analysis and Recognition (ICDAR), pp. 1156–1160. IEEE (2015)
17. Kim, S., Hori, T., Watanabe, S.: Joint CTC-attention based end-to-end speech recognition using multi-task learning. In: 2017 IEEE International Conference on Acoustics, Speech and Signal Processing (ICASSP), pp. 4835–4839. IEEE (2017)
18. Kuang, Z., et al.: MMOCR: A comprehensive toolbox for text detection, recognition and understanding. arXiv preprint arXiv:2108.06543 (2021)
19. Li, C., et al.: PP-OCRv3: More attempts for the improvement of ultra lightweight OCR system. arXiv preprint arXiv:2206.03001 (2022)
20. Li, H., Wang, P., Shen, C., Zhang, G.: Show, attend and read: A simple and strong baseline for irregular text recognition. In: Proceedings of the AAAI Conference on Artificial Intelligence. pp. 8610–8617 (2019)
21. Li, M., et al.: TrOCR: Transformer-based optical character recognition with pre-trained models. arXiv preprint arXiv:2109.10282 (2021)
22. Lin, T.-Y., et al.: Microsoft COCO: common objects in context. In: Fleet, D., Pajdla, T., Schiele, B., Tuytelaars, T. (eds.) ECCV 2014. LNCS, vol. 8693, pp. 740–755. Springer, Cham (2014). https://doi.org/10.1007/978-3-319-10602-1_48
23. Loshchilov, I., Hutter, F.: Decoupled weight decay regularization. arXiv preprint arXiv:1711.05101 (2017)
24. Nayef, N., et al.: ICDAR 2017 robust reading challenge on multi-lingual scene text detection and script identification - RRC-MLT. In: 2017 14th IAPR International Conference on Document Analysis and Recognition (ICDAR). vol. 01, pp. 1454–1459 (2017). https://doi.org/10.1109/ICDAR.2017.237
25. Rawls, S., Cao, H., Kumar, S., Natarjan, P.: Combining convolutional neural networks and LSTMs for segmentation free OCR. In: Proceedings of 2017 14th IAPR International Conference on Document Analysis and Recognition (ICDAR) (2017). https://doi.org/10.1109/ICDAR.2017.34
26. Singh, A., Pang, G., Toh, M., Huang, J., Galuba, W., Hassner, T.: TextOCR: towards large-scale end-to-end reasoning for arbitrary-shaped scene text. In: Proceedings of the IEEE/CVF Conference on Computer Vision and Pattern Recognition, pp. 8802–8812 (2021)
27. Smith, R., Antonova, D., Lee, D.S.: Adapting the tesseract open source OCR engine for multilingual OCR. In: Proceedings of the International Workshop on Multilingual OCR, pp. 1–8 (2009)
28. United Nations: Universal Declaration of Human Rights (Dec 1948)
29. Vaswani, A., et al.: Attention is all you need. In: Advances in Neural Information Processing Systems. vol. 30 (2017)

30. Veit, A., Matera, T., Neumann, L., Matas, J., Belongie, S.: COCO-text: Dataset and benchmark for text detection and recognition in natural images. arXiv preprint arXiv:1601.07140 (2016)
31. Watanabe, S., Hori, T., Kim, S., Hershey, J.R., Hayashi, T.: Hybrid CTC/attention architecture for end-to-end speech recognition. IEEE J. Sel. Top. Sig. Process. **11**(8), 1240–1253 (2017). https://doi.org/10.1109/JSTSP.2017.2763455
32. Wolf, T., et al.: Transformers: state-of-the-art natural language processing. In: Proceedings of the 2020 Conference on Empirical Methods in Natural Language Processing: System Demonstrations, pp. 38–45. Association for Computational Linguistics, Online (Oct 2020). https://doi.org/10.18653/v1/2020.emnlp-demos.6, https://aclanthology.org/2020.emnlp-demos.6

Multi-teacher Knowledge Distillation for End-to-End Text Image Machine Translation

Cong Ma[1,2], Yaping Zhang[1,2(✉)], Mei Tu[4], Yang Zhao[1,2], Yu Zhou[2,3], and Chengqing Zong[1,2]

[1] School of Artificial Intelligence, University of Chinese Academy of Sciences, Beijing 100049, People's Republic of China
{cong.ma,yaping.zhang,yang.zhao,cqzong}@nlpr.ia.ac.cn
[2] State Key Laboratory of Multimodal Artificial Intelligence Systems (MAIS), Institute of Automation, Chinese Academy of Sciences, Beijing 100190, People's Republic of China
yzhou@nlpr.ia.ac.cn
[3] Fanyu AI Laboratory, Zhongke Fanyu Technology Co., Ltd., Beijing 100190, People's Republic of China
[4] Samsung Research China - Beijing (SRC-B), Beijing, China
mei.tu@samsung.com

Abstract. Text image machine translation (TIMT) has been widely used in various real-world applications, which translates source language texts in images into another target language sentence. Existing methods on TIMT are mainly divided into two categories: the recognition-then-translation pipeline model and the end-to-end model. However, how to transfer knowledge from the pipeline model into the end-to-end model remains an unsolved problem. In this paper, we propose a novel Multi-Teacher Knowledge Distillation (MTKD) method to effectively distillate knowledge into the end-to-end TIMT model from the pipeline model. Specifically, three teachers are utilized to improve the performance of the end-to-end TIMT model. The image encoder in the end-to-end TIMT model is optimized with the knowledge distillation guidance from the recognition teacher encoder, while the sequential encoder and decoder are improved by transferring knowledge from the translation sequential and decoder teacher models. Furthermore, both token and sentence-level knowledge distillations are incorporated to better boost the translation performance. Extensive experimental results show that our proposed MTKD effectively improves the text image translation performance and outperforms existing end-to-end and pipeline models with fewer parameters and less decoding time, illustrating that MTKD can take advantage of both pipeline and end-to-end models. Our codes are available at: https://github.com/EriCongMa/MTKD_TIMT.

Keywords: Text Image Machine Translation · Knowledge Distillation · Machine Translation

G. A. Fink et al. (Eds.): ICDAR 2023, LNCS 14187, pp. 484–501, 2023.
https://doi.org/10.1007/978-3-031-41676-7_28

1   Introduction

Text image machine translation (TIMT) is a cross-modal generation task, which translates source language texts in images into target language sentences. Various real-world applications have been conducted for TIMT, such as digital document translation, scene text translation, handwritten text image translation, and so on. Existing TIMT systems are mainly constructed with a recognition-then-translation pipeline model [1,4,7,9,15], which first recognizes texts in images by a text image recognition (TIR) model [2,16,17,27,28], and then generates target language translation with a machine translation (MT) model [20,22,29, 30]. However, pipeline models have to train and deploy two separate models, leading to parameter redundancy and slow decoding speed. Meanwhile, errors in TIR model are further propagated by MT models, which causes more translation mistakes in the final translation results.

To address the shortcomings of pipeline models, end-to-end TIMT models are proposed with a more efficient architecture [14]. Although end-to-end models have fewer parameters and faster decoding speed, the end-to-end training data is limited compared with recognition or translation datasets, leading to inadequate training and limited translation performance of end-to-end models. As a result, how to explicitly incorporate external recognition or translation results has been studied by existing research [6,13]. Furthermore, transfer knowledge from TIR or MT models has been conducted to end-to-end TIMT models through feature transformation [18] and cross-modal mimic framework [5].

However, sub-modules in end-to-end TIMT models play quite different functions, which need different knowledge from various teacher models. Although existing methods explore to transfer knowledge from external models, how to introduce different knowledge into each sub-modules of the end-to-end TIMT model remains unsolved.

In this paper, we propose a novel multi-teacher knowledge distillation (MTKD) approach for end-to-end TIMT model, which is designed to transfer various types of knowledge into end-to-end TIMT model. Specifically, three sub-modules in end-to-end models are considered to optimize by distilling knowledge from different teacher models.

- Image encoder aims at extracting features of input images from pixel space to dense feature space, which has a similar function as the TIR image encoder. As a result, TIR image encoder is utilized as the teacher model for image encoder in end-to-end TIMT model to improve the image feature extraction.
- Sequential encoder in end-to-end TIMT model fuses the local image features into contextual features, which learns advanced semantic information of the sentences in text images. To guide semantic feature learning, MT sequential encoder offers the teacher guidance for TIMT sequential encoder to better map image features into semantic features.
- Decoder in end-to-end TIMT model generates target translation autoregressively, which has a similar function as the MT decoder. As so, the prediction distribution on target language vocabulary is utilized as the teacher distri-

bution to guide the decoder in end-to-end TIMT generate better prediction distribution.

By transferring different knowledge into corresponding sub-modules in end-to-end TIMT model, fine-grained knowledge distillation can better improve the translation quality of end-to-end TIMT models. In summary, our contributions are summarized as:

- We propose a novel multi-teacher knowledge distillation method for end-to-end TIMT model, which is carefully designed for fine-grained knowledge transferring to various sub-modules in end-to-end TIMT models.
- Various teacher knowledge distillation provides more improvements compared with single teacher guidance, indicating different sub-modules in end-to-end models need different knowledge information to better adapt corresponding functions.
- Extensive experimental results show our proposed MTKD method can effectively improve the translation performance of end-to-end TIMT models. Furthermore, MTKD based TIMT model also outperforms pipeline system with fewer parameters and less decoding time.

2   Related Work

2.1   Text Image Machine Translation

Text image machine translation models are mainly divided into pipeline and end-to-end models. Pipeline models deploy text image recognition and machine translation models respectively. Specifically, the source language text images are first fed into TIR models to obtain the recognized source language sentences. Second, the source language sentences are translated into the target language with the MT model. Various applications have been conducted with the pipeline TIMT architectures. Photos, scene images, document images, and manga pages are taken as the input text images. The TIR model recognizes the source language texts, and the MT model generates target language translation [1,3,4,7,9,23,25,26].

End-to-end TIMT models face the problem of end-to-end data scarcity and the performance is limited. To address the problem of data limitation, a multi-task learning method is proposed to incorporate external datasets [6,13,18]. Feature transformation module is proposed to bridge pre-trained TIR encoder and MT decoder [18]. The hierarchy Cross-Modal Mimic method is proposed to utilize MT model as a teacher model to guide the end-to-end TIMT student model [5].

2.2   Knowledge Distillation

Knowledge distillation has been widely used to distillate external knowledge into the student model to improve performance, speed up the training process, and

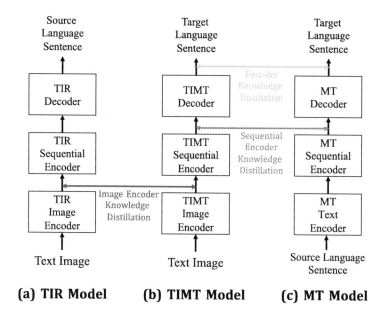

Fig. 1. Overall Diagram of (a) Text Image Recognition, (b) Text Image Machine Translation, (c) Machine Translation models and Multi-Teacher Knowledge Distillation.

decrease the parameter amounts in teacher models. Specifically, in sequence-to-sequence generation related tasks, token-level and sentence-level knowledge distillation have been proven effective in generation tasks [10,11]. Various tasks have been significantly improved through knowledge distillation method, like bilingual neural machine translation [19], multi-lingual translation [21], and speech translation [12].

To incorporate more knowledge into one student model, multiple teacher models are utilized in some studies to further transfer knowledge into student model. [21] proposed to use various teacher models in different training mini-batch to make the multilingual NMT model learn various language knowledge. DOPE is designed to incorporate multiple teacher models to guide different subnetworks of the student model to provide fine-grained knowledge like body, hand, and face segmentation information [24].

However, existing methods lack exploration in integrating various knowledge into end-to-end TIMT models. Our proposed multi-teacher knowledge distillation effectively addresses this problem by transferring different knowledge into various sub-modules to meet the corresponding functional characteristics of different modules.

3   Methodology

3.1   Problem Definition

The end-to-end TIMT model aims at translating source language texts in images into target language sentences. Let $\mathbf{I}$ be the source language text image and corresponding target language sentence is $\mathbf{Y}$ containing z tokens $\{y^1, y^2, ..., y^z\}$. The training object for the end-to-end TIMT model is to maximize the translation probability:

$$P(\mathbf{Y}|\mathbf{I}; \theta_{\text{TIMT}}) = \prod_{i=1}^{z} P(y^i|\mathbf{I}, \mathbf{Y}_{<i}) \tag{1}$$

where $\mathbf{Y}_{<i}$ represents the translation history at the i-th decoding step, and θ_{TIMT} denotes the parameters of end-to-end TIMT model.

Specifically, to generate target language translation, end-to-end TIMT model is divided into three sub-modules: image encoder, sequential encoder, and decoder as shown in Fig. 1 (b). Image encoder $\mathcal{I}$ extracts image features from pixel space and ResNet [8] is utilized as the image encoder in our work:

$$F_\mathcal{I} = \mathcal{I}(\mathbf{I}; \theta_\mathcal{I}) = \text{ResNet}(\mathbf{I}) \tag{2}$$

where $\mathbf{I} \in \mathbb{R}^{H \cdot W \cdot C}$ denotes the input text image, and H, W, C represent the height, width, and channel of input image respectively. $F_\mathcal{I} \in \mathbb{R}^{l_\mathcal{I} * c}$ denotes the image feature, and $l_\mathcal{I}, c$ represent length and channel of feature sequence respectively. Generally, image features encoded by convolutional network are $F'_\mathcal{I} \in \mathbb{R}^{h \cdot w \cdot c}$, where h, w, c represent the height, width, and channel of feature maps respectively. To meet the requirement of following sequential encoding, feature maps are resized to feature sequence by reducing height and width dimension into feature length: $l_\mathcal{I} = h \cdot w$. Thus, the output of image encoder is a feature sequence containing local information of input text image.

Sequential encoder $\mathcal{S}(\cdot)$ aims at encoding contextual semantic features given local features of input text image. Transformer encoder is utilized as the sequential encoder in this paper:

$$F_\mathcal{S} = \mathcal{S}(F_\mathcal{I}; \theta_\mathcal{S}) = \text{TransformerEncoder}(F_\mathcal{I}) \tag{3}$$

where $F_\mathcal{S} \in \mathbb{R}^{l_\mathcal{S} \cdot h_\mathcal{S}}$ represents the sequential features that contains contextual semantic information of the whole feature sequence. $l_\mathcal{S}, h_\mathcal{S}$ represent sequence length and hidden dimension of sequential features.

Finally, target language decoder $\mathcal{D}(\cdot)$ generates translation results autoregressively and transformer decoder is utilized in our work:

$$F_\mathcal{D} = \mathcal{D}(F_\mathcal{S}; \theta_\mathcal{D}) = \text{TransformerDecoder}(F_\mathcal{S}) \tag{4}$$

where $F_\mathcal{D} \in \mathbb{R}^{l_\mathcal{D} \cdot h_\mathcal{D}}$ represents the output of decoder. $l_\mathcal{D}, h_\mathcal{D}$ represent sequence length and hidden dimension of decoder features respectively. The final decoded word $\hat{y}^i_{\text{TIMT}}$ is calculated by:

$$\hat{y}_{\text{TIMT}}^i = \underset{j \in [1, |V_{\mathbf{Y}}|]}{\arg \max} \, P(\hat{y}_j^i | \mathbf{I}, \hat{\mathbf{Y}}_{<i}), \text{ where } P(\hat{y}_j^i | \mathbf{I}, \hat{\mathbf{Y}}_{<i}) \propto W_o F_{\mathcal{D}}^i \quad (5)$$

where $P(\hat{y}_j^i | \mathbf{I}, \hat{\mathbf{Y}}_{<i})$ denotes the probability that the decoder predicts the j-th word $\hat{y}_j^i$ in vocabulary at i-th decoding step. $W_o \in \mathbb{R}^{|V_{\mathbf{Y}}| \cdot h_{\mathcal{D}}}$ denotes a linear matrix that maps decoder features into target language words. $|V_{\mathbf{Y}}|, h_{\mathcal{D}}$ represent the size of target language vocabulary and the hidden dimension of decoder respectively. $F_{\mathcal{D}}^i$ means the i-th element of decoder feature $F_{\mathcal{D}}$, which represents the decoder information at position i. $\hat{\mathbf{Y}}_{<i}$ represents the translation history before i-th step. In summary, end-to-end TIMT model utilizes image encoder, sequential encoder, and target language decoder to generate target language translation results word by word.

To optimize the end-to-end TIMT model, the log-likelihood loss function is utilized:

$$\mathcal{L}_{\text{TIMT}} = - \sum_{(\mathbf{I}, \mathbf{Y}) \in \mathbf{D}_{\text{TIMT}}} \log P(\mathbf{Y} | \mathbf{I})$$

$$\log P(\mathbf{Y} | \mathbf{I}) = \sum_i^z \sum_j^{|V_{\mathbf{Y}}|} \mathbb{I}(\hat{y}_j^i = y^i) \log P(\hat{y}_j^i | \mathbf{I}, \hat{\mathbf{Y}}_{<i}) \quad (6)$$

where $\mathbb{I}(\hat{y}_j^i = y^i)$ is an indicator function which eques 1 when predicted word $\hat{y}^i$ is the same as the ground-truth y^i, otherwise it equals 0. z denotes the sentence length of target language ground-truth. $\mathbf{D}_{\text{TIMT}}$ represents the text image translation training dataset.

3.2   Architecture of Teacher Models

Different sub-modules in end-to-end TIMT model play quite different functions and need various knowledge guidance. Image encoder is utilized to extract local visual features from input text images, while a sequential encoder further encodes contextual semantic information from local visual features. Finally, a decoder is designed to generate translation results given sequential features. To incorporate various knowledge into sub-modules of end-to-end TIMT model, three teacher models are utilized to guide the optimization of image encoder, sequential encoder, and decoder respectively. Specifically, knowledge of extracting text image features is transferred from TIR encoder. MT sequential encoder provides the guidance of contextual semantic feature learning, while MT decoder distillates the target language generation knowledge into TIMT decoder.

Text Image Recognition Teacher Model. Considering image encoder extracts local visual features from input text images, which is consistent between TIMT and TIR tasks, TIR model is incorporated to provide guidance for image feature learning. In this paper, TIR models are also divided into three sub-modules as end-to-end TIMT model to better understand the information flow between teacher and student models. Similar to TIMT image encoder, TIR image encoder also aims at extracting local visual features of input text images:

$$F_{\mathcal{I}}^{\mathrm{TIR}} = \mathcal{I}^{\mathrm{TIR}}(\mathbf{I}; \theta_{\mathcal{I}}^{\mathrm{TIR}}) = \mathrm{ResNet}(\mathbf{I}) \tag{7}$$

where $F_{\mathcal{I}}^{\mathrm{TIR}}$ denotes the image features encoded by TIR image encoder $\mathcal{I}^{\mathrm{TIR}}(\cdot)$ and the dimension of $F_{\mathcal{I}}^{\mathrm{TIR}}$ is same as the image feature $F_{\mathcal{I}}$ of end-to-end TIMT model introduced in Sect. 3.1. $\theta_{\mathcal{I}}^{\mathrm{TIR}}$ represents the model parameters of TIR image encoder. The architecture of TIR image encoder is similar to TIMT image encoder, but these two models are trained with different supervised data.

TIR Sequential encoder is also designed to further extract contextual information by considering whole local visual features:

$$F_{\mathcal{S}}^{\mathrm{TIR}} = \mathcal{S}^{\mathrm{TIR}}(F_{\mathcal{I}}^{\mathrm{TIR}}; \theta_{\mathcal{S}}^{\mathrm{TIR}}) = \mathrm{TransformerEncoder}(F_{\mathcal{I}}^{\mathrm{TIR}}) \tag{8}$$

where $F_{\mathcal{S}}^{\mathrm{TIR}}, \mathcal{S}^{\mathrm{TIR}}(\cdot), \theta_{\mathcal{S}}^{\mathrm{TIR}}$ denote TIR sequential features, TIR sequential encoder, and parameters of TIR sequential encoder respectively.

Different from generating target language in TIMT decoder, TIR decoder predicts source language words autoregressively:

$$F_{\mathcal{D}}^{\mathrm{TIR}} = \mathcal{D}^{\mathrm{TIR}}(F_{\mathcal{S}}^{\mathrm{TIR}}; \theta_{\mathcal{D}}^{\mathrm{TIR}}) = \mathrm{TransformerDecoder}(F_{\mathcal{S}}^{\mathrm{TIR}}) \tag{9}$$

where $F_{\mathcal{D}}^{\mathrm{TIR}}, \mathcal{D}^{\mathrm{TIR}}(\cdot), \theta_{\mathcal{D}}^{\mathrm{TIR}}$ denote TIR decoder features, TIR decoder, and parameters of TIR decoder respectively. To further map TIR decoder feature into source language space, a transformation matrix is utilized to transform decoder feature into source language word:

$$\hat{x}_{\mathrm{TIR}}^i = \underset{j \in [1, |V_{\mathbf{X}}|]}{\arg\max} \, P(\hat{x}_j^i | \mathbf{I}, \hat{\mathbf{X}}_{<i}), \ \text{where } P(\hat{x}_j^i | \mathbf{I}, \hat{\mathbf{X}}_{<i}) \propto W_o^{\mathrm{TIR}} F_{\mathcal{D}}^{\mathrm{TIR}^i} \tag{10}$$

where $\hat{x}_j^i$ represents the j-th word in source language vocabulary at decoding position i, while $\hat{x}_{\mathrm{TIR}}^i$ represents the final predicted word of decoder at i-th decoding step. $W_o^{\mathrm{TIR}} \in \mathbb{R}^{|V_{\mathbf{X}}| \cdot h_{\mathcal{D}}^{\mathrm{TIR}}}$ denotes the transformation matrix from decoder feature space to source language space. $|V_{\mathbf{X}}|, h_{\mathcal{D}}^{\mathrm{TIR}}$ represent the size of source language vocabulary and feature dimension of TIR decoder respectively. $F_{\mathcal{D}}^{\mathrm{TIR}^i}$ denotes the TIR decoder feature at position i. $\hat{\mathbf{X}}_{<i}$ represents the recognition history before i-th decoding step.

The overall architecture of TIR and TIMT models is similar, but the supervised data is different. TIR model is trained with recognition data pair $< \mathbf{I}, \mathbf{X} >$, where $\mathbf{X}$ means the source language recognition label of input text image $\mathbf{I}$. While TIMT model is trained with text image translation pair $< \mathbf{I}, \mathbf{Y} >$, where $\mathbf{Y}$ means the target language translation of corresponding source language sentence $\mathbf{X}$. To optimize the parameters in TIR model, the log-likelihood loss is utilized similar to TIMT optimization:

$$\mathcal{L}_{\mathrm{TIR}} = - \sum_{(\mathbf{I}, \mathbf{X}) \in \mathbf{D}_{\mathrm{TIR}}} \log P(\mathbf{X}|\mathbf{I})$$

$$\log P(\mathbf{X}|\mathbf{I}) = \sum_i^z \sum_j^{|V_{\mathbf{X}}|} \mathbb{I}(\hat{x}_j^i = x^i) \log P(\hat{x}_j^i | \mathbf{I}, \hat{\mathbf{X}}_{<i}) \tag{11}$$

where $\hat{x}_j^i$ denotes the j-th word in source language vocabulary at i-th decoding step, while x^i represents the ground-truth word at i-th decoding step. z denotes the sentence length of ground-truth. $\mathbb{I}(\cdot)$ means the indicator function as introduced in Eq. (6). $\mathbf{D}_{\mathrm{TIR}}$ represents the text image recognition dataset.

Machine Translation Teacher Model. Different from cross-modal generation TIR and TIMT models, MT model is a text-to-text transformation network. Thus, the encoder of raw data is quite different from TIR and TIMT models. To obtain text features from source language sentence strings, an embedding layer based text encoder is utilized to map the input words into word embedding:

$$F_{\mathcal{T}}^{\mathrm{MT}} = \mathcal{T}^{\mathrm{MT}}(\mathbf{X}; \theta_{\mathcal{T}}^{\mathrm{MT}}) = \mathrm{Embedding}(\mathbf{X}) \tag{12}$$

where $F_{\mathcal{T}}^{\mathrm{MT}}, \mathcal{T}^{\mathrm{MT}}(\cdot), \theta_{\mathcal{T}}^{\mathrm{MT}}$ represent text features, MT text encoder, and parameters of MT text encoders respectively.

Word embedding only contains single word information rather than global semantic information. To better extract contextual semantic features, MT sequential encoder further encodes contextual information by considering all input words:

$$F_{\mathcal{S}}^{\mathrm{MT}} = \mathcal{S}^{\mathrm{MT}}(F_{\mathcal{T}}^{\mathrm{MT}}; \theta_{\mathcal{S}}^{\mathrm{MT}}) = \mathrm{TransformerEncoder}(F_{\mathcal{T}}^{\mathrm{MT}}) \tag{13}$$

where $F_{\mathcal{S}}^{\mathrm{MT}}, \mathcal{S}^{\mathrm{MT}}(\cdot), \theta_{\mathcal{S}}^{\mathrm{MT}}$ denote MT sequential feature, MT sequential encoder, and parameters of MT sequential encoder respectively. Similar to TIR and TIMT sequential encoder, transformer encoder is utilized to extract contextual semantic features given MT text features.

MT decoder generates target language translation word by word given MT sequential features:

$$F_{\mathcal{D}}^{\mathrm{MT}} = \mathcal{D}^{\mathrm{MT}}(F_{\mathcal{S}}^{\mathrm{MT}}; \theta_{\mathcal{D}}^{\mathrm{MT}}) = \mathrm{TransformerDecoder}(F_{\mathcal{S}}^{\mathrm{MT}}) \tag{14}$$

where $F_{\mathcal{D}}^{\mathrm{MT}}, \mathcal{D}^{\mathrm{MT}}(\cdot), \theta_{\mathcal{D}}^{\mathrm{MT}}$ represent MT decoder features, MT decoder, and parameters of MT decoder respectively. To further map MT decoder features into target language space, a transformation matrix is utilized to calculate the translation probability:

$$\hat{y}_{\mathrm{MT}}^i = \underset{j \in [1, |V_{\mathbf{Y}}|]}{\arg\max} \ P(\hat{y}_j^i | \mathbf{X}, \hat{\mathbf{Y}}_{<i}), \ \text{ where } P(\hat{y}_j^i | \mathbf{X}, \hat{\mathbf{Y}}_{<i}) \propto W_o^{\mathrm{MT}} F_{\mathcal{D}}^{\mathrm{MT}i} \tag{15}$$

where $\hat{y}_j^i$ represents the j-th word in target language vocabulary at i-th decoding step, while $\hat{y}_{\mathrm{MT}}^i$ represents the final predicted word of target language decoder at decoding position i. $\mathbf{X}, \hat{\mathbf{Y}}_{<i}$ denote source language sentence and translation history before i-th decoding step respectively. $W_o^{\mathrm{MT}} \in \mathbb{R}^{|V_{\mathbf{Y}}| \cdot h_{\mathcal{D}}^{\mathrm{MT}}}$ denotes the transformation matrix which maps MT decoder features into target language space. $|V_{\mathbf{Y}}|, h_{\mathcal{D}}^{\mathrm{MT}}$ denote the size of target language vocabulary and hidden dimension of MT decoder feature respectively.

$$\mathcal{L}_{\mathrm{MT}} = - \sum_{(\mathbf{X},\mathbf{Y})\in \mathbf{D}_{\mathrm{MT}}} \log P(\hat{\mathbf{Y}}|\mathbf{X})$$

$$\log P(\hat{\mathbf{Y}}|\mathbf{X}) = \sum_{i}^{z} \sum_{j}^{|V_{\mathbf{Y}}|} \mathbb{I}(\hat{y}_j^i = y^i) \log P(\hat{y}_j^i|\mathbf{X}, \hat{\mathbf{Y}}_{<i}) \qquad (16)$$

where $\hat{y}_j^i$ denotes the j-th word in target language vocabulary at i-th decoding step, while y^i represents ground-truth word at i-th decoding step. z denotes the sentence length of ground-truth. $\mathbb{I}(\cdot)$ means the indicator function as introduced in Eq. (6). $\mathbf{D}_{\mathrm{MT}}$ represents the text machine translation dataset.

From the comparison of TIR, MT, and TIMT architectures, they have similar and different functions. For example, TIR image encoder and TIMT image encoder have similar structure and functions. All the sequential encoders are similar in architecture and the functions all aim at extracting contextual semantic information. Furthermore, MT decoder and TIMT decoder are both designed to predict target language sentences, which has similar structure and function. As a result, sub-modules of TIMT model with similar architecture and function can as that of TIR or MT models can be improved by multi-teacher knowledge distillation.

3.3 Knowledge Distillation from TIR Image Encoder

TIMT image encoder and TIR image encoder both extract local visual features from input text images. Compared with TIMT task, TIR task has much more training data, thus TIR models can be better optimized to encode image features of text images. To address the data limitation of end-to-end TIMT task, knowledge distillation from TIR image encoder is proposed to transfer text image encoding knowledge into TIMT image encoder. As shown in Fig. 1, TIMT image encoder is optimized not only by end-to-end text image translation loss but also by the guidance from TIR image encoder. To align the TIMT image features with TIR image features, both token-level and sentence-level knowledge distillation are incorporated to guide TIMT image encoder to predict similar image features as TIR image features:

Token-Level Image Encoder Knowledge Distillation. TIMT and TIR image features are feature sequences as introduced in Sect. 3.1. To provide fine-grained guidance information, L2-Norm constraint is utilized to guide TIMT image encoder outputs:

$$\mathcal{L}_{\mathrm{TKD}}^{\mathcal{I}} = \frac{1}{B \cdot l_{\mathcal{I}}} \sum_{j}^{B} \sum_{i}^{l_{\mathcal{I}}} \|F_{\mathcal{I}}^{ij} - F_{\mathcal{I}}^{\mathrm{TIR}^{ij}}\|_2 \qquad (17)$$

where $\mathcal{L}_{\mathrm{TKD}}^{\mathcal{I}}$ denotes the token-level image encoder knowledge distillation loss function. $F_{\mathcal{I}}^{ij}, F_{\mathcal{I}}^{\mathrm{TIR}^{ij}}$ represent TIMT and TIR image features of j-th sample at position i respectively. $l_{\mathcal{I}}$ denotes the length of TIMT image feature sequence, and $l_{\mathcal{I}} = l_{\mathcal{I}}^{\text{TIR}}$ in our experiments, indicating the sequence length of TIMT and TIR image features are the same. B denotes the batch size.

Sentence-Level Image Encoder Knowledge Distillation. To provide sentence-level guidance, both TIMT and TIR global image features are calculated by average pooling:

$$\mathcal{L}_{\text{SKD}}^{\mathcal{I}} = \frac{1}{B} \sum_{j}^{B} \| \frac{1}{l_{\mathcal{I}}} \sum_{i}^{l_{\mathcal{I}}} F_{\mathcal{I}}^{ij} - \frac{1}{l_{\mathcal{I}}^{\text{TIR}}} \sum_{i}^{l_{\mathcal{I}}^{\text{TIR}}} F_{\mathcal{I}}^{\text{TIR}^{ij}} \|_2 \tag{18}$$

where $\mathcal{L}_{\text{SKD}}^{\mathcal{I}}$ represents the loss function of sentence-level image encoder knowledge distillation. By calculating the global image features, the optimization of TIMT image encoder is guided by the global alignment between TIMT and TIR image features.

Finally, the token-level and sentence-level image encoder knowledge distillation loss functions are fused to obtain image encoder knowledge distillation loss function $\mathcal{L}_{\text{KD}}^{\mathcal{I}}$, which provides multi-granularity knowledge distillation guidance information:

$$\mathcal{L}_{\text{KD}}^{\mathcal{I}} = \mathcal{L}_{\text{TKD}}^{\mathcal{I}} + \mathcal{L}_{\text{SKD}}^{\mathcal{I}} \tag{19}$$

3.4 Knowledge Distillation from MT Sequential Encoder

The sequential encoder is vital to TIMT task, because the contextual semantic features are important for cross-lingual generation. To improve the ability of TIMT sequential encoder, knowledge distillation from MT sequential encoder is incorporated to guide the optimization of TIMT sequential encoder as shown in Fig. 1. Similar to image encoder knowledge distillation, sequential encoder knowledge distillation also has token-level and sentence-level knowledge distillations:

Token-Level Sequential Encoder Knowledge Distillation. Similar to the token-level image encoder knowledge distillation, MT sequential features are regarded as the guidance for TIMT sequential features through L2-Norm constraint:

$$\mathcal{L}_{\text{TKD}}^{\mathcal{S}} = \frac{1}{B \cdot l_{\mathcal{S}}} \sum_{j}^{B} \sum_{i}^{l_{\mathcal{S}}} \| F_{\mathcal{S}}^{ij} - F_{\mathcal{S}}^{\text{MT}^{ij}} \|_2 \tag{20}$$

where $\mathcal{L}_{\text{TKD}}^{\mathcal{S}}$ represents sequential knowledge distillation loss function. $F_{\mathcal{S}}^{ij}, F_{\mathcal{S}}^{\text{MT}^{ij}}$ represent TIMT and MT sequential features of j-th sample at position i respectively. $l_{\mathcal{S}}$ denotes the length of TIMT sequential feature sequence, which is set the same as the length of MT sequential feature sequence $l_{\mathcal{S}}^{\text{MT}}$.

Sentence-Level Sequential Encoder Knowledge Distillation. To further provide global guidance of sequential feature learning, the sentence-level sequential encoder knowledge distillation is proposed by performing average pooling on TIMT and MT sequential features:

$$\mathcal{L}_{\text{SKD}}^{\mathcal{S}} = \frac{1}{B} \sum_{j}^{B} \| \frac{1}{l_{\mathcal{S}}} \sum_{i}^{l_{\mathcal{S}}} F_{\mathcal{S}}^{ij} - \frac{1}{l_{\mathcal{S}}^{\text{MT}}} \sum_{i}^{l_{\mathcal{S}}^{\text{MT}}} F_{\mathcal{S}}^{\text{MT}^{ij}} \|_2 \tag{21}$$

where $\mathcal{L}_{\text{SKD}}^{\mathcal{S}}$ denotes the sequential encoder knowledge distillation loss function. The length of TIMT and MT sequential features are the same ($l_{\mathcal{S}} = l_{\mathcal{S}}^{\text{MT}}$) as introduced in token-level sequential encoder knowledge distillation.

Overall sequential encoder knowledge distillation loss function $\mathcal{L}_{\text{KD}}^{\mathcal{S}}$ is obtained by combining token-level and sentence-level sequential encoder knowledge distillation:

$$\mathcal{L}_{\text{KD}}^{\mathcal{S}} = \mathcal{L}_{\text{TKD}}^{\mathcal{S}} + \mathcal{L}_{\text{SKD}}^{\mathcal{S}} \tag{22}$$

3.5   Knowledge Distillation from MT Decoder

Different from image and sequential encoder knowledge distillation, decoder knowledge distillation is proposed to align the predicted target language vocabulary distribution between TIMT and MT decoders. Token-level decoder knowledge distillation aims at aligning the prediction probability between TIMT and MT decoders at each decoding step, while sentence-level decoder knowledge distillation takes the MT predicted target language sentence as the ground-truth to calculate the decoding loss for the optimization of TIMT model.

Token-Level Decoder Knowledge Distillation. As introduced in Eq. (5), TIMT decoder predicts the j-th target language word at i-th decoding step with the probability of $P(\hat{y}_j^i | \mathbf{I}, \hat{\mathbf{Y}}_{<i}^{\text{TIMT}})$, while MT decoder generates the j-th target language word at i-th step with the probability of $P(\hat{y}_j^i | \mathbf{X}, \hat{\mathbf{Y}}_{<i}^{\text{MT}})$ as in Eq. (15). To align the decoding distribution, $\mathbf{I}$ and $\mathbf{X}$ are paired text images and corresponding source language text sentences. $\hat{\mathbf{Y}}_{<i}^{\text{TIMT}}$, $\hat{\mathbf{Y}}_{<i}^{\text{MT}}$ represent decoding history of TIMT and MT models respectively. The token-level decoder knowledge distillation loss is calculated by updating the vanilla cross-entropy loss:

$$\mathcal{L}_{\text{TKD}}^{\mathcal{D}} = - \sum_{i}^{z} \sum_{j}^{|V_{\mathbf{Y}}|} P(\hat{y}_j^i | \mathbf{X}, \hat{\mathbf{Y}}_{<i}^{\text{MT}}) \log P(\hat{y}_j^i | \mathbf{I}, \hat{\mathbf{Y}}_{<i}^{\text{TIMT}}) \tag{23}$$

where $\mathcal{L}_{\text{TKD}}^{\mathcal{D}}$ denotes the token-level decoder knowledge distillation loss. By transferring decoding knowledge from MT teacher decoder, the TIMT decoder is guided to have a similar predicted probability of target language words.

Sentence-Level Decoder Knowledge Distillation. To provide sentence-level decoding knowledge distillation, the MT model decoded target language sentences are utilized to replace original ground-truth sentences. Different from token-level decoder knowledge distillation, which is designed to align the decoding probability between TIMT and MT decoders, sentence-level decoder knowledge distillation aims at guiding the TIMT decoder to have similar translation results as MT decoder:

$$\mathcal{L}_{\text{SKD}}^{\mathcal{D}} = -\sum_{i}^{z}\sum_{j}^{|V_{\mathbf{Y}}|} \mathbb{I}(\hat{y}_j^i = \hat{y}_{\text{MT}}^i) \log P(\hat{y}_j^i | \mathbf{I}, \hat{\mathbf{Y}}_{<i}^{\text{TIMT}}) \tag{24}$$

where $\mathcal{L}_{\text{SKD}}^{\mathcal{D}}$ denotes sequence-level decoder knowledge distillation loss function. Different from the vanilla log-likelihood loss function, the ground-truth sentence is replaced as the MT prediction results. Thus the indicator function $\mathbb{I}(\hat{y}_j^i = \hat{y}_{\text{MT}}^i)$ equals 1 when the TIMT decoded word $\hat{y}_j^i$ is the same as the MT predicted word $\hat{y}_{\text{MT}}^i$. By incorporating both token-level and sentence-level decoder knowledge distillation, the overall loss function of decoder knowledge distillation is formulated as:

$$\mathcal{L}_{\text{KD}}^{\mathcal{D}} = \mathcal{L}_{\text{TKD}}^{\mathcal{D}} + \mathcal{L}_{\text{SKD}}^{\mathcal{D}} \tag{25}$$

The final loss function is the combination of end-to-end text image translation and knowledge distillation loss functions:

$$\begin{aligned} \mathcal{L}_{\text{ALL}} &= (1 - \lambda_{\text{KD}})\mathcal{L}_{\text{TIMT}} + \lambda_{\text{KD}}\mathcal{L}_{\text{KD}} \\ \mathcal{L}_{\text{KD}} &= \lambda_{\mathcal{I}}\mathcal{L}_{\text{KD}}^{\mathcal{I}} + \lambda_{\mathcal{S}}\mathcal{L}_{\text{KD}}^{\mathcal{S}} + \lambda_{\mathcal{D}}\mathcal{L}_{\text{KD}}^{\mathcal{D}} \end{aligned} \tag{26}$$

where $\lambda_{\text{KD}}, \lambda_{\mathcal{I}}, \lambda_{\mathcal{S}}, \lambda_{\mathcal{D}}$ represent the loss weight of overall knowledge distillation, image encoder knowledge distillation, sequential encoder knowledge distillation, and decoder knowledge distillation respectively.

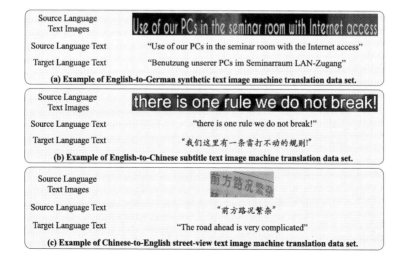

(a) **Example of English-to-German synthetic text image machine translation data set.**

(b) **Example of English-to-Chinese subtitle text image machine translation data set.**

(c) **Example of Chinese-to-English street-view text image machine translation data set.**

Fig. 2. Examples of synthetic, subtitle and street-view text image translation datasets.

4   Experiments

4.1   Datasets

To train the end-to-end TIMT model, the publicly available dataset released by [13] is utilized in our experiments. As shown in Fig. 2, this dataset contains samples from three domains: synthetic, subtitle, and street-view domains. The training and validation samples are all from synthetic domain, while samples in evaluation set are from all three domains. Three translation directions are conducted in this dataset: English-to-Chinese (EnCh), English-to-German (EnDe), and Chinese-to-English (ChEn) translation. There are 1,000,000 training samples, 2,000 validation samples, and 2,000 evaluation samples in synthetic domain. The subtitle test set contains 1,040 samples, while the street-view test set has 1,198 samples. To implement knowledge distillation, triple-aligned samples {**source language images, source language texts, target language texts**} are utilized to transfer the pre-trained knowledge from TIR and MT teacher models into the TIMT student model.

Table 1. Results of various knowledge distillation combinations on English-to-Chinese translation validation set. TKD and SKD represent using single token-level or sentence-level knowledge distillation loss. TKD+SKD means the fused token-level and sentence-level knowledge distillation are used for knowledge distillation loss function. BLEU Score is utilized to evaluate the translation performance.

No	λ_I	λ_S	λ_D	TKD	SKD	TKD+SKD
1	0	0	1	23.02	22.68	23.16
2	0	1	0	22.63	22.44	22.85
3	0	1	1	23.47	23.04	23.79
4	1	0	0	22.45	22.30	22.68
5	1	0	1	23.28	22.95	23.52
6	1	1	0	23.19	22.73	23.34
7	1	1	1	23.86	23.51	24.13

4.2   Experimental Setup

To provide a faire comparison with existing research on end-to-end TIMT task, a similar model architecture as [13] is utilized in our experiment. The TIMT image encoder is composed of TPS Net and Res Net, which extracts the image features from the raw input text images. The TIMT sequential encoder and decoder are 6-layer transformer encoder and 6-layer transformer decoder respectively, which is also the same as [13]. The MT model replaced the TIMT image encoder with an embedding layer based text encoder. The sequential encoder and decoder of the MT model are kept the same as the TIMT model. The preprocessing method and experimental setting are the same as [13]. For decoding results, sacre-BLEU [1] is calculated to evaluate the translation performance.

---

[1] https://github.com/mjpost/sacrebleu.

4.3   Results of Various Knowledge Distillation

Table 1 shows the results of various knowledge distillation (KD) combinations. Line No.1, No.2, and No.4 show the results of single-teacher KD. Single decoder KD (No.1) achieves the best single-teacher performance due to the strong guidance from decoding knowledge. Sequential encoder KD (No.2) outperforms image encoder KD (No.4), indicating semantic knowledge transferring is more important for TIMT task. For bi-teacher KD comparison, sequential encoder and decoder KD combination (No.3) performs well by incorporating semantic and decoding guidance. Finally, triple-teacher KD (No.7) achieves the best performance by transferring image encoder, sequential decoder, and decoder knowledge into end-to-end TIMT model, indicating incorporating accurate knowledge into various sub-modules is vital for performance improvements.

4.4   Comparison with Existing TIMT Methods

Compared with existing end-to-end TIMT models, MTKD has significant improvements by incorporating various knowledge into sub-modules of TIMT model. Table 2 shows the comparison between MTKD and existing TIMT models. TRBA [2] is a vanilla TIR model trained with translation dataset. CLTIR [6] proposed to train TIMT model with TIR multi-task learning. RTNet [18] bridges pre-trained TIR and MT models with feature transformer. METIMT [13] trains TIMT model with MT auxiliary task. MHCMM [5] is a mimic learning based method by introducing MT teacher for TIMT model. Different from existing research, MTKD incorporates both TIR and MT teachers into TIMT optimization. Meanwhile, various knowledge distillation is utilized to transfer accurate

Table 2. Comparison of existing end-to-end models with our proposed multi-teacher knowledge distillation (MTKD) method. MTKD utilizes the knowledge distillation setting of line No.7 in Table 1.

Architecture	Synthetic			Subtitle		Street
	EnCh	EnDe	ChEn	EnCh	ChEn	ChEn
Existing End-to-End Models						
TRBA [2]	9.61	7.36	4.77	12.12	5.18	0.36
CLTIR [6]	18.02	15.55	10.74	16.47	9.04	0.43
CLTIR+TIR [6]	19.44	16.31	13.52	17.96	11.25	1.74
RTNet [18]	18.91	15.82	12.54	17.63	10.63	1.07
RTNet+TIR [18]	19.63	16.78	14.01	18.82	11.50	1.93
MTETIMT [13]	19.25	16.27	13.16	17.73	10.79	1.69
MTETIMT+MT [13]	21.96	18.84	15.62	19.17	12.11	5.84
MHCMM [5]	22.08	18.97	15.66	19.24	12.12	5.87
Our Proposed Multi-Teacher Knowledge Distillation Method						
MTKD	**22.26**	**19.38**	**15.84**	**19.31**	**12.17**	**6.08**

498   C. Ma et al.

Table 3. Comparison of TIR+MT pipeline method with MTKD method on English-to-Chinese synthetic test set. Model size represents the parameter amount of the model. Decoding time means the time of predicting a sentence and the unit is second. BLEU score is utilized to evaluate the translation performance on valid and test set.

Architecture	Model Size↓	Decoding Time↓	Valid BLEU↑	Test BLEU↑
Pipeline	195.1M	0.33s	23.52	20.46
MTKD	121.9M	0.19s	24.13	22.26

knowledge into sub-modules of TIMT model. Finally, MTKD outperforms the existing best MHCMM model with 0.18 BLEU scores on average. Improvements in all three evaluation domains reveal the good generalization of MTKD.

4.5 Comparison with Pipeline Method

Table 3 shows the comparison of MTKD with the TIR+MT pipeline model. By transferring knowledge into TIMT model, MTKD has better translation performance, which effectively addresses the error propagation problems in pipeline model. With an end-to-end architecture, MTKD has fewer parameters than pipeline model. Meanwhile, MTKD has less decoding time than pipeline model, which is vital in real-world applications (Table 3).

4.6 Analysis of Hyper-parameter

The loss weight of knowledge distillation is a key hyper-parameter to balance the end-to-end TIMT loss and knowledge distillation losses. When $\lambda_{\mathrm{KD}} = 0$, the

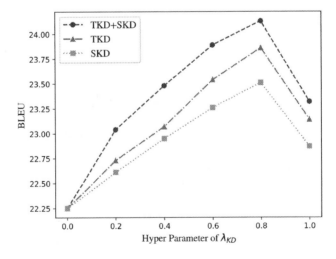

Fig. 3. Hyper-parameter analysis on the loss weight of knowledge distillation.

model is only optimized with end-to-end loss function and the performance is limited due to the end-to-end data scarcity and the difficulty of TIMT task. By incorporating KD loss, the performance is getting better and the optimal value for λ_{KD} is 0.8. When $\lambda_{KD} = 1$, the performance drops a bit, indicating end-to-end TIMT loss by guiding the model learns to predict as the ground-truth is also important for TIMT task.

5 Conclusion

In this paper, we propose a novel multi-teacher knowledge distillation (MTKD) method for end-to-end text image machine translation task. Three pre-trained teacher models are utilized to provide accurate knowledge for corresponding sub-modules in end-to-end TIMT model. By transferring various knowledge into sub-modules of TIMT model, the translation performance is significantly improved compared with existing methods. Meanwhile, token-level and sentence-level knowledge distillation are complementary for knowledge transferring, indicating that multi-granularity knowledge distillation is vital for TIMT improvements. Furthermore, MTKD based TIMT model outperforms pipeline models with a smaller model size and less decoding time, which has the advantages of both end-to-end and pipeline models. In the future, we will explore to transfer more knowledge into end-to-end TIMT model to further improve the translation performance.

Acknowledgement. This work has been supported by the National Natural Science Foundation of China (NSFC) grants 62106265.

References

1. Afli, H., Way, A.: Integrating optical character recognition and machine translation of historical documents. In: Proceedings of the Workshop on Language Technology Resources and Tools for Digital Humanities, LT4DH@COLING, pp. 109–116 (2016)
2. Baek, J., et al.: What is wrong with scene text recognition model comparisons? Dataset and model analysis. In: 2019 IEEE/CVF International Conference on Computer Vision, ICCV 2019, Seoul, Korea (South), October 27 - November 2, pp. 4714–4722 (2019)
3. Chang, Y., Chen, D., Zhang, Y., Yang, J.: An image-based automatic Arabic translation system. Pattern Recognit. **42**(9), 2127–2134 (2009)
4. Chen, J., Cao, H., Natarajan, P.: Integrating natural language processing with image document analysis: what we learned from two real-world applications. Int. J. Document Anal. Recognit. **18**(3), 235–247 (2015)
5. Chen, Z., Yin, F., Yang, Q., Liu, C.L.: Cross-lingual text image recognition via multi-hierarchy cross-modal mimic. IEEE Trans. Multimedia (TMM), pp. 1–13 (2022)
6. Chen, Z., Yin, F., Zhang, X., Yang, Q., Liu, C.: Cross-lingual text image recognition via multi-task sequence to sequence learning. In: 25th International Conference on Pattern Recognition (ICPR), pp. 3122–3129 (2020)

7. Du, J., Huo, Q., Sun, L., Sun, J.: Snap and translate using windows phone. In: 2011 International Conference on Document Analysis and Recognition (ICDAR), pp. 809–813. IEEE Computer Society (2011)

8. He, K., Zhang, X., Ren, S., Sun, J.: Deep residual learning for image recognition. In: 2016 IEEE Conference on Computer Vision and Pattern Recognition (CVPR), pp. 770–778 (2016)

9. Hinami, R., Ishiwatari, S., Yasuda, K., Matsui, Y.: Towards fully automated manga translation. In: The Thirty-Fifth AAAI Conference on Artificial Intelligence (AAAI) (2021)

10. Hinton, G.E., Vinyals, O., Dean, J.: Distilling the knowledge in a neural network. CoRR abs/1503.02531 (2015)

11. Kim, Y., Rush, A.M.: Sequence-level knowledge distillation. In: Proceedings of the 2016 Conference on Empirical Methods in Natural Language Processing, EMNLP 2016, Austin, Texas, USA, 1–4 November 2016, pp. 1317–1327. The Association for Computational Linguistics (2016)

12. Liu, Y., et al.: End-to-end speech translation with knowledge distillation. In: Interspeech 2019, 20th Annual Conference of the International Speech Communication Association, Graz, Austria, 15–19 September 2019, pp. 1128–1132. ISCA (2019)

13. Ma, C., et al.: Improving end-to-end text image translation from the auxiliary text translation task. In: 26th International Conference on Pattern Recognition, ICPR 2022, Montreal, QC, Canada, 21–25 August 2022, pp. 1664–1670. IEEE (2022)

14. Mansimov, E., Stern, M., Chen, M., Firat, O., Uszkoreit, J., Jain, P.: Towards end-to-end in-image neural machine translation. In: Proceedings of the First International Workshop on Natural Language Processing Beyond Text, pp. 70–74. Association for Computational Linguistics, Online (Nov 2020)

15. Shekar, K.C., Cross, M.A., Vasudevan, V.: Optical character recognition and neural machine translation using deep learning techniques. In: Saini, H.S., Sayal, R., Govardhan, A., Buyya, R. (eds.) Innovations in Computer Science and Engineering. LNNS, vol. 171, pp. 277–283. Springer, Singapore (2021). https://doi.org/10.1007/978-981-33-4543-0_30

16. Shi, B., Bai, X., Yao, C.: An end-to-end trainable neural network for image-based sequence recognition and its application to scene text recognition. IEEE Trans. Pattern Anal. Mach. Intell. **39**(11), 2298–2304 (2017)

17. Shi, B., Wang, X., Lyu, P., Yao, C., Bai, X.: Robust scene text recognition with automatic rectification. In: 2016 IEEE Conference on Computer Vision and Pattern Recognition, CVPR 2016, Las Vegas, NV, USA, 27–30 June 2016, pp. 4168–4176. IEEE Computer Society (2016)

18. Su, T., Liu, S., Zhou, S.: RTNet: an end-to-end method for handwritten text image translation. In: Lladós, J., Lopresti, D., Uchida, S. (eds.) ICDAR 2021. LNCS, vol. 12822, pp. 99–113. Springer, Cham (2021). https://doi.org/10.1007/978-3-030-86331-9_7

19. Sun, H., Wang, R., Chen, K., Utiyama, M., Sumita, E., Zhao, T.: Knowledge distillation for multilingual unsupervised neural machine translation. In: Proceedings of the 58th Annual Meeting of the Association for Computational Linguistics, ACL 2020, Online, 5–10 July 2020, pp. 3525–3535 (2020)

20. Sutskever, I., Vinyals, O., Le, Q.V.: Sequence to sequence learning with neural networks. In: Advances in Neural Information Processing Systems 27: Annual Conference on Neural Information Processing Systems 2014, 8–13 December 2014, Montreal, Quebec, Canada, pp. 3104–3112 (2014)

21. Tan, X., Ren, Y., He, D., Qin, T., Zhao, Z., Liu, T.: Multilingual neural machine translation with knowledge distillation. In: 7th International Conference on Learning Representations, ICLR 2019, New Orleans, LA, USA, 6–9 May 2019

22. Vaswani, A., et al.: Attention is all you need. In: Advances in Neural Information Processing Systems, pp. 5998–6008 (2017)

23. Watanabe, Y., Okada, Y., Kim, Y., Takeda, T.: Translation camera. In: Fourteenth International Conference on Pattern Recognition, ICPR 1998, Brisbane, Australia, 16–20 August 1998, pp. 613–617 (1998)

24. Weinzaepfel, P., Brégier, R., Combaluzier, H., Leroy, V., Rogez, G.: DOPE: distillation of part experts for whole-body 3D pose estimation in the wild. In: Computer Vision - ECCV 2020–16th European Conference, Glasgow, UK, 23–28 August 2020, Proceedings, Part XXVI. vol. 12371, pp. 380–397 (2020)

25. Wong, F., Chao, S., Chan, W.K.: Cyclops - snapshot translation system based on mobile device. J. Softw. **6**(9), 1664–1671 (2011)

26. Yang, J., Chen, X., Zhang, J., Zhang, Y., Waibel, A.: Automatic detection and translation of text from natural scenes. In: Proceedings of the IEEE International Conference on Acoustics, Speech, and Signal Processing, ICASSP 2002, 13–17 May 2002, Orlando, Florida, USA, pp. 2101–2104 (2002)

27. Zhang, Y., Nie, S., Liang, S., Liu, W.: Bidirectional adversarial domain adaptation with semantic consistency. In: Lin, Z. (ed.) PRCV 2019. LNCS, vol. 11859, pp. 184–198. Springer, Cham (2019). https://doi.org/10.1007/978-3-030-31726-3_16

28. Zhang, Y., Nie, S., Liang, S., Liu, W.: Robust text image recognition via adversarial sequence-to-sequence domain adaptation. IEEE Trans. Image Process. **30**, 3922–3933 (2021)

29. Zhao, Y., Xiang, L., Zhu, J., Zhang, J., Zhou, Y., Zong, C.: Knowledge graph enhanced neural machine translation via multi-task learning on sub-entity granularity. In: Proceedings of the 28th International Conference on Computational Linguistics, COLING 2020, 8–13 December 2020, pp. 4495–4505 (2020)

30. Zhao, Y., Zhang, J., Zhou, Y., Zong, C.: Knowledge graphs enhanced neural machine translation. In: Proceedings of the Twenty-Ninth International Joint Conference on Artificial Intelligence, IJCAI 2020, pp. 4039–4045 (2020)

Applications 2: Document Analysis and Systems

Multimodal Scoring Model for Handwritten Chinese Essay

Tonghua Su$^{(\boxtimes)}$ ⓘ, Jifeng Wang, Hongming You, and Zhongjie Wang ⓘ

School of Software, Harbin Institute of Technology, Harbin,
People's Republic of China
{thsu,rainy}@hit.edu.cn

Abstract. Essay writing plays a critical role in Chinese language skill teaching. With the smart education becomes a hot topic, the demand for automatic essay scoring (AES) has been emerging among teachers and students. Existing works frequently ignore the impact of the visual modal during the scoring process, such as writing quality in terms of neatness or legibility. This paper addresses the problem with a visual-textual integrating perspective and proposes a deep learning based multimodal AES. Specifically, implicit alignment algorithm is presented to cohere the distinct visual modal and text modal. Methods are tested on a large-scale dataset consisting of over 4000 essays including HSK publicly available samples. The results show that multi-modal AES reduce the MAE of scoring from 1.13 to 1.06, and the implicit alignment algorithm reduces it further to 1.01.

Keywords: Automated Essay Scoring · Multi-modal Learning · Implicit Alignment

1 Introduction

Essay teaching plays a critical role in teenagers' Chinese language education. Currently there exist problems for both students and teachers. From the perspective of students, it is difficult to improve the writing ability due to the lack of timely guidance and feedback. From the perspective of teachers, the dedicated essay scoring is a time-consuming work. Therefore, an automatic system for handwritten essay scoring is desirable for both industry and academics to improve the writers' essay writing abilities.

Existing works mostly tackle the task using natural language processing. The advent of pre-trained models represented by the bidirectional encoder representations from transformers (BERT) [3] has made a lot of progress. In recent years, a number of deep learning-based scoring methods based on text information have emerged. On the other hand, the industry has also attempt to create auxiliary teaching systems to reduce the huge burden work of teachers and improve students writing skills quickly and easily. However, there are still the following challenges and problems in achieving a fully usable scoring system.

© The Author(s), under exclusive license to Springer Nature Switzerland AG 2023
G. A. Fink et al. (Eds.): ICDAR 2023, LNCS 14187, pp. 505–519, 2023.
https://doi.org/10.1007/978-3-031-41676-7_29

We can see that exiting BERT-based essay assessing methods attribute to solely text modality. Indeed, teachers will not only pay attention to the content of the text itself, but also to the visual information, such as neatness of the handwriting and legibility of the calligraphy. It means that visual features play an indispensable role in the scoring process. Therefore, the scoring system needs to consider both textual modal and visual modal simultaneously. How to effectively integrate visual information into the grading model is also an issue worthy of study.

The paper proposes a multimodal scoring model based on regression to tackle the issue. To bridge the gaps between text and visual modalities,implicit alignment algorithm is designed. A pretrained full-page recognizer, named FPRNet [15], is adopted to facilitate the coupling of two modalities. As for visual modal, features such as the degree of neatness and the degree of writing legibility can be considered through this model. Experimental evaluations show that fusion based on implicit alignment outperforms the traditional manner.

2 Related Works

Here we summarize the general work of our predecessors into two parts. First of all, in terms of composition feature representation, the multimodal feature representation method can integrate the text information and image information of the essay, which is helpful for more accurate and comprehensive scoring of the essay. The current scoring systems mainly include deep learning methods based on pre-trained models and machine learning methods based on feature engineering, both of which only evaluate the content of the composition itself.

2.1 Multimodal Feature Text Representation

Multi-modal is one of the natural attributes of document text. For a deep understanding of text, not only the text content itself needs to be considered, but also the image information carrying the text content. How to effectively fuse features from multiple digital domains to achieve information complementarity is of great significance for the study of tasks based on multimodal features. The main goal of multimodal feature fusion is to reduce the heterogeneous difference between modalities and obtain multimodal representations while maintaining the integrity of each modality-specific semantics [4].

In recent years, the pre-training model represented by BERT has achieved great success. Many scholars have also followed the idea of pre-training and used a model structure similar to Transformer to carry out research on multi-modal representation. The model design used in this type of work can be divided into two types: 1) single-stream structure, for example, VisualBERT [11], Unicoder-V [9] and VL-BERT [3]; 2) dual-stream structure, such as ViBERT [16], Lxmert [17], The former maps all data in a unified semantic space, while the latter requires multi-modal information to be fused with different modal information through an independent encoder.

In the field of fusion of text features and visual features, LayoutLM [22] adopts the idea of single-stream structure to conduct research on document layout analysis, and provides more abundant information for downstream tasks through pre-training and natural text location information. The VSR [23] model uses a two-stream convolutional network to extract image visual features and text language features respectively. The semantic information is used to obtain the multimodal aggregation feature map. In scenarios of dense text understanding such as automatic document understanding and handwritten text review, how to align and fuse multimodal features is still an urgent problem to be solved.

2.2   Automated Scoring System

Automated essay scoring (AES) is an important research topic in the field of natural language processing. Typically, the automatic scoring task is modeled as a supervised machine learning problem. Project Essay Grade (PEG for short) [14] is the first automatic grading method in the field, which extracts some features from the text and uses linear regression to predict the score of the composition. Lonsdale et al. [12] attempted to use a linked grammar to score text, computed from the average sentence-level score obtained from the parser and style-based feature vectors, and used regression to predict the paper's score. At present, the Chinese-based automatic scoring system is still in the early stage of exploration. The literature [7] based on the HSK data set, the evaluation indicators are defined from the perspectives of composition complexity and accuracy, and the scoring is constructed using multiple linear regression.The literature [21] is based on the Chinese text processing for the first time, defining feature dimensions including dependency relationship and discourse structure analysis. The above work mainly adopts traditional machine learning methods and relies heavily on the selection of features, but it has gained wide recognition and use because of its interpretable modeling.

With the development of deep learning in recent years, many researchers have tried to use neural network models to automatically learn features. Tay et al. [19] believed that the coherence score of the article can be used as an important basis for evaluating the composition score, and based on this, they proposed the LSTM-based SkipFlow model, which can better capture the continuity characteristics between text sequences . The research of Nadeem et al. [13]showed that using pre-trained neural network can effectively improve the performance of essay scoring system. Continuing this idea based on deep learning, Li et al. [10]used the pre-trained BERT model for automatic scoring based on Chinese composition, and achieved good results. fruit. But in general, the method based on deep learning lacks good interpretability, and the dimension considered in the scoring is relatively single, so the scoring conclusion is not convincing.

From above literature study, it's obvious that the Chinese AES is still in the early stage of exploration.

3   Methods

This Chapter mainly introduces the proposed scoring method in detail. First, in Sect. 3.1 we model the problem. Section 3.2 introduces the scoring method based on multimodal fusion. As a comparison, Sect. 3.3 mainly introduces the scoring method based on linguistic features.

3.1   Problem Modeling

It's easy to see that the scoring problem can be modeled as a numerical regression problem. From the perspective of the label Y, theoretically, the essay score is a continuous metric space, and the difference between the predicted value and the real value $d = y_{true} - y_{pred}$ can directly measure the quantitative error between them. Therefore, it is preferable to model the scoring problem as a numerical regression problem rather than a classification problem.

But from the perspective of modeling, the essence of the classification model and the regression model are highly similar. Taking the neural network model as an example, assuming that the last hidden layer has m neurons, and the outputs is $\boldsymbol{h} = (h_1, h_2, ..., h_m)^T$. When it's used to deal with n classification problems, the hidden layer can be connected to n linear neurons, and then the model output can be converted into probabilities on n classes through normalization processing. The final output value is shown in Eq. 1:

$$\boldsymbol{h} = g(\boldsymbol{W}\boldsymbol{x} + \boldsymbol{c}), \tag{1}$$

where $\boldsymbol{W}$, $\boldsymbol{c}$ are built-in parameters of linear neurons, $g(.)$ is a normalized function represented by $softmax$. while handing regression problems, the hidden layer could be connected to n neurons to achieve dimension reduction, and connect them to one single neuron to normalize then get the output.

Therefore, it is also a feasible modeling scheme to discretize the continuous essay score value and transform it into a classification problem, and the continuous space hashing has been proved in tasks such as pose estimation, homography estimation, and image generation. Advantage. Therefore, in this paper will use two methods of classification and regression to model the essay scoring problem, and compare the scoring effects of the two modeling methods in the experiment in Sect. 4.

3.2   Multimodal Essay Scoring Method

Multimodal Feature Fusion Based on Implicit Alignment. Typically, the multimodal fusion architecture is usually a joint-based architecture, which maps multiple unimodal representation features into the same semantic space through operations such as addition and concat. As shown in Fig. 1a, the single-modal information is subjected to feature encoding, feature extraction, and then stitched into the same shared semantic space. Assuming that the visual features

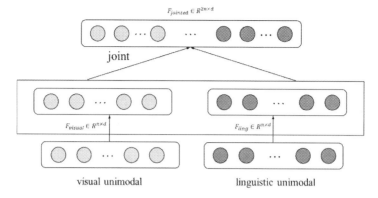

(a) Concat-based approach

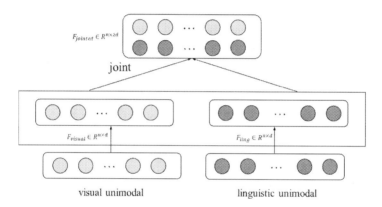

(b) Alignment-based approach

Fig. 1. Ways to joining two modalities

F_{visual} and semantic features F_{ling} are both hidden space vectors of the shape $[n, d]$, the fusion feature $F_{joint} \in R^{2n \times d}$.

Although the above approach can integrate multimodal features, the correlation between multimodal features needs to be captured by downstream modules, which is not conducive to the realization of information complementarity between multimodal features. Currently, some unsupervised methods based on display alignment have been proposed in the academic community to achieve fusion alignment between multimodal features. [18] proposes a dynamic time warping (DTW) algorithm based on dynamic programming for alignment between sequences. When aligning visual features and text sequences, a common practice is to obtain the corresponding high-dimensional visual features through the position of the text in the original image and the estimated receptive field. However,

510   T. Su et al.

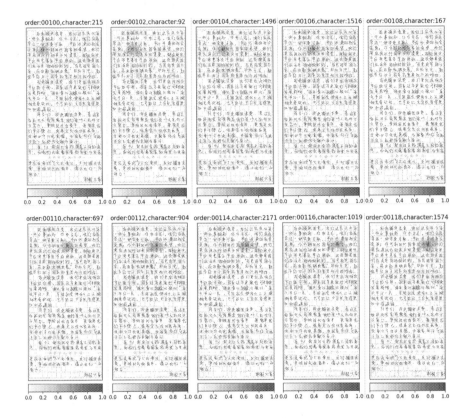

Fig. 2. Attention illustration of one sentence

this method needs to perceive the position of the text in the original image in advance, and thus it is not suitable for the handwritten essay scoring directly.

As for visual features, we consider the pretrained FPRNet model [15]. Combined with the characteristics of the upstream recognition model, since the decoder of the full-page handwriting recognition model FPRNet [15] adopts a dimension reduction method based on reshaping, the spatial correlation of high-dimensional visual features is largely preserved. We analyze the interpretability of the output results of the recognition model, as shown in Fig. 2, using the gradient map to represent the region of interest of each output Chinese character in the original image, it can be found that the semantic sequence recognized by FPRNet and the image features have excellent implicit alignment effect, we do not need to know the position of the Chinese characters in the original image, and we can obtain the high-dimensional image features corresponding to the characters without using the method of receptive field estimation. Using the serialized feature F_{seq} and the output result of CTC decoding, the alignment and fusion of images and text can be realized in the manner shown in Fig. 1b.

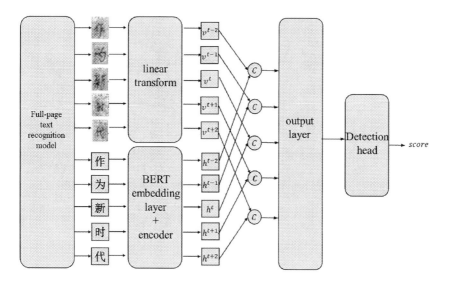

Fig. 3. Multimodal feature encoding method

Multimodal Scoring Model Architecture. The multimodal modeling method based on BERT transformation can realize direct alignment and feature fusion of multimodal features, as shown in Fig. 3. The feature vector $h \in R^{n \times d}$ is obtained after the existing text sequence passes through the pre-trained BERT embedding layer and encoding layer, where n is the sequence length and d is the feature dimension of the semantic feature. Use the visual features extracted by the text recognition model to screen, and concat the bit-by-bit visual features together to obtain the visual feature sequence $\mathbf{v} = W\mathbf{v} \in R_{n \times d}$ in order to compare it with the position-by-position splicing of semantic features realizes the direct alignment and fusion of multi-modal features and produces them.

In the selection of the detection head, we will use two modeling methods to design the detection head. When it is regarded as a classification task, the cross-entropy loss function is used as supervision, and the formula is shown in Eq. 2.

$$L = -\sum_{i=1}^{n} y_i log S(f_\theta(x_i)) \tag{2}$$

Where x is the classification label encoding of the model output after one hot encoding, $f(x)$ is the score belonging to a certain category, and S is the classification probability after passing the *softmax* activation function. When it is regarded as a regression task, the mean square error loss is used, and the sum of the squares of the error between the predicted value and the real value is used as the loss function, and its formula is shown in Eq. 3.

$$L = \frac{1}{n} \sum_{i=1}^{n} (y_i - f_\theta(x_i))^2 \tag{3}$$

Table 1. 90D linguistic features.

Feature category	Number of features	Example
word complexity	4	number of characters, number of phrases, ratio of class character and shape character complex word ratio
statement complexity	7	mean length of sentences, the mean length of T-units, the mean depth of the dependency trees
predicate collocations	23	ratio of noun-verb, ratio of adjective-adverb, ratio of adjective-noun
dependency structures	41	ratio of subject-predicate, ratio of inter-object, ratio of dynamic-complement, mean distance of subject-predicate, ratio of class symbol shape symbol in inter-object relation
logical expression level	15	ratio of level-1 constructions, ratio of level-2 constructions, ratio of level-3 constructions

The effects of different modeling methods on scoring will be shown in the experimental part.

3.3   Scoring Baseline Based on Linguistic Features

We describe a baseline based on linguistic features. The linguistics-based scoring problem is modeled as a standard machine learning problem. It takes the predicted text as input and aims to evaluate the text from the dimensions of linguistic features such as word complexity and sentence complexity.

We design 90D features for the baseline system to summarize the linguistic factors, as shown in Table 1. As the starting point, we consider the research on the linguistics area [8,20]. We also refer to existing Chinese & English automatic scoring systems [6,21] and their valuable experience.

In terms of word complexity, we introduce "New Chinese Proficiency Test (HSK) Vocabulary (2012 Revised Edition)" as a standard, HSK Level 4 or above vocabulary is regarded as complex words, and the proportion of complex statistical sentence-level language. In the process of counting the linguistic features in sentence-level, we introduced the pre-trained LTP [2] part-of-speech tagging model, dependency syntax analysis model and semantic dependency analysis model for feature extraction, and then consulted the original feature definition method in [5,21] for linguistic features' extraction, the relationship between some typical features and scores are shown in Fig. 4.

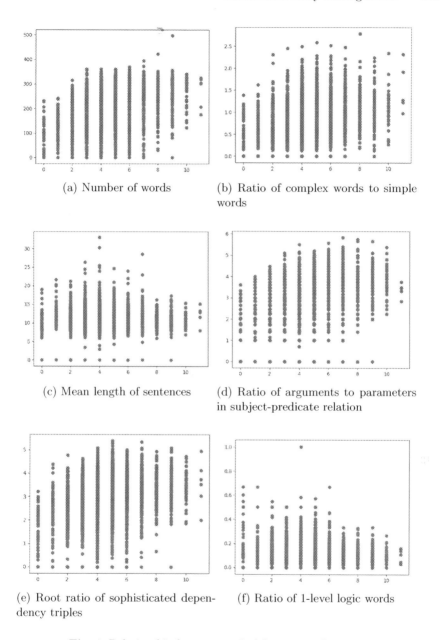

(a) Number of words

(b) Ratio of complex words to simple words

(c) Mean length of sentences

(d) Ratio of arguments to parameters in subject-predicate relation

(e) Root ratio of sophisticated dependency triples

(f) Ratio of 1-level logic words

Fig. 4. Relationship between typical features and scores

The x-axis in the scatter plot represents the score, the ordinate represents the value on the feature dimension, and each point represents a sample. Through the above-mentioned feature definition method, we can see that in most feature dimensions, samples with the same score are correlated in feature values. But we can also see that the sample discrimination on some features is not obvious enough. For example, the average sentence length of articles under different ratings is 10 to 15 words. Therefore, in addition to using the above five feature dimensions for class-by-class scoring, we also use stepwise regression and PCA to clean irrelevant features.

To select a best classification algorithms, we compared the effect of different methods for scoring tasks, including linear regression algorithm, random forest algorithm and ordered logistic regression algorithm. In Sect. 4, we will verify the specific performance of different algorithms on scoring tasks.

4   Experiments

4.1   Dataset

We compiled a data set for the purpose of model validation. We collected the HSK public available dataset [1] owned by Beijing Language and Culture University, which contains thousands of essay data from examinees of Chinese proficiency test within more than ten years, including essay picture, essay text and typos and sentence annotations etc.

After analysis, the scores of the HSK data set are scattered between [40,95] points, and the scores are all multiples of 5. The average sample score is 69.5 points, and the standard deviation is 10.9. Therefore, its essence is a score set with a full score of 20 points and a score interval of 1 point. According to the data distribution characteristics of the HSK data set, the self-owned data set was processed and normalized, so that the data from different data sources have the same mean and variance. On this basis, the training set, verification set and test set are divided by random sampling according to the ratio of 7:1:2.

4.2   Metric

In the selection of evaluation indicators, this paper adopts two standards for evaluating quality. A more intuitive way to measure performance is to use Mean Absolute Error (MAE) to intuitively record the absolute difference between the prediction results and expert scores. The average value of the error, its value is shown in formula 4.

$$MAX(X, h) = \frac{1}{m} \sum_{i=1}^{m} |h(x_i) - y_i| \tag{4}$$

where X represents the input sample, h represents the scoring function itself, m represents the number of samples in the input set, y_i represents the label value corresponding to the sample, and the smaller the absolute mean error, the better the scoring effect of the model.

Table 2. Comparison between modeling approaches.

Modeling approach	Learning rate	Loss function	MAE
classification	$2\,e$ -5	Cross Entropy	1.077
	$1\,e$-5	**Cross Entropy**	**1.015**
regression	$1\,e-5$	L2	1.100
	$1\,e-5$	L2	1.089
	$1\,e-5$	SmoothL1	1.100
	hierarchical learning rate	L2	1.162

4.3 Results

Modeling as Classification or Regression? As mentioned in Sect. 3.1, this paper designs two modeling methods of classification and regression to solve the scoring problem based on multimodal features. This section will show the performance difference between different modeling approaches.

In order to ensure the relative fairness of the comparison, in the model structure, the feature dimension is changed from the 512-dimensional feature output Dimensionality reduction to the category number dimension n. In the classification model, the model output and the label are directly calculated by the cross-entropy loss function; in the regression model, the linear layer is first used to output the regression value with sigmoid normalization, and the L2 loss and SmoothL1 loss are used as the model in the experiment. train.

In the setting of hyperparameters, since the essence of the multimodal scoring model is the fine-tuning of the BERT pre-training model, Therefore, a smaller learning rate is used for experiments, and a better hyperparameter configuration is verified through experiments. In terms of training strategy, in order to pursue a better effect of the regression model, this paper also tried to use layered training, setting a small learning rate of $1e-5$ for the BERT coding layer, and for the regression head Set a larger learning rate $1e-3$ for training. To sum up, the mean absolute error on the test set is used to measure the effect of the model. The experimental results of different modeling methods are shown in Table 2.

From the experimental results, it can be seen that modeling by classification tasks has achieved better results. In the test A mean absolute deviation of 1.015 was achieved on the set, outperforming regression modeling across multiple training configurations. Although intuitively the regression model is more in line with the needs of scoring scenarios, since the data set itself is a discrete score value, it naturally divides the continuous score range into segments, so the use of the cross-entropy loss function better captures the differences between classes. Applicable to the current scene requirements.

The Role of Multi-modal Architecture. In order to verify the gain of multimodal features to scoring models, this paper first constructs scoring models based on linguistic unimodal features and visual unimodal features. In the text

Table 3. Evaluation of the multimodal method.

Modality	Feature fusion method	MAE
visual unimodal	/	1.278
text unimodal	/	1.131
dual-modal	joint	1.063
dual-modal	**implicit alignment**	**1.015**

unimodal feature, use the pre-trained BERT encoder to connect to the learnable output layer and classifier; in the visual unimodal feature, use the serialized visual features extracted from the output of the recognition model as input and classify device for training. In order to verify the advantages of multimodal features based on implicit alignment, this paper also implements a multimodal feature fusion method based on concat. Visual features and linguistic features are spliced along the sequence direction and then passed to the classifier for scoring verification.

The four different modeling methods are all modeled as classification problems. Experiments are carried out using the cross-entropy loss function with the same optimizer and hyperparameters. The performance of the models obtained by the above-mentioned different experiments is shown in Table 3.

It can be seen from the experimental results that the serialized visual features and BERT-based linguistic features have preliminary scoring effects due to the introduced prior bias, and the average errors directly used for scoring are 1.278 and 1.131, respectively. The dual-modal features promote each other, and the use of dual-modal features can effectively improve the effect of the scoring model. Compared with the traditional concat-based feature fusion method, the method based on implicit alignment can better establish the correlation between the two modes, and further reduce the average error to 1.015, which proves the effectiveness of the fusion method based on implicit alignment.

Visualize the prediction results of the model based on multi-modal scoring, and count the error value between the model prediction score and the actual label score to obtain the scoring error distribution shown in Table 3. According to the table, in the test set, the model prediction value of 92.47 % of the samples is within the error range of two points above and below the true value, and the scoring model based on multi-modal features basically meets the usability requirements of the scoring system.

Comparison with Linguistic Features. This paper extracts linguistic features from the five dimensions of word complexity, sentence complexity, predicate collocation, dependency grammar and logical expression, and evaluates the writing quality from five aspects. In addition, this paper also uses the main program analysis method and the stepwise regression method to extract and reduce the dimensionality of features to obtain comprehensive linguistic features for scoring. In the experiment, we will respectively verify the effect of the feature definition method of the above five dimensions and the two fusion features in the composition scoring task.

As for the selection of the classification model, this paper also verified the performance of the linear regression model, the random forest model and the ordered logistic regression model on the scoring task through experiments. Taking the mean absolute error on the test set as the evaluation index, the performance of different classifiers on different feature selections is shown in the Fig. 5.

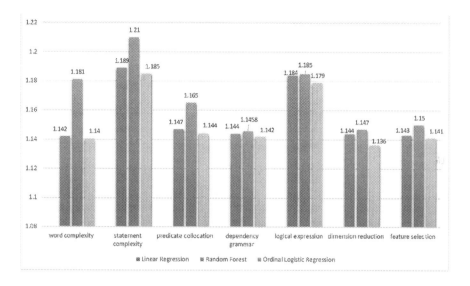

Fig. 5. Performance of different classifiers on different feature selections

From the perspective of feature selection, screening features from the perspective of word usage, predicate collocation, and dependency grammar can better score articles, while features in dimensions such as sentence complexity and logical expression lack advantages, which is similar to Fig. 4. The visualization results have similar performance, and the samples under different scores lack good separability in terms of logical word usage and average sentence length. On the whole, the best scoring performance is obtained by using dimensionality reduction data and using ordered logistic regression model training, with an average absolute error of 1.136 points, which is consistent with the BERT-based single-modal text scoring performance shown in Table 3. Reached a similar level, in line with overall expectations.

From the perspective of the model, the ordered logistic regression method generally achieves better performance. Compared with the random forest method, it has a 1%–5% improvement in the characteristics of each dimension. The ordered logistic regression algorithm models the task as a mixed task of regression and classification, which is naturally more in line with the characteristics of the scoring task, and its advantages have also been confirmed by experimental data. From the point of view of feature fusion, both dimensionality reduction using principal component analysis and feature screening using stepwise regression can improve the performance of the scoring model, and it is meaningful to add linguistic comprehensive scores to the integrated system.

5   Conclusion

Aiming at the problem of visual information negligence in automatic essay scoring, this paper proposes a multimodal scoring method based on implicit alignment. The recognition output is introduced to bridge between the text modality and the visual one. Validation was done on the data sets consisting more than 4000 samples including HSK publicly available corpus It shows that the multimodal feature fusion method based on implicit alignment is superior to the traditional single-modal method and joined multimodal method, and the MAE achieves an accuracy of 1.015. In addition, the method based on multidimensional linguistic features achieves the MAE of 1.036 which is slightly lower than the multimodal one. When the interpretation is preferable, linguistics-based features find its potential.

Acknowledgment. This work was supported by the National Key Research and Development Program of China (Grant No. 2020AAA0108003) and National Natural Science Foundation of China (Grant No. 62277011 and 61673140).

References

1. BLCU: HSK dynamic essay corpus ver 2.0 (in Chinese). https://hsk.blcu.edu.cn/. (26 Apr 2022)
2. Che, W., Feng, Y., Qin, L., Liu, T.: N-LTP: An open-source neural language technology platform for Chinese. In: Proceedings of NAACL-SD, pp. 42–49 (2021)
3. Devlin, J., Chang, M.W., Lee, K., Toutanova, K.: BERT: pre-training of deep bidirectional transformers for language understanding. In: Proceedings of NAACL-HLT, pp. 4171–4186 (2019)
4. He, J., Zhang, C., Li, X., Zhang, D.: Survey of research on multimodal fusion technology for deep learning (in Chinese). Comput. Eng. **46**(5), 1–11 (2020)
5. Hu, R., Xiao, H.: The construction of Chinese collocation knowledge bases and their application in second language acquisition. In: Applied Linguistics. vol. 1 (2019)
6. Huang, Z., Xie, J., Xun, E., et al.: Study of feature selection in HSK automated essay scoring. Comput. Eng. Appl. **6**, 118–122 (2014)
7. Huang, Z., Xie, J., Xun, E.: Study of feature selection in HSK automated essay scoring (in Chinese). Comput. Eng. Appl. **6**, 118–122 (2014)
8. Laufer, B., Nation, P.: Vocabulary size and use: lexical richness in L2 written production. Appl. Linguist. **16**(3), 307–322 (1995)
9. Li, G., Duan, N., Fang, Y., Gong, M., Jiang, D.: Unicoder-VL: a universal encoder for vision and language by cross-modal pre-training. In: Proceedings of the AAAI Conference on Artificial Intelligence. vol. 34, pp. 11336–11344 (2020)

10. Li, H., Dai, T.: Explore deep learning for Chinese essay automated scoring. In: Journal of Physics: Conference Series. vol. 1631, p. 012036. IOP Publishing (2020)

11. Li, L.H., Yatskar, M., Yin, D., Hsieh, C.J., Chang, K.W.: VisualBERT: A simple and performant baseline for vision and language. arXiv preprint arXiv:1908.03557 (2019)

12. Lonsdale, D., Strong-Krause, D.: Automated rating of ESL essays. In: Proceedings of the HLT-NAACL 03 Workshop on Building Educational Applications using Natural Language Processing, pp. 61–67 (2003)

13. Nadeem, F., Nguyen, H., Liu, Y., Ostendorf, M.: Automated essay scoring with discourse-aware neural models. In: Proceedings of the Fourteenth Workshop on Innovative use of NLP for Building Educational Applications, pp. 484–493 (2019)

14. Page, E.B.: The use of the computer in analyzing student essays. Int. Rev. Educ. **14**, 210–225 (1968)

15. Su, T., You, H., Liu, S., Wang, Z.: FPRNet: end-to-end full-page recognition model for handwritten Chinese essay. In: Porwal, U., Fornés, A., Shafait, F. (eds) Frontiers in Handwriting Recognition. ICFHR 2022. Lecture Notes in Computer Science. vol 13639. Springer, Cham (2022). https://doi.org/10.1007/978-3-031-21648-0_16

16. Su, W., et al.: VL-BERT: Pre-training of generic visual-linguistic representations. arXiv preprint arXiv:1908.08530 (2019)

17. Tan, H., Bansal, M.: LXMERT: Learning cross-modality encoder representations from transformers. arXiv preprint arXiv:1908.07490 (2019)

18. Tapaswi, M., Bäuml, M., Stiefelhagen, R.: Aligning plot synopses to videos for story-based retrieval. Int. J. Multimedia Inf. Retrieval **4**(1), 3–16 (2015)

19. Tay, Y., Phan, M., Tuan, L.A., Hui, S.C.: SkipFlow: incorporating neural coherence features for end-to-end automatic text scoring. In: Proceedings of the AAAI Conference on Artificial Intelligence. vol. 32 (2018)

20. Van Hout, R., Vermeer, A.: Comparing measures of lexical richness. Model. Assessing Vocabulary Knowl. **93**, 115 (2007)

21. Wang, Y., Hu, R.: A prompt-independent and interpretable automated essay scoring method for Chinese second language writing. In: Li, S. (ed.) CCL 2021. LNCS (LNAI), vol. 12869, pp. 450–470. Springer, Cham (2021). https://doi.org/10.1007/978-3-030-84186-7_30

22. Xu, Y., Li, M., Cui, L., Huang, S., Wei, F., Zhou, M.: LayoutLM: pre-training of text and layout for document image understanding. In: Proceedings of the 26th ACM SIGKDD International Conference on Knowledge Discovery & Data Mining, pp. 1192–1200 (2020)

23. Zhang, P., et al.: VSR: a unified framework for document layout analysis combining vision, semantics and relations. In: Lladós, J., Lopresti, D., Uchida, S. (eds.) ICDAR 2021. LNCS, vol. 12821, pp. 115–130. Springer, Cham (2021). https://doi.org/10.1007/978-3-030-86549-8_8

FCN-Boosted Historical Map Segmentation with Little Training Data

Josef Baloun[1,2]($\boxtimes$) (iD), Ladislav Lenc[1,2] (iD), and Pavel Král[1,2] (iD)

[1] Department of Computer Science and Engineering, Faculty of Applied Sciences,
University of West Bohemia, Pilsen, Czech Republic
{balounj,llenc,pkral}@kiv.zcu.cz
[2] NTIS - New Technologies for the Information Society, University of West Bohemia,
Pilsen, Czech Republic

Abstract. This paper deals with automatic image segmentation in poorly resourced areas. We concentrate on map content segmentation in historical maps as an example of such a domain. In such cases, conventional computer vision (CV) approaches fail in unexpected unique regions such as map content area exceeding the map frame, while deep learning methods lack boundary localization accuracy. Therefore, we propose an efficient approach that combines conventional CV techniques with deep learning and practically eliminates their drawbacks. To do so, we redefine the learning objective of a simple fully convolutional network to make the training easier and the model more robust even with few training samples. The presented method provides excellent results compared to more sophisticated but solely deep learning or traditional computer vision techniques as shown in "MapSeg" segmentation competition, where all other approaches were significantly outperformed. We further propose two additional approaches that improve the original method and set a new state-of-the-art result on the MapSeg dataset. The methods are further tested on an extended version of the Map Border dataset to show their robustness.

Keywords: Historical Map · Segmentation · Little Data

1 Introduction

Historical maps owned by various national archives and libraries are a rich source of information. They often contain valuable and precisely plotted geographical entities. The digitized materials then offer a great potential for many historical studies [4] and they are beneficial for geographical information systems (GIS) communities for example. The process of map vectorization is thus of a great interest.

In the last decades, such maps have been gradually digitized and a lot of the materials are already accessible in an electronic form. However, the digitization is only the first step in the processing of the maps. There is a number of tasks

G. A. Fink et al. (Eds.): ICDAR 2023, LNCS 14187, pp. 520–533, 2023.
https://doi.org/10.1007/978-3-031-41676-7_30

that have to be carried out in order to facilitate the search and fully utilize the materials which brings a number of research problems in the image processing field. The main emphasis is put on automatic approaches with good generalization abilities. However, this is problematic to achieve in some specific domains given the limited amount of annotated training data, which is almost always the case in the historical documents. It holds true in the case of historical maps as well. There are many differences between maps from different areas e.g. different colors, width of strokes, decorations and also map borders. Therefore, every collection is more or less unique. The manual annotation is time-consuming and therefore costly, which can explain also the lack of datasets in this domain.

The segmentation is an essential task which must be done after digitization. It allows further processing to focus only on the relevant area. Therefore, high demands are placed on the segmentation results. In this work, we concentrate on map content area detection according to the *Task 2* of the ICDAR 2021 Competition on Historical Map Segmentation [4] (*MapSeg*). The main goal is to provide a segmentation mask of a map content area (Fig. 1) and to remove features surrounding the actual map like map frame, legends and titles. Those elements are separated from the map content area by frames, however this is not the case for all of them. Moreover, the frames are frequently crooked or damaged and also exceeded by the map content area.

Although neural networks have remarkable learning capacity and achieved state-of-the-art results in many visual tasks including segmentation, their results still contain more or less errors in mask predictions given the low amount of training data. In the worst case, they do not work at all. On the other hand, it is also hard to handle document uniqueness with only conventional computer vision methods.

Therefore, we propose three methods with an easier learning objective that combines both conventional and deep learning approaches and lowers the effort needed for handcrafted features. Even though the methods utilize the simple FCN model, a significant improvement is achieved compared to much more sophisticated but purely deep learning or conventional computer vision techniques. We do not focus on different neural network architectures in this work, even though it may play a role. We instead focus on the combination of conventional and deep learning features in order to obtain the best results with a limited amount of data while minimizing the effort needed.

2   Related Works

Many methods were proposed for general image segmentation, e.g. fully convolutional networks (FCNs), where the input image and a segmentation mask are provided as a training sample. However, they rarely take into account the lack of data and the task specificity and only a few of them relate directly to maps. We first briefly report methods submitted to the MapSeg competition. Then we summarize methods solving map segmentation and analysis in general.

CMM [4] method is a representative of traditional approaches. The idea is to detect the contour lines and reconstruct them from the center of the image. It

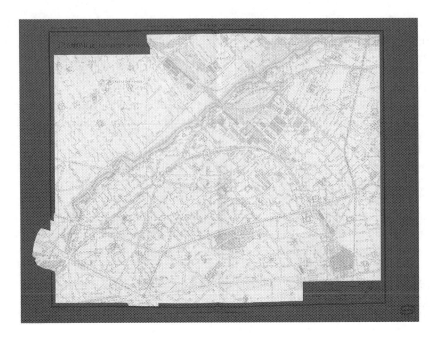

Fig. 1. MapSeg dataset sample: The map image overlayed with map content area ground-truth [5]

consists of several steps including the quasi-flat zone algorithm to eliminate map margin and a watershed to close the contour. *IRISA* [1] is another well performing traditional computer vision approach, that does not require any training. It relies on line segments. These segments are extracted from the image at various resolutions. Then, the grammatical rules are used to detect the map content area contour. *L3IRIS* [4] approach utilizes the state-of-the-art few-shot segmentation method HSNet [16].

The problem of map segmentation was solved for example in [14]. The authors proposed a method based on linear element features. A robust grid detection in historical maps relying on Hough transform is presented in [3]. The boom of neural networks and deep learning brought new possibilities for automatic segmentation and analysis of historical map resources. The potential usage of such methods was discussed in [10]. A method based on convolutional neural networks (*CNN*) was proposed in [15]. It uses an advanced guided watershed transform for obtaining superpixels [13]. A shallow CNN is then used for superpixel classification. A method for map segmentation utilizing handcrafted features, CNN and mathematical morphology was presented in [7]. A novel architecture for map segmentation was proposed in [9]. It has an encoder-decoder structure similarly as U-Net [19] and additionally uses cross-scale skip connections. A cadastre borders and important markers are detected in [12] utilizing an FCN or conventional computer vision techniques. A combination of deep learning and conventional computer vision methods is proposed in [8] to vectorize the historical maps.

Names of cities and other landscape features are detected using several object detection models and further processed in [11].

3   Map Content Area Segmentation

Since deep learning can deal with hardly definable specialties in the map documents, we employ a simple model as a feature extractor to predict the borders of the area. We have identified experimentally that predicting only border contours is a much easier learning objective than predicting the whole map content area (Fig. 2). If we train the FCN model to predict the whole map content area, we want to predict every pixel there as positive (e.g. roads, buildings or text). There are also similar conflicting objects outside the map content that we want to predict as negative (legends for example). On the other hand, when predicting only border contours, the network can focus for example on lines, transitions between "empty" and "non-empty" areas, border decorations or legends. The number of possible input variants for a positive pixel is therefore much lower than in the previous case. We found this helpful to train the network.

The predicted border contour can be closed and transformed into the map content area utilizing morphological operations for example. At the same time, the conventional computer vision approaches can improve the results in terms of localization accuracy. Therefore, the three main steps of the methods are *border prediction*, *image binarization* to improve localization details and *post-processing* to close the contour.

First, we describe the border prediction and image binarization steps. Since the post-processing step changes across the approaches, details are provided within each specific approach.

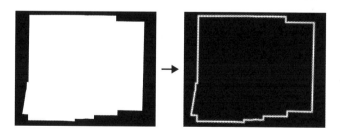

Fig. 2. Modified learning objective

3.1   Border Prediction

For the border prediction, we adapt a simplified U-Net-like FCN for general segmentation [2] as a feature extractor. As was shown in the paper, FCNs generalize
well and they can also deal with the small amount of training samples. Furthermore, the border contours often appear close to image borders. That information
can be utilized by the network thanks to the padding as discussed in the paper.

Compared to [2], the utilized network has half of the filters in the convolutional layers. The network's input is the whole down-sampled image and the
output is the predicted mask of the borders. Since the input images are large,
they are firstly eroded to propagate thin black lines and then down-sampled to
fit 1024 px rectangle. The reason for the down-sampling is a compromise between
network context capability, localization accuracy and computational costs. We
refer to [2] for further details of the architecture.

The ground-truth was automatically generated from the provided map content area masks in the original image resolution (Fig. 2). The value for each
pixel x was obtained using Gaussian function (Eq. 1), where $(x - b)$ stands for
the distance to the border and σ is set to 50.

$$f(x) = exp\left(-\frac{1}{2}\frac{(x-b)^2}{\sigma^2}\right) \tag{1}$$

The reason for that is to provide more true positives and decrease the possibility of discontinuities in predictions. The uncertainty, that the pixel does not
have to be strictly classified as the border or not, can also make classification
more robust as discussed in [18]. We also use image augmentation techniques
(mirroring, rotation and random distortion) to enlarge the training set.

3.2   Image Binarization

Since the borders are usually present in the input image, we found it useful to
use them directly in order to have as precise results as possible. Therefore we
adapt a recursive Otsu binarization method [17].

In a nutshell, the method firstly removes the background estimated by a
median filter. It is ideal to propagate thin lines and also to discard large homogeneous areas. This step also allows the method to deal with the brightness
inconsistency. After that, the image is recursively binarized using Otsu thresholding with hysteresis that reduces the amount of noise present in the binarized
image. A drawback is a significant amount of remaining noise in the result making it difficult to process.

3.3   UWB Method

The winning method of the MapSeg competition is depicted in Fig. 3. The input
image (Fig. 3a) is binarized (Fig. 3b) and the map content area borders (Fig. 3d)
are predicted in parallel. The prediction of map border is followed by post-
processing resulting in an estimated mask of the map content area (Fig. 3e).

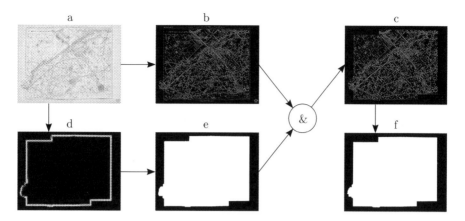

Fig. 3. Map content area segmentation process of UWB method: *(a)* input image, *(b)* binarized input image, *(c)* binarized image masked with estimated mask, *(d)* FCN border prediction, *(e)* estimated mask, *(f)* result

The estimated mask is then combined with binarized image utilizing logical and operation (Fig. 3c). Finally, the post-processing is repeated to obtain map content area mask (Fig. 3f).

The same post-processing is used for producing both masks Fig. 3e and 3f from the inputs Fig. 3d and Fig. 3c, respectively. It is similar to the morphological closing. It starts with dilation to fill eventual discontinuities. Then, the biggest connected component is selected and filled. Finally, the erosion is applied. The dilation and the erosion use the same rectangular kernel which is chosen with respect to the input images where map content borders usually follow horizontal and vertical lines.

A drawback of this method is the need to manually set the kernel. It is prone to improper settings and can cause fragmentation and errors if the contour is not properly closed. The post-processing also results in losing details as can be seen in Fig. 10. Therefore, we have proposed two improved versions of the method as described below.

3.4   BEW Method

To face the drawback of the baseline method, we use Border prediction, Euclidean distance transform and Watershed (*BEW*) according to Fig. 4. The method does not utilize any conventional features. Therefore, it can be used even if there is no possibility to extract the details conventionally.

A similar approach was presented in [8], but it fails if the contour is not properly closed. Therefore, we further extend the approach and use euclidean distance transform (Fig. 4c). In that case, the missing fragments are fixed as illustrated in Fig. 5.

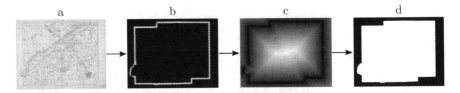

Fig. 4. Map content area segmentation process of BEW method: *(a)* input image, *(b)* FCN border prediction, *(c)* euclidean distance transform, *(d)* result using watershed

Fig. 5. From the left: detail of unclosed contour prediction, its euclidean distance transform, the result after watershed

3.5   BBEW Method

Optionally, the image binarization features can be used in order to improve localization accuracy as in the baseline method. The *BBEW* method uses Border prediction, Binarization, Euclidean distance transform and Watershed according to Fig. 6. The binarized image (Fig. 6b) is masked with predicted borders (Fig. 6d). Then, the euclidean distance transform is used to deal with unclosed contours. It provides a border distance matrix (Fig. 6c). Finally, the watershed is used to segment the map content area (Fig. 6f).

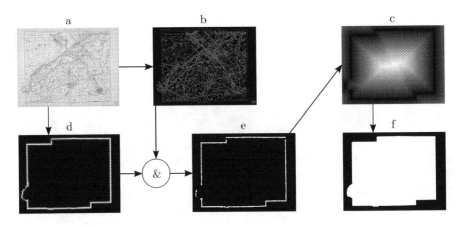

Fig. 6. Map content area segmentation process of BBEW method: *(a)* input image, *(b)* binarized input image, *(c)* euclidean distance transform, *(d)* FCN border prediction, *(e)* binarized image masked with predicted borders – dilated for visualization purposes, *(f)* result using watershed

4   Experimental Set-Up

In this section, we describe the dataset and evaluation criteria. For more details on the methods from Sect. 3, we refer to https://gitlab.kiv.zcu.cz/balounj/21_icdar_mapseg_competition, where the source codes and other related materials are freely available for non-commercial purposes.

4.1   MapSeg Competition Dataset

The map sheets constituting this dataset [5] are collected from 9 atlases of the City of Paris published between years 1894 and 1937. There are approximately 20 sheets for each year and the image resolution is very high (about 10000×10000 pixels).

The competition involved three tasks: Detect building blocks, Segment map content area and Locate graticule lines intersections. For each of these tasks, training, validation and test sets are available. For the second task, the sizes of train, validation and test sets are 26, 6 and 95 respectively. The dataset is available at https://zenodo.org/record/4817662.

4.2   Map Border Dataset

The Map Border dataset [12] consists of historical cadastral maps originating from the second half of the nineteenth century. It contains annotations for cadastre borders, important landmarks and other features for the task of border detection. We have extended the annotations with the map frame masks for the purposes of this work.

The image resolution is about 8400×6850 pixels. As illustrated in Fig. 7, the map sheets have different characteristics than the ones from MapSeg dataset.

We used 12 images for testing and 22 images for training and validation.

5   Evaluation Criteria

For the reported results, we follow the MapSeg competition [4] scenario and also use the provided evaluation tools [6]. For the map content area segmentation evaluation, the 95th percentile variant of Hausdorff distance $(d_{H_{0.95}})$ is used as error measure. The final measure is the average of all test image measures.

We consider the Hausdorff distance appropriate for the task, since it focuses on shapes and details at the borders. The result is not distorted since it is not affected by the large area as in Intersection over Union for example.

Fig. 7. Sample from Map Border dataset

6 Results

In this section, we report the obtained results and compare our *UWB*, *BEW* and *BBEW* methods with the best methods on the MapSeg dataset [5]. The proposed methods are further verified on the Map Border dataset.

As can be seen in Table 1, our methods surpassed the other methods by a significant margin. The proposed methods show excellent results even on the noisy historical map images.

Table 1. Final Hausdorff error ($d_{H_{0.95}}$ [px]) for map content area segmentation task

Method	MapSeg dataset	Map Border dataset
CMM	85	–
IRISA	112	–
L3IRIS	126	–
UWB (Ours)	19.0	25.5
BEW (Ours)	18.1	12.0
BBEW (Ours)	**12.0**	**9.5**

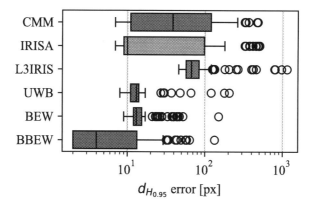

Fig. 8. Test images error distribution for map content area segmentation task on MapSeg dataset (Ours in red color)

The conventional approaches (*CMM* and *IRISA*) are very precise in the areas that contain the border contours, but they can hardly deal with other unique areas like map content exceeding the frame and some legends. This leads to higher variance in the errors in Fig. 8.

The deep learning approach (*L3IRIS*) has higher median value but smaller variance of errors. It usually catches the unique areas but it is missing the precision at the borders and the localization accuracy is not very convincing.

In the same figure, our approaches have small median value and also small variance of errors. They profit from both conventional and deep learning approaches and practically eliminate their drawbacks. On the other hand, the outliers in Fig. 8 are usually caused by wrongly predicted legends and are still present in each approach. This could be probably solved by extending the training part of the dataset, improved augmentation or further post-processing.

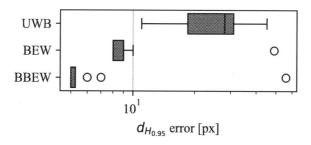

Fig. 9. Test images error distribution for map content area segmentation task on Map Border dataset

Fig. 10. Details of UWB method result

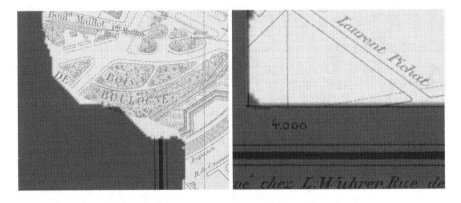

Fig. 11. Details of BEW method result

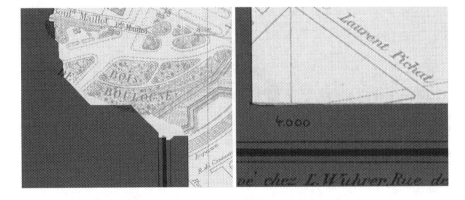

Fig. 12. Details of BBEW method result

The UWB method provides good results and allows to utilize the details provided by image binarization step. But naive post-processing has several drawbacks that can cause errors and loss of detail as in Fig. 10. As presented in Fig. 9, this issue is more evident on the results from Map Border dataset.

The BBEW method solves these drawbacks and provides as many details as possible while preventing the erroneous closing of the contour. It sets the new state-of-the-art result of $d_{H_{0.95}} = 12.0$. As illustrated in Fig. 12, the amount of detail is fascinating, especially in areas where border features can be obtained conventionally and further combined with deep learning features.

Interesting observation is a very good result of *BEW* method (Fig. 11). It indicates the advantage of the modified learning objective. *L3IRIS* utilizing the state-of-the-art few-shot segmentation method fails compared to *BEW* utilizing the simple general FCN segmentation model with modified learning objective and contour closing. It is obvious, that the border prediction is a much more suitable learning objective for the task and it has probably a bigger impact than the selection of the model.

7   Conclusions

In this paper, we faced the segmentation of historical maps. In such poorly resourced domain, common deep learning approaches often fail due to the lack of training data or even related data that could be used for transfer learning.

We have proposed three efficient map segmentation approaches that utilize an easier learning objective for a general FCN and post-processing to get the original objective. It allows them to work even with little training data while producing excellent results that can be optionally refined using conventional image binarization features.

The proposed methods are evaluated on the MapSeg dataset which was used in the ICDAR 2021 segmentation competition. We have shown that the proposed methods outperform significantly all other approaches and that the more suitable learning objective may have a bigger impact than the choice of a deep learning model. With *BBEW* method, we set the new state-of-the-art of $d_{H_{0.95}} = 12.0$ on MapSeg dataset. We further verified the methods on the Map Border dataset with corresponding results. Thus, the combination of deep learning with conventional computer vision techniques seems very promising, especially for poorly resourced domains such as historical documents.

Another contribution of our work consists in the availability of the source codes for the research purposes.

Acknowledgements. This work has been partly supported by the Grant No. SGS-2022-016 Advanced methods of data processing and analysis.

References

1. Aurelie, L., Jean, C.: Segmentation of historical maps without annotated data. In: The 6th International Workshop on Historical Document Imaging and Processing, pp. 19–24 (2021)

2. Baloun, J., Král, P., Lenc, L.: Chronseg: Novel dataset for segmentation of handwritten historical chronicles. In: Proceedings of the 13th International Conference on Agents and Artificial Intelligence - Volume 2: ICAART, pp. 314–322. INSTICC, SciTePress (2021). https://doi.org/10.5220/0010317203140322

3. Baloun, J., Lenc, L., Král, P.: Robust grid detection in historical map images. In: 2022 IEEE International Conference on Image Processing (ICIP), pp. 1931–1935 (2022). https://doi.org/10.1109/ICIP46576.2022.9897721

4. Chazalon, J., et al.: ICDAR 2021 competition on historical map segmentation. In: Lladós, J., Lopresti, D., Uchida, S. (eds.) ICDAR 2021. LNCS, vol. 12824, pp. 693–707. Springer, Cham (2021). https://doi.org/10.1007/978-3-030-86337-1_46

5. Chazalon, J., et al.: Icdar 2021 competition on historical map segmentation - dataset. online dataset (2021). https://doi.org/10.5281/zenodo.4817662

6. Chazalon, J., Edwin-Carlinet: icdar21-mapseg/icdar21-mapseg-eval: zenodo archival (2021). https://doi.org/10.5281/zenodo.4818400

7. Chen, Y., Carlinet, E., Chazalon, J., Mallet, C., Duménieu, B., Perret, J.: Combining deep learning and mathematical morphology for historical map segmentation. In: Lindblad, J., Malmberg, F., Sladoje, N. (eds.) DGMM 2021. LNCS, vol. 12708, pp. 79–92. Springer, Cham (2021). https://doi.org/10.1007/978-3-030-76657-3_5

8. Chen, Y., Carlinet, E., Chazalon, J., Mallet, C., Duménieu, B., Perret, J.: Vectorization of historical maps using deep edge filtering and closed shape extraction. In: Lladós, J., Lopresti, D., Uchida, S. (eds.) ICDAR 2021. LNCS, vol. 12824, pp. 510–525. Springer, Cham (2021). https://doi.org/10.1007/978-3-030-86337-1_34

9. Foroughi, F., Wang, J., Nemati, A., Chen, Z., Pei, H.: Mapsegnet: a fully automated model based on the encoder-decoder architecture for indoor map segmentation. IEEE Access **9**, 101530–101542 (2021)

10. Garcia-Molsosa, A., Orengo, H.A., Lawrence, D., Philip, G., Hopper, K., Petrie, C.A.: Potential of deep learning segmentation for the extraction of archaeological features from historical map series. Archaeol. Prospect. **28**(2), 187–199 (2021)

11. Lenc, L., Martínek, J., Baloun, J., Prantl, M., Král, P.: Historical map toponym extraction for efficient information retrieval. In: Uchida, S., Barney, E., Eglin, V. (eds.) DAS 2022. LNCS, vol. 13237, pp. 171–183. Springer, Cham (2022). https://doi.org/10.1007/978-3-031-06555-2_12

12. Lenc, L., Prantl, M., Martínek, J., Král, P.: Border detection for seamless connection of historical cadastral maps. In: Barney Smith, E.H., Pal, U. (eds.) ICDAR 2021. LNCS, vol. 12916, pp. 43–58. Springer, Cham (2021). https://doi.org/10.1007/978-3-030-86198-8_4

13. Li, Z., Chen, J.: Superpixel segmentation using linear spectral clustering. In: Proceedings of the IEEE Conference on Computer Vision and Pattern Recognition, pp. 1356–1363 (2015)

14. Liu, T., Miao, Q., Xu, P., Song, J., Quan, Y.: Color topographical map segmentation algorithm based on linear element features. Multimedia Tools Appl. **75**(10), 5417–5438 (2016)

15. Liu, T., Miao, Q., Xu, P., Zhang, S.: Superpixel-based shallow convolutional neural network (SSCNN) for scanned topographic map segmentation. Remote Sens. **12**(20), 3421 (2020)

16. Min, J., Kang, D., Cho, M.: Hypercorrelation squeeze for few-shot segmentation. CoRR abs/2104.01538 (2021). https://arxiv.org/abs/2104.01538

17. Nina, O., Morse, B., Barrett, W.: A recursive OTSU thresholding method for scanned document binarization. In: 2011 IEEE Workshop on Applications of Computer Vision (WACV), pp. 307–314. IEEE (2011)

18. Peterson, J.C., Battleday, R.M., Griffiths, T.L., Russakovsky, O.: Human uncertainty makes classification more robust. In: Proceedings of the IEEE/CVF International Conference on Computer Vision (ICCV), October 2019
19. Ronneberger, O., Fischer, P., Brox, T.: U-Net: convolutional networks for biomedical image segmentation. In: Navab, N., Hornegger, J., Wells, W.M., Frangi, A.F. (eds.) MICCAI 2015. LNCS, vol. 9351, pp. 234–241. Springer, Cham (2015). https://doi.org/10.1007/978-3-319-24574-4_28

MemeGraphs: Linking Memes to Knowledge Graphs

Vasiliki Kougia[1]([✉]), Simon Fetzel[1], Thomas Kirchmair[1], Erion Çano[1], Sina Moayed Baharlou[2], Sahand Sharifzadeh[3], and Benjamin Roth[1]

[1] Faculty of Computer Science, University of Vienna, Vienna, Austria
{vasiliki.kougia,simonf93,thomas.kirchmair,erion.cano,
benjamin.roth}@univie.ac.at
[2] Department of Electrical and Computer Engineering, Boston University,
Boston, MA, USA
baharlou@bu.edu
[3] Faculty of Computer Science, Ludwig Maximilians University of Munich,
Munich, Germany
sahand.sharifzadeh@gmail.com

Abstract. Memes are a popular form of communicating trends and ideas in social media and on the internet in general, combining the modalities of images and text. They can express humor and sarcasm but can also have offensive content. Analyzing and classifying memes automatically is challenging since their interpretation relies on the understanding of visual elements, language, and background knowledge. Thus, it is important to meaningfully represent these sources and the interaction between them in order to classify a meme as a whole. In this work, we propose to use scene graphs, that express images in terms of objects and their visual relations, and knowledge graphs as structured representations for meme classification with a Transformer-based architecture. We compare our approach with ImgBERT, a multimodal model that uses only learned (instead of structured) representations of the meme, and observe consistent improvements. We further provide a dataset with human graph annotations that we compare to automatically generated graphs and entity linking. Analysis shows that automatic methods link more entities than human annotators and that automatically generated graphs are better suited for hatefulness classification in memes.

Keywords: hate speech · internet memes · knowledge graphs · multimodal representations

1 Introduction

Internet memes are items such as images, videos, or twitter posts that are widely shared on social media and typically relate to several subjects, such as politics, social, news, and current internet trends.[1] Memes are a popular form of communication and they are often used as a means to express an opinion or stance

---

[1] Disclaimer: This paper contains examples of hateful content.

© The Author(s), under exclusive license to Springer Nature Switzerland AG 2023
G. A. Fink et al. (Eds.): ICDAR 2023, LNCS 14187, pp. 534–551, 2023.
https://doi.org/10.1007/978-3-031-41676-7_31

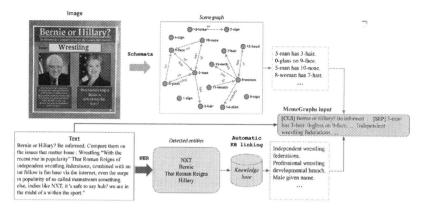

Fig. 1. The steps performed by the MemeGraphs method. The automatic augmentation consists of scene graphs generated automatically by a pre-trained model (Schemata [26]) and entities detected in the text by a pre-trained Named Entity Recognition (NER) model. Background knowledge for each entity is retrieved from a knowledge base (Wikidata). The final MemeGraphs input is created by concatenating these augmentations and adding them after the [SEP] token following the text of the meme in order to feed it to a Transformer for text classification.

in a humorous or sarcastic manner, but they can also be hateful and promote problematic content that is likely to hurt specific groups of people and hence be harmful to society in general [2]. Thus, analyzing trending memes can provide insight into people's reactions and opinions to important societal matters, as well as to the traits of different groups. This can help in tasks like filtering out harmful memes from internet platforms or extracting user opinions for socioeconomic studies. This work focuses on memes in the form of images with some form of superimposed text (sometimes referred to with the technical term *image macros*) with the goal of detecting hateful content.

Lately, there has been increased interest in deep learning models for analyzing memes and classifying them [10,11,23], for example in the context of the Hateful Memes competition organized by Facebook [13] and the shared task in the Workshop on Online Abuse and Harms (WOAH) 2021 that was the continuation of the first competition [21]. Another shared task in Semeval 2022 aimed at detecting misogyny in memes [11]. Alongside these shared tasks, the corresponding datasets were published with human annotations capturing important properties such as hatefulness and misogyny. Recent works use multimodal representation learning based on image features from Convolutional Neural Networks (CNNs), Transformer-based language models for the text [3,13,14] or multimodal (vision+language) Transformer models in order to classify memes [13]. Some works additionally incorporate specifically extracted image features such as race [33], person attributes [1,23], and automatic image captioning output [4,7]. However, none of these works have included the relations between objects in the form of scene graphs or background knowledge in the form of a text description for the objects depicted in the image.

In this paper, we address the issue of classifying hateful memes by performing an automated augmentation in order to represent the visual information and background knowledge (Fig. 1). We built on the MultiOFF dataset, which contains memes extracted from social media during the 2016 U.S. Presidential Election [28]. An off-the-shelf scene graph generation model was employed to produce scene graphs, which contain detected visual objects and relations between them for each meme, and a NER model to detect entities that we linked to background knowledge. The scene graphs were serialized as text resulting in a unified way to represent all the modalities expressed in a meme. This allows using a (unimodal) text classifier to classify the multimodal memes. Hence, we incorporated the automatically produced augmentations to classify the memes using a text-based Transformer model and show that they can improve its performance. The explicit representation of the image as serialized tokens also provides a more interpretable intermediate representation (compared to hidden layers in multimodal models such as ImgBERT [13,14]).

Furthermore, in order to examine how the results of the automatic augmentation would deviate from human ones, we performed manual augmentation. Two human evaluators corrected the automatically produced scene graphs and manually added background knowledge. We compare our automatic MemeGraphs method with models operating on the manually augmented data (only for training or for both training and inference) and find that automatic augmentations assist in achieving better results than manual ones.

Contribution. Our contributions can be summarized as follows:

– We propose MemeGraphs, a novel method for classifying memes utilizing scene graphs augmented with knowledge and providing insights for processing multimodal documents.
– We show that adding this kind of knowledge to a text-based Transformer model can improve its classification performance. Furthermore, we show that this yields improvements compared to a simple model only using learned representations to classify the memes, such as ImgBERT.
– We conduct extensive experiments with manual and automatic settings for obtaining this knowledge and show that the automatic setting of our MemeGraphs method provides more meaningful information.

In the following, we discuss related work and present our MemeGraphs method. Subsequently, in Sect. 4, we describe the models we implemented and report their results. Finally, we provide a qualitative analysis, including a discussion of the findings of human augmentation, and compare this with the automatic MemeGraphs method.[2]

2   Related Work

Combining text and image inputs is crucial for many tasks, e.g. for image search or (visual) question answering, relying on an image and its caption. Research

---

[2] The code and data for the MemeGraphs method are available on: https://github.com/vasilikikou/memegraphs.

has shown that adding images to text-based tasks (e.g., machine translation) improves the performance of the models [31]. However, the meaningful interpretation of text and image and, in particular, the relations between them, still remains challenging [5,17]. Commonly used approaches rely on Transformer models that are pre-trained on image+text pairs [6,12,18–20,27]. A step towards better scene understanding is to generate scene graphs [15]. Scene graphs provide structured knowledge about an image, e.g., objects, relations, and attributes. Recent works have shown that we can improve scene graph generation using message propagation between entities [29,32], and by employing background knowledge in form of knowledge graphs [26], texts [25], or using feedback connections. In [26], the authors proposed Schemata, a scene graph generation model consisting of two parts: the backbone module and the relational reasoning component. Additionally, Schemata uses feedback connections to further encourage the propagation of higher-level, class-based knowledge to each neighbor. The backbone is pre-trained on ImageNet [8] and the whole network is fine-tuned on Visual Genome [15] on the scene graph classification task. Scene graphs can help to achieve state-of-the-art results in several visual tasks [22,30]. Inspired by these approaches, we generate scene graphs to represent the visual information contained in memes by using the Schemata model (see Sect. 3).

A specific instance of vision and language tasks is memes classification. To address the need for automatic means that can detect hateful content in memes, datasets and models [3,23,28] were published in the last couple of years, and shared tasks [11,13,21] were organized to attract interest on this task. Methods that have been implemented for hateful memes detection can be grouped into three categories: 1. Unimodal methods that use either only the text or the image as input, 2. Multimodal approaches, where image embeddings from an image encoder are fed to a text model and both models are trained separately, and 3. Multimodal methods, consisting of vision+language Transformers, being pre-trained in a multimodal fashion. Current methods experiment with models from all three categories and focus on improving models from the third category by adding extra features [3,4,16,23]. These features can be visual attributes extracted from CNNs [1,13,16] (e.g. objects, entities or demographics), representations from CLIP [23,24], automatically generated captions [4,7], etc.

The Hateful Memes Challenge, hosted in 2020 by Facebook, was a binary classification task of hate detection [13].[3] Kiela et al. [13] created a dataset with 10,000 memes to which they added counterfactual examples in order to make the task more challenging for unimodal approaches. They experimented with several different settings and found that multimodal methods worked best. An extended version of the Hateful Memes Challenge was included as a shared task in the Workshop on Online Abuse and Harms (WOAH) [21]. The same dataset was used but it now included new fine-grained labels for two categories: protected category and attack type. In this shared task, multimodal approaches were dominant as well. A multimodal method introduced by [13] and subsequently also used for the shared task in WOAH [14] incorporated image embeddings as inputs to a

---

[3] https://www.drivendata.org/competitions/64/hateful-memes/.

text classifier.[4] This method belongs to the second category and is an early fusion approach meaning that the image embedding and the text embedding are concatenated before feeding them to the classifier. Different types of image and text components are employed in different works. Specifically, in ImgBERT [14], first, they feed the memes images to a convolutional neural network (CNN) and extract their embeddings. Then, they provide the text of the meme as input to BERT [9] and extract the [CLS] token representation. They concatenate the [CLS] token representation with the embedding of the meme's image and use the result as input to the classifier. During training only the text-based BERT part of ImgBERT is trained, while the image embeddings remain frozen.

Another dataset for detecting hateful content in memes is the MultiOFF dataset [28]. It contains memes that were extracted from social media during the 2016 U.S. presidential elections. The dataset was first shared on Kaggle and consisted of the image URL for each meme, its text and metadata, e.g., timestamp, author, likes, etc.[5] The authors obtained the images from the URLs and discarded any metadata. In total, this dataset contains 743 memes, which were annotated as hateful or non-hateful. In [28] the authors experimented with unimodal (text only) and multimodal (text and image) approaches, and the model with the highest F1 score was a CNN operating only on the text of the memes.

The above mentioned datasets focus on hate and offensive speech, but there are also datasets that cover other aspects of harmful content in memes. In [10], the authors focused on detecting propaganda in memes. They created and released a dataset with 950 memes extracted from Facebook groups, annotated for 22 different propaganda techniques. In their experiments they used existing unimodal and multimodal models and found that the latter, especially multimodally pre-trained Transformers perform best in their setting. Recently, a challenge called Multimedia Automatic Misogyny Identification (MAMI) focused on detecting misogyny and its exact form, i.e., stereotype, shaming, objectification and violence in memes [11]. In [23], they studied harm in memes and proposed a framework to detect harmful memes and the entities targeted. The authors also released their dataset with 7,096 memes in total about politics and COVID-19.

The existing challenge sets, resources and models show the importance of analyzing internet memes and the challenge to combine all the modalities that form a meme. However, current works focus on incorporating visual information in the form of individual features like the ones described above or automatically generated captions. We propose a novel approach to represent the visual content of memes using scene graphs, hence "translating" them into text form. Furthermore, current methods only extract entities from the images, but not from the captions or texts. We argue that often this is not sufficient (or feasible), since memes can also incorporate screenshots of text, as it is the case in the Multi-OFF dataset. Hence, we approach this problem by extracting the entities from the text in order to obtain more information. We further retrieve background

---

[4] The method was called Concat BERT in [13] and ImgBERT in [14]. Here we call it ImgBERT because we use their implementation.

[5] https://www.kaggle.com/datasets/SIZZLE/2016electionmemes.

knowledge for each extracted entity, and show that this approach is worthwhile to explore, since it allows for a more grounded and comprehensive automatic interpretation of memes.

3   MemeGraphs

In this section, we describe our method for automatic augmentation of a memes dataset, an approach we call MemeGraphs. Our proposed method consists of three steps: 1. Scene graph construction, 2. Knowledge linking to detected entities and 3. The construction of the final MemeGraphs input. The first two steps are performed automatically by using off-the-shelf models. Hence, the result of MemeGraphs are knowledge graphs representing the meme as a whole, which can be used to classify them, e.g., for hate detection. We build on the MultiOFF dataset, which contains memes extracted from social media during the 2016 U.S. Presidential Election [28]. In what follows, we present the three individual steps of MemeGraphs in detail and how the final result is constructed.

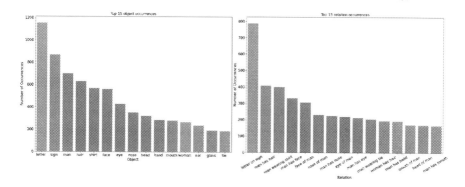

Fig. 2. The occurrences of the top 15 mostly detected objects (left) and relations (right) based on the automatic annotation.

3.1   Scene Graphs

Scene graphs provide information for an image in the form of a graph $SG = \{V, E\}$, where V are the objects depicted in the image (nodes) and E the relations between them (edges). Models that generate scene graphs output a set of relation triplets in the form of $\{object_1, relation, object_2\}$, which constitute the graph. In our task, we generate scene graphs for memes. Each meme m consists of two modalities $(I_m, T_m) \in I \times T$, where I is the set of images and T is the set of texts of the memes. We employ a scene graph generation model G, which takes as input a meme image and outputs a scene graph SG as:

$$SG_m = G(I_m) \qquad (1)$$

Towards this end, we apply Schemata, a pre-trained scene graph generation model [26] that was trained with a multi-task learning strategy to automatically predict scene graphs. We do not further fine-tune the model on the memes present in our data and we only consider the objects ranking amongst the top 16 objects for the scene graph according to the detection scores.[6] Hence, the scene graph generated for each meme is a set of triplets as follows:

$$SG_m \subseteq \{(object_i, relation, object_j)\}, \tag{2}$$

where $0 \leq i, j \leq 15$, $relation \in R$ and R is the set of all possible relations.

For 67 memes, Schemata did not produce any output. For the remaining 676 memes, 10,426 relations and 9,666 objects were detected in total. The number of unique objects and relations was 142 and 970 respectively. And while the number of objects detected in an image ranged from 2 to 16, the number of detected relations ranged from 1 to 40. The top 15 most frequently detected object and relation types are shown in Fig. 2.

3.2   Knowledge Linking

The second step of our MemeGraphs method is to obtain background knowledge for a given meme. In order to do that, first, we employ an NER model. This model detects named entities in a text, which can be person names, organizations, locations, etc. - depending on the task. It can provide useful information that assists to natural language understanding. Here, we feed the text of the meme as input to the NER model and get a list of extracted entities E as $E = NER(T_m)$, where T_m is the text of the meme. Each entity is then searched for in a knowledge base and its related information is retrieved. This way we obtain a text T_i for each entity E_i, so for a given meme we have:

$$KN_m = \{T_i\} = \{KB(E_i)\} \tag{3}$$

For the NER model we employ an off-the-shelf pre-trained Transformer model from spacy.[7] Then, we automatically search for each entity in Wikidata using the API, from which we obtain the description of a data entry for the corresponding entity.[8] Each data entry contains also other information besides the description that could be useful, e.g., the translation of the word in other languages, references to other databases, relevant images and relations e.g., "instance-of", "part-of", etc.

3.3   Knowledge Graph Input

The final result of the MemeGraphs augmentation is a serialized text-only representation of the meme that combines the information for all its modalities in one

---

[6] This number was chosen after observation of the scene graphs resulting from the memes in order to avoid having an overcrowded scene graph with multiple objects.

[7] https://spacy.io/universe/project/spacy-transformers.

[8] https://www.wikidata.org/wiki/Wikidata:Main_Page.

text. For the scene graphs we concatenate all the triplets that are detected in a meme (Eq. 2) and thus end up with a text $T_{sg,m}$. For the retrieved background knowledge we concatenate the texts of the KN for each meme (Eq. 3) and get a text $T_{kn,m}$. The final MemeGraphs input is the text of the meme followed by a [SEP] token and the concatenation of T_{sg} and T_{kn}.[9] This text can then be used as input to a classifier that will decide about the hatefulness of the meme. The complete process of the MemeGraphs method is depicted in Algorithm 1.

Algorithm 1: Outline of the MemeGraphs method

Data: a set of memes $\mathcal{M}$ consisting of images I and texts T.
Result: a set T_{kg} of texts representing the knowledge graphs of the memes.

1 // define a list to save the scene graphs for each meme;
2 $SG = \{\}$;
3 // apply the Schemata model;
4 **for** $m \in \mathcal{M}$ **do**
5 $\quad \lfloor \; SG_m = G(I_m)$;

6 // define a list to save the entities of each meme text;
7 $E = \{\}$;
8 // apply the NER model;
9 **for** $m \in \mathcal{M}$ **do**
10 $\quad \lfloor \; E_m = NER(T_m)$;

11 // retrieve information from the knowledge base for each entity;
12 $KN = \{\}$;
13 **for** $m \in \mathcal{M}$ **do**
14 $\quad$ **for** $e \in E_m$ **do**
15 $\quad \quad \lfloor \; KN_m = KB(e)$;

16 // define lists to save the concatenated texts for each meme;
17 $T_{sg}, T_{kn}, T_{kg} = \{\}, \{\}, \{\}$;
18 **for** $m \in \mathcal{M}$ **do**
19 $\quad$ $T_{sg,m} = concat(SG_m)$;
20 $\quad$ $T_{kn,m} = concat(KN_m)$;
21 $\quad$ $T_{kg,m} = concat(T_{sg,m}, T_{kn,m})$;
22 **return** T_{kg}

---

4 Benchmarking

4.1 Models

To evaluate the results of our automatic augmentation and how it can affect the performance of hateful memes classification, we employed a text-based Transformer with and without the MemeGraphs information as well as a multimodal

---

[9] All the texts are concatenated with a full stop.

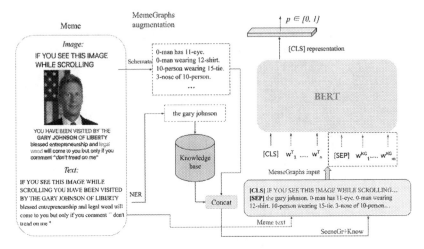

Fig. 3. The architecture of MemeGraphs[SceneGr+Know]. The scene graph produced by the Schemata model and the background knowledge for each entity are concatenated and given as input to BERT after the text of the meme and the [SEP] token.

model. Previous work has shown that using only the text of the memes for classification gives results highly competitive with multimodal methods [13,14,28]. On the other hand, unimodal image-based models have lowest results. We use a pre-trained BERT [9] model that takes as input the text of each meme (TxtBERT) [14] and different variants of this approach that include the MemeGraphs input. In order to feed the graphs as input to the model, we represent them as a text sequence (see Sect. 3), which is given as an extra text input after the [SEP] token (Fig. 3). In the model that we call MemeGraphs[SceneGr], this sequence contained only the scene graphs, which were represented by triplets of the detected objects and the relations between them (Eq. 2), e.g., "0-man has 11-eye. 0-man wearing 12-shirt.". In the model called MemeGraphs[Know], the text descriptions from the knowledge base corresponding to the detected entities are added as extra input (Eq. 3). For example, the second input sequence for the meme shown in Fig. 3 will be "American politician, businessman, and 29th Governor of New Mexico.". While in the MemeGraphs[SceneGr+Know] model, information from the whole knowledge graphs is added, comprising the scene graphs and the background knowledge concatenated with a full stop into one sequence (see Subsect. 3.3). In all the models, the [CLS] token is fed into a final linear layer with a sigmoid activation function that produced the probability of the meme being hateful.

In order to compare our MemeGraphs approach with a method that employs learned visual representations, we use ImgBERT, an early fusion multimodal model (see Sect. 2).

4.2   Experimental Setup

Each meme was labeled in the original dataset as offensive or non-offensive. The classification was based on the offensiveness labels that exist in the

Table 1. Test set precision (P), recall (R), and F1 scores averaged over twenty runs with different initializations and standard error of mean.

Model	P	R	F1
ImgBERT	0.394 ± 0.041	0.522 ± 0.080	0.408 ± 0.048
TxtBERT	**0.457±0.020**	0.516 ± 0.063	0.442 ± 0.032
MemeGraphs[SceneGr]	0.456 ± 0.010	0.577 ± 0.050	0.482 ± 0.019
MemeGraphs[Know]	0.446 ± 0.011	0.574 ± 0.045	**0.484±0.019**
MemeGraphs[SceneGr+Know]	0.451 ± 0.010	**0.583±0.063**	0.469 ± 0.035

Table 2. Test set precision (P), recall (R), and F1 scores from the model selected as best on the development set.

Model	P	R	F1
ImgBERT	0.389	**1.000**	0.560
TxtBERT	0.403	0.897	0.556
MemeGraphs[SceneGr]	0.398	0.845	0.541
MemeGraphs[Know]	0.396	0.690	0.503
MemeGraphs[SceneGr+Know]	**0.426**	0.948	**0.588**

MultiOFF dataset. The dataset is slightly imbalanced with 42% of the samples being offensive. For training and testing the models, we used the split provided by the authors of the dataset. The split consisted of training, validation, and test sets, which contained 445, 149, and 149 memes respectively. We employed the pre-trained BERT model provided by Hugging Face and fine-tuned it on our dataset.[10] To obtain image embeddings, we experimented with several pre-trained CNN models and based on the best results we chose DenseNet with 161 layers. The embeddings were extracted from the last pooling layer. The max length was set as the average token length of the training texts, depending on the input of each model. The models were trained using batch size 16, the weighted binary cross entropy loss, and AdamW optimizer with an initial learning rate of 2e−5. We used early stopping based on the validation loss with a patience of 3 epochs. We trained all models 20 times with different seeds and the training lasted between 4 and 6 epochs in each run. During inference, a threshold was used to determine if a meme is hateful or not based on the produced probability. This threshold was set to 0.5 for all the models.

4.3 Benchmarking Results

For each model, we obtained predictions for the test set from all twenty differently initialized runs. We evaluated each prediction set by calculating the

---

[10] https://huggingface.co/docs/Transformers/model_doc/bert#Transformers. BertForSequenceClassification.

F1 score defining the minority class (offensive class) as positive. In Table 1, we report the average F1 score over the twenty runs for each method and the corresponding standard error. In Table 2, the test score of the model that achieved the best score on the development set is shown. We observe that MemeGraphs[SceneGr+Know] outperforms the other models when predicting with the best checkpoint (Table 2). On the other hand, when looking at the average, the model with only the background knowledge as input achieves the best F1 score (MemeGraphs[Know]). We generally observe that the methods that incorporate our automatic augmentation (MemeGraphs) outperform the simple text-based fine-tuned BERT. Also, ImgBERT is outperformed showing that when the visual information is employed in the form of scene graphs the performance is improved, compared to when using only the image embeddings.

5 Analysis

5.1 Human Augmentation

Since the MemeGraphs augmentation is produced automatically it can potentially result in inaccuracies that can deteriorate the performance of the classification models. In order to examine this scenario and also the scene graphs themselves, which were produced by an off-the-shelf model, we performed human augmentation. The augmentation was conducted by two male students, who were doing a Master of Arts (M.A.) in Digital Humanities and the process lasted about 10 weeks. Several discussions took place before the augmentation started for the evaluators to get familiar with this study and understand its scope. They also carefully studied the guidelines before and during the process.

The goal of the human augmentation was to correct the automatically generated scene graphs and add background knowledge by linking the detected objects to a knowledge base (see Fig. 1). For each detected object or relation, the first step for the evaluators was to evaluate if it is correct or not and in case of an incorrect object to correct it. The second step was to link each object to its entry in Wikidata. For example, the detected object "man" was linked to the entry for "man". The objects represent generic types, e.g., "man", "woman" etc., but in some cases, a specific instance of the object type might be shown, which we call entity, e.g., the woman depicted is Hillary Clinton (as shown in Fig. 1). Then, the evaluators searched for the entry of each detected object and its entity (if existing) in Wikidata and added the corresponding links. The evaluators worked towards achieving high precision and in order to limit the scope of the augmentation, no new objects or relations were added.

After a first round of augmenting a small sub-sample (around 10%), preliminary guidelines were created. These guidelines described the process mentioned in the previous paragraph in simple and brief steps. A few systematic difficulties for the Schemata algorithm were encountered in this first round that can be summarized as follows:

– The text of the meme appearing on the image was often detected as a "sign" and some parts of the text as "letter". However, this text is not actually a sign and is not part of the image.
– In some cases, the same object was detected multiple times.
– Some memes in the dataset were screenshots of text, so there was not any useful visual information extracted from them.
– In a few cases, there was a specific entity depicted in the meme, but the corresponding object type was not detected.

The above-mentioned difficulties were discussed with the evaluators and the guidelines were revised to include clear instructions for those cases. Based on that and other observations of the evaluators about the data, the following final guidelines were defined:

– **Meme text or sign.** A distinction between a meme's text and an actual sign in the image, e.g., a sticker or a banner with text, was made. Detected objects that were referring to the meme text were discarded.
– **Multiple object detection.** In cases where the same object type appeared multiple times, the annotators would inspect the bounding boxes in the images to verify whether it is the same object. If yes, then only one occurrence was kept.
– **Screenshots.** Memes that are screenshots of text and do not contain visual information were disregarded from the augmentation process.
– **Missed objects.** The human augmentation is based on the automatically generated scene graphs and thus, no new objects or their corresponding entities were added.

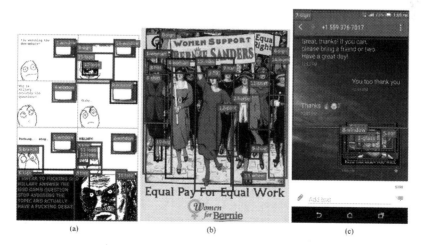

Fig. 4. Examples of memes after the automatic augmentation. In sub-figures (a) and (b), we see cases where the detected object is incorrect (e.g., "10-tree" in (a) and "13-wheel" in (b)), but the correct object is not clear to define. In sub-figure (c) we see that the same object, i.e., "0-face" can depict two entities.

- **Bounding boxes.** The bounding boxes that were drawn on the image during the automatic augmentation are not used in our study, hence they were not changed in the human augmentation.
- **Incorrect objects.** For each detected object, the evaluators determined if its type was correct or not. If not, then the correct type for this object was indicated, when it was possible. In cases where the object did not exist in the image at all, it was simply removed.
- **Relation correction.** For each relation, the evaluators determined if it was correct or not. If the type of objects that the relation referred to was incorrect, then, the type was replaced with the correct one. Following the same approach as with the objects, no new relations were added neither was an object that was not detected added to a relation.
- **Knowledge base.** Each object was mapped to a specific entity when possible by adding a link to Wikidata.

After the guidelines were finalized, the entire dataset was augmented. The inter-annotator agreement and Cohen's kappa regarding the correctness of the detected objects were 83.84% and 0.60 respectively, while for the correctness of the detected relations it was 78.05% and 0.53. The first evaluator found that 12.95% of the automatically detected objects needed correction. The second evaluator found a larger percentage of incorrect objects around 21.86%. Similar outcomes were observed for the relations, where the first evaluator found 22.13% of the relations to be incorrect, while the second found 31.48%. The number of objects found incorrect by both evaluators was 1,127 and for these cases, they agreed on the correct object 282 times. This low agreement shows that deciding about the exact type of objects shown in an image is a difficult task. The relations found incorrect by both evaluators were 1,820. Regarding the knowledge base linking, more inconsistencies were found and the evaluators added the same link in 4,314 out of 9,666 cases. In Fig. 4, we see examples of cases that caused the low agreement. Sub-figures (a) and (b) contain detected objects that both evaluators agreed are incorrect (e.g., "10-tree" in (a) and "13-wheel" in (b)). However, they did not add the same correct object. For example, the object "13-wheel" in (b), was corrected as "shoe" by one evaluator and as "foot" by the other. Both of these corrections can be considered valid. Regarding adding the links to Wikidata, similar uncertainties can be found. In cases of disagreement, we chose the correct object or link for the final dataset by the following heuristic: for each annotated alternative, we counted its occurrence in the part of the dataset that had 100% agreement, i.e., the more frequent object or link was finally chosen. For the entity links, there are cases that the evaluators added a different link, but both of the links were correct. In sub-figure (c), we see such a case in which "0-face" depicts both "Bernie Sanders" and "Yoda". Hence, both links were added to the final augmentations. In cases where the evaluators disagreed regarding the correctness of an object or relation, they were removed.

To conclude our analysis based on human augmentation, we used the results of the aforementioned process as inputs to the models described in Sect. 4.1. We experimented with two different settings for training and testing the models.

First, we used the results of the manual augmentation, which are the corrected scene graphs and manually linked knowledge, both for training and inference (manual/manual). Second, we combined the two augmentations and trained on the manual ones, and tested on the automatic ones (manual/automatic). In this case, the automatic augmentations are the scene graphs corrected automatically based on the manual corrections of the training data. We kept in the development and test scene graphs only the objects that were manually marked as correct at least once in the training scene graphs and removed the rest (i.e., the ones that the evaluators found were always falsely detected by the scene graph model). The knowledge base information consisted of the descriptions of the automatically detected text entities (See Subsect. 3.2). We used the same experimental setup as for the MemeGraphs method described in Subsect. 4.2. The average results over the 20 runs are shown in Table 3 and the score of the models performing best on the development set in Table 4. We observe that in both settings MemeGraphs[SceneGr+Know] achieves the best F1 score in Table 4, similar to the fully automatic setting (Table 2). On the other hand, when looking at the average, the TxtBERT with only the scene graphs as input obtains the best score (MemeGraphs[SceneGr])in both the manual/manual and the manual/automatic settings.

Table 3. Precision (P), recall (R) and F1 scores averaged over twenty runs and standard error of mean on the test set for each setting.

Model	Manual/manual			Manual/automatic		
	P	R	F1	P	R	F1
ImgBERT	0.394 ± 0.041	0.522 ± 0.080	0.408 ± 0.048	0.394 ± 0.041	0.522 ± 0.080	0.408 ± 0.048
TxtBERT	$\mathbf{0.457 \pm 0.020}$	0.516 ± 0.063	0.442 ± 0.032	0.457 ± 0.020	0.516 ± 0.063	0.442 ± 0.032
MemeGraphs[SceneGr]	0.410 ± 0.024	$\mathbf{0.604 \pm 0.051}$	$\mathbf{0.474 \pm 0.029}$	0.404 ± 0.024	$\mathbf{0.592 \pm 0.050}$	$\mathbf{0.467 \pm 0.029}$
MemeGraphs[Know]	0.408 ± 0.024	0.507 ± 0.050	0.434 ± 0.030	$\mathbf{0.460 \pm 0.050}$	0.309 ± 0.079	0.254 ± 0.049
MemeGraphs[SceneGr+Know]	0.436 ± 0.031	0.560 ± 0.051	0.443 ± 0.033	0.368 ± 0.030	0.331 ± 0.079	0.280 ± 0.045

Table 4. Precision (P), recall (R) and F1 scores on the test set of the best model based on the development set for each setting.

Model	Manual/manual			Manual/automatic		
	P	R	F1	P	R	F1
ImgBERT	0.389	**1.000**	0.560	0.389	**1.000**	0.560
TxtBERT	0.403	0.897	0.556	0.403	0.897	0.556
MemeGraphs[SceneGr]	0.419	0.845	0.560	0.417	0.828	0.555
MemeGraphs[Know]	0.400	0.655	0.497	0.406	0.931	0.565
MemeGraphs[SceneGr+Know]	**0.424**	0.862	**0.568**	**0.422**	0.931	**0.581**

5.2   Discussion

We observe that overall our proposed method outperforms its competitors in terms of F1 score in all settings (fully automatic, manual/manual, manual/automatic). Between the different settings, we see that the models with the fully automatic setting have the best scores (MemeGraphs[Know] in Table 1 and MemeGraphs[SceneGr+Know] in Table 2), even though manual augmentations would be expected to be more accurate than the automatic ones. Furthermore, not only the best score is achieved by the MemeGraphs[Know] model in the fully automatic setting, but this model's performance is improved in this setting compared to the manual ones. The scene graphs infused model also achieves better results in the fully automatic setting, showing that the automatically produced scene graphs are accurate enough and no manual correction is needed. In the manual/automatic setting, all the models performed worse compared to the other settings. This fact shows that models trained on manual annotations are not able to generalize in the automatic setting. This holds true especially in the MemeGraphs[Know] model, since in the automatic setting no information for the type of the objects exists in the input. To gain insights into that behavior, we analyze the different challenges that were faced in the manual and the automatic augmentation and compare their results.

During the manual augmentation, both correcting the scene graphs and adding background knowledge were found challenging. Regarding the scene graphs, memes contain complex information and images, which made the correction of objects difficult for the evaluators (see Subsect. 5.1). Background knowledge for specific entities was also difficult to add for three main reasons: 1. the evaluators may not know the person depicted in the image, 2. many memes were screenshots of posts, so there was no actual visual information, and 3. meme texts often refer to entities that are not shown in the image. This resulted in detecting entities for only 409 memes out of 743. The automatic entity detection, on the other hand, was based on the text, which assisted in overcoming the three aforementioned challenges and extracted entities for all the memes.

Regarding the automatic augmentation based on the entities detected by the NER model, the main challenge consists of linking the detected entities to the knowledge base. Even though the model managed to extract entities from all the texts showing that we can obtain rich information, linking them to the knowledge base was not easy. Many times the entities appear in the text with only their first name, e.g., "Hillary", or their first and last name concatenated, e.g., "donaldtrump", or are detected alongside some other word, e.g., "green Bernie". When these entities are searched for in the knowledge base, the results might not be accurate and contain data entries for many entities from which we choose the first as the most related one. However, this can lead to adding a link to the wrong entity.

6   Conclusion

In order to understand memes, it is necessary to correctly interpret the image, and the text and to connect it with appropriate general background knowledge (outside of the meme). In this work, we introduced models infused with scene graphs and world knowledge retrieved from WikiData. As a foundational representation, scene graphs were automatically generated, which relate the most important objects in the meme image to each other. Typed objects from the scene graph and named entities from the text were extracted automatically and linked to WikiData. This structured information (scene graph and information from WikiData) was then serialized as a sequence of tokens and concatenated with the original text from the meme for classification with a Transformer language model. We found that adding the graph representation and knowledge from Wikidata improved performance on hateful meme detection compared to classification on text alone, and compared to a multimodal model based on pre-trained image embeddings in addition to text. We also provide a dataset with human corrections of the automatically generated graphs, and an analysis that shows that the (uncorrected) automatic graphs and the corrected ones perform similarly well for hatefulness detection with our approach.

Acknowledgements. This research was funded by the Deutsche Forschungsgemein-schaft (DFG, German Research Foundation) - RO 5127/2-1 and the Vienna Science and Technology Fund (WWTF)[10.47379/VRG19008]. We thank Christos Bintsis for participating in the manual augmentation. We also thank Matthias Aßenmacher and the anonymous reviewers for their valuable feedback.

References

1. Aggarwal, P., Liman, M.E., Gold, D., Zesch, T.: VL-BERT+: detecting protected groups in hateful multimodal memes. In: Proceedings of the 5th Workshop on Online Abuse and Harms (WOAH 2021), pp. 207–214. Online, August 2021. https://doi.org/10.18653/v1/2021.woah-1.22
2. Henrique Luz de Araujo, P., Roth, B.: Checking HateCheck: a cross-functional analysis of behaviour-aware learning for hate speech detection. In: Proceedings of NLP Power! The First Workshop on Efficient Benchmarking in NLP, pp. 75–83. Dublin, Ireland (2022)
3. Behera, P., Mamta, Ekbal, A.: Only text? Only image? Or both? Predicting sentiment of internet memes. In: Proceedings of the 17th International Conference on Natural Language Processing (ICON), pp. 444–452. Indian Institute of Technology Patna, Patna, India (2020). https://aclanthology.org/2020.icon-main.60
4. Blaier, E., Malkiel, I., Wolf, L.: Caption enriched samples for improving hateful memes detection. In: Proceedings of the 2021 Conference on Empirical Methods in Natural Language Processing, pp. 9350–9358. Online and Punta Cana, Dominican Republic (2021). https://doi.org/10.18653/v1/2021.emnlp-main.738
5. Chen, S., Aguilar, G., Neves, L., Solorio, T.: Can images help recognize entities? A study of the role of images for multimodal NER. In: Proceedings of the 2021 EMNLP Workshop W-NUT: The Seventh Workshop on Noisy User-Generated Text, pp. 87–96. Online and Punta Cana, Dominican Republic (2021)

6. Chen, Y.-C., et al.: UNITER: UNiversal image-TExt representation learning. In: Vedaldi, A., Bischof, H., Brox, T., Frahm, J.-M. (eds.) ECCV 2020. LNCS, vol. 12375, pp. 104–120. Springer, Cham (2020). https://doi.org/10.1007/978-3-030-58577-8_7

7. Das, A., Wahi, J.S., Li, S.: Detecting hate speech in multi-modal memes. arXiv preprint: arXiv:2012.14891 (2020)

8. Deng, J., Dong, W., Socher, R., Li, L.J., Li, K., Fei-Fei, L.: ImageNet: a large-scale hierarchical image database. In: IEEE Conference on Computer Vision and Pattern Recognition, pp. 248–255. Miami Beach, FL, USA (2009)

9. Devlin, J., Chang, M.W., Lee, K., Toutanova, K.: BERT: pre-training of deep bidirectional transformers for language understanding. In: Proceedings of the 2019 Conference of the North American Chapter of the Association for Computational Linguistics: Human Language Technologies, Volume 1 (Long and Short Papers), pp. 4171–4186. Minneapolis, Minnesota, USA (2019)

10. Dimitrov, D., et al.: Detecting propaganda techniques in memes. In: Proceedings of the 59th Annual Meeting of the Association for Computational Linguistics and the 11th International Joint Conference on Natural Language Processing (Volume 1: Long Papers), pp. 6603–6617. Online (2021). https://doi.org/10.18653/v1/2021.acl-long.516

11. Fersini, E., et al.: SemEval-2022 task 5: Multimedia automatic misogyny identification. In: Proceedings of the 16th International Workshop on Semantic Evaluation (SemEval-2022), pp. 533–549. Seattle, United States (2022)

12. Gan, Z., Chen, Y.C., Li, L., Zhu, C., Cheng, Y., Liu, J.: Large-scale adversarial training for vision-and-language representation learning. arXiv:2006.06195 (2020)

13. Kiela, D., et al.: The hateful memes challenge: detecting hate speech in multimodal memes. Adv. Neural. Inf. Process. Syst. **33**, 2611–2624 (2020)

14. Kougia, V., Pavlopoulos, J.: Multimodal or text? Retrieval or BERT? Benchmarking classifiers for the shared task on hateful memes. In: Proceedings of the 5th Workshop on Online Abuse and Harms (WOAH 2021), pp. 220–225. Online (2021). https://doi.org/10.18653/v1/2021.woah-1.24

15. Krishna, R., et al.: Visual genome: connecting language and vision using crowd-sourced dense image annotations. Int. J. Comput. Vision **123**(1), 32–73 (2017)

16. Lee, R.K.W., Cao, R., Fan, Z., Jiang, J., Chong, W.H.: Disentangling hate in online memes. In: Proceedings of the 29th ACM International Conference on Multimedia, pp. 5138–5147 (2021)

17. Li, J., Ataman, D., Sennrich, R.: Vision matters when it should: sanity checking multimodal machine translation models. In: Proceedings of the 2021 Conference on Empirical Methods in Natural Language Processing, pp. 8556–8562. Online and Punta Cana, Dominican Republic (2021)

18. Li, L.H., Yatskar, M., Yin, D., Hsieh, C.J., Chang, K.W.: VisualBERT: a simple and performant baseline for vision and language. arXiv:1908.03557 (2019)

19. Li, X., et al.: OSCAR: object-semantics aligned pre-training for vision-language tasks. In: Vedaldi, A., Bischof, H., Brox, T., Frahm, J.-M. (eds.) ECCV 2020. LNCS, vol. 12375, pp. 121–137. Springer, Cham (2020). https://doi.org/10.1007/978-3-030-58577-8_8

20. Lu, J., Batra, D., Parikh, D., Lee, S.: VilBERT: pretraining task-agnostic visiolinguistic representations for vision-and-language tasks. arXiv:1908.02265 (2019)

21. Mathias, L., et al.: Findings of the WOAH 5 shared task on fine grained hateful memes detection. In: Proceedings of the 5th Workshop on Online Abuse and Harms (WOAH 2021), pp. 201–206. Online (2021). https://doi.org/10.18653/v1/2021.woah-1.21

22. Mozes, M., Schmitt, M., Golkov, V., Schütze, H., Cremers, D.: Scene graph generation for better image captioning? arXiv:2109.11398 (2021)
23. Pramanick, S., Sharma, S., Dimitrov, D., Akhtar, M.S., Nakov, P., Chakraborty, T.: MOMENTA: a multimodal framework for detecting harmful memes and their targets. In: Findings of the Association for Computational Linguistics: EMNLP 2021, pp. 4439–4455. Punta Cana, Dominican Republic (2021). https://doi.org/10.18653/v1/2021.findings-emnlp.379
24. Radford, A., et al.: Learning transferable visual models from natural language supervision. In: International Conference on Machine Learning, pp. 8748–8763 (2021)
25. Sharifzadeh, S., Baharlou, S.M., Schmitt, M., Schütze, H., Tresp, V.: Improving scene graph classification by exploiting knowledge from texts. Proc. AAAI Conf. Artif. Intell. **36**(2), 2189–2197 (2022)
26. Sharifzadeh, S., Baharlou, S.M., Tresp, V.: Classification by attention: scene graph classification with prior knowledge. In: Proceedings of the Thirty-Fifth AAAI Conference on Artificial Intelligence (AAAI-21), pp. 5025–5033. Online (2021)
27. Su, W., et al.: Vl-BERT: pre-training of generic visual-linguistic representations. arXiv:1908.08530 (2019)
28. Suryawanshi, S., Chakravarthi, B.R., Arcan, M., Buitelaar, P.: Multimodal meme dataset (MultiOFF) for identifying offensive content in image and text. In: Proceedings of the Second Workshop on Trolling, Aggression and Cyberbullying, pp. 32–41. Marseille, France (2020). https://aclanthology.org/2020.trac-1.6
29. Yang, J., Lu, J., Lee, S., Batra, D., Parikh, D.: Graph R-CNN for scene graph generation. In: Ferrari, V., Hebert, M., Sminchisescu, C., Weiss, Y. (eds.) ECCV 2018. LNCS, vol. 11205, pp. 690–706. Springer, Cham (2018). https://doi.org/10.1007/978-3-030-01246-5_41
30. Yang, X., Tang, K., Zhang, H., Cai, J.: Auto-encoding scene graphs for image captioning. In: Proceedings of the IEEE/CVF Conference on Computer Vision and Pattern Recognition, pp. 10685–10694. Long Beach, CA, USA (2019)
31. Yin, Y., Meng, F., Su, J., Zhou, C., Yang, Z., Zhou, J., Luo, J.: A novel graph-based multi-modal fusion encoder for neural machine translation. In: Proceedings of the 58th Annual Meeting of the Association for Computational Linguistics (ACL), pp. 3025–3035. Online and Punta Cana, Dominican Republic (2020)
32. Zellers, R., Yatskar, M., Thomson, S., Choi, Y.: Neural motifs: scene graph parsing with global context. In: Proceedings of the IEEE Conference on Computer Vision and Pattern Recognition, pp. 5831–5840 (2018)
33. Zhu, R.: Enhance multimodal transformer with external label and in-domain pretrain: hateful meme challenge winning solution. arXiv:2012.08290 (2020)

Author Index